SECOND EDITION

THEORY AND APPLICATIONS OF HIGHER-DIMENSIONAL HADAMARD MATRICES

SECOND EDITION

THEORY AND APPLICATIONS OF HIGHER-DIMENSIONAL HADAMARD MATRICES

YI XIAN YANG
XIN XIN NIU
CHENG QING XU

CRC Press is an imprint of the
Taylor & Francis Group, an **informa** business

CRC Press
Taylor & Francis Group
6000 Broken Sound Parkway NW, Suite 300
Boca Raton, FL 33487-2742

First issued in paperback 2019

ISBN-13: 978-1-4398-1807-7 (hbk)
ISBN-13: 978-0-367-38440-1 (pbk)

Library of Congress Cataloging-in-Publication Data

Yang, Yi Xian.
Theory and applications of higher-dimensional Hadamard marices / Yi Xian Yang. -- 2nd ed. / Xin Xin Niu, Cheng Qing Xu.
p. cm.
Includes bibliographical references and index.
ISBN 978-1-4398-1807-7
1. Hadamard matrices. I. Niu, Xin Xin. II. Xu, Cheng Qing. III. Title.

QA166.4.Y36 2010
512.9'434--dc22 2009038241

Visit the Taylor & Francis Web site at
http://www.taylorandfrancis.com

and the CRC Press Web site at
http://www.crcpress.com

Preface to the Second Edition

Time is flying! Eight years have passed since the publication of the first edition of my book, *Theory and Applications of Higher-Dimensional Hadamard Matrices.*

During the past eight years, my group and I have been applying the Hadamard related theory to many engineering projects and lucky to get quite a few good results, especially in the design and analysis of perfect digital signals and arrays for communications, radars, cryptology, and information security. Thus when the editor of Science Press invited me to republish the book, my colleague and I were very happy to revise the book.

Compared with the first edition, a new part (Part IV) with two new chapters (Chapter 7 and Chapter 8) has been added. In this new version, we concentrate on higher dimensional Hadamard matrices and their applications in telecommunications and information security. This revised book is naturally divided into four parts according to the dimensions of Hadamard matrices processed and applications.

The first part concentrates on the classical 2-dimensional Walsh and Hadamard matrices. Fast algorithms, updated constructions, existence results and their generalized forms are presented for Walsh and Hadamard matrices. Some useful dyadic operation tools are presented, e.g., dyadic additions and dyadic groups. New Hadamard designs based on dyadic addition sets and difference family are also stated here.

The second part deals with the lower-dimensional cases, e.g., 3-, 4-, and 6-dimensional Walsh and Hadmard matrices and transforms. One of the aims of this part is to make it easier to smoothly move from 2-dimensional cases to the general higher-dimensional cases. This part concentrates on the 3-dimensional Hadamard and Walsh matrices. Constructions based upon direct multiplication, and upon recursive methods, and perfect binary arrays are also introduced. Another important topic of this part is the existence and construction of 3-dimensional Hadamard matrices of orders $4k$ and $4k + 2$, respectively, and a group of transforms based on 2-, 3-, 4-, and 6-dimensional Walsh-Hadamard matrices and their corresponding fast algorithms. Different new sequences and arrays are introduced, e.g., the optical orthogonal codes, periodic complementary binary array family, dyadic complementary sequence family, and Bent complementary functions.

The third part is the key part, which investigates the n-dimensional Hadamard matrices of order 2, which have been proved equivalent to the well-known H-Boolean functions and the perfect binary arrays of order 2. This equivalence motivates a group of perfect results about the enumeration of higher-dimensional Hadamard matrices of order 2. Applications of these matrices to feed forward networking, stream cipher, Bent functions, and error correcting codes are presented in turn. After introducing the definitions of the regular, proper, improper, and generalized higher-dimensional Hadamard matrices, many theorems about their existence and constructions are presented. Perfect binary arrays, generalized perfect arrays, and the orthogonal designs are also used to construct new higher-dimensional Hadamard matrices. The other content of this part is the Boolean approach of Hadamard matrices, including the correlation immunity of Boolean functions and Boolean substitutions, stream and block ciphers based on Hadamard matrices, the applications of Hadamard matrices in error-correcting codes, and image coding.

The fourth part states some examples of applications of Hadamard-related ideas to the designs and analysis of 1-dimensional sequences (Chapter 7) and 2-dimensional arrays (Chapter 8). Specifically, enumerations, constructions and correlation immunities of Boolean functions, corresponding to the n-dimensional matrix of order 2, with cryptographic significance are presented in Section 7.1. The correlation functions of Bent-like sequences (including the Gold-Geometric sequences, Generalized Geometric sequences, and p-ary d-form sequences) are calculated in Section 7.2. Hadamard-like difference sets are used in Section 7.3 to construct sequences pairs with mismatched filtering which are, in fact, semi-arrays with two-level autocorrelation functions and good cross-correlations. A few unusual sequences analyses are listed in Section 7.4–especially the implementations of Boolean functions by neural networks, the calculations of linear complexities of sequences represented by the truth tables of Boolean functions, the estimations of upper and lower bounds of periodic ambiguity functions of EQC-Based TFHC, and the properties of auto-, cross- and triple-correlations of sequences. Constructions, correlations, and enumerations of Costas arrays are established in Section 8.1. Section 8.2 concentrates on the parameters, bounds and constructions of optical orthogonal codes, which are families of arrays with good auto- and cross-correlations. Besides the applications stated in Part 4, the theory and ideas of Hadamard matrices can be used in many other areas of communications and information security.

Many open problems in the study of the theory of higher-dimensional Hadamard matrices are also listed in the book. We hope that these research problems will motivate further developments.

The authors would like to give their thanks to the following research foundations, by which this book has been supported: (1) National Basic Research Program of China (973 Program) (No. 2007CB310704, 2007CB311203); (2) National Natural Science Foundation of China (No. 60673098, 90718001, 60821001, 60731160626); (3) Specialized Research Fund for the Doctoral Program of Higher Education (No. 20070013005, 20070013007); (4) The 111 Project (No. B08004); (5) The National 863 (No. 2007AA01Z430).

Yixian Yang

Preface to the First Edition

Just over one hundred years ago, in 1893, Jacques Hadamard found binary (± 1) matrices of orders 12 and 20 whose rows (resp. columns) were pairwise orthogonal. These matrices satisfy the determinantal upper bound for binary matrices. Hadamard actually proposed the question of seeking the maximal determinant of matrices with entries on the unit circle, but his name has become associated with the question concerning real (binary) matrices. Hadamard was not the first person to study these matrices. For example, Sylvester had found, in 1857, such row (column) pairwise orthogonal binary matrices of all orders of powers of two. Nevertheless, Hadamard proved that binary matrices with a maximal determinant could exist only for orders 1, 2, and $4t$, t a positive integer.

With regard to the practical applications of Hadamard matrices, it was Hall, Baumert, and Golomb who sparked the interest in Hadamard matrices over the past 30 years. They made use of the Hadamard matrix of order 32 to design an eight-bit error-correcting code for two reasons. First, error-correcting codes based on Hadamard matrices have good error correction capability and good decoding algorithms. Second, because Hadamard matrices are (± 1)-valued, all the computer processing can be accomplished using additions and subtractions rather than multiplication.

Walsh matrices are the simplest and the most popular special kinds of Hadamard matrices. Walsh matrices are generated by sampling the Walsh functions, which are families of orthogonal complete functions. Based on the Walsh matrices, a very efficient orthogonal transform, called Walsh-Hadamard transform, was developed. The Walsh-Hadamard transform is now playing a more and more important role in signal processing and image coding.

Shlichta discovered in 1971 that there exist higher-dimensional binary arrays which possess a range of orthogonality properties. In particular, Shlichta constructed 3-dimensional arrays with the property that any sub-array obtained by fixing one index is a 2-dimensional Hadamard matrix. The study of higher-dimensional Hadamard matrices was mainly motivated by another important paper of Shlichta, "Higher-Dimensional Hadamard Matrices," which was published in *IEEE Trans. on Inform.*, in 1979. Since then a lot of papers on the existence, construction, and enumeration of higher-dimensional Hadamard matrices have been reported. For example, Hammer and Seberry found, in 1982, that higher-dimensional orthogonal designs can be used to construct higher-dimensional Hadamard matrices. To the author's knowledge much of the research achievements on higher-dimensional Hadamard matrices have been accomplished by Agaian, De Launey, Hammer, Seberry, Yixian Yang, Horadam, Shlichta, Jedwab, Lin, Chen, and others. Many new papers have been published, thus none can collect together all of the newest results in this area.

The book divides naturally into three parts according to the dimensions of Hadamard matrices processed.

The first part, Chapter 1 and Chapter 2, concentrates upon the classical 2-dimensional cases. Because quite a few books (or chapters in them) have been published which introduce the progress of (2-dimensional) Hadamard matrices, we prefer to present an introductory survey rather than to restate many known long proofs. Chapter 1 introduces Walsh matrices and Walsh transforms, which have been widely used in engineering fields. Fast algorithms for Walsh transforms and various useful properties of Walsh matrices are also stated. Chapter 2 is about (2-dimensional) Hadamard matrices, especially their construction, existence, and their generalized forms. The updated strongest Hadamard construction theorems presented in this chapter are helpful for readers to understand how difficult it is to prove or disprove the famous Hadamard conjecture.

The second part, Chapters 3 and 4, deals with the lower-dimensional cases, e.g., 3-, 4-, and 6-dimensional Walsh and Hadmard matrices and transforms. One of the aims of this part is to make it easier to smoothly move from 2-dimensional cases to the general higher-dimensional cases. Chapter 3 concentrates on the 3-dimensional Hadamard and Walsh matrices. Constructions based upon direct multiplication, and upon recursive methods, and perfect binary arrays are introduced. Another important topic of this chapter is the existence and construction of 3-dimensional Hadamard matrices of orders $4k$ and $4k+2$, respectively. Chapter 4 introduces a group of transforms based on 2-, 3-, 4-, and 6-dimensional Walsh-Hadamard matrices and their corresponding fast algorithms. The algebraic theory of higher-dimensional Walsh-Hadamard matrices is presented also.

Finally, the third part, which is the key part of the book, consists of the last two chapters (Chapter 5 and 6). To the author's knowledge, the contents in this part (and the previous second part) have never been included in any published books. This part is divided into chapters according to the orders of the matrices (arrays) processed. Chapter 5 investigates the n-dimensional Hadamard matrices of order 2, which have been proved equivalent to the well known H-Boolean functions and the perfect binary arrays of order 2. This equivalence motivates a group of perfect results about the enumeration of higher-dimensional Hadamard matrices of order 2. Applications of these matrices to feed forward networking, stream cipher, Bent functions and error correcting codes are presented in turn. Chapter 6, which is the longest chapter of the book, aims at introducing Hadamard matrices of general dimension and order. After introducing the definitions of the regular, proper, improper, and generalized higher-dimensional Hadamard matrices, many theorems about their existence and constructions are presented. Perfect binary arrays, generalized perfect arrays, and the orthogonal designs are also used to construct new higher-dimensional Hadamard matrices. The last chapter of the book is a concluding chapter of questions, which includes a list of open problems in the study of the theory of higher-dimensional Hadamard matrices. We hope that these research problems will motivate further developments.

In order to satisfy readers with this special interest, we list, at the end of each chapter, as many up-to-date references as possible.

I would like to thank my supervisors, Professors Zhenming Hu and Jiongpang Zhou, for their guidance during my academic years at the Information Security Center of Beijing University of Posts and Telecommunications (BUPT). During my research years in higher-dimensional Hadamard matrices I benefited from Professors De Launey, Hammer, Seberry, Horadam, Shlichta, Jedwab. My thanks go to many of their papers, theses and communications. I was attracted into the area of higher-dimensional Hadamard matrices by Shlichta's paper, “Higher-Dimensional Hadamard Matrices” published in *IEEE Trans. on Inform. Theory.* My first journal paper was motivated by Hammer and Seberry’s paper “Higher-Dimensional Orthogonal Designs and Applications” published in *IEEE Trans. on Inform. Theory.* It is Dr. Jedwab’s wonderful Ph.D thesis “Perfect arrays, barker arrays and difference sets” that motivated me to finish the first book on higher-dimensional Hadamard matrices. One of my main aims in this book is to motivate other authors to begin to publish more books on higher-dimensional Hadamard matrices and their applications, so that the readers in other areas can know what has been done in the area of higher-dimensional Hadamard matrices.

I specially thank my wife, Xinxin Niu, and my son, Mulong Yang, for their support. It is not hard to imagine how much they have sacrificed in family life during the past years. I would like to dedicate this book to my wife and son. Finally, I also dedicate this book to my parents, Mr. Zhongquan Yang and Mrs. De Lian Wei, for their love.

Yixian Yang

Contents

Preface to the Second Edition

Preface to the First Edition

Part I 2-Dimensional Cases

Chapter 1 Walsh Matrices · · · 3

1.1 Walsh Functions and Matrices · · · 3

1.1.1 Definitions · · · 3

1.1.2 Ordering · · · 8

1.2 Orthogonality and Completeness · · · 11

1.2.1 Orthogonality · · · 12

1.2.2 Completeness · · · 12

1.3 Walsh Transforms and Fast Algorithms · · · 14

1.3.1 Walsh-Ordered Walsh-Hadamard Transforms · · · 14

1.3.2 Hadamard-Ordered Walsh-Hadamard Transforms · · · 16

Bibliography · · · 17

Chapter 2 Hadamard Matrices · · · 19

2.1 Definitions · · · 20

2.1.1 Hadamard Matrices · · · 20

2.1.2 Hadamard Designs · · · 23

2.1.3 Williamson Matrices · · · 28

2.2 Construction · · · 34

2.2.1 General Constructions · · · 34

2.2.2 Amicable Hadamard Matrices · · · 37

2.2.3 Skew Hadamard Matrices · · · 40

2.2.4 Symmetric Hadamard Matrices · · · 42

2.3 Existence · · · 44

2.3.1 Orthogonal Designs and Hadamard Matrices · · · 44

2.3.2 Existence Results · · · 49

Bibliography · · · 52

Part II Lower-Dimensional Cases

Chapter 3 3-Dimensional Hadamard Matrices · · · 57

3.1 Definitions and Constructions · · · 57

3.1.1 Definitions · · · 57

3.1.2 Constructions Based on Direct Multiplications · · · 59

3.1.3 Constructions Based on 2-Dimensional Hadamard Matrices · · · 63

3.2 3-Dimensional Hadamard Matrices of Order $4k+2$ · · · 65

3.3 3-Dimensional Hadamard Matrices of Order $4k$ · · · 67

3.3.1 Recursive Constructions of Perfect Binary Arrays · · · 67

3.3.2 Quasi-Perfect Binary Arrays ········ 72
3.3.3 3-Dimensional Hadamard Matrices Based on PBA(2^m, 2^m) and PBA(3.2^m, 3.2^m) ········ 76
3.4 3-Dimensional Walsh Matrices ········ 80
3.4.1 Generalized 2-Dimensional Walsh Matrices ········ 80
3.4.2 3-Dimensional Walsh Matrices ········ 84
3.4.3 3-Dimensional Pan-Walsh Matrices ········ 89
3.4.4 Analytic Representations ········ 91
Bibliography ········ 97
Chapter 4 Multi-Dimensional Walsh-Hadamard Transforms ········ 99
4.1 Conventional 2-Dimensional Walsh-Hadamard Transforms ········ 99
4.1.1 2-Dimensional Walsh-Hadamard Transforms ········ 99
4.1.2 Definitions of 4-Dimensional Hadamard Matrices ········ 101
4.2 Algebraic Theory of Higher-Dimensional Matrices ········ 103
4.3 Multi-Dimensional Walsh-Hadamard Transforms ········ 114
4.3.1 Transforms Based on 3-Dimensional Hadamard Matrices ········ 114
4.3.2 Transforms Based on 4-Dimensional Hadamard Matrices ········ 115
4.3.3 Transforms Based on 6-Dimensional Hadamard Matrices ········ 118
Bibliography ········ 121

Part III General Higher-Dimensional Cases

Chapter 5 n-Dimensional Hadamard Matrices of Order 2 ········ 125
5.1 Constructions of 2^n Hadamard Matrices ········ 125
5.1.1 Equivalence between 2^n Hadamard Matrices and H-Boolean Functions ········ 126
5.1.2 Existence of H-Boolean Functions ········ 127
5.1.3 Constructions of H-Boolean Functions ········ 135
5.2 Enumeration of 2^n Hadamard Matrices ········ 140
5.2.1 Classification of 2^4 Hadamard Matrices ········ 140
5.2.2 Enumeration of 2^5 Hadamard Matrices ········ 147
5.2.3 Enumeration of General 2^n Hadamard Matrices ········ 148
5.3 Applications ········ 153
5.3.1 Strict Avalanche Criterion and H-Boolean Functions ········ 153
5.3.2 Bent Functions and H-Boolean Functions ········ 158
5.3.3 Reed-Muller Codes and H-Boolean Functions ········ 162
Bibliography ········ 166
Chapter 6 General Higher-Dimensional Hadamard Matrices ········ 169
6.1 Definitions, Existences and Constructions ········ 169
6.1.1 n-Dimensional Hadamard Matrices of Order $2k$ ········ 170
6.1.2 Proper and Improper n-Dimensional Hadamard Matrices ········ 183
6.1.3 Generalized Higher-Dimensional Hadamard Matrices ········ 187
6.2 Higher-Dimensional Hadamard Matrices Based on Perfect Binary Arrays ········ 190
6.2.1 n-Dimensional Hadamard Matrices Based on PBAs ········ 190

6.2.2 Construction and Existence of Higher-Dimensional PBAs ········ 192
6.2.3 Generalized Perfect Arrays ········ 201
6.3 Higher-Dimensional Hadamard Matrices Based on Orthogonal Designs ········ 221
6.3.1 Definitions of Orthogonality ········ 221
6.3.2 Higher-Dimensional Orthogonal Designs ········ 225
6.3.3 Higher-Dimensional Hadamard Matrices from Orthogonal Designs ········ 228
Bibliography ········ 235

Part IV Applications to Signal Design and Analysis

Chapter 7 Design and Analysis of Sequences ········ 239
7.1 Sequences of Cryptographic Significance ········ 239
7.1.1 Enumerating Boolean Functions of Cryptographic Significance ········ 239
7.1.2 Constructing Boolean Functions of Cryptographic Significance ········ 251
7.1.3 Correlation Immunity of Boolean Functions ········ 261
7.1.4 Entropy Immunity of Feedforward Networks ········ 267
7.2 Correlation Functions of Geometric Sequences ········ 276
7.2.1 On the Correlation Functions of a Family of Gold-Geometric Sequences ········ 276
7.2.2 On the Correlation Functions of a Family of Generalized Geometric Sequences ········ 289
7.2.3 On the Correlation Functions of p-Ary d-Form Sequences ········ 306
7.3 Sequence Pairs with Mismatched Filtering ········ 315
7.3.1 Binary Sequences Pairs with Two-Level Autocorrelation Functions (BSPT) ·· 315
7.3.2 Difference Set Pairs ········ 320
7.3.3 Construction of BSPTs ········ 324
7.3.4 Periodic Complementary Binary Sequence Pairs ········ 332
7.4 Sequence Unusual Analysis ········ 340
7.4.1 Boolean Neural Network Design ········ 340
7.4.2 Linear Complexity and Random Sequences with Period 2^n ········ 347
7.4.3 Periodic Ambiguity Functions of EQC-Based TFHC ········ 352
7.4.4 Auto-, Cross-, and Triple Correlations of Sequences ········ 359
Bibliography ········ 368
Chapter 8 Design and Analysis of Arrays ········ 371
8.1 Costas Arrays ········ 372
8.1.1 Correlations of Costas Arrays ········ 372
8.1.2 Algebraically Constructed Costas Arrays ········ 383
8.1.3 Enumeration Limitation of Costas Arrays ········ 389
8.2 Optical Orthogonal Codes ········ 396
8.2.1 Parameters Bounds of Optical Orthogonal Codes ········ 396
8.2.2 Truncated Costas Optical Orthogonal Codes ········ 403
Bibliography ········ 417
Concluding Questions ········ 419
Index ········ 421

Part I
2-Dimensional Cases

Chapter 1

Walsh Matrices

Walsh matrices are the simplest and the most popular special kind of Hadamard matrices, which are defined as the (± 1)-valued orthogonal matrix. Walsh matrices are generated by sampling the Walsh functions, which are families of orthogonal complete functions. The orders of Walsh matrices are always equal to 2^n, where n is a non-negative integer. If the $+1$s in a Walsh matrix are replaced by -1s and -1s by 1s, then a good error correcting code with Hamming distance $m/2$, where m is the order of the matrix, is constructed. Walsh matrices are widely used in communications, signal processing, and physics, and have an extensive and widely scattered literature. This chapter concentrates on the definitions, generations and ordering of Walsh matrices, and on Walsh transforms with fast algorithms.

1.1 Walsh Functions and Matrices

Walsh functions belong to the class of piecewise constant basis functions which were developed in the nineteen twenties and have played an important role in scientific and engineering applications. The foundations of the field of Walsh functions were laid by Rademacher (in 1922), Walsh (in 1923), Fine (in 1945), Paley (in 1952), and Kaczmarz and Steinhaus (in 1951). The engineering approach to the study and utilization of these functions was originated by Harmuth (in 1969), who introduced the concept of sequency to represent the associated, generalized frequency defined as one half the mean rate of zero crossings. Possible applications of Walsh functions to signal multiplexing, bandwidth compression, digital filtering, pattern recognition, statistical analysis, function approximation, and others are suggested and extensively examined.

1.1.1 Definitions

In order to define the Walsh functions, we introduce, at first, a family of important orthogonal (but incomplete) functions which are called Rademacher functions[1]:

$$\mathrm{RAD}(n,t) = \mathrm{sign}[\sin(2^n \pi t)], \quad n = 0, 1, \cdots, \tag{1.1}$$

where $\mathrm{sign}[x]$ is the sign function of x, i.e., $\mathrm{sign}[x] = 1$ if $x > 0$ and $\mathrm{sign}[x] = -1$ if $x < 0$.

Clearly, Rademacher functions are derived from sinusoidal functions which have identical zero crossing positions. Rademacher functions have two arguments n and t such that $R(n,t)$ has 2^{n-1} periods of square wave over a normalized time base $0 \leqslant t \leqslant 1$. The amplitudes of the functions are $+1$ and -1. The first function $R(0,t)$ is equal to one for the entire interval $0 \leqslant t \leqslant 1$. The next and subsequent functions are square waves having odd symmetry.

Rademacher functions are periodic with period 1, i.e.,

$$\text{RAD}(n,t) = \text{RAD}(n,t+1).$$

They are also periodic over shorter intervals such that

$$\text{RAD}(n, t + m2^{1-n}) = \text{RAD}(n,t), \quad n = 1, 2, \cdots; \quad m = \pm 1, \pm 2, \cdots.$$

Rademacher functions can also be generated using the recurrence relation

$$\text{RAD}(n,t) = \text{RAD}(1, 2^{n-1}t)$$

with

$$\text{RAD}(1,t) = \begin{cases} 1, & t \in [0, 1/2), \\ -1, & t \in [1/2, 1). \end{cases}$$

In order to define the Walsh functions we note that each integer n, $0 \leqslant n \leqslant 2^m - 1$, has a unique binary extension of the form

$$n = \sum_{k=0}^{m-1} n_k 2^k, \quad \text{where } n_k = 0 \text{ or } 1. \tag{1.2}$$

Then the n-th Paley-ordered Walsh function is defined by

$$\text{Wal}_{\text{P}}(n,t) = \prod_{k=0}^{m-1} [\text{RAD}(k+1,t)]^{n_k}. \tag{1.3}$$

For example, because $7 = 1 \times 2^2 + 1 \times 2^1 + 1 \times 2^0$,

$$\begin{aligned} \text{Wal}_{\text{P}}(7,t) &= \text{RAD}(3,t)\text{RAD}(2,t)\text{RAD}(1,t) \\ &= \text{sign}[\sin(2^3 \pi t)] \times \text{sign}[\sin(2^2 \pi t)] \times \text{sign}[\sin(2^1 \pi t)]. \end{aligned}$$

Thus Walsh functions form an ordered set of rectangular waveforms taking only two amplitude values +1 and −1. Unlike the Rademacher functions the Walsh rectangular waveforms do not have unit mark-space ratio. Like the sine and cosine functions, two arguments are required for complete definition, a time period, t, and an ordering number, n, related to frequency in a way which is described later.

The Walsh functions can also be defined by their time argument. In fact, each non-negative real number t, $0 \leqslant t < 1$, can be uniquely decomposed as

$$t = \sum_{k=1}^{\infty} t_k 2^{-k}, \quad \text{with } t_k = 0 \text{ or } 1. \tag{1.4}$$

Then the Rademacher functions are derived by

$$\begin{cases} \text{RAD}(0,t) = 1, \\ \text{RAD}(k,t) = (-1)^{t_k}, \quad k = 1, 2, \cdots. \end{cases} \tag{1.5}$$

Hence by setting Equation (1.5) into Equation (1.3), we have the following equivalent definition of the Paley-ordered Walsh functions[1]:

$$\text{Wal}_{\text{P}}(n,t) = (-1)^{\sum_{k=0}^{m-1} n_k t_{k+1}}. \tag{1.6}$$

A straightforward consequence of Equation (1.6) is the following identity:

Lemma 1.1.1 [2–4] *Let q and n be two non-negative integers. Thus*

$$\mathrm{Wal_P}(n,t)\mathrm{Wal_P}(q,t) = \mathrm{Wal_P}(q \oplus n, t), \tag{1.7}$$

where $q \oplus n$ is the dyadic summation of q and n, i.e., $q \oplus n = k$ if and only if their binary extensions $q = (q_0, q_1, \cdots, q_{m-1})$, $n = (n_0, n_1, \cdots, n_{m-1})$, and $k = (k_0, k_1, \cdots, k_{m-1})$ satisfy $(n_i + q_i) \bmod 2 = k_i$ for all $0 \leqslant i \leqslant m-1$.

Thus the set of Paley-ordered Walsh functions forms an Abelian group under the multiplication operation.

Let k, $0 \leqslant k \leqslant 2^m - 1$, be an integer with its binary extension being $k = \sum_{i=0}^{m-1} k_i 2^i$. Then, by Equation (1.6), the discrete sampling of a Walsh function $\mathrm{Wal_P}(n,t)$ at the point $t = k/2^m$ is

$$W_\mathrm{P}(n,k) = (-1)^{\sum_{i=0}^{m-1} n_i k_{m-1-i}}. \tag{1.8}$$

Therefore by sampling the continuous Walsh functions with the unit space $1/2^m$ we have the following (± 1)-valued matrix of size $2^m \times 2^m$:

$$W_\mathrm{P} = [W_\mathrm{P}(n,k)] = \left[(-1)^{\sum_{i=0}^{m-1} n_i k_{m-1-i}}\right]. \tag{1.9}$$

This matrix is called the Paley-ordered Walsh matrix.

For example, if $m = 3$, then the corresponding Paley-ordered Walsh matrix is

$$W_\mathrm{P} = \begin{bmatrix} + & + & + & + & + & + & + & + \\ + & + & + & + & - & - & - & - \\ + & + & - & - & + & + & - & - \\ + & + & - & - & - & - & + & + \\ + & - & + & - & + & - & + & - \\ + & - & + & - & - & + & - & + \\ + & - & - & + & + & - & - & + \\ + & - & - & + & - & + & + & - \end{bmatrix}.$$

Theorem 1.1.1 [2–4] *Let $W_\mathrm{P} = [W_\mathrm{P}(n,k)]$, $0 \leqslant n, k \leqslant 2^m - 1$, be a Paley-ordered Walsh matrix. Then*

(1) *$W_\mathrm{P}(n,k) = W_\mathrm{P}(k,n)$ for all $0 \leqslant n, k \leqslant 2^m - 1$, i.e., the Paley-ordered Walsh matrix is symmetrical;*

(2) *$W_\mathrm{P}(n,k)W_\mathrm{P}(n,q) = W_\mathrm{P}(n, k \oplus q)$ for all $0 \leqslant n, k, q \leqslant 2^m - 1$, i.e., the set of columns of the Paley-ordered Walsh matrix is also closed under the bit-wise multiplication;*

(3) *The matrix W_P is (± 1)-valued and orthogonal. In other words, W_P is a Hadamard matrix.*

Proof The first two statements are direct consequences of Equation (1.9). The third statement is owed to the second statement and the following identity:

$$\sum_{k=0}^{2^m-1} W_\mathrm{P}(n,k) = \sum_{k=0}^{2^m-1} (-1)^{\sum_{i=0}^{m-1} n_i k_{m-1-i}}$$

$$
\begin{aligned}
&= \sum_{k_0=0}^{1} \cdots \sum_{k_{m-1}=0}^{1} (-1)^{\sum_{i=0}^{m-1} n_i k_{m-1-i}} \\
&= \prod_{i=0}^{m-1} \sum_{k_i=0}^{1} (-1)^{n_i k_{m-1-i}} \\
&= 0 \text{ (provided that } n_i \neq 0 \text{ for some } i).
\end{aligned}
$$

In other words, except for the all ones row (the 0-th row), the rows of the matrix W_{P} are balanced by 1 and -1. □

The set of Walsh function series produced by Equation (1.5) or equivalently by Equation (1.3) can also be obtained in several other different ways, each of which has its own particular advantages. The methods considered in the following context are:

(1) By means of a difference equation;

(2) Through the Hadamard matrices.

Both of these derivations are, of course, mathematical processes for which computational algorithms can be developed and the series produced using the digital computer or obtained directly by using digital logic.

From Difference Equations[2–5] This method gives the function directly in sequency order. Sequency is a term used for describing a periodic repetition rate which is independent of waveform. It is defined as "one half of the average number of zero crossings per unit time interval." From this we see that frequency can be regarded as a special measure of sequency applicable to sinusoidal waveforms only. Applying the definition of sequency to periodic and aperiodic function, we obtain:

(1) The sequency of a periodic function equals one half the number of sign changes per period;

(2) The sequency of an aperiodic function equals one half the number of sign changes per unit time, if this limit exists.

Assume that the normalized time base is $0 \leqslant t \leqslant 1$. A given sequency-ordered Walsh function is defined from its preceding harmonic function so that we commence with a definition of $\mathrm{Wal}(0,t) = 1$ within the time-base and 0 outside the time-base. Then the entire set of sequency-ordered Walsh functions $\{\mathrm{Wal}(n,t) : n = 0, 1, \cdots\}$ can be obtained by an iterative process. The difference equation is given as

$$
\mathrm{Wal}(2j+i,t) = (-1)^{\lfloor j/2 \rfloor + i}[\mathrm{Wal}(j,2t) + (-1)^{j+i}\mathrm{Wal}(j,2(t-1/2))], \tag{1.10}
$$

where $i = 0$ or 1, $j = 0, 1, \cdots$ and $\lfloor x \rfloor$ is the floor function.

For $N = 2^m$ equally spaced discrete points, Equation (1.10) can be written as

$$
\mathrm{Wal}(2j+i,n) = (-1)^{\lfloor j/2 \rfloor + i}[\mathrm{Wal}(j,2n) + (-1)^{j+i}\mathrm{Wal}(j,2(n-N/2))], \tag{1.11}
$$

where $0 \leqslant n \leqslant 2^m - 1$.

Commencing with the known $\mathrm{Wal}(0,t) = 1$ within the time base (i.e., $j = 0$, and $i = 1$) for $n \leqslant N/2$ its value for $\mathrm{Wal}(j,2n)$ will be 1 and for $n > N/2$ the function falls outside the time base and will be 0. Similarly to $\text{Wal}(j, 2(n - N/2))$ for $n < N/2$ the function again falls outside the time-base and will become 0, whilst for $n > N/2$ the value is 1. The sign of these functions will be modified by the factors $(-1)^{j+i}$ and $(-1)^{\lfloor j/2 \rfloor + i}$ in accordance with Equation (1.11).

The operation of difference Equation (1.10) may be considered as equivalent to compressing the previous Walsh function $\text{Wal}(j, 2n)$ into the left hand part of the time base by selection of alternate points and, after left adjustment, adding to these on the right hand side a similar valued set of points but all having an opposite sign.

From Walsh-Hadamard Matrices[2–4,6] A Walsh-Hadamard matrix is a (± 1)-valued square array with its rows (and columns) orthogonal to one another. The smallest non-trivial Walsh-Hadamard matrix is

$$H_2 = \begin{bmatrix} 1 & 1 \\ 1 & -1 \end{bmatrix}.$$

The other higher matrices of size $2^m \times 2^m$ can be obtained from the recursive relationship

$$H_{2^m} = \begin{bmatrix} H_{2^{m-1}} & H_{2^{m-1}} \\ H_{2^{m-1}} & -H_{2^{m-1}} \end{bmatrix}, \tag{1.12}$$

i.e., the direct product of $H_{2^{m-1}}$ and H_2 (in general, the direct product of two matrices A and B, represented by $A \otimes B$, is another matrix obtained by replacing the (i, j)-th entry a_{ij} by the matrix $a_{ij}B$). For example,

$$H_4 = H_2 \otimes H_2 = \begin{bmatrix} H_2 & H_2 \\ H_2 & -H_2 \end{bmatrix} = \begin{bmatrix} 1 & 1 & 1 & 1 \\ 1 & -1 & 1 & -1 \\ 1 & 1 & -1 & -1 \\ 1 & -1 & -1 & 1 \end{bmatrix}$$

and

$$H_8 = H_4 \otimes H_2 = \begin{bmatrix} H_4 & H_4 \\ H_4 & -H_4 \end{bmatrix}$$

$$= \begin{bmatrix} 1 & 1 & 1 & 1 & 1 & 1 & 1 & 1 \\ 1 & -1 & 1 & -1 & 1 & -1 & 1 & -1 \\ 1 & 1 & -1 & -1 & 1 & 1 & -1 & -1 \\ 1 & -1 & -1 & 1 & 1 & -1 & -1 & 1 \\ 1 & 1 & 1 & 1 & -1 & -1 & -1 & -1 \\ 1 & -1 & 1 & -1 & -1 & 1 & -1 & 1 \\ 1 & 1 & -1 & -1 & -1 & -1 & 1 & 1 \\ 1 & -1 & -1 & 1 & -1 & 1 & 1 & -1 \end{bmatrix}.$$

When the n-th row of the matrix H_{2^m} is denoted by a function $\text{Wal}_h(n, t)$, $0 \leqslant n \leqslant 2^m - 1$, we obtain the Hadamard-ordered Walsh function series.

Replacing each row of the matrix in Equation (1.12) by its equivalent Paley-ordered Walsh functions, we can form a series of functions which will indicate the ordering obtained through this derivation. For example, from the above matrix H_8, it is easy to see that

$$\text{Wal}_h(0, t) = \text{Wal}_\text{P}(0, t), \quad \text{Wal}_h(1, t) = \text{Wal}_\text{P}(4, t),$$

$$\text{Wal}_h(2,t) = \text{Wal}_\text{P}(2,t), \quad \text{Wal}_h(3,t) = \text{Wal}_\text{P}(6,t),$$

$$\text{Wal}_h(4,t) = \text{Wal}_\text{P}(1,t), \quad \text{Wal}_h(5,t) = \text{Wal}_\text{P}(5,t),$$

$$\text{Wal}_h(6,t) = \text{Wal}_\text{P}(3,t), \quad \text{Wal}_h(7,t) = \text{Wal}_\text{P}(7,t).$$

The relationship between these Walsh-Hadamard matrices and a sampled set Paley-ordered Walsh functions is now clear. They simply express a Walsh function series having positive phasing and arranged in bit-reversed Paley order. This ordering is sometimes referred to as the "lexicographic ordering."

1.1.2 Ordering

From the last subsection we know that the generation of a Walsh function series can be carried out in a number of ways, and that the order of the functions produced can be different from each other. So it is necessary now to consider the ordering of Walsh functions in some details. There are three main ordering conventions in common use:

Sequency Order[1] This is Walsh's original order for his function $\text{Wal}(n,t)$, see Equations (1.10) and (1.11). In this order the components are arranged in ascending order of zero crossings. It is directly related to frequency, in which we find that Fourier components are also arranged in increasing harmonic number (zero-crossings divided by two). Sequency order is closest to our practical experience with other orthogonal functions (e.g., sinusoidal functions).

Paley Order[1] This is the order obtained by generation from successive Rademacher functions, see Equation (1.3). The Paley ordering has certain analytical and computational advantages and is used for most mathematical discussions. The Paley-ordered functions may be defined as the eigenfunctions of a logical differential operator and this definition is of great value in the mathematical development of the theory.

Hadamard Order[1] This ordering follows the Walsh-Hadamard matrices, see Equation (1.12). It is the ordering that is obtained if one computes fast Walsh transforms without sorting in the manner of the Cooley-Tukey fast Fourier transform algorithm. Hence it is computationally advantageous.

Paley order and Hadamard order are used in theoretical mathematical work in image transmission and for computational efficiency. Sequency order is favored for communications and signal processing work such as spectral analysis and filtering.

In order to show the relationships amongst the Walsh functions of different orders, we now introduce the concept of "binary to gray code" conversion and "gray code to binary" conversion as follows:

Binary to gray code conversion: let $0 \leqslant n \leqslant 2^m - 1$ be an integer with its binary extension being

$$(n_{m-1}, n_{m-2}, \cdots, n_0).$$

Its binary to gray code is

$$(g_{m-1}, g_{m-2}, \cdots, g_0),$$

where $g_{m-1} = n_{m-1}$ and $g_i = (n_i + n_{i+1})\text{mod}2$, $0 \leqslant i \leqslant m-2$.

Gray code to binary conversion: converting gray code to binary, we start with the left most digit and move to the right, making $n_i = g_i$ if the number of 1s preceding g_i is even, and making $n_i = 1 - g_i$ if the number of 1s preceding g_i is odd. During this process, zero 1s is treated as an even number of 1s.

Theorem 1.1.2 [1] *Let t, $0 \leqslant t < 1$, be a number and $t = \sum_{k=1}^{\infty} t_k 2^{-k}$, where $t_k = 0$ or 1. And let $(g_{m-1}, g_{m-2}, \cdots, g_0)$ be the binary-to-gray code of $(n_{m-1}, n_{m-2}, \cdots, n_0)$. Then the sequency-ordered Walsh functions can be equivalently defined by*

$$\mathrm{Wal}(n,t) = (-1)^{\sum_{k=0}^{m-1} g_k t_{k+1}} = (-1)^{\sum_{k=0}^{m-1} (n_k + n_{k+1}) t_{k+1}}. \tag{1.13}$$

Proof The functions defined by Equation (1.13) satisfy the recursive relationship in Equations (1.10) and (1.11). □

From Theorem 1.13 and Equation (1.6) we find that the Walsh functions in Paley order and sequency order are related by

$$\mathrm{Wal_P}(n,t) = \mathrm{Wal}(b(n),t), \tag{1.14}$$

where $b(n)$ represents the gray code to binary conversion of the integer n.

Theorem 1.1.3 [1] *The function* $\mathrm{Wal}(n,t)$ *has n zero-crossings. Or, equivalently, the sequency s_n of* $\mathrm{Wal}(n,t)$ *is given by*

$$S_n = \begin{cases} 0, & n = 0, \\ n/2, & n = even, \\ (n+1)/2, & n = odd. \end{cases}$$

Proof By induction on m: the case of $m = 0$ is trivial. Assume that the theorem is true for $n < 2^m$. It is now necessary to prove that the theorem is true for $n < 2^{m+1}$.

If $2^m \leqslant n < 2^{m+1}$ then $n_{m+1} = 1$. Hence by Equation (1.13) we have:

$$\begin{aligned} \mathrm{Wal}(n,t) &= (-1)^{\sum_{k=0}^{m} (n_k + n_{k+1}) t_{k+1}} \\ &= (-1)^{\left[\sum_{k=0}^{m-2} (n_k + n_{k+1}) t_{k+1} + n_{m-1} t_m\right] + n_m t_m + n_m t_{m-1}} \\ &= (-1)^{n_m (t_m + t_{m+1})} \mathrm{Wal}(n - 2^m, t) \\ &= (-1)^{(t_m + t_{m+1})} \mathrm{Wal}(n - 2^m, t). \end{aligned} \tag{1.15}$$

While the $(-1)^{t_m + t_{m+1}}$ produces a zero-crossing only if (t_m, t_{m+1}) is changed from $(0,0)$ to $(0,1)$ or from $(1,0)$ to $(1,1)$. In other words, the zero-crossing points are of the form $t = k/2^{m+1}$, where k is odd. Thus $(-1)^{t_m + t_{m+1}}$ has 2^m zero-crossing points.

From Equation (1.13) the function $\mathrm{Wal}(n - 2^m, t)$ is unchanged at the zero-crossing points of $(-1)^{t_m + t_{m+1}}$. Thus, by Equation (1.15), the zero-crossing points of $(-1)^{t_m + t_{m+1}}$ are also those of $\mathrm{Wal}(n,t)$.

Outside the zero-crossing points of $(-1)^{t_m + t_{m+1}}$, by the induction assumption and Equation (1.15), the function $\mathrm{Wal}(n - 2^m, t)$ has $n - 2^m$ zero-crossing points.

Thus, the total zero-crossing points of $\mathrm{Wal}(n,t)$ is equal to $2^m + (n - 2^m) = n$. □

An alternative notation of $\mathrm{Wal}(n,t)$ is to classify the Walsh functions in terms of even and odd waveform symmetry, viz.,

$$\begin{cases} \mathrm{Wal}(2k,t) = \mathrm{Cal}(k,t), & \text{called the Walsh cosine,} \\ \mathrm{Wal}(2k-1,t) = \mathrm{Sal}(k,t), & \text{called the Walsh sine,} \end{cases} \tag{1.16}$$

where $0 \leqslant k \leqslant 2^{m-1}$. This classification defines two further Walsh series having close similarities with the cosine and sine series. It can be indicated that the normalized Walsh functions are symmetrical about their mid or zero time point. Defining the range of the function as $-1/2 \leqslant t \leqslant 1/2$, the functions are either directly symmetrical (Cal functions) or inversely symmetrical (Sal functions). In the latter case the ones found in the left hand side are mirrored by zeros in the right-hand side and vice versa. Thus, a symmetry relationship can be stated as $\text{Wal}(n,t) = \text{Wal}(t,n)$.

From Equations (1.5) and (1.13) we obtain the following equivalent definition of sequency-ordered Walsh functions:

$$\text{Wal}(n,t) = \prod_{k=0}^{m-1} \text{RAD}(k+1,t)^{g_k}, \tag{1.17}$$

where $g=(g_{m-1}, g_{m-2}, \cdots, g_0)$ is the binary to gray code of the integer $n=(n_{m-1}, n_{m-2}, \cdots, n_0)$.

By sampling the functions in Equation (1.17) with the unit space $1/2^m$, we get the sequency-ordered Walsh matrix:

$$W = [W(n,i)] = \left[(-1)^{\sum_{k=0}^{m-1}(n_k+n_{k+1})i_{m-1-k}}\right], \tag{1.18}$$

where $0 \leqslant n, i \leqslant 2^m - 1$. This matrix W is also a (± 1)-valued orthogonal matrix of size $2^m \times 2^m$.

Similarly, the Hadamard-ordered Walsh functions are equal to

$$\text{Wal}_h(n,t) = \prod_{k=0}^{m-1} \text{RAD}(k+1,t)^{n_{m-1-k}}, \tag{1.19}$$

and its matrix form is

$$W_h = [W_h(n,i)] = \left[(-1)^{\sum_{k=0}^{m-1} n_k i_k}\right], \tag{1.20}$$

where $0 \leqslant n, i \leqslant 2^m - 1$. This matrix W_h is also a (± 1)-valued orthogonal matrix of size $2^m \times 2^m$.

From Equations (1.19) and (1.17), it is clear that the Hadamard-ordered Walsh functions are related to those in sequency order by the relation

$$\text{Wal}_h(n,t) = \text{Wal}(b(\langle n \rangle), t), \tag{1.21}$$

where $\langle n \rangle$ is obtained by the bit-reversal of n, and $b(\langle n \rangle)$ is the gray code to binary conversion of $\langle n \rangle$.

Similarly, from Equations (1.19) and (1.3), we have

$$\text{Wal}_h(n,t) = \text{Wal}_\text{P}(\langle n \rangle, t). \tag{1.22}$$

Up to now, the relationships amongst Walsh functions and matrices of the Paley, sequency, and Hadamard orders have been completed by Equations (1.14), (1.21), and (1.22).

Besides the Paley order, Hadamard order, and sequency order, there are many other orders for Walsh functions. It has been proved that the universal order of Walsh functions, denoted by $\text{UW} = [\text{UW}(i,j)]$, $0 \leqslant i, j \leqslant 2^n - 1$, can be defined by

$$\text{UW}(i,j) = (-1)^{iAj'} \tag{1.23}$$

for some non-singular matrix $A = [A(i,j)]$ of size $n \times n$, where $\boldsymbol{i} := (i_0, i_1, \cdots, i_{n-1})$ and $\boldsymbol{j} := (j_0, j_1, \cdots, j_{n-1})$ are the binary expressions of the integers i and j, respectively.

For example, if

$$A = \begin{bmatrix} 0 & 0 & 0 & 0 & \cdots & 0 & 0 & 1 & 1 \\ 0 & 0 & 0 & 0 & \cdots & 0 & 1 & 1 & 0 \\ 0 & 0 & 0 & 0 & \cdots & 1 & 1 & 0 & 0 \\ \vdots & \vdots & \vdots & \vdots & & \vdots & \vdots & \vdots & \vdots \\ 0 & 1 & 1 & 0 & \cdots & 0 & 0 & 0 & 0 \\ 1 & 1 & 0 & 0 & \cdots & 0 & 0 & 0 & 0 \end{bmatrix},$$

$$\begin{bmatrix} 0 & 0 & 0 & \cdots & 0 & 0 & 1 \\ 0 & 0 & 0 & \cdots & 0 & 1 & 0 \\ 0 & 0 & 0 & \cdots & 1 & 0 & 0 \\ \vdots & \vdots & \vdots & & \vdots & \vdots & \vdots \\ 0 & 0 & 1 & \cdots & 0 & 0 & 0 \\ 0 & 1 & 0 & \cdots & 0 & 0 & 0 \\ 1 & 0 & 0 & \cdots & 0 & 0 & 0 \end{bmatrix}$$

or

$$\begin{bmatrix} 1 & 0 & 0 & \cdots & 0 & 0 & 0 \\ 0 & 1 & 0 & \cdots & 0 & 0 & 0 \\ 0 & 0 & 1 & \cdots & 0 & 0 & 0 \\ \vdots & \vdots & \vdots & & \vdots & \vdots & \vdots \\ 0 & 0 & 0 & \cdots & 1 & 0 & 0 \\ 0 & 0 & 0 & \cdots & 0 & 1 & 0 \\ 0 & 0 & 0 & \cdots & 0 & 0 & 1 \end{bmatrix},$$

then the Walsh functions in Equation (1.23) are sequency-ordered, Paley-ordered, and Hadamard-ordered, respectively.

Because the number of non-singular matrices, in GF(2) and of size $n \times n$, is enumerated by

$$S(n) =: \begin{cases} \prod_{i=1}^{n/2} (2^{n+1} - 2^{2i}), & n \text{ even}, \\ \prod_{i=0}^{(n-1)/2} (2^n - 2^{2i}), & n \text{ odd}, \end{cases}$$

there are at all $S(n)$ different orders for the Walsh functions defined by binary matrices of size $n \times n$.

1.2 Orthogonality and Completeness

The mathematical techniques of studying functions, signals, and systems through series expansions in orthogonal complete sets of basis functions are now a standard tool in all branches of science and engineering. This section introduces the orthogonality and completeness of Walsh functions and matrices.

1.2.1 Orthogonality

A family of real-valued functions $S_n(t)$, $n = 0, 1, \cdots$, is said to be orthogonal with weight K, a non-negative constant, over the interval $0 \leqslant t \leqslant T$ if each pair of its member functions, say $S_n(t)$ and $S_m(t)$, are orthogonal with each other, i.e.,

$$\int_0^T K S_n(t) S_m(t) \mathrm{d}t = \begin{cases} K, & \text{if } n = m, \\ 0, & \text{if } n \neq m. \end{cases} \tag{1.24}$$

If the constant K is one, then the family is normalized and is called an orthonormal family of functions. An orthogonal non-normalized family can always be converted into an orthonormal family.

Theorem 1.2.1 [1–4,7] *The family of Walsh functions* $\{\mathrm{Wal_P}(n, t) : 0 \leqslant n \leqslant 2^m - 1, 0 \leqslant t < 1\}$ *is orthonormal over the interval* $[0, 1)$. *In other words,*

$$\int_0^1 \mathrm{Wal_P}(u, t) \mathrm{Wal_P}(v, t) \mathrm{d}t = 0, \quad 0 \leqslant u \neq v \leqslant 2^m - 1 \tag{1.25}$$

and

$$\int_0^1 [\mathrm{Wal_P}(n, t)]^2 = 1, \quad n = 0, 1, 2, \cdots. \tag{1.26}$$

Proof The Equation (1.26) is trivial, because of the identity $[\mathrm{Wal_P}(n,\ t)]^2 = 1$. The Equation (1.25) is based on the facts that (1) the non-trivial rows of the discrete sampling of $\mathrm{Wal_P}(n, t)$, or equivalently the Walsh matrix W_p, are balanced by 1s and -1s; and (2) the functions $\{\mathrm{Wal_P}(n, t) : 0 \leqslant n \leqslant 2^m - 1, 0 \leqslant t < 1\}$ are of waveforms with their space being larger than or equal to $1/2^m$. □

From Theorem 1.2.1 and Equations (1.14), (1.21), and (1.22), it is easy to see that the Walsh functions in both Hadamard and sequency orders are also orthonormal over the interval $[0, 1)$.

1.2.2 Completeness

Let $f(t)$ be a signal time function defined over a time interval $(0, T)$, and $S_n(t)$, $n = 0, 1, \cdots$ be a family of orthogonal functions. When we use the following series

$$\sum_{n=0}^{N-1} C_n S_n(t), \quad N = 1, 2, \cdots$$

to represent the function $f(t)$, the resultant mean square approximation error (MSE) is

$$\mathrm{MSE} = \int_0^T \left[f(t) - \sum_{n=0}^{N-1} C_n S_n(t) \right]^2 \mathrm{d}t. \tag{1.27}$$

The family $S_n(t)$, $n = 0, 1, \cdots$ is called a complete orthogonal function series, if the MSE in Equation (1.27) monotonically decreases to zero as the integer N increases to infinite, where the coefficient C_n, $n = 0, 1, \cdots$ is defined by

$$C_n = \frac{1}{T} \int_0^T f(t) S_n(t) \mathrm{d}t. \tag{1.28}$$

A function series is called incomplete if it is not complete. For example, the Rademacher functions defined by Equation (1.1) are incomplete.

A complete orthogonal function series $S_n(t)$ is always a closed series, or equivalently, there exists no $f(t)$ that satisfies both $0 < \int_0^T f^2(t)\mathrm{d}t < \infty$ and $\int_0^T f(t)S_n(t)\mathrm{d}t = 0$ for each integer n. More generally, a function series is said to be complete (or closed) if no function exists which is orthogonal to every function of the series unless the integral of the square of the function is itself zero.

One of the most important complete orthogonal function series defines the trigonometric functions sine and cosine by

$$S_n(t) = \sqrt{2}\cos(2\pi nt) \quad \text{or} \quad \sqrt{2}\sin(2\pi nt). \tag{1.29}$$

Thus, every function can be expressed as a sum of sinusoidal components (Fourier series), viz.,

$$f(t) = \frac{a_0}{2} + \sum_{k=1}^{\infty}[a_k \cos(k\omega_0 t) + b_k \sin(k\omega_0 t)].$$

Similarly to the sine-cosine functions, we have the following:

Theorem 1.2.2 [1–4,7] *The family of Walsh functions* $\{\mathrm{Wal_P}(n,t) : 0 \leqslant n \leqslant 2^m - 1, 0 \leqslant t < 1\}$ *is complete and orthogonal. In other words, if* $f(t)$ *is a function satisfying both* $f(t) \in L^2[0,1)$, *i.e.,* $f^2(t)$ *is integrable over* $[0,1)$, *and*

$$\int_0^1 f(t)\mathrm{Wal_P}(n,t)\mathrm{d}t = 0, \quad n = 0, 1, 2, \cdots,$$

then the $f(t)$ *is a zero function.*

This theorem together with Equations (1.14), (1.21), and (1.22) implies that the Walsh functions in both Hadamard and sequency orders are also complete orthogonal.

Moreover, it has been proved that:

Theorem 1.2.3 [1–4,7] *Every function* $f(t) \in L^2[0,1)$ *can be decomposed as*

$$\begin{aligned} f(t) &= a_0\mathrm{Cal}(0,t) + \sum_{k=1}^{\infty}[a_k\mathrm{Cal}(k,t) + b_k\mathrm{Sal}(k,t)] \\ &= \sum_{k=0}^{\infty}[a_k\mathrm{Cal}(k,t) + b_{k+1}\mathrm{Sal}(k+1,t)], \end{aligned}$$

where the coefficients a_k *and* b_k *are defined by*

$$a_k = \int_0^1 f(t)\mathrm{Cal}(k,t)\mathrm{d}t, \quad k = 0, 1, 2, \cdots$$

and

$$b_k = \int_0^1 f(t)\mathrm{Sal}(k,t)\mathrm{d}t, \quad k = 1, 2, 3, \cdots.$$

The Parseval Identity is now stated as:

Theorem 1.2.4 *If $f(t) \in L^2[0,1)$, then*

$$\int_0^1 [f(t)]^2 \mathrm{d}t = \sum_{k=0}^{\infty} [a_k^2 + b_{k+1}^2],$$

where a_k and b_k are the same as those in Theorem 1.2.3.

1.3 Walsh Transforms and Fast Algorithms[8]

The famous Fourier transforms are based on the orthogonal and complete families of functions, sine and cosine functions. Walsh matrices are sampled from another orthogonal and complete families of functions called Walsh functions, thus a great deal of research in the field of Walsh functions has been devoted to the study and development of easy and fast algorithms for computing Walsh transforms. The sampling principle in the framework of Walsh signals and transforms was studied by Kak (1970) and Cheng and Johnson (1973). Their sequency-based sampling theorem states that: "a causal time function $x(t)$ sequency-band limited to $B = 2^k$ zeros per second, can be uniquely reconstructed from its samples at every $T = 1/B$ seconds for all positive time". The simplest version of a Walsh transform which provides the coefficients in Hadamard ordering, was given by Whechel and Guinn (1968). They actually derived the so-called fast Hadamard transform (FHT) by using the recursive structure of the Hadamard matrix and applying a modified form of the Cooley-Tukey FFT algorithm. In 1970, Henderson showed how to circumvent the difficulty of obtaining the Walsh coefficients in bit-reversed order when the data are arranged in the Hadamard forward order. Andrews and Caspari (1970) presented a generalized approach for fast digital Walsh-Hadamard spectral analysis using the direct multiplication of matrices. Another method for computing the Walsh-Hadamard transform and the R-transform was presented in 1970 by Ulman, where the transform coefficients were produced in increasing sequency order.

A real time technique for computing the digital Walsh-Hadamard transform of two-dimensional pictures was developed by Alexandridis and Klinger (1972) by using the matrix formulation. A hardware implementation for real-time, parallel Walsh-Hadamard transformation was also provided. In 1972, another sequency-ordered fast Walsh transform (FWT) computation algorithm was presented. This algorithm was based on the Cooley-Tukey type of FHT and required an equal computational effort as the FHT.

This section is devoted to the study of the Walsh-Hadamard transform, which is perhaps the most well known of the non-sinusoidal orthogonal transforms. Fast algorithms to compute these transforms are developed and the notion of Walsh spectra is introduced.

1.3.1 Walsh-Ordered Walsh-Hadamard Transforms[8]

For details about the contents of this subsection the readers are recommended to [1].

Before considering the development of Walsh-Hadamard transforms, it is instructive to study some aspects of representing a given continuous signal $f(t)$ in the form of a Walsh series. For the purposes of discussion we will assume that $f(t)$ is defined on the half-open unit interval $t \in [0,1)$. From Theorem 1.2.3, it is known that every integrable function $f(t)$

may be represented by the following Walsh series

$$f(t) = a_0 + a_1 \mathrm{Wal}(1,t) + a_2 \mathrm{Wal}(2,t) + \cdots, \tag{1.30}$$

where $t \in (0,1)$ and for each integer n the coefficient a_n is

$$a_n = \int_0^1 f(t)\mathrm{Wal}(n,t)dt. \tag{1.31}$$

In other words, we have the following pair of transforms:

(1) The forward transform is

$$f(t) = \sum_{n=0}^{\infty} F_n \mathrm{Wal}(n,t); \tag{1.32}$$

(2) The reverse transform is

$$F_n = \int_0^1 f(t)\mathrm{Wal}(n,t)dt. \tag{1.33}$$

If the integration in Equation (1.33) is replaced by summation through the use of the trapezium rule of $N = 2^m$ sampling points f_i, then we obtain the following finite discrete Walsh transform pair

$$F_n = \frac{1}{N}\sum_{i=0}^{N-1} f_i \mathrm{Wal}(n,i), \quad 0 \leqslant n \leqslant N-1; \quad \text{(forward transform)} \tag{1.34}$$

and

$$f_i = \sum_{n=0}^{N-1} F_n \mathrm{Wal}(n,i), \quad 0 \leqslant i \leqslant N-1; \quad \text{(inverse transform)} \tag{1.35}$$

Let $W = [\mathrm{Wal}(n,i)]$, $0 \leqslant n, i \leqslant N-1$, be the sequency-ordered Walsh matrix of size $N \times N$. Then the transforms in Equations (1.34) and (1.35) can be equivalently stated as

$$F = \frac{1}{N} fW \tag{1.36}$$

and

$$f = FW, \tag{1.37}$$

where $f := (f_0, \cdots, f_{N-1})$ and $F := (F_0, \cdots, F_{N-1})$. The transforms defined by Equations (1.36) and (1.37) are called the Walsh-ordered Walsh-Hadamard transforms (WHT_W). It is evident from this definition that N^2 additions and subtractions are required to compute the WHT_W coefficients F. In the following text, we will develop a fast algorithm which yields the F in only $N \log_2 N$ additions and subtractions.

The fast Walsh-ordered Walsh-Hadamard transform (FWHT_W) is illustrated in the following four steps:

Step 1. Bit-reverse the input and order it in ascending index order. If $f =: (f_0, \cdots, f_{N-1})$ is the given data sequence, then we denote the bit-reversed sequence in ascending index order by $f' =: (f'_0, \cdots, f'_{N-1})$;

Step 2. Define a reversal, which is illustrated by:

without a reversal, we would have

$$\left.\begin{aligned} f'_{k+1}(s) &= f'_k(s) + f'_k(s+2) \\ f'_{k+1}(s+1) &= f'_k(s+1) + f'_k(s+3) \end{aligned}\right\} \text{additions}$$

and

$$\left.\begin{aligned} f'_{k+1}(s+2) &= f'_k(s) - f'_k(s+2) \\ f'_{k+1}(s+3) &= f'_k(s+1) - f'_k(s+3) \end{aligned}\right\} \text{subtractions.}$$

However, with a reversal we have

$$\left.\begin{aligned} f'_{k+1}(s) &= f'_k(s) - f'_k(s+2) \\ f'_{k+1}(s+1) &= f'_k(s+1) - f'_k(s+3) \end{aligned}\right\} \text{subtractions}$$

and

$$\left.\begin{aligned} f'_{k+1}(s+2) &= f'_k(s) + f'_k(s+2) \\ f'_{k+1}(s+3) &= f'_k(s+1) + f'_k(s+3) \end{aligned}\right\} \text{additions.}$$

Step 3. Define a block, which is a group of additions/subtractions which are disconnected from their neighbors, above or below;

Step 4. List the rules for locating the blocks where reversals occur.

1.3.2 Hadamard-Ordered Walsh-Hadamard Transforms

For details of the contents of this subsection the readers are recommended to the paper [8].

Let H_N be the Walsh-Hadamard matrix of size $N \times N = 2^n \times 2^n$ that is recursively constructed by Equation (1.12). The forward Hadamard-ordered Walsh-Hadamard transform (WHT_H) of the data sequence $g = (g_0, g_1, \cdots, g_{N-1})$ is defined by

$$G =: (G_0, \cdots, G_{N-1}) = \frac{1}{N} g H_N. \tag{1.38}$$

Because the Walsh-Hadamard matrix H_N is symmetric (i.e., $H'_N = H_N$) and orthogonal (i.e., $H'_N H_N = N I_N$) and the inverse matrix of H_N is proportional to itself (i.e., $(H_N)^{-1} = H_N/N$), the inverse transform of Equation (1.38) is

$$g = G H_N, \tag{1.39}$$

which is called the inverse Hadamard-ordered Walsh-Hadamard transform (IWHT_H).

The other equivalent definition of the Walsh-Hadamard matrix H_N, $N = 2^n$, is

$$H_N = \overbrace{H_2 \otimes H_2 \otimes \cdots \otimes H_2}^{n},$$

where $\otimes$ stands for the direct multiplication of matrices.

Thus the matrix H_N can be decomposed as the multiplication of the following n sparse matrices

$$H_N = \prod_{k=0}^{n-1} (I_{2^k} \otimes H_2 \otimes I_{2^{n-1-k}}). \tag{1.40}$$

Setting Equation (1.40) into (1.38) and (1.39), a fast Hadamard-ordered Walsh-Hadamard transform is obtained which requires only $N \log_2 N$ additions and subtractions.

Bibliography

[1] Ahmed N and Rao K R. *Orthogonal Transforms for Digital Signal Processing.* Berlin: Springer-Verlag, 1975.

[2] Beauchamp K G. *Applications of Walsh and Related Functions.* London: Academic Press, 1984.

[3] Tzafestas S G. *Walsh Functions in Signal and Systems Analysis and Design.* New York: Van Nostrand Reinhold Co., 1985.

[4] Yang Y X and Lin X D. *Coding and Cryptography.* Beijing: PPT Press, 1992.

[5] Song H Y and Golomb S W. On the existence of cyclic Hadamard difference sets. *IEEE Trans. IT*, 1994, 40(4): 1266-1270.

[6] Gotsman C. A note on functions governed by Walsh expressions. *IEEE Trans. on Inform. Theory*, 1991, 37(3): 694-695.

[7] Beauchamp K G. *Walsh Functions and Their Applications.* London: Academic Press, 1975.

[8] Banta E D. Coding and decoding nonlinear block codes using logical hadamard transform. *IEEE Trans. IT*, 1978, 24(4): 761-763.

Chapter 2

Hadamard Matrices

Hadamard matrices were originally investigated as a family of orthogonal matrices by Sylvester in 1857. In 1893, Hadamard proved that if $A = [A_{i,j}]$, $0 \leqslant i, j \leqslant n-1$, is a matrix of size $n \times n$ (or called "of order n"), then

$$|\det A|^2 \leqslant \prod_{i=0}^{n-1} \sum_{j=0}^{n-1} |A_{i,j}|^2 . \tag{2.1}$$

Furthermore, Hadamard discovered that if $|A_{i,j}| \leqslant 1$ and the equality (2.1) is satisfied, then the matrix A is (± 1)-valued and its size m is equal to 1, 2, or $4t$. Subsequently matrices satisfying the Equality (2.1) came to be called Hadamard matrices, which were further studied by Scarpis in 1898. The next major milestone was owed to Paley, in 1933. In 1944 and 1947, Williamson obtained more results of considerable interest. With regard to the practical applications of Hadamard matrices, Hall, Baumert, and Golomb working at the U.S. Jet Propulsion Laboratories (JPL) sparked the interest in Hadamard matrices over the past 30 years[1]. In the sixties the JPL was working toward building the Mariner and Voyager space probes to visit Mars and the other planets of the solar system. In order to obtain high quality color pictures of the backs of the planets, the space probes have to take three black and white pictures through red, green, and blue filters. Each picture is then divided into a 1000×1000 array of black and white pixels. Each pixel is graded on a scale of 1 to 16. These grades are then used to choose a code-word in an eight error-correcting code based on the Hadamard matrix of order 32. The code word is transmitted to the earth. After error-correction, the three black and white pictures are reconstructed and then a computer is used to obtain the colored pictures. Hadamard matrices are used for these code words for two reasons. First, error-correcting codes based on Hadamard matrices have good error correction capability and good decoding algorithms. Second, because Hadamard matrices are (± 1)-valued, all the computer processing can be accomplished using additions and subtractions rather than multiplication.

Up to now, besides more and more applications, much of the research work is devoted to the proof of the following famous Hadamard conjecture:

There exists at least one Hadamard matrix of size 1×1, 2×2 *and* $4t \times 4t$, *for each positive integer* t.

This chapter concentrates on the basic definitions, constructions and existence results about the Hadamard matrices. Many references are listed at the end of this chapter so that interested readers can trace the up-to-date results related to Hadamard matrices.

2.1 Definitions

2.1.1 Hadamard Matrices

Although the Hadamard matrices can be defined by the Equality (2.1), in the following text we will use another equivalent and more intuitive definition:

Definition 2.1.1 [2] *Let $H = [H_{i,j}]$, $0 \leqslant i, j \leqslant n-1$ be a (± 1)-valued matrix of size $n \times n$. Then H is called a Hadamard matrix if and only if*

$$HH' = nI_n, \tag{2.2}$$

where H' stands for the transpose of H and I_n is the unit matrix of order n.

In other words, a Hadamard matrix is an orthogonal (± 1)-valued matrix. Thus $H'H = HH'$. Up to now many Hadamard matrices have been constructed. For example, the Hadamard matrices of order less than or equal to 200 have been proved to exist for the following orders[3]: 4, 8, 12, 16, 20, 24, 28, 32, 36, 40, 44, 48, 52, 56, 60, 64, 68, 72, 76, 80, 84, 88, 92, 96, 100, 104, 108, 112, 116, 120, 124, 128, 132, 136, 140, 144, 148, 152, 156, 160, 164, 168, 172, 176, 180, 184, 192, 196, 200.

It is obvious that the set of Hadamard matrices is closed under the following five transforms:

(1) Permuting the rows;
(2) Permuting the columns;
(3) Multiplying some rows by -1;
(4) Multiplying some columns by -1;
(5) Transposing.

By using these five transforms in suitable order, every Hadamard matrix can be changed to a Hadamard matrix such that its first row (resp. column) is all 1 which is called row-normal (resp. column-normal).

Definition 2.1.2 [4] *A Hadamard matrix is called normal iff it is both row-normal and column-normal.*

For example, the Hadamard-ordered Walsh matrices introduced in the last chapter are normal Hadamard matrices of order 2^m, $m = 1, 2, \cdots$. If H is a normal Hadamard matrix of order $4n$, then every row (column) except the first has $2n$ + 1s in each row (column); further, n − 1s in any row (column) overlap n − 1s in each other row (column).

An important necessary condition for Hadamard matrices is that their sizes should be the multiple of 4, precisely,

Theorem 2.1.1 [1] *A matrix of size $m \times m$, $m > 2$, is a Hadamard matrix only if $4 \mid m$.*

Proof Let the matrix be $A = [A_{ij}]$. Because of the orthogonality we have

$$\sum_{i=0}^{m-1} (A_{1i} + A_{2i})(A_{1i} + A_{3i}) = \sum_{i=0}^{m-1} A_{1i}^2 = m. \tag{2.3}$$

Whilst, on the other hand,

$$A_{1i} + A_{2i} = \pm 2, 0 \quad \text{and} \quad A_{1i} + A_{3i} = \pm 2, 0.$$

Thus, the summation in Equation (2.3) is of the form $4t$, because each of its terms is a multiple of 4. □

Clearly, the famous Hadamard conjecture is the inverse of this theorem. Another subclass of Hadamard matrices is the regular subclass defined by:

Definition 2.1.3 [5] *A Hadamard matrix H is called regular if there exists some integer, say t, such that*

$$HJ = tJ.$$

Equivalently, the sum of each row of the regular Hadamard matrix H is equal to the same number t.

For example[1], the following matrix H is a regular Hadamard matrix of order 16 with $t = -4$:

$$H = \begin{bmatrix}
-1 & 1 & 1 & 1 & 1 & 1 & -1 & -1 & -1 & -1 & 1 & -1 & -1 & -1 & -1 & -1 \\
1 & -1 & 1 & -1 & -1 & 1 & 1 & 1 & -1 & -1 & -1 & -1 & -1 & 1 & -1 & -1 \\
1 & 1 & -1 & -1 & -1 & 1 & -1 & -1 & 1 & 1 & -1 & -1 & -1 & -1 & 1 & -1 \\
1 & -1 & -1 & -1 & 1 & -1 & 1 & -1 & 1 & 1 & 1 & -1 & -1 & -1 & -1 & -1 \\
1 & -1 & -1 & 1 & -1 & -1 & -1 & 1 & -1 & 1 & 1 & -1 & 1 & -1 & -1 & -1 \\
1 & 1 & 1 & -1 & -1 & -1 & -1 & -1 & -1 & -1 & -1 & 1 & 1 & -1 & -1 & 1 \\
-1 & 1 & -1 & 1 & -1 & -1 & -1 & 1 & 1 & -1 & -1 & 1 & -1 & 1 & -1 & -1 \\
-1 & 1 & -1 & -1 & 1 & -1 & 1 & -1 & -1 & 1 & -1 & -1 & 1 & 1 & -1 & -1 \\
-1 & -1 & 1 & 1 & -1 & -1 & 1 & -1 & -1 & 1 & -1 & 1 & -1 & -1 & 1 & -1 \\
-1 & -1 & 1 & -1 & 1 & -1 & -1 & 1 & 1 & -1 & -1 & -1 & 1 & -1 & 1 & -1 \\
1 & -1 & -1 & 1 & 1 & -1 & -1 & -1 & -1 & -1 & -1 & -1 & -1 & 1 & 1 & 1 \\
-1 & -1 & -1 & 1 & -1 & 1 & 1 & -1 & 1 & -1 & -1 & -1 & 1 & -1 & -1 & 1 \\
-1 & -1 & -1 & -1 & 1 & 1 & -1 & 1 & -1 & 1 & -1 & 1 & -1 & -1 & -1 & 1 \\
-1 & 1 & -1 & -1 & -1 & -1 & 1 & 1 & -1 & -1 & 1 & -1 & -1 & -1 & 1 & 1 \\
-1 & -1 & 1 & -1 & -1 & -1 & -1 & -1 & 1 & 1 & 1 & -1 & -1 & 1 & -1 & 1 \\
-1 & -1 & -1 & -1 & -1 & 1 & -1 & -1 & -1 & -1 & 1 & 1 & 1 & 1 & 1 & -1
\end{bmatrix}.$$

A Hadamard matrix H is called a symmetric Hadamard matrix iff $H' = H$, exactly one of the top left corners of H and $-H$ is 1. Rename by B the matrix with 1 as its top left corner. Multiplying by -1 those columns of B where the top most row is -1ed. Similarly, multiplying by -1 those rows of B where the left most column is -1ed. Then the matrix B is changed into the following normal form

$$A = \begin{pmatrix} 1 & e \\ e' & C \end{pmatrix}, \tag{2.4}$$

where $e = (1, 1, \cdots, 1)$ is the all-ones vector of length $m - 1$. The matrix C in Equation (2.4) is called the kernel of the symmetric Hadamard matrix H.

Lemma 2.1.1 *Let C be the kernel of a symmetric Hadamard matrix H of order m. Then*

$$C' = C; \quad CJ_{m-1} = J_{m-1}C = -J_{m-1}; \quad C^2 = mI_{m-1} - J_{m-1}. \tag{2.5}$$

Proof Without loss of generality, we assume that

$$H = \begin{pmatrix} 1 & e \\ e' & C \end{pmatrix}. \tag{2.6}$$

Because $H = H'$, we have $C = C'$. From the equation

$$\begin{aligned}
mI_m &= HH' \\
&= \begin{pmatrix} 1 & e \\ e' & C \end{pmatrix} \begin{pmatrix} 1 & e \\ e' & C \end{pmatrix} \\
&= \begin{pmatrix} m & e + eC \\ e' + Ce' & e'e + C^2 \end{pmatrix},
\end{aligned}$$

we obtain the equations

$$eC = -e, \quad C^2 = mI_{m-1} - e'e = mI_{m-1} - J_{m-1}.$$

The lemma follows. □

Definition 2.1.4 *A Hadamard matrix H of order m is called a skew Hadamard matrix iff*

$$H = S + I \text{ and } S' = -S. \tag{2.7}$$

Lemma 2.1.2 *Let H be a skew Hadamard matrix of order m, see Equation* (2.7). *Then*

$$SS' = (m-1)I. \tag{2.8}$$

Proof The lemma follows the identity:

$$\begin{aligned} mI &= HH' \\ &= (S+I)(S'+I) \\ &= SS' + I. \end{aligned}$$

□

The condition $S' = -S$ implies that all of the diagonal elements of S are zero. Multiplying -1s to some rows and columns, the matrix S can be changed to

$$\begin{pmatrix} 0 & e \\ -e' & W \end{pmatrix}. \tag{2.9}$$

The matrix W in Equation (2.9) is called the kernel of the skew Hadamard matrix H.

Lemma 2.1.3 *Let W be the kernel of a skew Hadamard matrix H of order m. Then*

$$WW' = (m-1)I_{m-1} - J_{m-1} \tag{2.10}$$

and

$$WJ_{m-1} = 0, \quad W' = -W. \tag{2.11}$$

Proof Without loss of generality we assume that

$$H = I + \begin{pmatrix} 0 & e \\ -e' & W \end{pmatrix} =: I + S.$$

The equation $S' = -S$ implies $W' = -W$. From Lemma 2.1.2 we have

$$\begin{aligned} (m-1)I &= \begin{pmatrix} 0 & e \\ -e' & W \end{pmatrix} \begin{pmatrix} 0 & -e \\ e' & W' \end{pmatrix} \\ &= \begin{pmatrix} ee' & eW' \\ We' & e'e + WW' \end{pmatrix}. \end{aligned}$$

Hence

$$We' = 0, \quad J_{m-1} + WW' = (m-1)I_{m-1}.$$

The lemma follows. □

Definition 2.1.5 *Let M be a skew Hadamard matrix, and N a symmetric Hadamard matrix of the same order. The pair M and N are called amicable Hadamard matrices if*

$$MN = NM'. \tag{2.12}$$

The Equation (2.12) implies that MN is also a symmetric matrix.

Lemma 2.1.4 *If $M = S + I$ and N are a pair of amicable Hadamard matrices, then*

$$SN = NS'. \tag{2.13}$$

Proof Because of $MN = (S + I)N = SN + N$, and $NM' = N(S' + I) = NS' + N$, the proof is finished by Equation (2.12). □

2.1.2 Hadamard Designs

Let $S = \{s_1, s_2, \cdots, s_v\}$ be a v-element set and $B = \{B_1, B_2, \cdots, B_b\}$ a family consisted of b subsets of S. The family B is called a (b, v, r, k, λ)-balanced incomplete block design, or a (b, v, r, k, λ)-BIBD in short, if and only if all of the following three conditions are satisfied:

(1) Each subset B_i, $1 \leqslant i \leqslant b$, contains the same number, k, of elements of S;

(2) Each element $s \in S$ appears in exactly r subsets of B;

(3) Each pair $\{s_i, s_j\}$ is contained in just λ subsets of B.

The family B can also be uniquely described by its associated matrix $A = [A(i,j)]$, which is a $(0,1)$-valued matrix of size $b \times v$ and for each $1 \leqslant i \leqslant b$, $1 \leqslant j \leqslant v$,

$$A(i,j) = \begin{cases} 1, & \text{if } s_j \in B_i, \\ 0, & \text{if } s_j \notin B_i. \end{cases} \tag{2.14}$$

In particular, if $b = v$ and $r = k$, then a (b, v, r, k, λ)-BIBD is called a symmetrical balanced incomplete block design, or shortly, a (v, k, λ)-design. It is known that the associated matrix of every (v, k, λ)-design satisfies the necessary conditions presented in the following lemma.

Lemma 2.1.5 *If $A = [A(i,j)]$ is the associated matrix of some (v, k, λ)-design, then,*

$$\lambda(v-1) = k(k-1); \tag{2.15}$$

$$A'A = (k-\lambda)I_v + \lambda J_v; \tag{2.16}$$

$$Aw_v = kw_v; \tag{2.17}$$

$$\frac{(v+\lambda)}{2} \geqslant k \geqslant \sqrt{\lambda v}. \tag{2.18}$$

If v is even, then

$$k - \lambda = c^2 \tag{2.19}$$

for some integer c, where I_v is the unit matrix of order v; J_v the all-one matrix of order v; and w_v the all-one column vector of length v.

One of the useful corollaries of Lemma 2.1.5 is

Corollary 2.1.1 *Let $A = [A(i,j)]$ be the associated matrix of some (v,k,λ)-design. Then,*

$$A'A = (k-\lambda)I_v + \lambda J_v; \tag{2.20}$$

$$AA' = (k-\lambda)I_v + \lambda J_v; \tag{2.21}$$

$$J_v A = kJ_v; \tag{2.22}$$

$$AJ_v = kJ_v. \tag{2.23}$$

Lemma 2.1.6 *Let $A = [A(i,j)]$ be a $(0,1)$-valued non-singular matrix of order v. Then Equation (2.20) is true iff Equation (2.21) is true. Moreover, the A is an associate matrix of some (v,k,λ)-design iff Equation (2.20) (or (2.21)) is satisfied.*

One of the most important such designs is the Hadamard design, which is defined by

Definition 2.1.6 *A $(4t-1, 2t-1, t-1)$-design is called a Hadamard design.*

It will be clear that Hadamard designs and Hadamard matrices are uniquely generated by each other. In fact,

Theorem 2.1.2 *There exists a $(4t-1, 2t-1, t-1)$-Hadamard design if and only if there exists a Hadamard matrix of order $4t$.*

Proof At first, we prove that each Hadamard matrix of order $4t$ produces a $(4t-1, 2t-1, t-1)$-Hadamard design.

Without the loss of generality we assume that $H = [H_{ij}]$ is a normal Hadamard matrix of order $4t$. Let A_1 be the sub-matrix of H which is produced by canceling the first row and column of H. Thus A_1 is of order $4t-1$. Let A be the matrix defined by

$$A = \frac{1}{2}(A_1 + J), \tag{2.24}$$

where J is the all-one matrix of order $4t-1$. In other words, A is transformed from A_1 by changing -1 to 0.

Let $e = (1, 1, \cdots, 1)$ be an all-one vector of length $4t-1$. Then the Equation (2.2) is equivalent to

$$\begin{pmatrix} 1 & e \\ e' & A_1 \end{pmatrix} \begin{pmatrix} 1 & e \\ e' & A_1' \end{pmatrix} = \begin{pmatrix} 4t & e + eA_1' \\ e' + A_1 e' & e'e + A_1 A_1' \end{pmatrix} = 4tI. \tag{2.25}$$

Because of $e'e = J_{4t-1}$, we have

$$\begin{cases} A_1 A_1' = 4tI - J, \\ A_1 J = -J, \\ JA_1' = -J. \end{cases} \tag{2.26}$$

Thus the following two equations are obtained:

$$\begin{aligned} AA' &= \frac{1}{4}(A_1 + J)(A_1' + J) \\ &= tI + (t-1)J \end{aligned} \tag{2.27}$$

and

$$\begin{aligned} AJ &= \frac{1}{2}(A_1 + J)J \\ &= \frac{1}{2}(A_1 J + J^2) \\ &= (2t-1)J. \end{aligned} \tag{2.28}$$

Equation (2.27) implies that $\det A \neq 0$, for nontrivial t. Thus from Equations (2.27) and (2.28) we know that A is the associated matrix of some $(4t-1, 2t-1, t-1)$-design, which results in the existence of a $(4t-1, 2t-1, t-1)$-Hadamard design.

We then try to prove that a Hadamard matrix of order $4t$ can be constructed by a $(4t-1, 2t-1, t-1)$-Hadamard design.

If there exists a $(4t-1, 2t-1, t-1)$-Hadamard design with its associated matrix A, then from Corollary 2.1.1 we know that the Equations (2.27) and (2.28) are satisfied. Let

$$A_1 = 2A - J. \tag{2.29}$$

Then the (± 1)-valued matrix A_1 satisfies

$$\begin{aligned} A_1 J &= (2A - J)J \\ &= 2AJ - (4t-1)J \\ &= -J, \end{aligned}$$

$$JA_1' = (A_1 J)' = (-J)' = -J$$

and

$$\begin{aligned} A_1 A_1' &= (2A - J)(2A' - J) \\ &= 4AA' - 2AJ - 2JA' + J^2 \\ &= 4tI - J. \end{aligned}$$

Hence the matrix $\begin{pmatrix} 1 & e \\ e' & A_1 \end{pmatrix}$ is the desired Hadamard matrix of order $4t$, where e is the all-one row of length $4t-1$. □

Besides Theorem 2.1.2 there are some other relationships between (v, k, λ)-designs and Hadamard matrices. For example, by making use of the design theory we can prove the following theorems, which provide us with some necessary and/or sufficient conditions of regular Hadamard matrices.

Theorem 2.1.3 *If H is a regular Hadamard matrix of order $4m$, then there exists a positive integer s such that $m = s^2$ and*

$$HJ = JH = \pm 2sJ.$$

Proof Because H is regular,

$$HH' = H'H = 4mI \qquad \text{and} \qquad HJ = tJ, \tag{2.30}$$

for some integer t. Let A be the $(0,1)$-valued matrix defined by

$$A := \frac{1}{2}(H + J).$$

Then, by using Equation (2.30), we have

$$\begin{aligned} AA' &= \frac{1}{4}(H + J)(H' + J) \\ &= mI + (m + t/2)J \end{aligned} \quad (2.31)$$

and

$$AJ = (2m + t/2)J. \quad (2.32)$$

Hence t is even. Note that $\det A \neq 0$. Then from Lemma 2.1.6, it is known that the matrix A is the associated matrix of some $(4m, 2m + t/2, m + t/2)$-design. Hence Equation (2.15) becomes

$$(m + t/2)(4m - 1) = (2m + t/2)(2m + t/2 - 1)$$

or equivalently

$$m = (t/2)^2 =: s^2 \qquad \text{and} \qquad t = \pm 2s. \quad (2.33)$$

On the other hand, Equation (2.30) and the identity $JH'H = 4mJ$ imply

$$JH = \frac{4m}{tJ} = tJ.$$

□

Theorem 2.1.4 *There exists a $(4u^2, 2u^2 \pm u, u^2 \pm u)$-design if and only if there exists a regular Hadamard matrix of order $4u^2$.*

Proof On the one hand, if H is a regular Hadamard matrix of order $4u^2$, then by Equation (2.33) we have $t = \pm 2u$. Equations (2.31), (2.32), and Lemma 2.1.6 imply that the matrix $A := (H + J)/2$ is the associated matrix of some $(4u^2, 2u^2 \pm u, u^2 \pm u)$-design.

On the other hand, if there is a $(4u^2, 2u^2 \pm u, u^2 \pm u)$-design with its associated matrix A, then

$$AA' = u^2 I + (u^2 \pm u)J$$

and

$$AJ = JA = (2u^2 \pm u)J.$$

Let H be the (± 1)-valued matrix defined by $H =: 2A - J$. Then

$$HH' = (2A - J)(2A' - J) = 4u^2 I$$

and

$$HJ = JH = \pm 2uJ.$$

Therefore H is the regular Hadamard matrix of order $4u^2$. □

Definition 2.1.7 *Let v and k be two positive integers, and $D = \{a_1, a_2, \cdots, a_k\}(\text{mod} v)$ a set of k different* (mod v) *integers. The set D is called a (v, k, λ)-cyclic difference set, abbreviated as (v, k, λ)-CDS, iff for each $d \not\equiv 0$ (mod$v$), there are exactly λ pairs (a_i, a_j) such that*

$$d \equiv a_i - a_j (\text{mod} v).$$

For example, let $v = 11$ and $k = 5$. Then the set $D = \{1, 3, 4, 5, 9\}$ is a $(11, 5, 2)$-CDS.

From the following lemma it is clear that the cyclic difference sets are, in fact, subclasses of (v, k, λ)-designs.

Lemma 2.1.7 *A set $D = \{a_1, a_2, \cdots, a_k\}$ (mod v) of k different (mod v) integers is a (v, k, λ)-CDS if and only if the family $\mathcal{B} = \{B_1, B_2, \cdots, B_v\}$ forms a (v, k, λ)-design, where $B_1 := D$, $B_2 := \alpha(B_1)$, $\cdots$, $B_v =: \alpha(B_{v-1})$, and α the one-to-one mapping from the ring Z_v to itself defined by*

$$\alpha : i \longrightarrow i + 1 (\text{mod} v).$$

A direct corollary of Lemma 2.1.7 and Theorem 2.1.2 is the following

Corollary 2.1.2 *If there is a $(4t - 1, 2t - 1, t - 1)$-CDS, then there exists a Hadamard matrix of order $4t$.*

Two of the important cyclic difference sets are stated in the following lemma.

Lemma 2.1.8 (1) *Let p be a prime, e a positive integer, and $p^e = 4t - 1$. Let G be the additive group of the field $\mathrm{GF}(p^e)$. Then the set $Q =: \{g^2 : g \in \mathrm{GF}(p^e)\}$ forms a $(4t - 1, 2t - 1, t - 1)$-CDS in G;*

(2) *Let p and q be two different primes such that $p^e = q^f - 2$. Let G be the group $G =: \{(a, b) : a \in \mathrm{GF}(p^e), b \in \mathrm{GF}(q^f)\}$ with its addition $+$ and production $.$ defined by*

$$(a, b) + (c, d) =: (a + c, b + d); \text{ and } (a, b).(c, d) = (ac, bd).$$

Let $D =: D_1 \cup D_2 \cup D_3$, where

$$D_1 =: \{(a, b) : a = x^2 \neq 0, b = y^2 \neq 0, x \in \mathrm{GF}(p^e), y \in \mathrm{GF}(q^f)\},$$

$$D_2 =: \{(c, d) : c \text{ and } d \text{ are non-quadratic in } \mathrm{GF}(p^e) \text{ and } \mathrm{GF}(q^f), \text{ respectively } \}$$

and

$$D_3 =: \{(g, 0) : g \in \mathrm{GF}(p^e)\}.$$

Then the set D is a $(p^e q^f, (p^e q^f - 1)/2, (p^e q^f - 3)/4)$-CDS in the group G.

Applying Lemma 2.1.8 to Theorem 2.1.2, the following theorem is obtained.

Theorem 2.1.5 *There exist Hadamard matrices of orders m if $m = p^e + 1 \equiv 0 (\text{mod} 4)$ or $m = p^e q^f + 1$, $q^f = p^e + 2$.*

Definition 2.1.8 [6] *Let V be an additive Abelian group of order v. Let $S_1, S_2, \cdots, S_n$, $S_i = \{s_{i1}, s_{i2}, \cdots, s_{ik_i}\}$, $1 \leqslant i \leqslant n$, be subsets of V. If, for each nonzero element $g \in V$, the equation*

$$g = s_{ij} - s_{im}$$

has exactly λ solutions, then $S_1, S_2, \cdots, S_n$ is called an n-$\{v; k_1, \cdots, k_n; \lambda\}$ supplementary difference sets (SDS). If $k_1 = k_2 = \cdots = k_n = k$, we abbreviate it as n-$(v; k; \lambda)$ SDS.

2.1.3 Williamson Matrices

Theorem 2.1.6 *If there are four $m \times m$ (± 1)-valued matrices A_1, A_2, A_3, and A_4 such that the following three conditions are satisfied:*

C1. *They are symmetric;*

C2. $A_iA_j = A_jA_i$, *for all* $1 \leqslant i, j \leqslant 4$;

C3. *The following equation is true:*

$$A_1^2 + A_2^2 + A_3^2 + A_4^2 = 4mI_m. \tag{2.34}$$

Then the matrix H defined by

$$H = \begin{pmatrix} A_1 & A_2 & A_3 & A_4 \\ -A_2 & A_1 & -A_4 & A_3 \\ -A_3 & A_4 & A_1 & -A_2 \\ -A_4 & -A_3 & A_2 & A_1 \end{pmatrix} \tag{2.35}$$

is a Hadamard matrix of order $4m$.

This theorem can be proved by direct verification. The four matrices satisfying the conditions C1, C2 and C3 are called Williamson matrices. Thus constructions of Williamson matrices imply the constructions of Hadamard matrices. In the following text we concentrate on Williamson matrices.

A matrix $B = [B(i, j)]$, $0 \leqslant i, j \leqslant m - 1$ is called cyclic if it is recursively produced by the first row by

$$B(i, j) = B(0, (j - i) \bmod m).$$

This kind of matrix is denoted by $B = \text{circ}(B(0, 0), B(0, 1), \cdots, B(0, m - 1))$.

Let $U = \text{circ}(0, 1, 0, \cdots, 0)$ be a cyclic matrix. And let A, B, C, D be four (± 1)-valued matrices defined respectively by

$$A = a_0I + a_1U + \cdots + a_{m-1}U^{m-1}, \tag{2.36}$$

$$B = b_0I + b_1U + \cdots + b_{m-1}U^{m-1}, \tag{2.37}$$

$$C = c_0I + c_1U + \cdots + c_{m-1}U^{m-1} \tag{2.38}$$

and

$$D = d_0I + d_1U + \cdots + d_{m-1}U^{m-1}, \tag{2.39}$$

where $a_i, b_i, c_i, d_i = \pm 1$, $0 \leqslant i \leqslant m - 1$.

If the coefficients a_i, b_i, c_i, and d_i, $1 \leqslant i \leqslant m - 1$, satisfy

$$a_{m-i} = a_i; \quad b_{m-i} = b_i; \quad c_{m-i} = c_i; \quad d_{m-i} = d_I, \tag{2.40}$$

then the matrices A, B, C, and D are all symmetric, i.e., the condition C1 in Theorem 2.1.6 is satisfied. It is not difficult to find that the condition C2 is also satisfied. Thus we obtain the following

Theorem 2.1.7 *The matrices A, B, C, and D, respectively, defined by Equations* (2.36)–(2.39) *are Williamson matrices if Equations* (2.40) *and* (2.34) *are simultaneously satisfied.*

Let m be odd integer. Without loss of generality we assume that

$$a_0 = b_0 = c_0 = d_0 = 1. \tag{2.41}$$

Then the matrices A, B, C, and D, respectively, defined by Equations (2.36)–(2.39) can be rewritten as

$$A = P_1 - N_1; \qquad B = P_2 - N_2; \qquad C = P_3 - N_3 \quad \text{and} \quad D = P_4 - N_4, \tag{2.42}$$

where

$$\begin{cases} P_1 = \sum_{a_i=1} U^i, & N_1 = \sum_{a_i=-1} U^i, \\ P_2 = \sum_{b_i=1} U^i, & N_2 = \sum_{b_i=-1} U^i, \\ P_3 = \sum_{c_i=1} U^i, & N_3 = \sum_{c_i=-1} U^i, \\ P_4 = \sum_{d_i=1} U^i, & N_4 = \sum_{d_i=-1} U^i. \end{cases} \tag{2.43}$$

Because of the identity

$$P_i + N_i = J, \quad 1 \leqslant i \leqslant 4, \tag{2.44}$$

the Equation (2.34) is transformed to

$$\sum_{i=1}^{4} (2P_i - J)^2 = 4mI. \tag{2.45}$$

Note that each U^i is a permutation matrix. Then $U^i J = J U^i = J$. Hence

$$P_i J = J P_i = p_i J, \quad 1 \leqslant i \leqslant 4, \tag{2.46}$$

where p_i represents the number of terms in P_i, $1 \leqslant i \leqslant 4$.

By making use of Equation (2.46), the Equation (2.45) can be rewritten as

$$4\sum_{i=1}^{4} P_i^2 - 4\sum_{i=1}^{4} P_i J + \sum_{i=1}^{4} J^2 = 4mI,$$

which is equivalent to

$$\sum_{i=1}^{4} P_i^2 = \left(\sum_{i=1}^{4} p_i - m\right) J + mI. \tag{2.47}$$

If Equation (2.40) is satisfied, then because of $a_0 = 1$, $m =$ odd, we know that p_1, p_2, p_3, and p_4 are all odd integers. Thus Equation (2.47) implies

$$\sum_{a_i=1} U^{2i} + \sum_{b_i=1} U^{2i} + \sum_{c_i=1} U^{2i} + \sum_{d_i=1} U^{2i} \equiv J + I \ (\text{mod} 2), \tag{2.48}$$

which implies further that $a_i + b_i + c_i + d_i = \pm 2$, for each $1 \leqslant i \leqslant m-1$. Therefore, we have proved the following lemma.

Lemma 2.1.9 *If m is odd and the matrices A, B, C, and D, respectively, defined by Equations* (2.36)–(2.39), *satisfy Equations* (2.40), (2.41), *and* (2.34), *then the coefficients a_i, b_i, c_i, and d_i satisfy $a_i + b_i + c_i + d_i = \pm 2$, for each $1 \leqslant i \leqslant m-1$.*

Let

$$\begin{cases} W_1 = (A+B+C-D)/2, \\ W_2 = (A+B-C+D)/2, \\ W_3 = (A-B+C+D)/2, \\ W_4 = (-A+B+C+D)/2 \end{cases} \tag{2.49}$$

or equivalently,

$$\begin{cases} A = (W_1+W_2+W_3-W_4)/2, \\ B = (W_1+W_2-W_3+W_4)/2, \\ C = (W_1-W_2+W_3+W_4)/2, \\ D = (-W_1+W_2+W_3+W_4)/2. \end{cases} \tag{2.50}$$

Then Equation (2.34) is equivalent to

$$W_1^2 + W_2^2 + W_3^2 + W_4^2 = 4mI_m. \tag{2.51}$$

In conclusion, we have

Lemma 2.1.10 *Let m be odd. There exist matrices A, B, C, and D, described in Lemma* 2.1.9 *if and only if there exist symmetric matrices W_1, W_2, W_3, and W_4 satisfying Equation* (2.51) *and the following*

(1) *For $1 \leqslant i \leqslant 4$,*

$$W_i = I \pm 2U^j \pm 2U^s \pm \cdots \pm 2U^t, \quad 1 \leqslant j < s < \cdots < t \leqslant m-1; \tag{2.52}$$

(2) *For each j, $1 \leqslant j \leqslant m-1$, the U^j appears once in just one of W_1, W_2, W_3, and W_4.*

If the matrices W_1, W_2, W_3, and W_4 described in Lemma 2.1.10 are constructed, then Williamson matrices A, B, C, and D can be generated by Equation (2.50). Hence a Hadamard matrix H is obtained by Equation (2.35).

Note that the characteristic equation of the cyclic matrix U is $\lambda^m - 1 = 0$, which has m different solutions. Thus there exists a non-singular matrix, say S, that

$$U = SVS^{-1}, \tag{2.53}$$

where V is a diagonal matrix $V = \mathrm{dia}(r_1, \cdots, r_m)$, and $r_1, \cdots, r_m$ is the m different solutions. Setting Equation (2.53) into Equation (2.52), one obtains

$$W_i = S(I \pm 2V^j \pm 2V^s \pm \cdots \pm 2V^t)^{-1}.$$

Let $\overline{W_i} = I \pm 2V^j \pm 2V^s \pm \cdots \pm 2V^t$. Then Equation (2.51) is changed to

$$\overline{W_1}^2 + \overline{W_2}^2 + \overline{W_3}^2 + \overline{W_4}^2 = 4mI_m. \tag{2.54}$$

Because 1 is a solution of the characteristic equation, without loss of generality we assume that $r_j = 1$. Then by comparing the (j,j)-th elements of both left and right hand sides of Equation (2.54), we have

$$(1 \pm 2 \pm \cdots \pm 2)^2 + (1 \pm 2 \pm \cdots \pm 2)^2 + (1 \pm 2 \pm \cdots \pm 2)^2 + (1 \pm 2 \pm \cdots \pm 2)^2 = 4m. \tag{2.55}$$

Lemma 2.1.11 (Lagrange) *Every positive integer n can be decomposed into the sum of four squares, i.e., $n = a^2 + b^2 + c^2 + d^2$. Moreover, if m is a positive odd, then the number $4m$ can be decomposed into the summation of four odd squares.*

In conclusion, Williamson matrices can be constructed by the following steps:

Step 1. Decompose $4m$ into the sum of four odd squares:

$$4m = a^2 + b^2 + c^2 + d^2, \tag{2.56}$$

where a, b, c, and d are odd.

Step 2. Compare Equations (2.55) with (2.56) to find the matrices W_i, $1 \leqslant i \leqslant 4$, that satisfy both Lemma 2.1.10 and Equation (2.51);

Step 3. Define the matrices A, B, C, and D from Equation (2.50);

Step 4. Construct the Hadamard matrix H by Equation (2.35).

In order to illustrate the above steps, we introduce a few examples here:

Example 2.1.1 Let $u^i + U^{23-i} =: E_i$, $1 \leqslant i \leqslant 11$. Then because of

$$92 = 4 \times 23 = 9^2 + 3^2 + 1^2 + 1^2 \tag{2.57}$$

and

$$\begin{aligned} 92I = &(I + 2E_2 + 2E_6)^2 + (I - 2E_3 + 2E_1 - 2E_{10})^2 \\ &+(I + 2E_5 - 2E_7)^2 + (I + 2E_{11} - 2E_8 + 2E_9 - 2E_4)^2, \end{aligned}$$

the matrices W_i can be defined by $W_1 = I + 2E_2 + 2E_6$, $W_2 = I - 2E_3 + 2E_1 - 2E_{10}$, $W_3 = I + 2E_5 - 2E_7$, and $W_4 = I + 2E_{11} - 2E_8 + 2E_9 - 2E_4$. Hence Williamson matrices of order 23 can be constructed. Therefore there exists a Hadamard matrix of order 92.

Example 2.1.2 Because of $4 \times 43 = 172 = 13^2 + 1^2 + 1^2 + 1^2$ and

$$\begin{aligned} 172I = &(I + 2Y_0 - 2Y_2)^2 + (I + 2Y_3 - 2Y_1)^2 \\ &+(I + 2Y_4 - 2Y_6)^2 + (I + 2Y_5)^2, \end{aligned}$$

where

$$Y_i =: E_{3^i} + E_{3^{7+i}} + E_{3^{14+i}} \quad \text{and} \quad E_j =: U^j + U^{43-j}.$$

Hence Williamson matrices A, B, C, and D of order 43 can be constructed by $W_1 =: I + 2Y_0 - 2Y_2$, $W_2 =: I + 2Y_3 - 2Y_1$, $W_3 =: I + 2Y_4 - 2Y_6$, and $W_4 =: I + 2Y_5$. Therefore there exists a Hadamard matrix of order 172.

Example 2.1.3 Because of $116 = 4 \times 29 = 9^2 + 5^2 + 3^2 + 1^2$ and

$$\begin{aligned} 116I = &(I + 2E_2 - 2E_4 + 2E_6 - 2E_9 - 2E_{11} + 2E_{12})^2 \\ &+(I - 2E_3 - 2E_5 + 2E_7 - 2E_8 + 2E_{10})^2(I + 2E_1)^2 \\ &+(I + 2E_{13} + 2E_{14})^2, \end{aligned}$$

where $E_i =: U^i + U^{29-i}$. Hence Williamson matrices A, B, C, and D of order 29 can be constructed by $W_1 =: I + 2E_2 - 2E_4 + 2E_6 - 2E_9 - 2E_{11} + 2E_{12}$, $W_2 =: I - 2E_3 - 2E_5 +$

$2E_7 - 2E_8 + 2E_{10}$, $W_3 =: I + 2E_1$, and $W_4 =: I + 2E_{13} + 2E_{14}$. Therefore there exists a Hadamard matrix of order 116.

The following theorem presents us more Hadamard matrices constructed by cyclic Williamson matrices.

Theorem 2.1.8 *If A, B, C, and D are cyclic Williamson matrices of order m, then the matrix H defined by*

$$\begin{pmatrix}
A & A & A & B & -B & C & -C & -D & B & C & -D & -D \\
A & -A & B & -A & -B & -D & D & -C & -B & -D & -C & -C \\
A & -B & -A & A & -D & D & -B & B & -C & -D & C & -C \\
B & A & -A & -A & D & D & D & C & C & -B & -B & -C \\
B & -D & D & D & A & A & A & C & -C & B & -C & B \\
B & C & -D & D & A & -A & C & -A & -D & C & B & -B \\
D & -C & B & -B & A & -C & -A & A & B & C & D & -D \\
-C & -D & -C & -D & C & A & -A & -A & -D & B & -B & -B \\
D & -C & -B & -B & -B & C & C & -D & A & A & A & D \\
-D & -B & C & C & C & B & B & -D & A & -A & D & -A \\
C & -B & -C & C & D & -B & -D & -B & A & -D & -A & A \\
-C & -D & -D & C & -C & -B & B & B & D & A & -A & -A
\end{pmatrix}$$

is a Hadamard matrix of order $12m$.

In order to construct more Hadamard matrices, we introduce here a new concept in the following definition.

Definition 2.1.9 [6] *A $(0, \pm 1)$-valued matrix $W =: W(p, k)$, $p \geqslant k$, of order p is called a weighting matrix of order p and weight k if*

$$WW' = kI_p.$$

Thus, if $p = k$, a $W(p, p)$ is a Hadamard matrix of order p.

Lemma 2.1.12 [6] *Let p be a prime power. Then there exists a weighting matrix $W = W(p+1, p)$ satisfying $W' = (-1)^{(p-1)/2}W$. Furthermore, if $p \equiv 3 \pmod 4$, then $W + I_p$ is a Hadamard matrix of order $p + 1$.*

Proof Arrange the elements of $\mathrm{GF}(p) = \{a_0, a_1, \cdots, a_{p-1}\}$ so that $a_0 = 0$, and $a_{p-i} = -a_i$, $1 \leqslant i \leqslant p-1$. Define a matrix $Q = [Q(i,j)]$ by

$$Q(i,j) = \chi(a_j - a_i) =: \begin{cases} 0, & \text{if } i = j, \\ 1, & \text{if } a_j - a_i = y^2 \text{ for } y \in \mathrm{GF}(p), \\ -1, & \text{otherwise.} \end{cases}$$

Then we have the next identities:

$$QQ' = pI - J, \qquad QJ = JQ = 0, \qquad Q' = (-1)^{(p-1)/2}Q,$$

since exactly half of $a_1, \cdots, a_{p-1}$ are squares, -1 is a square for $p \equiv 1$ (mod4) but not for $p \equiv 3$ (mod4), and $\sum_y \chi(y)\chi(y+c) = \sum_y \chi(y^2)\chi(1+cy^{-1}) = \sum_{x\neq 1}\chi(x) = -1$.

Let e be the all-one vector of length p. Then the matrix

$$W = \begin{bmatrix} 0 & e \\ (-1)^{(p-1)/2}e' & Q \end{bmatrix}$$

is the required weighting matrix $W(p+1, p)$.

If $p \equiv 3$ (mod4), then $W + I_p$ is a Hadamard matrix of order $p+1$. □

Theorem 2.1.9[6] *Let $q \equiv 1$(mod4) be a prime power. Then there exists a weighting matrix $W(q+1, q)$ of the form*

$$S = \begin{bmatrix} A & B \\ B & -A \end{bmatrix},$$

where A and B are cyclic and A is of zero diagonal.

Proof Let α be a primitive element of GF(q^2). Let V be a basis of the vector space GF(q^2) over GF(q). Based on this basis, we can define a matrix v by

$$(v) := \frac{1}{2}\begin{bmatrix} \alpha^{q-1} + \alpha^{1-q} & (\alpha^{q-1} - \alpha^{1-q})\alpha^{(q+1)/2} \\ (\alpha^{q-1} - \alpha^{1-q})\alpha^{-(q+1)/2} & \alpha^{q-1} + \alpha^{1-q} \end{bmatrix}.$$

This matrix v satisfies $\det(v) = 1$, and has two eigenvalues $\alpha_1 =: \alpha^{q-1}$ and $\alpha_2 =: \alpha^{1-q}$. Note that $\alpha_1^{(q+1)/2} = \alpha_2^{(q+1)/2} \in$ GF(q) and no $k < (q+1)/2$ satisfies $\alpha_1^k \in$ GF(q) or $\alpha_2^k \in$ GF(q). Hence v acts on the projective line PG$(1, q)$ as a permutation with period $(q+1)/2$ and without any fixed point. Thus the points of PG$(1, q)$ can be divided into two sets of transitivity, say v_1 and v_2, each containing $(q+1)/2$ points.

Let w be another matrix defined by

$$(w) := \begin{bmatrix} 0 & \alpha^{q+1} \\ 1 & 0 \end{bmatrix}.$$

Then $\chi\det(w) = -\chi(-1)$, $wv = vw$, and w has two eigenvalues $\beta_1 := \alpha^{(q+1)/2}$ and $\beta_2 := -\alpha^{(1+q)/2}$ satisfying $\beta_1^2 = \beta_2^2 \in$ GF(q). Hence w acts on PG$(1, q)$ as a permutation of period 2 which maps each point in v_1 into v_2. For each i, $1 \leqslant i \leqslant (q+1)/2$, the mapping $v^i w$ has no eigenvalue in GF(q).

Represent the $q+1$ points of PG$(1, q) = \{x_0, x_1, \cdots, x_q\}$ by the following $q+1$ vectors in V: x, $v(x)$, $v^2(x)$, $\cdots$, $v^{(q-1)/2}$, $w(x)$, $vw(x)$, $\cdots$, $v^{(q-1)/2}w(x)$.

Let $S = [\chi\det(x_i, x_j)]$. Observe that each linear mapping $u : V \to V$ satisfies $\det(u(x), u(y)) = \det(w\det(x, y))$, $x, y \in V$. Thus

$$\begin{aligned} \det(v^i w(x), v^j w(x)) &= \det(w)\det(v^i(x), v^j(x)) \\ &= \det(w)\det(x, v^{j-i}(x)); \\ \det(v^i(x), v^j w(x)) &= -\det(v^i w(x), v^j(x)) \\ &= \det(v^j(x), v^i w(x)) \end{aligned}$$

and

$$\det(v^i(x), v^j(x)) = -\det(v^{(q+1)/2+i}, v^j(x)).$$

In other words, this matrix S is the required matrix. The theorem follows. □

Theorem 2.1.10 [6] *Let $p \equiv 1$ (mod4) be a prime power. Then there are Williamson matrices of order $p(p+1)/2$.*

Proof Let A and B be the matrices in Theorem 2.1.9. Let Q be the matrix that appeared in Lemma 2.1.12. Then the required Williamson matrices X_1, X_2, X_3, and X_4 are

$$\begin{aligned} X_1 &= (I \otimes J) + (A \otimes (I + Q)), \\ X_2 &= B \otimes (I + Q), \\ X_3 &= (I \otimes J) + (A \otimes (I - Q)), \\ X_4 &= B \otimes (I - Q). \end{aligned}$$

□

2.2 Construction

2.2.1 General Constructions

We have seen in the last chapter that direct multiplication (or Kronecker production) is a very useful approach for the construction of Walsh matrices. The following theorem shows that this approach works for Hadamard matrices too.

Theorem 2.2.1 *Let A and B be Hadamard matrices of orders m and n, respectively. Then their direct product $A \otimes B$ is a Hadamard matrix of order mn.*

Proof At first the matrix $A \otimes B$ is clearly (± 1)-valued. Secondly,

$$\begin{aligned} (A \otimes B)(A \otimes B)' &= (A \otimes B)(A' \otimes B') \\ &= (AA') \otimes (BB') \\ &= (mI_m) \otimes (nI_n) \\ &= (mn)I_{mn}. \end{aligned}$$

The theorem follows. □

A simple, but useful, consequence of this theorem is

Corollary 2.2.1 *Let H_i, $1 \leqslant i \leqslant t$, be a Hadamard matrix of order m_i. Then $H_1 \otimes H_2 \otimes \cdots \otimes H_t$ is a Hadamard matrix of order $\prod_{i=1}^{t} m_i$.*

For example, Hadamard matrices of order 2^n can be formed as the direct product of some of order 2.

Theorem 2.2.2 *Let $H = U + I$ be a skew Hadamard matrix of order h, $M = W + I$ and $N = N'$ be a pair of amicable Hadamard matrices of order m. Let X, Y, Z be three (± 1)-valued matrices of order l and*

$$\begin{aligned}
(XY')' &= XY', \\
(YZ')' &= YZ', \\
(ZX')' &= ZX', \\
XX' &= aI + (l-a)J, \\
YY' &= cI + (l-c)J, \\
ZZ' &= (l+1)I - J,
\end{aligned}$$

where $(m-1)c = m(l-h+1) - a$. *Then the matrix*

$$D =: U \otimes N \otimes Z + I_h \otimes W \otimes Y + I_h \otimes I_m \otimes X \tag{2.58}$$

is a Hadamard matrix of order hlm.

Proof On one hand, the matrix D defined by Equation (2.58) is clearly (± 1)-valued.

On the other hand, by the assumptions on H, M, and N, the following identities are obtained:

$$\begin{aligned}
U' &= -U, \\
UU' &= (h-1)I_h, \\
W' &= -W, \\
WW' &= (m-1)I_m, \\
MN &= NM', \\
MM' &= N^2 = mI_m, \\
WN &= NW'.
\end{aligned}$$

Thus we have

$$\begin{aligned}
DD' = {} & UU' \otimes N^2 \otimes ZZ' + I_h \otimes WW' \otimes YY' + I_h \otimes I_m \otimes XX' \\
& + U' \otimes WN \otimes YZ' + U \otimes NW' \otimes XY' + U' \otimes N \otimes XZ' \\
& + U \otimes N \otimes ZX' + I_h \otimes W' \otimes XY' + I_h \otimes W \otimes YX' \\
= {} & UU' \otimes N^2 \otimes ZZ' + I_h \otimes WW' \otimes YY' + I_h \otimes I_m \otimes XX' \\
& + (U' + U) \otimes WN \otimes YZ' + (U' + U) \otimes N \otimes ZX' \\
& + I_h \otimes (W + W') \otimes XY' \\
= {} & (h-1)I_h \otimes mI_m \otimes ((l+1)I_l - J_l) + I_h \otimes I_m \otimes (c(m-1)I_l \\
& + (l-c)(m-1)J_l) + I_h \otimes I_m \otimes (aI_l + (l-a)J_l) \\
= {} & I_{mh} \otimes (m(h-1)(l+1) + c(m-1) + a)I_l \\
& + I_{mh} \otimes (-m(h-1) + (l-c)(m-1) + l - a)J_l \\
= {} & mlhI_{mlh}.
\end{aligned}$$

In other words, we have proved that D is a Hadamard matrix of order mhl. □

Corollary 2.2.2 *If there exist:*

(1) *A skew Hadamard matrix of order* h;

(2) *Two* (± 1)*-valued matrices* Y *and* Z *of order* l *such that*

$$(YZ')' = YZ', \tag{2.59}$$

$$YY' = cI + (l - c)J \tag{2.60}$$

and

$$ZZ' = (l+1)I - J, \tag{2.61}$$

where $c = l - h + 1$. Then there exists a Hadamard matrix of order lh.

Proof This corollary is, in fact, a special case of Theorem 2.2.2 for $X = Y$, $a = c$ and $m = 1$. □

Corollary 2.2.3 *If there exists a skew Hadamard matrix of order h, then there exists a Hadamard matrix of order $h(h-1)$.*

Proof Let H be a skew Hadamard matrix of order h, and $Z - I_{h-1}$ the kernel of H. Then, by Lemma 2.1.3, Z is a (± 1)-valued matrix satisfying

$$(Z - I_{h-1})(Z - I_{h-1})' = (h-1)I_{h-1} - J_{h-1}, \tag{2.62}$$

$$(Z - I_{h-1})J_{h-1} = 0, \tag{2.63}$$

$$(Z - I_{h-1})' = -(Z - I_{h-1}). \tag{2.64}$$

Therefore,

$$Z' = -Z + 2I_{h-1} \quad \text{(by Eq. (2.64))}, \tag{2.65}$$

$$J_{h-1}Z' = J_{h-1} = ZJ_{h-1} \quad \text{(by Eq. (2.63))}, \tag{2.66}$$

$$ZZ' = hI_{h-1} - J_{h-1} \quad \text{(by Eq. (2.65) + (2.62))}. \tag{2.67}$$

Note that

$$J_{h-1}J_{h-1} = (h-1)J_{h-1}. \tag{2.68}$$

Thus if we choose $Y = J$, $l = h - 1$, and $c = 0$, then Equations (2.66), (2.68), and (2.67) reduce to Equations (2.59), (2.60), and (2.61), respectively. The proof is finished by Corollary 2.2.2. □

Corollary 2.2.4 *A Hadamard matrix of order $h(h+3)$ can be constructed from a skew Hadamard matrix of order h and a symmetric Hadamard matrix of order $h+4$.*

Proof Let

$$A = \begin{pmatrix} 1 & e \\ e' & C \end{pmatrix}$$

be a symmetric Hadamard matrix of order $h+4$, where C is the kernel. Thus C is also symmetric and

$$C^2 = (h+4)I_{h+3} - J_{h+3} \quad \text{and} \quad CJ_{h+3} = J_{h+3}C = -J_{h+3}. \tag{2.69}$$

Let $Y = J_{h+3} - 2I_{h+3}$. Then Y is a symmetric (± 1)-valued matrix and

$$\begin{aligned} YC' &= (J_{h+3} - 2I_{h+3})C \\ &= -J_{h+3} - 2C \\ &= C(J_{h+3} - 2I_{h+3}) \\ &= CY', \end{aligned} \tag{2.70}$$

$$YY' = Y^2 = (h-1)J_{h+3} + 4I_{h+3}. \tag{2.71}$$

If we choose $Z = C$, $c = 4$, and $l = h + 3$, then Equations (2.70), (2.71), and (2.69) reduce Equations (2.59), (2.60), and (2.61), respectively. The proof is finished by Corollary 2.2.2. □

2.2.2 Amicable Hadamard Matrices

We have known that amicable Hadamard matrices are useful in constructing skew Hadamard matrices. In fact, the truth of the conjecture "amicable Hadamard matrices exist for every order 2 and $4n$, $n \geqslant 1$", would imply the two conjectures "skew Hadamard matrices exist for every order 2 and $4n$, $n \geqslant 1$" and "symmetric Hadamard matrices exist for every order 2 and $4n$, $n \geqslant 1$". Infinite families of amicable Hadamard matrices can be constructed by the following theorems.

Theorem 2.2.3 *Let $t \geqslant 0$, $l \geqslant 0$, $e_i \geqslant 1$, p_i a prime satisfying $(p_i)^{e_i} \equiv 3(\text{mod}4)$, for $1 \leqslant i \leqslant l$. If*

$$m = 2^t \prod_{i=1}^{l} ((p_i)^{e_i} + 1), \tag{2.72}$$

then there exist amicable Hadamard matrices of order m.

Proof The amicable Hadamard matrices are constructed as follows:

Case 1. $m = 1$, i.e., $t = l = 0$. Then $M = N = (1)$ are a pair of amicable Hadamard matrices of order 1.

Case 2. $m = 2$, i.e., $t = 1$, $l = 0$. Then $M = \begin{pmatrix} 1 & 1 \\ -1 & 1 \end{pmatrix}$ and $N = \begin{pmatrix} 1 & 1 \\ 1 & -1 \end{pmatrix}$ are a pair of amicable Hadamard matrices of order 2.

Case 3. $m = p^e + 1$, i.e., $t = 0$, $l = 1$, $p_1 = p$, and $e_1 = e$. Let $q = p^e$, and $\chi(z)$ be the quadratic characteristic function over $\text{GF}(q) - \{0\}$, i.e.,

$$\chi(z) = \begin{cases} 1, & \text{if } z = a^2, \text{ for some non-zero } a \in \text{GF}(q), \\ 0, & \text{if } z = 0, \\ -1, & \text{otherwise.} \end{cases}$$

Number the elements of $\text{GF}(q)$ by $\text{GF}(q) = \{z_1, z_2, \cdots, z_q\}$ and form a matrix $Q = [Q(i,j)]$, $1 \leqslant i, j \leqslant q$, by $Q(i,j) = \chi(z_j - z_i)$. Let

$$A = \begin{pmatrix} 0 & e \\ -e' & Q \end{pmatrix}, \tag{2.73}$$

where e is a row vector of length q. Because p is odd, we can rearrange GF(q) so that $z_1 = 0$, and

$$z_{q+2-i} = -z_i, \quad 2 \leqslant i \leqslant q. \tag{2.74}$$

The equation $p^e \equiv 3 \pmod 4$ implies that the matrix A in Equation (2.73) satisfies $A' = -A$. Thus

$$AA' = \begin{pmatrix} q & eQ' \\ QQ' & J_q + QQ' \end{pmatrix} = \begin{pmatrix} q & 0 \\ 0 & qI_q \end{pmatrix} = qI_m.$$

Let $M =: A + I$. Then M is (± 1)-valued and $MM' = AA' + A + A' + I = (q+1)I = mI$. Hence, M is a skew Hadamard matrix of order m.

Let

$$W = \begin{pmatrix} 0 & 0 & \cdots & 0 & 1 \\ 0 & \cdots & 0 & 1 & 0 \\ 0 & \cdots & 1 & 0 & 0 \\ \vdots & \vdots & \vdots & \vdots & \vdots \\ 1 & 0 & \cdots & 0 & 0 \end{pmatrix} \quad \text{and} \quad V = \begin{pmatrix} 1 & 0 \\ 0 & W \end{pmatrix}. \tag{2.75}$$

Thus

$$W' = W, \qquad W^2 = I_{q-1}, \qquad V' = V \quad \text{and} \quad V^2 = I_q. \tag{2.76}$$

Let N be a (± 1)-valued matrix of order $q+1$ defined by

$$N =: \begin{pmatrix} 1 & 0 \\ 0 & -V \end{pmatrix} M = \begin{pmatrix} 1 & 0 \\ 0 & -V \end{pmatrix} + \begin{pmatrix} 0 & e \\ e' & -VQ \end{pmatrix},$$

because $M = A + I$. Note that the $(1,j)$-th element of VQ is $\chi(z_j - z_1) = \chi(z_j + z_1)$; moreover the (i,j)-th element of VQ is

$$\chi(z_j - z_{q+2-i}) = \chi(z_j - (-z_i)) = \chi(z_j + z_i),$$

in other words, the matrix VQ is symmetric, therefore the N is symmetric.

From Equation (2.76), we have

$$NN' = \begin{pmatrix} 1 & 0 \\ 0 & -V \end{pmatrix} MM' \begin{pmatrix} 1 & 0 \\ 0 & -V \end{pmatrix} = m \begin{pmatrix} 1 & 0 \\ 0 & V^2 \end{pmatrix} = mI_m.$$

Thus N is a Hadamard matrix of order m. Furthermore,

$$MN = MN' = MM' \begin{pmatrix} 1 & 0 \\ 0 & -V \end{pmatrix} = m \begin{pmatrix} 1 & 0 \\ 0 & -V \end{pmatrix},$$

and

$$NM' = \begin{pmatrix} 1 & 0 \\ 0 & -V \end{pmatrix} MM' = m \begin{pmatrix} 1 & 0 \\ 0 & -V \end{pmatrix},$$

i.e., $MN = NM'$. Therefore M and N are indeed a pair of amicable Hadamard matrices of order $q+1$.

Now it is necessary to prove that if there exist simultaneously pairs of amicable Hadamard matrices of orders m and h, then there exist a pair of amicable Hadamard matrices of order mh.

In fact, if $M_h = S_h + I_h$ and N_h (resp., $M_m = S_m + I_m$ and N_m) is a pair of amicable Hadamard matrices of order h (resp. m), where S_h and S_m are anti-symmetric, while N_m and N_h are symmetric.

Let $M_{hm} =: I_h \otimes M_m + S_h \otimes N_m$, and $N_{hm} := N_h \otimes N_m$. Then M_{hm} and N_{hm} are Hadamard matrices with N_{hm} being symmetric.

Because of

$$\begin{aligned} M_{hm} - I_{hm} &= I_h \otimes (S_m + I_m) + S_h \otimes N_m - I_{hm} \\ &= I_h \otimes S_m + S_h \otimes N_m, \end{aligned}$$

we know that $M_{hm} - I_{hm}$ is anti-symmetric. By Lemma 2.1.4 we have

$$\begin{aligned} M_{hm}N_{hm} &= (I_h \otimes M_m + S_h \otimes N_m)(N_h \otimes N_m) \\ &= N_h \otimes M_m N_m + S_h N_h \otimes N_m^2 \\ &= N_h \otimes N_m M'_m + N_h S'_h \otimes N_m^2 \\ &= (N_h \otimes N_m)(I_h \otimes M'_m + S'_h \otimes N_m) \\ &= N_{hm} M'_{hm}. \end{aligned}$$

Up to now we have proved that M_{hm} and N_{hm} are a pair of amicable Hadamard matrices. □

Theorem 2.2.4 *If A, B, C, and D are four (± 1)-valued matrices satisfying*

$$\begin{gathered} C = I + U, \quad U' = -U, \\ A' = A, \quad B' = B, \quad D' = D, \\ AA' + BB' = CC' + DD' = 2(v+1)I - 2J, \\ eA' = eB' = eC' = eD' = e, \\ AB = BA, \quad CD = DC', \end{gathered}$$

where e is the all ones vector of length v. Let

$$N = \begin{pmatrix} 1 & 1 & e & e \\ 1 & -1 & -e & e \\ e' & -e' & A & -B \\ e' & e' & -B & -A \end{pmatrix}$$

and

$$M = \begin{pmatrix} 1 & 1 & e & e \\ -1 & 1 & e & -e \\ -e' & -e' & C & D \\ -e' & e' & -D & C \end{pmatrix}.$$

Then N is a symmetric Hadamard matrix of order $2(v+1)$, and M is a skew Hadamard matrix of order $2(v+1)$. In addition, if both $AC' - BD$ and $BC' + AD$ are symmetric, them M and N form a pair of amicable Hadamard matrices of order $2(v+1)$.

Proof It can be proved easily by direct verification. □

The other important result about amicable Hadamard matrices is

Theorem 2.2.5 *Let $t = p^e \equiv 1(\text{mod}4)$; $q = g^f = 2t+1$, where p and g are primes, and e and f positive integers. Then there exist amicable Hadamard matrices of order $2(t+1)$.*

In conclusion, it has been proved that there exist amicable Hadamard matrices of the following orders[1]:

Family 1. The order of 2^t, where $t \geqslant 0$ an integer;

Family 2. The order of $p^r + 1$, where $p^r \equiv 3(\text{mod}4)$;

Family 3. The order of $(p-1)^u + 1$, where p is the order of normalized amicable Hadamard matrices, $u > 0$ an odd integer;

Family 4. The order of $2(q+1)$, where $2q+1$ is a prime power, and q a prime satisfying $q \equiv 1(\text{mod}4)$;

Family 5. The order of $(|\, t\,| + 1)(q+1)$, where q is a prime power satisfying $q \equiv 5\ (\text{mod}8) = s^2 + 4t^2$, $s \equiv 1\ (\text{mod}4)$, and $|\, t\,| + 1$ is the order of amicable orthogonal designs of type $(1+|t|; (1,|t|); ((|t|+1)/2, (|t|+1)/2))$ (the definition of amicable orthogonal design will be introduced in Definition 2.3.1);

Family 6. The order of $2^r(q+1)$, where q is a prime power satisfying

$$q \equiv 5(\text{mod}8) = s^2 + 4(2^r - 1)^2, \qquad s \equiv 1(\text{mod}4),$$

r a positive integer;

Family 7. The order of $2(q+1)$, where $q \equiv 5(\text{mod}8)$;

Family 8. The order of S, where S is the product of two amicable Hadamard matrices.

2.2.3 Skew Hadamard Matrices

Some of the most powerful methods for constructing Hadamard matrices depend on the existence of skew Hadamard matrices. It is not an easy task to find skew Hadamard matrices. For example, the Kronecker product of skew Hadamard matrices is not a skew Hadamard matrix, although it has been proved that if h_1 and h_2 are the orders of amicable Hadamard matrices, then there are amicable Hadamard matrices of order h_1h_2. Fortunately, many constructions for amicable Hadamard matrices are also valid for the skew Hadamard matrices.

The following theorems generate some new skew Hadamard matrices by amicable Hadamard matrices and skew Hadamard matrices.

Theorem 2.2.6 *If there exists a skew Hadamard matrix of order h and amicable Hadamard matrices of order m, then there also exist skew Hadamard matrices of order mh.*

Proof Let $S = I_h + \overline{S}$ be a skew Hadamard matrix of order h. Then

$$\overline{S}' = -\overline{S}, \quad \overline{S}\overline{S}' = (h-1)I_h. \tag{2.77}$$

Let $M = I_m + \overline{M}$ and N is a pair of amicable Hadamard matrices of order m. Then

$$MN = NM'. \tag{2.78}$$

Let $K = I_h \otimes M + \overline{S} \otimes N$, which is a (± 1)-valued matrix of order mh. From Equations (2.77) and (2.78), we know that

$$\begin{aligned} KK' &= (I_h \otimes M + \overline{S} \otimes N)(I_h \otimes M' - \overline{S} \otimes N) \\ &= I_h \otimes (MM') - \overline{S}^2 \otimes N^2 \\ &= I_h \otimes mI_m + (h-1)I_h \otimes mI_m \\ &= mhI_{mh} \end{aligned}$$

and

$$\begin{aligned} (K - I_{mk})' &= (I_h \otimes \overline{M} + \overline{S} \otimes N)' \\ &= I_h \otimes M' + \overline{S}' \otimes N \\ &= -I_h \otimes \overline{M} - \overline{S} \otimes N \\ &= -(K - I_{mh}). \end{aligned}$$

Thus we have proved that K is a skew Hadamard matrix of order mh. □

The other important results about skew Hadamard matrices are

Theorem 2.2.7 *Let p be a prime and $p^r = 2m+1 \equiv 5 \pmod 8$. Then there exists a skew Hadamard matrix of order $4(m+1)$.*

Theorem 2.2.8 [6] *If A, B, C, D are square circulant matrices of order m, R is the back diagonal $(0,1)$ matrix, then if A is a skew Hadamard matrix and if*

$$AA' + BB' + CC' + DD' = 4mI,$$

then the following matrices

$$\begin{bmatrix} A & BR & CR & DR \\ -BR & A & -D'R & C'R \\ -CR & D'R & A & -B'R \\ -DR & -C'R & B'R & A \end{bmatrix} \quad or \quad \begin{bmatrix} A & BR & CR & DR \\ -BR & A & D'R & -C'R \\ -CR & -D'R & A & B'R \\ -DR & C'R & -B'R & A \end{bmatrix}$$

are skew Hadamard matrices of order $4m$.

In conclusion, it has been proved that there exist skew Hadamard matrices of the following orders[1]:

Family 1. The order of $2^t \prod k_i$, where $k_i = p_i^{r_i} + 1 \equiv 0 \pmod 4$, p_i prime, r_i and t positive integers;

Family 2. The order of $(p-1)^u + 1$, where u is a positive odd integer, and p is the order of another skew Hadamard matrix;

Family 3. The order of $2(q+1)$, where $q \equiv 5 \pmod 8$ is a prime power;

Family 4. The order of $2(q+1)$, where $q = p^t$ is a prime power with $p \equiv 5 \pmod 8$ and $t \equiv 2 \pmod 4$;

Family 5. The order of $2^s(q+1)$, where $q = p^t$ is a prime power, $p \equiv 5 \pmod 8$, $t \equiv 2 \pmod 4$, and $s \geqslant 1$ an integer;

Family 6. The order of $4m$, where m is an odd integer in $[3, 31]$ or $m \in \{37, 39, 43, 49, 65, 67, 93, 113, 121, 127, 129, 133, 157, 163, 181, 217, 219, 241, 267\}$;

Family 7. The order of $n(n-1)(m-1)$, where m and n are the orders of amicable Hadamard matrices such that $(m-1)n/m$ is the order of another skew Hadamard matrix;

Family 8. The order of $mn(n-1)$, where n is the order of amicable orthogonal designs of types $((1, n-1); (n))$ and mn is the order of an orthogonal design of type $(1, m, mn-m-1)$(see Definition 2.3.1);

Family 9. The order of $4(q+1)$, where $q \equiv 9 (\text{mod} 16)$ is a prime power;

Family 10. The order of $(|\, t \,|+1)(q+1)$, where $q = s^2+4t^2 \equiv 5(\text{mod} 8)$ is a prime power, and $|\, t \,|+1$ is the order of another skew Hadamard matrix;

Family 11. The order of $4(1+q+q^2)$, where q is a prime power such that (1) $1+q+q^2 \equiv 3, 5$, or $7(\text{mod} 8)$ is a prime power; or (2) $3+2q+2q^2$ is a prime power;

Family 12. The order of $2^t q$, where $q = s^2+4r^2 \equiv 5(\text{mod} 8)$ is a prime power, and an orthogonal design of type $(2^t; 1, a, b, c, c+ \mid r \mid)$ exists where $1+a+b+2c+ \mid r \mid = 2^t$ and $a(q+1)+b(q-4) = 2^t$;

Family 13. The order of hm, where h is the order of another skew Hadamard matrix, and m is the order of a pair of amicable Hadamard matrices.

2.2.4 Symmetric Hadamard Matrices

One of the two matrices in a pair of amicable Hadamard matrices is symmetric. Thus Theorems 2.2.3 and 2.2.5, respectively, imply the following:

Theorem 2.2.9 *Let $t \geqslant 0$, $l \geqslant 0$, $e_i \geqslant 1$, p_i a prime satisfying $(p_i)^{e_i} \equiv 3 \pmod 4$, for $1 \leqslant i \leqslant l$. If*

$$m = 2^t \prod_{i=1}^{l} ((p_i)^{e_i}+1), \tag{2.79}$$

then there exist symmetric Hadamard matrix of order m.

Theorem 2.2.10 *Let $t = p^e \equiv 1(\text{mod} 4)$; $q = g^f = 2t+1$, where p and g are primes, e and f positive integers. Then there exists symmetric Hadamard matrix of order $2(t+1)$.*

Theorem 2.2.11 *Let p be a prime, l a positive integer, and $q = p^l \equiv 1(\text{mod} 4)$. Then there exists a symmetric Hadamard matrix of order $2(q+1)$.*

Proof Let G, q, and Q be the same as those in the proof of Theorem 2.2.3. Let

$$P = \begin{pmatrix} 0 & e \\ e' & Q \end{pmatrix}, \quad N = P + I_{q+1}. \tag{2.80}$$

Then the diagonal of Q is all-zero, $Q' = Q$, $QQ' = qI - J$, $QJ = JQ = 0$. Thus N is a symmetric (± 1)-valued matrix, and

$$P^2 = \begin{pmatrix} q & eQ \\ Qe' & J+Q^2 \end{pmatrix} = qI_{q+1}. \tag{2.81}$$

Let

$$H = \begin{pmatrix} -N & P-I \\ P-I & N \end{pmatrix}, \tag{2.82}$$

which is a symmetric (± 1)-valued matrix of order $2(q+1)$. From Equation (2.81), we have

$$\begin{aligned} HH' &= \begin{pmatrix} (P+I)^2 + (P-I)^2 & 0 \\ 0 & (P+I)^2 + (P-I)^2 \end{pmatrix} \\ &= 2\begin{pmatrix} (P^2+I) & 0 \\ 0 & (P^2+I) \end{pmatrix} \\ &= 2(q+1)I_{2(q+1)}. \end{aligned}$$

In other words, the matrix H is the symmetric Hadamard matrix of order $2(q+1)$.

□

Lemma 2.2.1 *Let H_1 and H_2 be symmetric Hadamard matrices of orders m_1 and m_2, respectively. Then $H_1 \otimes H_2$ is a symmetric Hadamard matrix of order $m_1 m_2$.*

This lemma will be used in the proofs of the following theorems.

Theorem 2.2.12 *Let $t \geqslant 0$, $l \geqslant 0$, $e_i \geqslant 1$, p_i an odd prime and*

$$m = 2^t \prod_{i=1}^{l} ((p_i)^{e_i} + 1). \tag{2.83}$$

If the number of p_i satisfying $(p_i)^{e_i} \equiv 1 \pmod 4$ is upper bounded by t, then there exists a symmetric Hadamard matrix of order m.

Proof This theorem follows the direct multiplication of the Hadamard matrices presented in Equation (2.82), in Theorem 2.2.3 and some $\begin{pmatrix} 1 & 1 \\ 1 & -1 \end{pmatrix}$. □

Theorem 2.2.13 *Let p be a prime, e a positive integer, and $q = p^e \equiv 1 \pmod 4$. Then there exists a symmetric Hadamard matrix of order $2q(q+1)$.*

Proof Let P and N be those matrices of order $q+1$ defined by Equation (2.80). Let

$$X = \begin{pmatrix} 1 & 1 \\ 1 & -1 \end{pmatrix} \otimes J_q$$

and

$$\begin{aligned} Y &= \begin{pmatrix} Q+I & Q-I \\ Q-I & -Q-I \end{pmatrix} \\ &= \begin{pmatrix} 1 & 1 \\ 1 & -1 \end{pmatrix} \otimes Q + \begin{pmatrix} 1 & -1 \\ -1 & -1 \end{pmatrix} \otimes I_q, \end{aligned}$$

where Q is the matrix which appears in the matrix P. Then

$$X' = X, \quad Y' = Y, \quad X^2 = 2qI_2 \otimes J_q, \tag{2.84}$$

$$\begin{aligned} Y^2 &= 2I_2 \otimes Q^2 + 2I_2 \otimes I_q \\ &= 2I_2 \otimes ((q+1)I_q - J_q), \end{aligned} \tag{2.85}$$

$$\begin{aligned} XY + YX &= 2I_2 \otimes JQ + 2\begin{pmatrix} 0 & -1 \\ 1 & 0 \end{pmatrix} \otimes J_q + 2I_2 \otimes QJ \\ &+2\begin{pmatrix} 0 & 1 \\ -1 & 0 \end{pmatrix} = 0. \end{aligned} \tag{2.86}$$

Let $H = I_{q+1} \otimes X + P \otimes Y$. Then H is a (± 1)-valued symmetric matrix. From Equations (2.84)–(2.86), we have

$$\begin{aligned} HH' &= I_{q+1} \otimes X^2 + P^2 \otimes Y^2 \\ &= 2q(q+1)I_{2q(q+1)}. \end{aligned}$$

In other words, this H is a symmetric Hadamard matrix of order $2q(q+1)$. □

Theorem 2.2.14 *Let $t \geqslant l \geqslant 0$, p_i a prime, and $p_i^{e_i} \equiv 1 \pmod 4$, $1 \leqslant i \leqslant l$. Let*

$$m = 2^t \prod_{i=1}^{l} (p_i^{e_i}(p_i^{e_i} + 1)).$$

Then there exists a symmetric Hadamard matrix of order m.

Proof This theorem is the direct consequence of Lemma 2.2.1 and Theorem 2.2.13 and the known symmetric Hadamard matrix $\begin{pmatrix} 1 & 1 \\ 1 & -1 \end{pmatrix}$. □

2.3 Existence

2.3.1 Orthogonal Designs and Hadamard Matrices

A $(0, \pm x_1, \cdots, \pm x_l)$-valued matrix, X, of order n is called an orthogonal design of order n and type $(s_1, \cdots, s_l)$, s_i positive integer, if and only if

$$XX' = \left(\sum_{i=1}^{l} s_i x_i^2\right) I_n.$$

Alternatively, X is an orthogonal matrix with each row containing exactly s_i entries of the type $\pm x_i$, $1 \leqslant i \leqslant l$.

It is obvious that Hadamard matrices of order n are (± 1)-valued orthogonal designs of type (n). Some other orthogonal designs of special interest are[7]:

(1) The following matrix X is an orthogonal design of order 4 and type $(1, 1, 1, 1)$:

$$X = \begin{bmatrix} A & B & C & D \\ -B & A & -D & C \\ -C & D & A & -B \\ -D & -C & B & A \end{bmatrix} \text{ or } \begin{bmatrix} A & B & C & D \\ -B & A & D & -C \\ -C & -D & A & B \\ -D & C & -B & A \end{bmatrix}.$$

(2) The following matrix X is an orthogonal design of order 8 and type $(1, 1, 1, 1, 1, 1, 1, 1)$:

$$X = \begin{bmatrix} A & B & C & D & E & F & G & H \\ -B & A & D & -C & F & -E & -H & G \\ -C & -D & A & B & G & H & -E & -F \\ -D & C & -B & A & H & -G & F & -E \\ -E & -F & -G & -H & A & B & C & D \\ -F & E & -H & G & -B & A & -D & C \\ -G & H & E & -F & -C & D & A & -B \\ -H & -G & F & E & -D & -C & B & A \end{bmatrix}.$$

In addition orthogonal designs of order 12 and type $(3, 3, 3, 3)$, order 24 and type $(3, 3, 3, 3, 3, 3, 3, 3)$, order 20 and type $(5, 5, 5, 5)$, and order 36 and type $(9, 9, 9, 9)$ can also be constructed.

The following existence result about orthogonal designs will be used in the construction of Hadamard matrices.

Lemma 2.3.1 [6, 7] *There exist orthogonal designs of type $(1, m-1, nm-n-m)$ and order $2^t = (m-1)n$.*

Definition 2.3.1 [6, 7] *Let $(0, \pm x_1, \cdots, \pm x_s)$-valued X (resp. $(0, \pm y_1, \cdots, \pm y_t)$-valued Y) be an orthogonal design of order n and type $(u_1, \cdots, u_s)$ (resp., $(v_1, \cdots, v_t)$). If $XY' = YX'$ is satisfied, the pair X and Y are called amicable orthogonal designs. Sometimes this pair are also called amicable orthogonal designs of type $[(u_1, \cdots, u_s); (v_1, \cdots, v_t)]$ and order n.*

For example[6],

$$X = \begin{bmatrix} x_1 & x_2 \\ x_2 & -x_1 \end{bmatrix} \quad \text{and} \quad Y = \begin{bmatrix} y_1 & y_2 \\ -y_2 & y_1 \end{bmatrix} \tag{2.87}$$

are amicable orthogonal designs of type $[(1, 1); (1, 1)]$ and order 2.

$$X = \begin{bmatrix} x_1 & x_2 & x_3 & x_3 \\ -x_2 & x_1 & x_3 & -x_3 \\ x_3 & x_3 & -x_1 & -x_2 \\ x_3 & -x_3 & x_2 & -x_1 \end{bmatrix} \quad \text{and} \quad Y = \begin{bmatrix} y_1 & y_2 & y_3 & y_3 \\ y_2 & -y_1 & y_3 & -y_3 \\ -y_3 & -y_3 & y_2 & y_1 \\ -y_3 & y_3 & y_1 & -y_2 \end{bmatrix} \tag{2.88}$$

are amicable orthogonal designs of the type $[(1, 1, 2); (1, 1, 2)]$ and order 2[6].

Comparing the definition of amicable Hadamard matrices with Definition 2.3.1, the following lemma is obtained.

Lemma 2.3.2 [6, 7] *Let $M = I + S$ and N be amicable Hadamard matrices of order n. Then (1) M and N are amicable orthogonal designs of type $[(n); (n)]$ and order n; (2) S and N are amicable orthogonal designs of type $[(1, n-1); (n)]$ and order n.*

Proof The first part is directly owed to Definitions 2.3.1 and 2.1.5. The second part is obtained by Lemmas 2.1.4 and 2.1.2. □

It has been proved that

Lemma 2.3.3 [6, 7] *For each integer t, there exist amicable orthogonal designs of types*

$$((1,1,2,4,\cdots,2^t);(2^t,2^t)) \quad \text{and} \quad ((1,2^{t+1}-1);(2^t,2^t)).$$

Let $G = \{z_1, z_2, \cdots, z_t\}$ be an additive Abelian group of order t. Let ψ and ϕ be two mappings from G to a commutative ring. Define two matrices $M = [m(i,j)]$ and $N = [n(i,j)]$ by

$$m(i,j) = \psi(z_j - z_i) \qquad \text{and} \qquad n(i,j) = \phi(z_j + z_i), \quad 1 \leqslant i,j \leqslant t.$$

These matrices M and N are called type 1 and type 2 matrices, respectively. Here the words "type 1" illustrate the way the elements of G are ordered and which one of the following two functions ψ or ϕ is used.

Let X be a subset of G and $0 \notin X$. If ψ and ϕ are defined by

$$\psi(x) = \begin{cases} a, & x = 0, \\ b, & x \in X, \\ c, & \text{otherwise}, \end{cases} \quad \text{and} \quad \phi(x) = \begin{cases} d, & x = 0, \\ e, & x \in X, \\ f, & \text{otherwise}. \end{cases}$$

Then the M (resp., N) is called type 1 (a,b,c) (resp., type 2 (d,e,f)) incidence matrix generated by X.

If the restriction $0 \notin X$ is dropped, and

$$\psi(x) = \phi(x) = \begin{cases} 1, & x \in X, \\ -1, & \text{otherwise}, \end{cases}$$

then the M (resp., N) is called type 1 (resp., type 2) ± 1 incidence matrix generated by X, and if

$$\psi(x) = \phi(x) = \begin{cases} 1, & x \in X \\ 0, & \text{otherwise}, \end{cases}$$

then the M (resp., N) is called type 1 (resp., type 2) $(0,1)$ incidence matrix generated by X.

Definition 2.3.2 [7, 8] *Let α be a primitive element of* GF(q)*, where $q = p^k = ef + 1$ is a prime power. Let $G = \langle x \rangle$, the Abelian group generated by x. The following subsets, C_i, $0 \leqslant i \leqslant e-1$, of* GF$(q)$ *are called the cyclotomic classes, where*

$$C_i =: \{\alpha^{es+i} : 0 \leqslant s \leqslant f-1\}, \quad 0 \leqslant i \leqslant e-1.$$

Hence these cyclotomic classes are disjointed from each other and their union is exactly the group G itself.

Theorem 2.3.1 [6, 7] *Let q be a prime power and $q \equiv 5 \pmod 8$, $q = s^2 + 4t^2$, $s \equiv 1 \pmod 4$. If there are amicable orthogonal designs of type $[(1, 2r-1);(r,r)]$ and order $2r = |t|+1$, then there exist amicable Hadamard matrices of order $(|t|+1)(q+1)$.*

Proof Let C_i be the cyclotomic classes in Definition 2.3.2. Then the sets

$$C_0 \cap C_1, \overbrace{C_0 \cap C_2, C_0 \cap C_2, \cdots, C_0 \cap C_2}^{|t|}$$

are $(|t|+1)-\{q;(q-1)/2;(|t|+1)(q-3)/4\}$ supplementary difference sets(SDS) satisfying

$$x \in C_0 \cap C_1 \implies -x \notin C_0 \cap C_1$$

and

$$y \in C_0 \cap C_2 \implies -y \in C_0 \cap C_2.$$

In addition, the sets

$$\overbrace{C_0 \cap C_2, \cdots, C_0 \cap C_2}^{(|t|)/2} \overbrace{C_1 \cap C_3, \cdots, C_1 \cap C_3}^{(|t|)/2}$$

are $(|t|+1)$-$\{q;(q-1)/2;(|t|+1)(q-3)/4\}$ supplementary difference sets(SDS) satisfying

$$y \in C_0 \cap C_2 \implies -y \in C_0 \cap C_2$$

and

$$z \in C_1 \cap C_3 \implies -z \in C_1 \cap C_3.$$

Let A be the type 1 (± 1) incidence matrix generated by $C_0 \cap C_1$, and B (resp. C) the type 2 (± 1) incidence matrix generated by $C_0 \cap C_2$ (resp. $C_1 \cap C_3$). Then

$$AJ = BJ = CJ = -J, \quad (A+I)' = -(A+I), \quad B' = B, \quad C' = C$$

and

$$\begin{aligned} AA' + \mid t \mid BB' &= (\mid t \mid +1)/2(BB' + CC') \\ &= (\mid t \mid +1)(q+1)I - (\mid t \mid +1)J. \end{aligned}$$

Let $P = x_0 U + x_1 V$ and $Q = x_3 X + x_4 Y$ be the amicable orthogonal designs of order $2r$ and type $((1, 2r-1); (r, r))$. Without loss of generality it can be assumed that $U = I$, $V' = -V$, $X' = X$, and $Y' = Y$, in fact, if otherwise simultaneously multiplying P and Q by some matrix W. Let e be the all ones vector of length q. Then the required amicable Hadamard matrices are

$$E = \begin{bmatrix} U+V & (U+V) \otimes e \\ (-U+V) \otimes e' & U \otimes (-A) + V \otimes B \end{bmatrix}$$

and

$$F = \begin{bmatrix} X+Y & (X+Y) \otimes e \\ (X+Y) \otimes e' & X \otimes C + Y \otimes D \end{bmatrix}.$$

□

Corollary 2.3.1 [6, 7] *Let q be a prime power, $q \equiv 5$ (mod8), and $q = s^2 + 4t^2$, $s \equiv 1$ (mod4). If $|\, t \,| = 2^r - 1$, r a positive integer, then there exist amicable Hadamard matrices of order $2^r(q+1)$.*

Proof This is a simple consequence of Theorem 2.3.1 and Lemma 2.3.3. □

Corollary 2.3.2 [7, 9] *Let q be a prime power, $q \equiv 5$ (mod8), and $q = s^2+4$, or $q = s^2+36$, $s \equiv 1$ (mod4). Then there exist amicable Hadamard matrices of orders $2(s^2+5)$ or $4(s^2+37)$.*

Definition 2.3.3 [7, 10] *A set of m (± 1)-valued matrices $A_1, A_2, \cdots, A_m$ of order n are called suitable matrices for the orthogonal design of type $(s_1, s_2, \cdots, s_m)$ if the following two equations are satisfied:*

(1) $A_i A_j' = A_j A_i$, $1 \leqslant i, j \leqslant m$

and

(2) $\sum_{i=1}^{m} s_i A_i A_i' = (\sum s_i) n I_n$.

A simple observation of suitable matrices and orthogonal designs is

Theorem 2.3.2 [7, 11] *Let X be a $(0, \pm x_1, \pm x_2)$-valued orthogonal design of type $(1, m-1, m(h-1))$ and order mh. Let A_1, A_2 are suitable matrices of order n. Then replacing the variable x_i in X by the matrix A_i, $1 \leqslant i \leqslant 2$, a Hadamard matrix of order mhn is generated.*

Furthermore, the following results are also true:

Theorem 2.3.3 [7, 11] *Suppose that there exist m suitable matrices of order n and an orthogonal design of type $(s_1, s_2, \cdots, s_m)$ and order $\sum s_i$. Then there exists a Hadamard matrix of order $(\sum s_i)m$.*

Theorem 2.3.4 [7, 12] *Suppose that there exists an orthogonal design of order $4t$ and type (t, t, t, t) and that there exist 4 suitable matrices A, B, C, D of order m satisfying*

$$AA' + BB' + CC' + DD' = 4mI_m.$$

Then there exists a Hadamard matrix of order $4m$.

Corollary 2.3.3 [7, 13] *Let n be the order of Hadamard matrix H. If there exists an orthogonal design, say D, of type $(1, m-1, nm-n-m)$ and order $n(m-1)$, then there exists a Hadamard matrix of order $n(n-1)(m-1)$.*

Proof Without loss of generality, we assume

$$H = \begin{bmatrix} 1 & e \\ -e' & P \end{bmatrix},$$

where e is the all one row of length $n-1$. Then the matrix P satisfies

$$PJ = J \quad \text{and} \quad PP' = nI - J.$$

The proof is finished by replacing the variables of the orthogonal design D by the matrices P, J, P, respectively. □

Corollary 2.3.4 [7, 8] *There exists a Hadamard matrix of order $2^s(2^s-1)(2^t-1)$ for each pair of nonnegative integers s and t.*

Proof It can be proved by an obvious consequence of Lemma 2.3.1 and Corollary 2.3.3. □

Definition 2.3.4 *A weighting matrix $W = W(p, p-1)$ is called a symmetric conference matrix if $W' = W$ and $p \equiv 2$ (mod4).*

Theorem 2.3.5 [7, 10] *If there exists a Hadamard matrix H, of order $k > 1$, a symmetric conference matrix C, of order n, a pair of amicable orthogonal designs, M, N, of order m and type $((1, m-1); (m/2, m/2))$, and suitable matrices of order p, then there exists a Hadamard matrix of order $nmkp$.*

Proof Let P be a matrix defined by

$$P = \begin{bmatrix} 0 & -1 \\ 1 & 0 \end{bmatrix} \otimes I_{k/2}.$$

Then the matrix $R := C \otimes H \otimes N + I \otimes PH \otimes M$ is an orthogonal design of type $(k, (m-1)k, (n-1)mk/2, (n-1)nk/2)$ and order nkm. The proof is finished by replacing the variables in this orthogonal design by the suitable matrices. □

2.3.2 Existence Results

Most of the following existence results are based on the orthogonal designs. First we restate a well known theorem in number theory.

Theorem 2.3.6 [7, 11] *Let x and y be two positive co-prime integers, and N an integer satisfying $N > (x-1)(y-1)$. Then there exist non-negative integers a and b such that $N = ax + by$.*

A direct consequence of this theorem is

Corollary 2.3.5 [7, 11] *Let $v \geqslant 9$ be an odd integer, $x = v+1$, and $y = v-3$. Then there exist non-negative integers a, b, and t such that $a(v+1) + b(v-3) = 2^t$.*

Proof Let $g = \gcd(v+1, v-3)$. Then $g = 1, 2$, or 4. Let m be the smallest number of the form $m = 2^k$ such that

$$m > [(v+1)/g - 1][(v-3)/g - 1].$$

Then by Theorem 2.3.6, there exist integers a and b such that

$$a(v+1)/g + b(v-3)/g = m.$$

The corollary follows. □

The first existence result is

Lemma 2.3.4 [7, 6] *Let $p \geqslant 9$ be a prime and $p \equiv 3 \pmod 4$. Then there exists an integer t such that a Hadamard matrix of order $2^s p$ can always be obtained if $s \geqslant t$.*

Proof Let $x = p+1$ and $y = p-3$. Then, by Corollary 2.3.5, there exist a and b satisfying $ax + by = 2^t =: n$ for some t. Let D be an orthogonal design of the variables x_1, x_2 and x_3, of order 2^t and type $(a, b, n-a-b)$. The required Hadamard matrix is obtained by replacing each variable x_1 by the matrix J, each variable x_2 by $J - 2I$ and each variable x_3 by the back circulant $(\pm)$-valued matrix $B = (Q+1)R$, where R is the back diagonal matrix and $Q = [Q_{ij}]$ is defined by

$$Q_{ij} = \begin{cases} 0, & \text{if } i = j, \\ 1, & \text{if } j - i = y^2 \text{ for some } y \in \mathrm{GF}(p), \\ -1, & \text{otherwise.} \end{cases}$$

□

Lemma 2.3.5 [6, 7] *Let $p \geqslant 9$ be a prime and $p \equiv 1 \pmod 4$. Then there exists an integer t such that a Hadamard matrix of order $2^s p$ can always be obtained if $s \geqslant t+1$.*

This lemma can be proved in the same way as that of Lemma 2.3.4. In fact, let F be an orthogonal design of the variables x_1, x_2, x_3 and x_4, of order $2^{t+1} =: n$ and type $(2a, 2b, n-a-b, n-a-b)$. Then the required matrix is obtained by replacing each variable x_1 by J, each variable x_2 by $J-2I$, and the variables x_3 and x_4 by $X = I+Q$ and $Y = I-Q$, respectively.

The previous two lemmas complete the proof of the existence of Hadamard matrices of order $2^t \cdot p$ for all prime p except $p = 2, 3, 5$, and 7.

The existence of Hadamard matrices of order $4m$ and $4n$ implies the existence of that of order $4 \times 4mn = 16mn$ by direct multiplication. The following result reduces the order to 8mn in the resulting Hadamard matrix.

Lemma 2.3.6 [7, 10] *Let H and E be Hadamard matrices of orders $4m$ and $4n$, respectively. Then there is a Hadamard matrix of order $8mn$.*

Proof Write the H and E as

$$H = \begin{bmatrix} P & Q \\ R & S \end{bmatrix} \quad \text{and} \quad E = \begin{bmatrix} K & L \\ M & N \end{bmatrix}.$$

Because of $HH' = 4mI$ and $EE' = 4nI$, we have

$$\begin{aligned} &PP' + QQ' = RR' + SS' = 2mI, \\ &PR' + QS' = 0 = RP' + SQ', \\ &KK' + LL' = MM' + NN' = 2nI, \\ &KM' + LN' = 0 = MK' + NL'. \end{aligned}$$

The resulting Hadamard matrix of order $8mn$ is

$$\begin{bmatrix} \frac{1}{2}(P+Q) \otimes K + \frac{1}{2}(P-Q) \otimes M & \frac{1}{2}(P+Q) \otimes L + \frac{1}{2}(P-Q) \otimes N \\ \frac{1}{2}(R+S) \otimes K + \frac{1}{2}(R-S) \otimes M & \frac{1}{2}(R+S) \otimes L + \frac{1}{2}(R-S) \otimes N \end{bmatrix}. \qquad \square$$

For example, we have known that there are Hadamard matrices of orders 20 and 12. Thus Lemma 2.3.6 guarantees the existence of a Hadamard matrix of order $8 \times 5 \times 3 = 120$.

Lemma 2.3.7 [6, 7] *There exist Hadamard matrices of order $2^t p$ for $p = 3, 5, 7$.*

Proof The required Hadamard matrices can be formed by applying Lemma 2.3.6 to the known Hadamard matrices of orders 12, 20, 28, and 2^t. □

Theorem 2.3.7 [7, 8] *Let q be a positive integer. Then there exists $t = t(q)$ such that a Hadamard matrix of order $2^s q$ exists for every $s \geqslant t$.*

Proof Apply Lemmas 2.3.4, 2.3.5 and/or 2.3.7 to each prime factor of q. The proof of the theorem is finished by the direct multiplication of Hadamard matrices. □

A strengthened form of Lemma 2.3.6 can be stated as

Theorem 2.3.8 [7, 10] *Suppose that there are Hadamard matrices of orders $4m$, $4n$, $4p$, and $4q$. Then there exists a Hadamard matrix of order $16\,mnpq$.*

A natural generalization of a Hadamard matrix is the so-called complex Hadamard matrix, which is defined by

Definition 2.3.5 [7, 13] *A matrix C of order $2n$ with elements ± 1, $\pm i$ that satisfies $CC^* = 2nI$ is called a complex Hadamard matrix, where C^* is the transpose of the conjugate matrix of C.*

The following result is similar to Lemma 2.3.6.

Theorem 2.3.9 [7, 13] *Suppose that there are a complex Hadamard matrix of order $2n$ and a Hadamard matrix of order $4m$. Then there exists a Hadamard matrix of order $8mn$.*

Some of the other strongest existence results are listed in the following[1, 5, 7]:

Class 1. Let $p \equiv 3(\mathrm{mod}4)$ be a prime power. Then there exists a Hadamard matrix of order $p+1$;

Class 2. Let $p \equiv 1(\mathrm{mod}4)$ be a prime power. Then there exists a Hadamard matrix of order $2(p+1)$;

Class 3. Suppose that there is a Hadamard matrix of order n. Then there is a symmetric regular Hadamard matrix with constant diagonal of order n^2;

Class 4. Let q be any positive integer. Then there exists a Hadamard matrix of order $2^t q$ for every $t \geqslant [2\log_2(q-3)]+1$;

Class 5. Let p and $p+2$ be twin prime powers. Then there exists a $t \leqslant [\log_2(p+3)(p-1)(p^2+2p-7)]-2$ so that there is a Hadamard matrix of order $2^t p(p+2)$;

Class 6. Let $p+1$ be the order of a symmetric Hadamard matrix. Then there exists a $t \leqslant [\log_2(p-3)(p-7)]-2$ so that there is a Hadamard matrix of order $2^t p$;

Class 7. Let pq be an odd natural number. Suppose that an orthogonal design of order $2^s p$ and type $(2^r a, 2^r b, 2^r c)$ exists, $s \geqslant s_0$, $2^{s-r}p = a+b+c$. Then there exists a Hadamard matrix of order $2^t \cdot p \cdot q$;

Class 8. Let q be a positive integer. Then there exists a $t = t(q)$ so that there is a Hadamard matrix of order $2^s 3q$ for all $s \geqslant t$;

Class 9. There is a Hadamard matrix of order $8 \cdot 49 \cdot 3^t = 342 \cdot 3^t$ for all $t \geqslant 0$;

Class 10. If $n \equiv 3(\mathrm{mod}4)$ is a prime power, there is a Hadamard matrix of order $n^2(n+1)9^t$ for all $t \geqslant 0$;

Class 11. There exist Hadamard matrices of orders $4 \cdot 3^t$, $4 \cdot 5^t$, $4 \cdot 13^t$, $4 \cdot 17^t$, $4 \cdot 29^t$, $4 \cdot 37^t$, $4 \cdot 41^t$, $4 \cdot 53^t$, $4 \cdot 61^t$, $4 \cdot 101^t$, $t \geqslant 0$; $4 \cdot g^{4i}$, $4 \cdot g^{4i+1}$, $4 \cdot g^{4i+2}$, $8 \cdot g^{4i+3}$, $i \geqslant 0$, $g = 7, 11$; and $4 \cdot p^k$ whenever $p = 1 + 2^a \cdot 10^b \cdot 26^c$ is prime, $a, b, c \geqslant 0$.

Up to now, much progress on Hadamard matrices has been achieved. For the updated papers in this area the readers are recommended to see references [14–71].

Bibliography

[1] Craigen R, Wallis W. Hadamard matrices: 1893-1993 // *Proc. of the 24th Southeastern Int. Conf. On Combinatorics, Graph Theory, and Computing*, Boca Raton, FL, 1993. *Congr. Numer*, 1993, 97: 99-129.

[2] Agaian S S. Hadamard matrices and their applications // *Lecture Notes in Mathematics 1168.* Berlin: Springer-Verlag, 1985.

[3] Assmus E. Hadamard matrices and their designs: a coding-theoretic approach. *Trans. Amer. Math. Soc.*, 1990, 330(1): 269-293.

[4] Balasubramanian K. Computer generation of Hadamard matrices. *J. Comput. Chem.*, 1993, 14(5): 603-619.

[5] Craigen R and Kharaghani H. On the existence of regular Hadamard matrices. 21*th Manitoba Conf. On Numerical Math. and Comput., Congr., Numer.*, 1994, 99: 277-283.

[6] Seberry J and Yamada M. Amicable Hadamard matrices and amicable orthogonal designs. *Utilitas Math.*, 1991, 40: 179-192.

[7] Geramita A and Seberry J. *Orthogonal Designs.* New York: Marcel Dekker, Inc., 1979.

[8] Seberry J. Existence of SBIBD($4k^2, 2k^2 \pm k, k^2 \pm k$) and Hadamard matrices with maximal excess. *Australas J. Combin.*, 1991, 4: 87-91.

[9] Seberry J. SBIBD($4k^2, 2k^2 + k, k^2 + k$) and Hadamard matrices of order $4k^2$ with maximum excess are equivalent. *Graphs Combin.*, 1989, 5(4): 373-383.

[10] Seberry J, Whiteman A L. New Hadamard matrices and conference matrices obtained via Mathon's construction. *Graphs Combin*, 1988, 4(4): 355-377.

[11] Seberry J and Yamada M. Hadamard matrices, sequences, and block designs // *Contemporary Design Theory: a Collection of Surveys.* Edited by Dinitz J H and Douglas R Stinson. John Wiley and Sons, Inc., 1992: 431-560.

[12] Seberry J. A new construction for Williamson-type matrices. *Graphs Combin.*, 1986, 2(1): 81-87.

[13] Seberry J and Yamada M. On the multiplication theorem of Hadamard natrices of generalized quaternion type using M-structures. *J. Combin. Math. Combin. Comput.*, 1993, 13: 97-106.

[14] Balasubramanian K. Characterization of Hadamard matrices. *Molecular Phys.*, 1993, 78(5): 1309-1329.

[15] Bussemaker F and Tonchev V D. New extremal double-even codes of length 56 derived from Hadamard matrices of order 28. *Discrete Math.*, 1989, 76: 45-59.

[16] Bussemaker F and Tonchev V D. Extremal double-even codes of length 40 derived from Hadamard matrices of order 20. *Discrete Math.*, 1990, 82: 317-321.

[17] Cantian L. et al. Codes from Hadamard matrices and profiles of Hadamard matrices. *J Combin. Math. Combin. Comput.*, 1992, 12: 57-64.

[18] Chan W. Necessary conditions for Menon difference sets. *Des. Codes Cryptogr.*, 1993, 3(2): 147-154.

[19] Cohen G, et al. A survey of base sequences, disjoint complementary sequences and OD(*4t; t,t,t,t*). *J. Combin. Math. Combin. Comput.*, 1989, 5: 69-103.

[20] Craigen R. Product of four Hadamard matrices. *J. of Combin., Ser. A*, 1992, 59: 318-320.

[21] Craigen R. Constructing Hadamard matrices with orthogonal Pairs. *Ars. Combin.*, 1992, 33: 57-64.

[22] Craigen R. Equivalence classes of inverse orthogonal and unit Hadamard matrices. *Bull. Austral., Math., Soc.*, 1991, 44(1): 109-115.

[23] Craigen R. Embedding rectangular matrices in Hadamard matrices. *Linear and Multilinear Algebra*, 1991, 29(2): 91-92.

[24] Davis J and Jedwab J. A summary of Menon difference sets. *Congr. Numer.*, 1993, 93: 203-207.

[25] Dokvic D. Williamson matrices of order 4n for n=33, 35, 39. *Discrete Math.*, 1993, 115: 267-271.

[26] Dokvoick D. Williamson matrices of order 4×29 and 4×31. *J. of Combin. Ser. A*, 1992, 59: 309-311.

[27] Dokvic D. Ten Hadamard matrices of order 1852 of Goethals-Seidel Type. *European J. Combin.*, 1992, 13(4): 245-248.

[28] Dokvic D. Construction of some new Hadamard matrices. *Bull. Austral., Math., Soc.*, 1992, 45(2): 327-332.

[29] Dragomir D. Ten new orders for Hadamard matrices of skew type. *Univ. Beograd., Publ. Elektrotehn, Fak., Ser., Mat.*, 1992, 3: 47-59.

[30] Gluck D. Hadamard difference sets in groups of order 64. *J. of Combin., Ser. A*, 1989, 51: 138-140.

[31] Holzmann W H and Kharaghani H. On the access of Hadamard matrices. *Congr. Numer*, 1993, 92: 257-260.

[32] Kharaghani H. A construction for Hadamard matrices. *Discrete Math.*, 1993, 120: 115-120.

[33] Kimura H. New Hadamard matrix of order 24. *Graphs Combin.*, 1989, 5(3): 235-242.

[34] Koukouvinos C and Kounias S. Construction of some Hadamard matrices with maximum excess. *Discrete Math*, 1990, 85: 295-300.

[35] Koukouvinos C, Seberry J. Hadamard matrices of order 8(mod16) with maximum excess. *Discrete Math.*, 1991, 92: 173-176.

[36] Koukouvinos C and Kounias S. Hadamard matrices of the Williamson type of order $4m$, $m = pq$, an exhaustive search for $m = 33$. *Discrete Math.*, 1988, 68: 45-57.

[37] Kounias S and Farmakis N. On the excess of Hadamard matrices. *Discrete Math.*, 1988, 68: 59-69.

[38] Koukouvinos C. On the excess of Hadamard matrices. *Utilitas Math.*, 1994, 45: 97-101.

[39] Koukouvinos C and Seberry J. Constructing Hadamard matrices from orthogonal designs. *Australas, J. Combin*, 1992, 6: 267-278.

[40] Koukouvinos C and Seberry J. Construction of new Hadamard matrices with maximal excess and infinity many new SBIBD$(4k^2, 2k^2+k, k^2+k)$ // *Graphs, Matrices, and Designs*, 255-267, *Lecture Notes in Pure and Appl. Math.*, 139. New York: Dekker, 1993.

[41] Koukouvinos C, Kounias S and Seberry J. Further results On base sequences, disjoint complementary sequences, $OD(4t; t, t, t, t)$ and the excess of Hadamard matrices. *Ars. Combin.*, 1990, 30: 241-255.

[42] Koukouvinos C, Kounias S and Seberry J. Further Hadamard matrices with maximal excess and new SBIBD$(4k^2, 2k^2 + k, k^2 + k)$. *Utilitas Math.*, 1989, 36: 135-150.

[43] Kraemer R. Proof of a conjecture on Hadamard 2-groups. *J. of Combin., Ser. A*, 1993, 63, 1-10.

[44] Launey W. A product of twelve Hadamard matrices. *Australas., J. Combin.*, 1993, 7: 123-127.

[45] Lin C, Wallis W. Symmetric and Skew equivalence of Hadamard matrices. *Congr. Numer.*, 1991, 85: 73-79.

[46] Lin C, Wallis W. Profiles of Hadamard matrices of order 24. *Congr. Numer*, 1988, 66: 93-102.

[47] Mcfarland R L. Necessary conditions for Hadamard difference sets // Ray-Chaudhuri D ed. *Coding Theory and Design Theory, Part II.* Berlin: Springer, 1990: 257-272.

[48] Mcfarland R L. Subdifference sets of Hadamard difference sets. *J. Combin. Theory, Ser.* A, 1990, 54: 112-122.

[49] Mcfarland R L. Necessary conditions for Hadamard difference sets // *Coding Theory and Design Theory. Part II*, 257-272. *IMA Vol. Math. Appl.* 21. New York: Springer, 1990.

[50] Meisner D. Families of Menon difference sets // Combinatorics'90 (Gaeta, 1990), 365-380. *Ann., Discrete Math.*, 52. Amsterdam: North-Holland, 1992.

[51] Meisner D. On a construction of regular Hadamard matrices. *Math., Appl.*, 1992, 3(4): 233-240.

[52] Miyamoto M. A construction of Hadamard matrices. *J. of Combin., Ser. A*, 1991, 57: 86-108.

[53] Momura K. Spin models constructed from Hadamard matrices. *J. Combin. Theory, Ser.*, 1994, 68(2): 251-261.

[54] Mullin R C and Wevrick D. Singular points in pair covers and their relation to Hadamard designs. *Discrete Math.*, 1991, 92: 221-225.

[55] Noboru I. Note on Hadamard groups of quadratic residue type. *Hokkaido Math. J.*, 1993, 22(3): 373-378.

[56] Noboru I. Nearly triply regular Hadamard designs and tournaments. *Math. J. Okayama. Univ.*, 1990, 32: 1-5.

[57] Smith K W. Non-Abelian Hadamard difference sets. *J. of Combinatorial Theory, Ser. A*, 1995, 70(1): 144-156.

[58] Sole P, Ghafoor A. and Sheikh S A. The covering Radius of Hadamard codes in odd graphs. *Discrete Applied Math.*, 1992, 37/38: 501-510.

[59] Spence E. Classification of Hadamard matrices of order 24 and 28. *Discrete Mathematics*, 1995, 140(1-3): 185-243.

[60] Tonchev V D. Self-dual codes and Hadamard matrices. *Discrete Applied Math.*, 1991, 33(1-3): 235-240.

[61] Wallis W. Hadamard Matrices // *Combinatorial and Graph-Theoretical Problems in Linear Algebra* (*Minneapolis, MN,* 1991), 235-243. *IMA Vol. Math. Appl.*, 50. New York: Springer, 1993.

[62] Xia M. Some infinite classes of special Williamson matrices and difference sets. *J. of Combinatorial Theory, Ser. A*, 1992, 61: 230-242.

[63] Xia M and Xia T. Hadamard matrices constructed from supplementary difference sets in the class F_1. *J. of Combin. Des.*, 1994, 2(5): 325-339.

[64] Xia M. An infinite classes of supplementary difference sets and Williamson matrices. *J. of Combin., Ser.A*, 1991, 58: 310-317.

[65] Xia M. Some infinite classes of Williamson matrices and weighting matrices. *Australas J. Combin.*, 1992, 6: 107-110.

[66] Xia M. Hadamard Matrices, Combinatorial Designs and Applications // (Hungshan, 1988), 179-181, *Lecture Notes in Pure and Appl. Math.*, 126. New York: Dekker, 1990.

[67] Yamada M. Some new series of Hadamard matrices. *J. Austral. Math. Soc. Ser. A*, 1989, 46(3): 371-383.

[68] Yang Y X and Lin X D. *Coding and Cryptography.* Beijing: PPT Press, 1992.

[69] Yamada M. Hadamard matrices of generalized quaternion type. *Discrete Math.*, 1991, 87: 187-196.

[70] Zhu L. Equivalence classes of Hadamard matrices of order 32. *Congr. Numer*, 1993, 95: 179-182.

[71] Zhu L. An infinite family of complex amicable Hadamard matrices. *Bull. Inst. Combin. Appl.*, 1991, 1: 37-40.

Part II

Lower-Dimensional Cases

Chapter 3

3-Dimensional Hadamard Matrices

3.1 Definitions and Constructions

3.1.1 Definitions[2]

It is known that a 2-dimensional Hadamard matrix $H = [H(i,j)]$, $0 \leqslant i,j \leqslant m-1$, of order $m \times m$ is in fact a binary (± 1)-valued orthogonal matrix satisfying $HH' = mI_m$. In other words, a 2-dimensional Hadamard matrix is a matrix such that its $(2-1)$-dimensional layers, in each axis-normal orientation, are orthogonal to each other, i.e., the $(2-1)$-dimensional layers $(H(0,a),\cdots,H(m-1,a)$; $(H(0,b),\cdots,H(m-1,b)$ in y-axis orientation and the $(2-1)$-dimensional layers $(H(a,0),\cdots,H(a,m-1)$; $(H(b,0),\cdots,H(b,m-1)$ in x-axis orientation satisfy $H(i,j) = -1$ or 1 and

$$\sum_{i=0}^{m-1} H(i,a)H(i,b) = \sum_{j=0}^{m-1} H(a,j)H(b,j) = m\delta_{ab}, \tag{3.1}$$

where $\delta_{ab} = 1$ iff $a = b$, otherwise $\delta_{ab} = 0$.

Similar to the definition of 2-dimensional Hadamard matrices, a 3-dimensional Hadamard matrix of order $m \times m$ can be defined as a (± 1)-valued matrix $H = [H(i,j,k)]$, $0 \leqslant i,j,k \leqslant m-1$, such that its $(3-1)$-dimensional layers, in each axis-normal orientation, are orthogonal to each other, i.e., the $(3-1)$-dimensional layers $[H(i,j,a) : 0 \leqslant i,j \leqslant m-1]$ and $[H(i,j,b) : 0 \leqslant i,j \leqslant m-1]$ in z-axis orientation are orthogonal to each other; the $(3-1)$-dimensional layers $[H(i,a,k) : 0 \leqslant i,k \leqslant m-1]$ and $[H(i,b,k) : 0 \leqslant i,k \leqslant m-1]$ in y-axis orientation are orthogonal to each other; and the $(3-1)$-dimensional layers $[H(a,j,k) : 0 \leqslant j,k \leqslant m-1]$ and $[H(b,j,k) : 0 \leqslant j,k \leqslant m-1]$ in x-axis orientation are also orthogonal to each other. In other words, a three-dimensional Hadamard matrix is one in which all parallel two-dimensional layers, in all axis-normal orientations, are orthogonal to each other. That is, the three-dimensional matrix, $H = [H(i,j,k)]$, $0 \leqslant i,j,k \leqslant m-1$, of order m is Hadamardian if all $H(i,j,k) = -1$ or 1 and if the following are satisfied[1, 2]:

$$\begin{aligned}\sum_{i=0}^{m-1}\sum_{j=0}^{m-1} H(i,j,a)H(i,j,b) &= \sum_{j=0}^{m-1}\sum_{k=0}^{m-1} H(a,j,k)H(b,j,k)\\ &= \sum_{i=0}^{m-1}\sum_{k=0}^{m-1} H(i,a,k)H(i,b,k)\\ &= m^2\delta_{ab}.\end{aligned} \tag{3.2}$$

A much stronger definition of a 3-dimensional Hadamard matrix is defined by those matrices in which all of the 2-dimensional layers, in all axis-normal orientations, are in themselves (two-dimensional) Hadamard matrices[1, 2]. In other words,

$$\sum_{i=0}^{m-1} H(i,a,r)H(i,b,r) = \sum_{j=0}^{m-1} H(a,j,r)H(b,j,r) = m\delta_{ab}, \quad r = \text{any value of } k\ , \tag{3.3}$$

i.e., the 2-dimensional layers in z-normal orientation are 2-dimensional Hadamard matrices.

$$\sum_{i=0}^{m-1} H(i,q,a)H(i,q,b) = \sum_{k=0}^{m-1} H(a,q,k)H(b,q,k) = m\delta_{ab}, \quad q = \text{any value of } j, \tag{3.4}$$

i.e., the 2-dimensional layers in y-normal orientation are 2-dimensional Hadamard matrices.

$$\sum_{j=0}^{m-1} H(p,j,a)H(p,j,b) = \sum_{k=0}^{m-1} H(p,a,k)H(p,b,k) = m\delta_{ab}, \quad p = \text{any value of } i, \tag{3.5}$$

i.e., the 2-dimensional layers in x-normal orientation are 2-dimensional Hadamard matrices.

Matrices satisfying Equations (3.3), (3.4), and (3.5) are precisely called "absolutely proper" three-dimensional Hadamard matrices. Matrices satisfying Equation (3.2) but not all of Equations (3.3)–(3.5) are called "improper." Thus, a matrix satisfying Equations (3.2), (3.3), and (3.4) but not (3.5) is called "improper in the x-direction." Matrices satisfying Equations (3.2), (3.3) and (3.4) are called "proper" in the z-direction. Three-dimensional Hadamard matrices which are improper in every axial direction are called absolutely improper.

Each three-dimensional matrix $A = [A(i,j,k)]$, $0 \leqslant i,j,k \leqslant m-1$ can be described by its two-dimensional layers in each direction, e.g. the m layers $[A(i,j,0)]$, $[A(i,j,1)]$, $[A(i,j,2)]$, $\cdots$, $[A(i,j,m-1)]$ in the z-direction.

Thus it is easy to verify that the two 3-dimensional matrix $A = [A(i,j,k)]$, $0 \leqslant i,j,k \leqslant 1$, described by

$$[A(i,j,0)] = \begin{bmatrix} 1 & 1 \\ 1 & -1 \end{bmatrix} \quad \text{and} \quad [A(i,j,1)] = \begin{bmatrix} -1 & 1 \\ 1 & 1 \end{bmatrix}$$

is an absolutely proper three-dimensional Hadamard matrix[1, 2].

The 3-dimensional matrix $B = [B(i,j,k)]$, $0 \leqslant i,j,k \leqslant 1$ described by

$$[B(i,j,0)] = \begin{bmatrix} -1 & 1 \\ 1 & 1 \end{bmatrix} \quad \text{and} \quad [B(i,j,1)] = \begin{bmatrix} -1 & 1 \\ -1 & -1 \end{bmatrix}$$

is improper in x-direction. In fact, its x-direction layers are

$$[B(0,j,k)] = \begin{bmatrix} -1 & -1 \\ 1 & 1 \end{bmatrix} \quad \text{and} \quad [B(1,j,k)] = \begin{bmatrix} 1 & 1 \\ 1 & -1 \end{bmatrix}.$$

One of them, $[B(0,j,k)]$, is not a 2-dimensional Hadamard matrix[1, 2].

The 3-dimensional matrix $C = [C(i,j,k)]$, $0 \leqslant i,j,k \leqslant 1$ described by

$$[C(i,j,0)] = \begin{bmatrix} 1 & 1 \\ 1 & -1 \end{bmatrix} \quad \text{and} \quad [C(i,j,1)] = \begin{bmatrix} 1 & -1 \\ -1 & -1 \end{bmatrix}$$

is an absolutely proper three-dimensional Hadamard matrix[1, 2].

The 3-dimensional matrix $D = [D(i,j,k)]$, $0 \leqslant i,j,k \leqslant 1$ described by

$$[D(i,j,0)] = \begin{bmatrix} -1 & 1 \\ 1 & 1 \end{bmatrix} \quad \text{and} \quad [D(i,j,1)] = \begin{bmatrix} 1 & -1 \\ 1 & 1 \end{bmatrix}$$

is improper in the x-direction. In fact, its x-direction layers are

$$[D(0,j,k)] = \begin{bmatrix} -1 & 1 \\ 1 & -1 \end{bmatrix} \quad \text{and} \quad [D(1,j,k)] = \begin{bmatrix} 1 & 1 \\ 1 & 1 \end{bmatrix},$$

which are not 2-dimensional Hadamard matrices[1, 2].

3.1.2 Constructions Based on Direct Multiplications

For the details of the contents of this subsection the readers are recommended to see [1].

Direct multiplication (or Kronecker production) of two matrices A and B is the matrix C constructed from A and B by replacing each element $A(i,j)$ by the submatrix $A(i,j)B$. Exactly, if the given two matrices are

$$A = \begin{bmatrix} A(0,0) & A(0,1) & \cdots & A(0,m-1) \\ A(1,0) & A(1,1) & \cdots & A(1,m-1) \\ \vdots & \vdots & & \vdots \\ A(n-1,0) & A(n-1,1) & \cdots & A(n-1,m-1) \end{bmatrix}$$

and

$$B = \begin{bmatrix} B(0,0) & B(0,1) & \cdots & B(0,M-1) \\ B(1,0) & B(1,1) & \cdots & B(1,M-1) \\ \vdots & \vdots & & \vdots \\ B(N-1,0) & B(N-1,1) & \cdots & B(N-1,M-1) \end{bmatrix},$$

then the direct multiplication of A and B is another matrix $C = [C(i,j)]$ of order $(nN) \times (mM)$ defined by

$$C = \begin{bmatrix} A(0,0)B & A(0,1)B & \cdots & A(0,m-1)B \\ A(1,0)B & A(1,1)B & \cdots & A(1,m-1)B \\ \vdots & \vdots & & \vdots \\ A(n-1,0)B & A(n-1,1)B & \cdots & A(n-1,m-1)B \end{bmatrix}.$$

This definition is denoted by $C = A \otimes B$, or equivalently, the general elements of the direct multiplication matrix C are

$$C(i,j) = A\left(\left\lfloor \frac{i}{N} \right\rfloor, \left\lfloor \frac{j}{M} \right\rfloor\right) B([i]_N, [j]_M),$$
$$\text{for } 0 \leqslant i \leqslant (nN-1),\ 0 \leqslant j \leqslant (mM-1), \tag{3.6}$$

where $\lfloor x \rfloor$, the floor function, denotes the largest integer not larger than x, and $[i]_K \equiv i \bmod K$, the remainder of i divided by K. Thus $i = \lfloor \frac{i}{N} \rfloor N + [i]_N$ and $j = \lfloor \frac{j}{M} \rfloor M + [i]_M$.

The concept of direct multiplication defined by Equation (3.6) can be generalized to three-dimensional cases as follows[3]:

Let

$$A = [A(i,j,k)], \quad 0 \leqslant i \leqslant n-1,\ 0 \leqslant j \leqslant m-1,\ 0 \leqslant k \leqslant p-1$$

and

$$B = [B(i,j,k)], \quad 0 \leqslant i \leqslant N-1,\ 0 \leqslant j \leqslant M-1,\ 0 \leqslant k \leqslant P-1$$

are two three-dimensional matrices of orders $n \times m \times p$ and $N \times M \times P$, respectively. Then the direct product of A and B is a three-dimensional matrix

$$C = A \otimes B = [C(i,j,k)], \quad 0 \leqslant i \leqslant Nn-1, \quad 0 \leqslant j \leqslant Mm-1,\ 0 \leqslant k \leqslant pP-1$$

of order $(nN) \times (mM) \times (pP)$ defined by

$$C(i,j,k) = A\left(\left\lfloor \frac{i}{N} \right\rfloor, \left\lfloor \frac{j}{M} \right\rfloor, \left\lfloor \frac{k}{P} \right\rfloor\right) B([i]_N, [j]_M, [k]_P), \tag{3.7}$$

for $0 \leqslant i \leqslant (nN-1)$, $0 \leqslant j \leqslant (mM-1), 0 \leqslant k \leqslant pP-1$. Thus the matrix C is constructed from its mother matrices A and B by replacing each element $A(i,j,k)$ by a 3-dimensional sub-matrix $A(i,j,k)B$.

Direct multiplication is an important construction approach for 2-dimensional Hadamard matrices. In fact, we know that if A and B are two 2-dimensional Hadamard matrices, then so is their direct multiplication matrix $C = A \otimes B$. This kind of construction is still valid for 3-dimensional cases. Specifically, we have the following theorems.

Theorem 3.1.1[1−3] *The direct multiplication of two three-dimensional Hadamard matrices is also a three-dimensional Hadamard matrix. Furthermore, the product matrix is proper in those directions in which both the parent matrices are proper.*

Proof Let $A = [A(i,j,k)]$ and $B = [B(i,j,k)]$ be two 3-dimensional Hadamard matrices of order n and m, respectively. Their direct production $C = A \otimes B$ is of order (nm). Thus, by Equation (3.7), we have

$$\begin{aligned}
&\sum_{i=0}^{nm-1} \sum_{j=0}^{nm-1} C(i,j,a)C(i,j,b) \\
&= A\left(\left\lfloor \frac{i}{m} \right\rfloor, \left\lfloor \frac{j}{m} \right\rfloor, \left\lfloor \frac{a}{m} \right\rfloor\right) \cdot B([i]_m, [j]_m, [a]_m) \\
&\cdot A\left(\left\lfloor \frac{i}{m} \right\rfloor, \left\lfloor \frac{j}{m} \right\rfloor, \left\lfloor \frac{b}{m} \right\rfloor\right) \cdot B([i]_m, [j]_m, [b]_m)
\end{aligned}$$

$$
\begin{aligned}
&= \sum_{u=0}^{n-1}\sum_{v=0}^{n-1}\sum_{p=0}^{m-1}\sum_{q=0}^{m-1} A\left(u,v,\left\lfloor\frac{a}{m}\right\rfloor\right)\cdot A\left(u,v,\left\lfloor\frac{b}{m}\right\rfloor\right)\cdot B(p,q,[a]_m)\cdot B(p,q,[b]_m)\\
&= \left[\sum_{u=0}^{n-1}\sum_{v=0}^{n-1} A\left(u,v,\left\lfloor\frac{a}{m}\right\rfloor\right)\cdot A\left(u,v,\left\lfloor\frac{b}{m}\right\rfloor\right)\right]\\
&\quad\cdot\left[\sum_{p=0}^{m-1}\sum_{q=0}^{m-1} B(p,q,[a]_m)\cdot B(p,q,[b]_m)\right]\\
&= (nm)^2\delta_{a,b},
\end{aligned}
$$

where the last equation is owed to the properties that (1) A and B are themselves 3-dimensional Hadamard matrices, and (2) $a \neq b$ implies that $\left\lfloor\frac{a}{m}\right\rfloor \neq \left\lfloor\frac{b}{m}\right\rfloor$ or $[a]_m \neq [b]_m$.

Similarly, it can be proved that

$$
\begin{aligned}
\sum_{j=0}^{nm-1}\sum_{k=0}^{nm-1} C(a,j,k)C(b,j,k) &= \sum_{i=0}^{nm-1}\sum_{k=0}^{nm-1} C(i,a,k)C(i,b,k)\\
&= (nm)^2\delta_{a,b}.
\end{aligned}
$$

Therefore Equation (3.2) is satisfied by C, i.e., C is indeed a 3-dimensional Hadamard matrix.

Every layer of C in the x- (resp., y- or z-) direction is the direct multiplication of two layers of the parents in the x- (resp., y- or z-) direction. Thus C is proper in those directions in which both the parent matrices are proper. □

Corollary 3.1.1 [1–3] *Three-dimensional Hadamard matrices of order 2^t can be generated from $t-1$ successive direct multiplication among three-dimensional Hadamard matrices of order 2.*

Three-dimensional Hadamard matrices of order m^2 can also be generated by the direct multiplication of three two-dimensional Hadamard matrices of order m in different orientations.

For example, the 2-dimensional Hadamard matrix

$$
A = \begin{bmatrix} 1 & 1 \\ 1 & -1 \end{bmatrix}
$$

can be respectively treated as a 3-dimensional matrix $A_1 = [A_1(i,j,k)]$, $k = 0$, $0 \leqslant i,j \leqslant 1$, of order $2\times 2\times 1$ such that $[A_1(i,j,0)] = [A(i,j)]$, a 3-dimensional matrix $A_2 = [A_2(i,j,k)]$, $j = 0$, $0 \leqslant i,k \leqslant 1$, of order $2\times 1\times 2$ such that $[A_2(i,0,k)] = [A(i,k)]$, and a 3-dimensional matrix $A_3 = [A_3(i,j,k)]$, $i = 0$, $0 \leqslant j,k \leqslant 1$, of order $1\times 2\times 2$ such that $[A_3(0,j,k)] = [A(j,k)]$. The direct multiplication of these three matrices is

$$
C = [C(i,j,k)] = (A_1 \otimes A_2) \otimes A_3, \quad 0 \leqslant i,j,k \leqslant 3.
$$

Its layers in the z-direction are

$$
C(i,j,0) = \begin{bmatrix} 1 & 1 & 1 & 1 \\ 1 & 1 & 1 & 1 \\ 1 & 1 & -1 & -1 \\ 1 & 1 & -1 & -1 \end{bmatrix}, \quad C(i,j,1) = \begin{bmatrix} 1 & -1 & 1 & -1 \\ 1 & -1 & 1 & -1 \\ 1 & -1 & -1 & 1 \\ 1 & -1 & -1 & 1 \end{bmatrix}
$$

and

$$C(i,j,2)=\begin{bmatrix}1&1&1&1\\-1&-1&-1&-1\\1&1&-1&-1\\-1&-1&1&1\end{bmatrix},\quad C(i,j,3)=\begin{bmatrix}1&-1&1&-1\\-1&1&-1&1\\1&-1&-1&1\\-1&1&1&-1\end{bmatrix}.$$

It can be verified that the above direct multiplication matrix $C=(A_1\otimes A_2)\otimes A_3$ is a 3-dimensional Hadmard matrix of order $4\times 4\times 4$.

In general, we have the following theorem.

Theorem 3.1.2 [1–3] *Let $A=[A(i,j)]$ be a 2-dimensional Hadamard matrix of order $m\times m$, and let $A_1=[A_1(i,j,k)]$ (resp., $A_2=[A_2(i,j,k)]$, and $A_3=[A_3(i,j,k)]$) be the 3-dimensional matrix of order $m\times m\times 1$ (resp., $m\times 1\times m$, and $1\times m\times m$) that are produced from the matrix $A=[A(i,j)]$ by $A_1(i,j,0)=A(i,j)$ (resp., $A_2(i,0,k)=A(i,k)$, and $A_3(0,j,k)=A(j,k)$). Then the direct multiplication $C=(A_1\otimes A_2)\otimes A_3$ is a 3-dimensional Hadamard matrix of order $m^2\times m^2\times m^2$.*

Proof By the definition of direct multiplication, the general formula of $C(i,j,k)$, $0\leqslant i,j,k\leqslant m^2-1$, is

$$C(i,j,k)=A\left(\left\lfloor\frac{i}{m}\right\rfloor,\left\lfloor\frac{j}{m}\right\rfloor\right)A\left([i]_m,\left\lfloor\frac{k}{m}\right\rfloor\right)A([j]_m,[k]_m).$$

Therefore

$$\begin{aligned}&\sum_{i,j=0}^{m^2-1}C(i,j,a)C(i,j,b)\\ =&\sum_{i,j=0}^{m^2-1}A\left([i]_m,\left\lfloor\frac{a}{m}\right\rfloor\right)A([j]_m,[a]_m)\\ &\cdot A\left([i]_m,\left\lfloor\frac{b}{m}\right\rfloor\right)A([j]_m,[b]_m)\quad\text{for }(A(p,q))^2=1\\ =&m^2\sum_{i=0}^{m-1}A\left(i,\left\lfloor\frac{a}{m}\right\rfloor\right)\cdot A\left(i,\left\lfloor\frac{b}{m}\right\rfloor\right)\sum_{j=0}^{m-1}A(j,[a]_m)A(j,[b]_m)\\ =&m^4\delta_{a,b}.\end{aligned}$$

The last equation is due to the facts that: (1) The matrix A is itself a Hadamard matrix; and (2) $a\neq b$ implies $[a]_m\neq[b]_m$ or $\left\lfloor\frac{a}{m}\right\rfloor\neq\left\lfloor\frac{b}{m}\right\rfloor$.

By the same way, it can be proved that

$$\sum_{j,k=0}^{m^2-1}C(a,j,k)C(b,j,k)=\sum_{i,k=0}^{m^2-1}C(i,a,k)C(i,b,k)=m^4\delta_{a,b}.$$

Therefore Equation (3.2) is satisfied by C, i.e., C is indeed a 3-dimensional Hadamard matrix. □

The 3-dimensional Hadamard matrices constructed by Theorem 3.1.2 are absolutely improper in all of the possible directions. A more general form of Theorem 3.1.2 will be presented in the third part of the book.

3.1.3 Constructions Based on 2-Dimensional Hadamard Matrices

Besides the direct multiplications, other methods for generating 3-dimensional Hadamard matrices are at present restricted to special cases involving propriety or at least some degree of correlation within the two-dimensional layers[1−3].

For example, if the 2-dimensional layers in the y-direction are themselves 2-dimensional Hadamard matrices, or if the 2-dimensional layers in the x-direction are themselves 2-dimensional Hadamard matrices, then the 2-dimensional layers in the z-direction are orthogonal to each other.

Thus the equation

$$\sum_{i=0}^{m-1}\sum_{j=0}^{m-1} H(i,j,a)H(i,j,b) = m^2\delta_{ab}$$

is satisfied if either

$$\sum_{i=0}^{m-1} H(i,j,a)H(i,j,b) = m\delta_{ab} \tag{3.8}$$

or

$$\sum_{j=0}^{m-1} H(i,j,a)H(i,j,b) = m\delta_{ab}. \tag{3.9}$$

Similarly, if the 2-dimensional layers in the z-direction are themselves 2-dimensional Hadamard matrices, or if the 2-dimensional layers in the y-direction are themselves 2-dimensional Hadamard matrices, then the 2-dimensional layers in the x-direction are orthogonal to each other.

Thus the equation

$$\sum_{j=0}^{m-1}\sum_{k=0}^{m-1} H(a,j,k)H(b,j,k) = m^2\delta_{ab}$$

is satisfied if either

$$\sum_{j=0}^{m-1} H(a,j,k)H(b,j,k) = m\delta_{ab} \tag{3.10}$$

or

$$\sum_{k=0}^{m-1} H(a,j,k)H(b,j,k) = m\delta_{ab}. \tag{3.11}$$

If the 2-dimensional layers in the x-direction are themselves 2-dimensional Hadamard matrices, or if the 2-dimensional layers in the z-direction are themselves 2-dimensional Hadamard matrices, then the 2-dimensional layers in the y-direction are orthogonal to each other.

Thus the equation

$$\sum_{i=0}^{m-1}\sum_{k=0}^{m-1} H(i,a,k)H(i,b,k) = m^2\delta_{ab}$$

is satisfied if either

$$\sum_{i=0}^{m-1} H(i,a,k)H(i,b,k) = m\delta_{ab} \tag{3.12}$$

or

$$\sum_{k=0}^{m-1} H(i,a,k)H(i,b,k) = m\delta_{ab}. \tag{3.13}$$

Each of these equations specifies a possible orientation for orthogonality of either rows or columns within the two-dimensional layers in one direction. A three-dimensional Hadamard matrix exists whenever there are orthogonalities having at least one correlation vector in each axial direction, e.g.[1−3],

Equations(3.8) + (3.10) + (3.12);

Equations(3.8) + (3.10) + (3.13);

Equations(3.8) + (3.11) + (3.12);

Equations(3.8) + (3.11) + (3.13);

Equations(3.9) + (3.10) + (3.12);

Equations(3.9) + (3.10) + (3.13);

Equations(3.9) + (3.11) + (3.12);

Equations(3.9) + (3.11) + (3.13).

Note that certain pairs of these equations also specify that all layers in one direction are two-dimensional Hadamard matrices, e.g., Equations (3.10) +(3.12) is equivalent to Equation (3.3).

From these relations, it follows that a three-dimensional Hadamard matrix is specified[1−3], if

(a) all layers in one direction are two-dimensional Hadamard matrices which are orthogonal to each other (e.g., Equations (3.8)+(3.10)+(3.12));

(b) all layers in two directions are Hadamard matrices (e.g., Equations (3.9)+(3.13)+(3.10)+(3.12)); or

(c) in any direction, all layers are orthogonal in at least one layer direction so that collectively there is at least one correlation vector in each axial direction (e.g., Equations (3.8)+(3.10)+(3.13)).

By making use of these rules, we can construct three-dimensional Hadamard matrices proper in at least two directions, if we have been given a two-dimensional Hadamard matrix. In fact, if the rows (or columns) of a two-dimensional Hadamard matrix are cyclically permuted, then the resultant Hadamard matrix is orthogonal to the mother matrix. Therefore the set of successive cyclic row-permutations of a given two-dimensional Hadamard matrix $[h(i,j)]$ of order $n \times n$, i.e.,

$$H(i,j,k) = h(i,(j+k) \bmod n) \tag{3.14}$$

form the successive layers of a three-dimensional matrix which satisfies both rules (a) and (b). This matrix is therefore a three-dimensional Hadamard matrix proper in at least two directions.

To sum up, it has been found possible to construct the following three-dimensional Hadamard matrices[1−3]:

- Absolutely proper 3-dimensional Hadamard matrices of order 2^t, plus a variety of partially proper and improper ones, by direct multiplication of 2^3 matrices;
- Absolutely improper 3-dimensional Hadamard matrices of order m^2 by successive direct multiplication of two-dimensional m^2 matrices in different orientations;

- Two-directional proper 3-dimensional Hadamard matrices of order m by cyclic permutation of the rows of any m^2 two-dimensional matrix;
- In some cases, absolutely proper matrices of order m from mirror-symmetrical m^2 matrices by assuming threefold symmetry.

3.2 3-Dimensional Hadamard Matrices of Order $4k + 2$

One of the most important necessary conditions for a 2-dimensional Hadamard matrix is that its order n satisfies $n = 4k$ or $n = 2$. In the following section, we will show, by example, that the order n of some 3-dimensional Hadamard matrix can take the form of $n = 4k + 2$, $k \geqslant 1$.

Theorem 3.2.1 *If $H = [H(i,j,k)]$, $0 \leqslant i,j,k \leqslant n-1$, is a 3-dimensional Hadamard matrix of order n, then this n must be an even integer.*

Proof $H(i,j,k) = \pm 1$ implies $(H(i,j,k))^2 = 1$. The orthogonality of the 3-dimensional Hadamard matrix implies the following equations

$$\sum_{i=0}^{n-1}\sum_{j=0}^{n-1} H(i,j,0)H(i,j,1) = 0 \tag{3.15}$$

and

$$\sum_{i=0}^{n-1}\sum_{j=0}^{n-1} H(i,j,0)H(i,j,0) = n^2. \tag{3.16}$$

Therefore

$$\begin{aligned}
&\sum_{i=0}^{n-1}\sum_{j=0}^{n-1} H(i,j,0)[H(i,j,0) + H(i,j,1)] \\
&= \sum_{i=0}^{n-1}\sum_{j=0}^{n-1} H(i,j,0)H(i,j,0) + \sum_{i=0}^{n-1}\sum_{j=0}^{n-1} H(i,j,0)H(i,j,1) \\
&= n^2 + 0 \\
&= n^2.
\end{aligned} \tag{3.17}$$

On the other hand,

$$H(i,j,0) + H(i,j,1) = (\pm 1) + (\pm 1) = \text{even}.$$

Thus the number n^2 must be even, and so is n itself. □

The constructions stated in the last section produce 3-dimensional Hadamard matrices of order $4k$ only. Theorem 3.2.1 motivates us to try to find those of order $4k+2$, $k \geqslant 1$. The smallest such order is $4 \times 1 + 2 = 6$, i.e., the case of $k = 1$. This subsection will show an example of a 3-dimensional Hadamard matrix of order 6 which is based on the concept of perfect binary array, defined by

Definition 3.2.1 [4, 5] *A (± 1)-valued matrix $S = [A(i,j)]$, $0 \leqslant i \leqslant n-1$, $0 \leqslant j \leqslant m-1$, is called a (2-dimensional) perfect binary array of order $n \times m$, abbreviated as* PBA(n,m), *if and only if its 2-dimensional cyclic autocorrelation is a δ-function, i.e.,*

$$R_A(s,t)=\sum_{i=0}^{n-1}\sum_{j=0}^{m-1}A(i,j)A(i+s,j+t)=0 \text{ if } (s,t)\neq(0,0), \tag{3.18}$$

where $(i+s)$ *and* $(j+t)$ *refer to* $(i+s)\bmod n$ *and* $(j+t) \bmod m$, *respectively.*

For example[6–8], the matrix

$$A=\begin{bmatrix} - & + & + & + & + & - \\ + & - & + & + & + & - \\ + & + & - & + & + & - \\ + & + & + & - & + & - \\ + & + & + & + & - & - \\ - & - & - & - & - & + \end{bmatrix} \tag{3.19}$$

is a PBA(6, 6)[9]. Where '$-$', and '$+$' refer to -1, and $+1$, respectively.

Theorem 3.2.2 *If* $A=[A(i,j)]$, $0\leqslant i,j\leqslant m-1$, *is a* PBA$(m,m)$, *then the matrix* A *produces a 3-dimensional Hadamard matrix, say* $B=[B(i,j,k)]$, $0\leqslant i,j,k\leqslant m-1$, *of order* m, *by*

$$B(i,j,k)=A(k+i,k+j), \quad 0\leqslant i,j,k\leqslant m-1, \tag{3.20}$$

where $i+k$ *and* $j+k$ *refer to* $(i+k)\bmod m$ *and* $(j+k)\bmod m$, *respectively.*

Proof Let $0\leqslant a\neq b\leqslant m-1$. Then

$$\begin{aligned}\sum_{i=0}^{m-1}\sum_{j=0}^{m-1}B(i,j,a)B(i,j,b)&=\sum_{i=0}^{m-1}\sum_{j=0}^{m-1}A(i+a,j+a)A(i+b,j+b)\\ &=\sum_{i=0}^{m-1}\sum_{j=0}^{m-1}A(i,j)A(i+(b-a),j+(b-a))\\ &=0.\end{aligned}$$

The last equation is owed to the property that the matrix $A=[A(i,j)]$ is itself a PBA.

$$\begin{aligned}\sum_{i=0}^{m-1}\sum_{j=0}^{m-1}B(i,a,k)B(i,b,k)&=\sum_{i=0}^{m-1}\sum_{j=0}^{m-1}A(i+k,a+k)A(i+k,b+k)\\ &=\sum_{i=0}^{m-1}\sum_{j=0}^{m-1}A(i,j)A(i,j+(b-a))\\ &=0.\end{aligned}$$

Similarly, it can be proved that

$$\sum_{i=0}^{m-1}\sum_{j=0}^{m-1}B(a,j,k)B(b,j,k)=0.$$

Thus the matrix $B=[B(i,j,k)]$ satisfies Equation (3.2). □

Applying Theorem 3.2.2 to the PBA(6, 6) in Equation (3.19), we obtain a 3-dimensional Hadamard matrix of order 6 which is described by its layers in the z-direction as follows

$$B(i,j,0) = \begin{bmatrix} - & + & + & + & + & - \\ + & - & + & + & + & - \\ + & + & - & + & + & - \\ + & + & + & - & + & - \\ + & + & + & + & - & - \\ - & - & - & - & - & + \end{bmatrix}, \quad B(i,j,1) = \begin{bmatrix} - & + & + & + & - & + \\ + & - & + & + & - & + \\ + & + & - & + & - & + \\ + & + & + & - & - & + \\ - & - & - & - & + & - \\ + & + & + & + & - & - \end{bmatrix},$$

$$B(i,j,2) = \begin{bmatrix} - & + & + & - & + & + \\ + & - & + & - & + & + \\ + & + & - & - & + & + \\ - & - & - & + & - & - \\ + & + & + & - & - & + \\ + & + & + & - & + & - \end{bmatrix}, \quad B(i,j,3) = \begin{bmatrix} - & + & - & + & + & + \\ + & - & - & + & + & + \\ - & - & + & - & - & - \\ + & + & - & - & + & + \\ + & + & - & + & - & + \\ + & + & - & + & + & - \end{bmatrix},$$

$$B(i,j,4) = \begin{bmatrix} - & - & + & + & + & + \\ - & + & - & - & - & - \\ + & - & - & + & + & + \\ + & - & + & - & + & + \\ + & - & + & + & - & + \\ + & - & + & + & + & - \end{bmatrix}, \quad B(i,j,5) = \begin{bmatrix} + & - & - & - & - & - \\ - & - & + & + & + & + \\ - & + & - & + & + & + \\ - & + & + & - & + & + \\ - & + & + & + & - & + \\ - & + & + & + & + & - \end{bmatrix}.$$

Lemma 3.2.1 [5] *For each positive integer b there exists at least one* $\mathrm{PBA}(2.3^b, 2.3^b)$.

This lemma is, in fact, a direct corollary of a theorem in Chapter 6 on *General Higher Dimensional Hadamard Matrices*. Thus we omit here the proof.

Lemma 3.2.1 together with Theorem 3.2.2 reduces the following result.

Theorem 3.2.3 *For each positive integer b there exists at least one 3-dimensional Hadamard matrix of order $2 \cdot 3^b$.*

The above 3-dimensional Hadamard matrix of order 6 is clearly an example of a matrix of Theorem 3.2.3. It is very strange that except for the above 3-dimensional Hadamard matrices, no other 3-dimensional Hadamard matrix of order $4k+2 \neq 2.3^b$, $k > 1$, $b \geqslant 0$, has been found. Thus it is an open problem to find such 3-dimensional Hadamard matrices.

3.3 3-Dimensional Hadamard Matrices of Order $4k$

Many 3-dimensional Hadamard matrices of order $4k$ have been constructed in the first section. This section concentrates on the construction of 2-dimensional PBAs. Thus 3-dimensional Hadamard matrices are found by Theorem 3.2.2. The contents of this section are mainly due to [5].

3.3.1 Recursive Constructions of Perfect Binary Arrays

We start this subsection with the following necessary condition on the orders of perfect binary arrays.

Theorem 3.3.1 *Let $A = [A(i,j)]$, $0 \leqslant i,j \leqslant m-1$, be a* $\mathrm{PBA}(m,m)$. *Then the m must be even.*

Proof By the definition of PBA, we have

$$\sum_{i=0}^{m-1}\sum_{j=0}^{m-1} A(i,j)A(i+1,j+1)=0.$$

Because each term $A(i,j)A(i+1,j+1)$ in the left hand is ± 1, so m^2 should be even, otherwise the summation in the left hand can't be zero. Hence the order m is even. □

From Theorem 3.3.1, we know that the order of PBA(m,m) should satisfy $m = 2, 4, 6, 8, 10, 12, \cdots$. We have shown a PBA$(6,6)$ in the last subsection. Here are a few other known PBAs of small orders:

(1) PBA(2, 2):

$$A=\begin{bmatrix} + & + \\ + & - \end{bmatrix};$$

(2) PBA(4, 4):

$$A=\begin{bmatrix} + & + & + & - \\ + & + & + & - \\ + & + & + & - \\ - & - & - & + \end{bmatrix};$$

(3) PBA(8, 8)[10]:

$$A=\begin{bmatrix} - & + & + & + & - & + & + & + \\ + & - & + & + & - & + & - & - \\ + & + & + & - & + & + & + & - \\ + & + & - & - & - & - & + & + \\ - & - & + & - & - & - & + & - \\ + & + & + & - & - & - & - & + \\ + & - & + & + & + & - & + & + \\ + & - & - & + & - & + & + & - \end{bmatrix};$$

(4) PBA(12, 12)[6]:

$$A=\begin{bmatrix} + & - & + & - & + & - & - & + & - & - & - & - \\ + & - & - & - & + & - & - & - & - & - & + & + \\ + & + & - & + & + & + & + & - & - & + & - & - \\ - & + & - & + & - & + & + & - & + & + & + & + \\ + & - & - & - & + & - & - & - & - & - & + & + \\ + & + & - & + & + & + & + & - & - & + & - & - \\ + & - & + & - & + & - & - & + & - & - & - & - \\ - & + & + & + & - & + & + & + & + & + & - & - \\ + & + & - & + & + & + & + & - & - & + & - & - \\ + & - & + & - & + & - & - & + & - & - & - & - \\ + & - & - & - & + & - & - & - & - & - & + & + \\ - & - & + & - & - & - & - & + & + & - & + & + \end{bmatrix}.$$

In addition, some other PBAs of small size, e.g., PBA(16, 16), PBA(24, 24), PBA(32, 32), and PBA(64, 64), have been constructed[11]. These small PBAs are useful in the general recursive constructions.

Definition 3.3.1 [5, 8] *Let* $A = [A(i,j)]$, $B = [B(i,j)]$, $0 \leqslant i \leqslant s-1$, $0 \leqslant j \leqslant t-1$, *be two* (± 1)*-valued arrays. The column interleaving* $ic(A,B)$ *of* A *with* B *is the* $s \times (2t)$ (± 1)*-valued array* $C = [C(i,j)]$ *given by*

$$C(i,2j) = A(i,j) \quad \text{and} \quad C(i,2j+1) = B(i,j),$$
$$0 \leqslant i \leqslant s-1 \text{ and } 0 \leqslant j \leqslant t-1.$$

The row interleaving $ir(A,B)$ *of* A *with* B *is the* $(2s) \times t$ (± 1)*-valued array* $D = [d(i,j)]$ *given by*

$$d(2i,j) = A(i,j) \quad \text{and} \quad d(2i+1,j) = B(i,j), \quad 0 \leqslant i \leqslant s-1 \text{ and } 0 \leqslant j \leqslant t-1.$$

The 2-dimensional cyclic cross-correlation between two matrices, say A and B, of the same size, say $n \times m$, is another matrix $R_{AB} = [R_{AB}(u,v)]$, $0 \leqslant u \leqslant n-1$, $0 \leqslant v \leqslant m-1$, defined by[5]

$$R_{AB}(u,v) = \sum_{i=0}^{n-1} \sum_{j=0}^{m-1} A(i,j)B(i+u,j+v),$$

where $i+u \equiv (i+u) \bmod n$ and $j+v \equiv (j+v) \bmod m$. If $A = B$, then $R_{AA}(u,v)$ is the autocorrelation and abbreviated as $R_A(u,v)$.

By Definition 3.3.1 and the definition of 2-dimensional cyclic auto- and cross-correlations we have the following identities:

Lemma 3.3.1 [5, 8] *Let* $A = [A(i,j)]$, *and* $B = [B(i,j)]$, $0 \leqslant i \leqslant s-1$, $0 \leqslant j \leqslant t-1$, *be two* (± 1)*-valued arrays. Let* $C = ic(A,B)$ *and* $D = ir(A,B)$. *Then*

$$\begin{aligned} R_C(u,2v) &= R_A(u,v) + R_B(u,v), \\ R_C(u,2v+1) &= R_{AB}(u,v) + R_{AB}(s-u,t-v-1), \\ R_D(2u,v) &= R_A(u,v) + R_B(u,v), \\ R_D(2u+1,v) &= R_{AB}(u,v) + R_{AB}(s-u-1,t-v) \end{aligned}$$

for all $0 \leqslant u \leqslant s-1$ *and* $0 \leqslant v \leqslant t-1$.

Theorem 3.3.2 [5, 8] *Let* $A = [A(i,j)]$, $B = [B(i,j)]$, $0 \leqslant i \leqslant s-1$, $0 \leqslant j \leqslant t-1$, *be two* (± 1)*-valued arrays. Any two of the following imply the third:*

(1) $C = ic(A,B)$ *is a* PBA$(s,2t)$;

(2) $D = ir(A,B)$ *is a* PBA$(2s,t)$;

(3) $R_{AB}(u,v) = R_{AB}(u+1,v-1)$ *for all* $0 \leqslant u \leqslant s-1$, $0 \leqslant v \leqslant t-1$.

Proof If the first two equations are satisfied, then

$$R_C(u,2v+1) = R_D(2u+1,v)(=0)$$

for all $0 \leqslant u \leqslant s-1$, $0 \leqslant v \leqslant t-1$. Hence by Lemma 3.3.1,

$$R_{AB}(s-u,t-v-1) = R_{AB}(s-u-1,t-v)$$

for all $0 \leqslant u \leqslant s-1$, $0 \leqslant v \leqslant t-1$. Replacing $s-u-1$ by u and $t-v$ by v, we obtain the third equation.

Conversely, if the third equation holds, then by Lemma 3.3.1,

$$R_C(u, 2v) = R_D(2u, v) \text{ and } R_C(u, 2v+1) = R_D(2u+1, v)$$

for all $0 \leqslant u \leqslant s-1$, $0 \leqslant v \leqslant t-1$. Hence the first equation holds if and only if the second equation holds. □

Definition 3.3.2 [5, 7] *Let $A = [A(i,j)]$, $B = [B(i,j)]$, $0 \leqslant i \leqslant s-1$, $0 \leqslant j \leqslant t-1$, be two (± 1)-valued arrays. A and B are called complementary if*

$$R_A(u,v) + R_B(u,v) = 0 \text{ for all } (u,v) \neq (0,0)$$

and uncorrelated if

$$R_{AB}(u,v) = 0 \text{ for all } u, v.$$

Now, we are ready to construct PBAs C and D by interleaving appropriate arrays A and B.

Theorem 3.3.3 [5, 8] *Let $A = [A(i,j)]$, $B = [B(i,j)]$, $0 \leqslant i \leqslant s-1$, $0 \leqslant j \leqslant t-1$, be two (± 1)-valued arrays. Then $C = ic(A,B)$ is a* PBA$(s, 2t)$ *(resp., $D = ir(A,B)$ is a* PBA$(2s,t)$) *if and only if A and B are complementary arrays such that*

$$R_{AB}(u,v) + R_{AB}(s-u, t-v-1) = 0 \text{ for all } u, v$$

$$(\text{resp., } R_{AB}(u,v) + R_{AB}(s-u-1, t-v) = 0 \text{ for all } u, v).$$

Proof This theorem follows immediately from Definition 3.3.2 and Lemma 3.3.1. □

Corollary 3.3.1 [5, 8] *Let $A = [A(i,j)]$, $B = [B(i,j)]$, $0 \leqslant i \leqslant s-1$, $0 \leqslant j \leqslant t-1$, be two (± 1)-valued arrays which are complementary and uncorrelated. Then $C = ic(A,B)$ is a* PBA$(s, 2t)$ *and $D = ir(A,B)$ is a* PBA$(2s,t)$.

Let $A = [A(i,j)]$, $B = [B(i,j)]$, $0 \leqslant i \leqslant s-1$, $0 \leqslant j \leqslant t-1$, be two (± 1)-valued arrays. Define a $(2s) \times t$ binary array $E = [E(i,j)] = \begin{bmatrix} A \\ B \end{bmatrix}$ by

$$E(i,j) = A(i,j) \text{ and } E(i+s, j) = B(i,j) \text{ for all } 0 \leqslant i \leqslant s-1,\ 0 \leqslant j \leqslant t-1,$$

and a $s \times (2t)$ binary array $F = [F(i,j)] = [AB]$ by

$$F(i,j) = A(i,j) \text{ and } F(i, j+t) = B(i,j) \text{ for all } 0 \leqslant i \leqslant s-1,\ 0 \leqslant j \leqslant t-1.$$

The following lemma states how to construct uncorrelated binary arrays.

Lemma 3.3.2 [5, 8, 12] *Let $A = [A(i,j)]$, $B = [B(i,j)]$, $0 \leqslant i \leqslant s-1$, $0 \leqslant j \leqslant t-1$, be two (± 1)-valued arrays. Then*

(1) $A' =: \begin{bmatrix} A \\ A \end{bmatrix}$ *and* $B' =: \begin{bmatrix} B \\ -B \end{bmatrix}$ *are uncorrelated;*

(2) $A'' =: [A\ A]$ *and* $B'' =: [B\ \ -B]$ *are uncorrelated.*

This lemma can be proved by directly verification.

Definition 3.3.3 [7, 5, 12] *Let* $B = [B(i,j)]$, $0 \leqslant i \leqslant s-1$, $0 \leqslant j \leqslant t-1$, *be a* (± 1)*-valued array. B is called row-wise quasiperfect if* $B' =: \begin{bmatrix} B \\ -B \end{bmatrix}$ *satisfies* $R_{B'}(u,v) = 0$ *for all* $(u,v) \neq (0,0)$ *or* $(s,0)$.

B is called column-wise quasi-perfect if $B'' =: [B - B]$ *satisfies* $R_{B''}(u,v) = 0$ *for all* $(u,v) \neq (0,0)$ *or* $(0,t)$.

We write RQPBA(s,t) *(resp.,* CQPBA(s,t)*) for an* $s \times t$ *array which is either row-wise or column-wise quasi-perfect.*

Note that B is row-wise quasi-perfect if and only if B^{T} is column-wise quasi-perfect. Note also that if $A = [A(i,j)]$, $B = [B(i,j)]$ and $C = [C(i,j)]$, where $B(i,j) = (-1)^i A(i,j)$ and $C(i,j) = (-1)^j A(i,j)$, then

(1) for s odd, A is a PBA(s,t) if and only if B is a RQPBA(s,t);

(2) for t odd, A is a PBA(s,t) if and only if C is a CQPBA(s,t).

Lemma 3.3.3 [5, 7] *Let A be a* PBA(s,t) *and B a quasi-perfect binary array. If B is row-wise quasi-perfect, then* $A' =: \begin{bmatrix} A \\ A \end{bmatrix}$ *is complementary to* $B' =: \begin{bmatrix} B \\ -B \end{bmatrix}$. *If B is column-wise quasi-perfect, then* $A'' =: [A\ A]$ *is complementary to* $B'' =: [B\ \ -B]$.

Proof By Definition 3.3.3, we have

$$R_{B'}(u,v) = \begin{cases} 0, & \text{for all } (u,v) \neq (0,0), (s,0), \\ -2st, & \text{for } (u,v) = (s,0), \end{cases}$$

and

$$R_{A'}(u,v) = \begin{cases} 0, & \text{for all } (u,v) \neq (0,0), (s,0), \\ 2st, & \text{for } (u,v) = (s,0). \end{cases}$$

Hence

$$R_{A'}(u,v) + R_{B'}(u,v) = 0 \text{ for all } (u,v) \neq (0,0).$$

Similarly

$$R_{A''}(u,v) + R_{B''}(u,v) = 0 \text{ for all } (u,v) \neq (0,0).$$

□

Theorem 3.3.4 [5, 7] *If there exist a* PBA(s,t), *A, and a quasi-perfect binary array, B, of size* $s \times t$, *then there exist a* PBA$(2s, 2t)$. *Moreover, if the quasi-perfect binary array B is row-wise quasi-perfect, there also exists a* PBA$(4s, t)$, *and if B is column-wise quasi-perfect, there also exists a* PBA$(s, 4t)$.

Proof If B is row-wise quasi-perfect, then, by Lemmas 3.3.2 and 3.3.3, $A' =: \begin{bmatrix} A \\ A \end{bmatrix}$ and $B' =: \begin{bmatrix} B \\ -B \end{bmatrix}$ are complementary uncorrelated $(2s) \times t$ (± 1)-valued arrays. Hence

$ic(A', B')$ is a PBA$(2s, 2t)$ and $ir(A', B')$ is a PBA$(4s, t)$, by Corollary 3.3.1.

Similarly, if B is column-wise quasi-perfect and $A'' =: [A\ A]$ and $B'' =: [B\ \ -B]$, then $ir(A'', B'')$ is a PBA$(2s, 2t)$ and $ic(A'', B'')$ is a PBA$(s, 4t)$. □

It should be remarked that with the definitions used in the above proof, if a PBA takes one of the forms constructed, namely, $ic(A', B')$ or $ir(A', B')$ (resp., $ic(A'', B'')$ or $ir(A'', B'')$) for any PBAs, A and B, of size $s \times t$, then A is perfect and B is row-wise (resp. column-wise) quasi-perfect. This follows easily from Lemma 3.3.1 and the following identities: for all $0 \leqslant u \leqslant s-1$, $0 \leqslant v \leqslant t-1$, $(u, v) \neq (0, 0)$,

$$R_{A'}(u+s, v) = R_{A'}(u, v) \text{ and}$$
$$R_{B'}(u+s, v) = -R_{B'}(u, v)$$
$$(\text{reps. } R_{A''}(u, v+t) = R_{A''}(u, v) \text{ and}$$
$$R_{B''}(u, v+t) = -R_{B''}(u, v)).$$

3.3.2 Quasi-Perfect Binary Arrays

In order to make use of Theorem 3.3.4 to construct PBAs, we need as many quasi-perfect binary arrays as possible. This subsection aims at constructing quasi-perfect and doubly quasi-perfect binary arrays[5, 7].

Definition 3.3.4 [5, 7] *Let $A = [A(i, j)]$ and $B = [B(i, j)]$, $0 \leqslant i \leqslant s-1$, $0 \leqslant j \leqslant t-1$, be (± 1)-valued arrays. Let $A' =: \begin{bmatrix} A \\ -A \end{bmatrix}$, $B' =: \begin{bmatrix} B \\ -B \end{bmatrix}$, and $A'' =: [A\ \ -A]$, $B'' =: [B\ \ -B]$. A and B are called row-wise quasi-complementary if*

$$R_{A'}(u, v) + R_{B'}(u, v) = 0 \textit{ for all } (u, v) \neq (0, 0), (s, 0),$$

and column-wise quasi-complementary if

$$R_{A''}(u, v) + R_{B''}(u, v) = 0 \textit{ for all } (u, v) \neq (0, 0), (0, t).$$

A and B are called row-wise quasi-uncorrelated if A' and B' are uncorrelated, and column-wise quasi-uncorrelated if A'' and B'' are uncorrelated.

Theorem 3.3.5 [5, 7] *Let $A = [A(i, j)]$ and $B = [B(i, j)]$, $0 \leqslant i \leqslant s-1$, $0 \leqslant j \leqslant t-1$, be (± 1)-valued arrays.*

(1) *Let $A' =: \begin{bmatrix} A \\ -A \end{bmatrix}$, $B' =: \begin{bmatrix} B \\ -B \end{bmatrix}$. Then $C = ic(A, B)$ (resp., $D = ir(A, B)$) is a RQPBA$(s, 2t)$ (resp., RQPBA$(2s, t)$) if and only if A and B are row-wise quasi-complementary arrays such that*

$$R_{A'B'}(u, v) + R_{A'B'}(2s-u, t-v-1) = 0 \textit{ for all } u, v$$

$$(\textit{resp.,}\ \ R_{A'B'}(u, v) + R_{A'B'}(2s-u-1, t-v) = 0 \textit{ for all } u, v).$$

(2) *Let* $A'' =: [A \ -A]$, $B'' =: [B \ -B]$. *Then* $C = ic(A, B)$ (*resp.*, $D = ir(A, B)$) *is a* CQPBA$(s, 2t)$ (*resp.*, CQPBA$(2s, t)$) *if and only if* A *and* B *are column-wise quasi-complementary arrays such that*

$$R_{A''B''}(u, v) + R_{A''B''}(s - u, 2t - v - 1) = 0 \text{ for all } u, v$$

$$(\textit{resp.}, \quad R_{A''B''}(u, v) + R_{A''B''}(s - u - 1, 2t - v) = 0 \text{ for all } u, v).$$

Proof The first statement follows by applying Lemma 3.3.1 to $C' =: \begin{bmatrix} C \\ -C \end{bmatrix} = ic(A', B')$ $\left(\text{resp., } D' =: \begin{bmatrix} D \\ -D \end{bmatrix}\right)$.

The second statement follows by applying Lemma 3.3.1 to $C'' =: [C \ -C] = ic(A'', B'')$ $\left(\text{resp., } D'' =: [D \ -D]\right)$. □

Corollary 3.3.2 [5, 7, 12] *Let* A *and* B *be row-wise (resp., column-wise) quasi-complementary, row-wise (resp., column-wise) quasi-uncorrelated* $s \times t$ (± 1)*-valued arrays. Then* $C = ic(A, B)$ *is a row-wise (resp., column-wise) quasi-perfect binary array of size* $s \times (2t)$ *and* $D = ir(A, B)$ *is a row-wise (resp., column-wise) quasi-perfect binary array of size* $(2s) \times t$.

The following lemma shows how to construct quasi-uncorrelated binary arrays.

Lemma 3.3.4 [5, 7, 12] *Let* $A = [A(i, j)]$ *and* $B = [B(i, j)]$, $0 \leqslant i \leqslant s - 1$, $0 \leqslant j \leqslant t - 1$, *be* (± 1)*-valued arrays. Then*

(1) $A' =: \begin{bmatrix} A \\ A \end{bmatrix}$ *and* $B' =: \begin{bmatrix} B \\ -B \end{bmatrix}$ *are column-wise quasi-uncorrelated; and*

(2) $A'' =: [A \ A]$ *and* $B'' =: [B \ -B]$ *are row-wise quasi-uncorrelated.*

Proof The first statement follows by applying the first statement of Lemma 3.3.2 to $[A \ -A]$ and $[B \ -B]$.

The second statement follows by applying the second statement of Lemma 3.3.2 to $\begin{bmatrix} A \\ -A \end{bmatrix}$ and $\begin{bmatrix} B \\ -B \end{bmatrix}$. □

Definition 3.3.5 [5, 7] *Let* $C = [C(i, j)]$, $0 \leqslant i \leqslant s - 1$, $0 \leqslant j \leqslant t - 1$, *be a* (± 1)*-valued array.* C *is called doubly quasi-perfect, written by* DQPBA(s, t), *if and only if the cyclic autocorrelation of the matrix*

$$C' = \begin{bmatrix} C & -C \\ -C & C \end{bmatrix}$$

satisfies $R_{C'}(u, v) = 0$ *for all* $(u, v) \neq (0, 0), (s, 0), (0, t), (s, t)$.

Note that C is doubly quasi-perfect if and only if C^{T} is doubly quasi-perfect. Note also that if $A = [A(i, j)]$, $B = [B(i, j)]$ and $C = [C(i, j)]$ where $B(i, j) = (-1)^i C(i, j)$ and $A(i, j) = (-1)^j C(i, j)$, then

(1) For s odd, C is a DQPBA(s, t) if and only if B is a CQPBA(s, t);

(2) For t odd, C is a DQPBA(s, t) if and only if A is a RQPBA(s, t).

Lemma 3.3.5 [5, 7] *Let* B *be a quasi-perfect binary array of size* $s \times t$ *and* C *a* DQPBA(s, t).

(1) *If B is row-wise quasi-perfect, then $[B\ B]$ and $[C\ \ -C]$ are row-wise quasi-complementary; and,*

(2) *If B is column-wise quasi-perfect, then* $\begin{bmatrix} B \\ B \end{bmatrix}$ *and* $\begin{bmatrix} C \\ -C \end{bmatrix}$ *are column-wise quasi-complementary.*

Proof The proof is similar to that of Lemma 3.3.3. □

Theorem 3.3.6 [5, 7] *If there exist a row-wise (resp., column-wise) quasi-perfect binary array, say B, of size $s \times t$, and a DQPBA(s,t), say C, then there exist a row-wise (resp., column-wise) quasi-perfect binary array of size $(2s) \times (2t)$ and a RQPBA$(s, 4t)$ (resp., CQPBA$(4s, t)$).*

Proof The proof is similar to that of Theorem 3.3.4. In fact, if B is row-wise quasi-perfect, then $ir\,([B\ B], [C\ \ -C])$ is a RQPBA$(2s, 2t)$ and $ic\,([B\ B], [C\ \ -C])$ is a RQPBA$(s, 4t)$.

If B is column-wise quasi-perfect, then

$$ic\left(\begin{bmatrix} B \\ B \end{bmatrix}, \begin{bmatrix} C \\ -C \end{bmatrix}\right) \text{ is a CQPBA}(2s, 2t)$$

and

$$ir\left(\begin{bmatrix} B \\ B \end{bmatrix}, \begin{bmatrix} C \\ -C \end{bmatrix}\right) \text{ is a CQPBA}(4s, t).$$ □

We note that if a row-wise (resp., column-wise) quasi-perfect binary array takes one of the forms constructed in the above proof, namely $ir\,([B\ B], [C\ \ -C])$ or $ic\,([B\ B], [C\ \ -C])$ $\left(\text{resp., } ic\left(\begin{bmatrix} B \\ B \end{bmatrix}, \begin{bmatrix} C \\ -C \end{bmatrix}\right) \text{ or } ir\left(\begin{bmatrix} B \\ B \end{bmatrix}, \begin{bmatrix} C \\ -C \end{bmatrix}\right)\right)$ for any $s \times t$ binary arrays B and C, then B is row-wise (resp., column-wise) quasi-perfect and C is doubly quasi-perfect.

Definition 3.3.6 [5, 7] *Let $A = [A(i,j)]$ and $B = [B(i,j)]$, $0 \leqslant i \leqslant s-1$, $0 \leqslant j \leqslant t-1$, be (± 1)-valued arrays and let c be an integer. B is called the row-wise c-shear of A if $ct \equiv 0 \pmod{s}$ and*

$$B(i,j) = A(i - cj, j), \quad \textit{for all } i, j,$$

or B is called the column-wise c-shear of A if $cs \equiv 0 (\text{mod} t)$ and

$$B(i,j) = A(i, j - ci), \quad \textit{for all } i, j$$

(we regard the sums $i - cj$ and $j - ci$ to be reduced modulo s and t, respectively).

The condition on c is necessary for the c-shear to be well-defined. Note that if B is the row-wise (resp., column-wise) c-shear of A, then A is the row-wise (resp., column-wise) $(-c)$-shear of B. With this definition in mind, it is easy to verify the following lemma.

Lemma 3.3.6 [5, 7] *Let $A = [A(i,j)]$, $0 \leqslant i \leqslant s-1$, $0 \leqslant j \leqslant t-1$, be a (± 1)-valued array.*

(1) *If A' is the row-wise c-shear of A, then*

$$R_{A'}(u,v) = R_A(u - cv, v) \ \textit{for all } u, v;$$

(2) *If A'' is the column-wise c-shear of A, then*

$$R_{A''}(u,v) = R_A(u, v - cu) \text{ for all } u, v.$$

Corollary 3.3.3 [5, 7] *Let A be an $(rs) \times (rt)$ binary array for some positive integer r. Let B be the c-shear of A (row-wise or column-wise). Then*

$$R_A(u,v) = 0 \text{ for } u \not\equiv 0(\text{mod} s) \text{ or } v \not\equiv 0(\text{mod} t)$$

if and only if

$$R_B(u,v) = 0 \text{ for } u \not\equiv 0(\text{mod} s) \text{ or } v \not\equiv 0(\text{mod} t).$$

Proof This corollary follows from Lemma 3.3.6, together with the necessary condition on c for the c-shear to be well-defined. □

Definition 3.3.7 [5, 8] *Let B be a $(2s) \times (2t)$ binary array. Let $B' = [B'(i,j)]$ be the row-wise (resp., column-wise) c-shear of B and define an $s \times t$ binary array $A' = [A'(i,j)]$ by $A'(i,j) = B'(i,j)$, for all $0 \leqslant i \leqslant s-1$, $0 \leqslant j \leqslant t-1$. A' is called the row-wise (resp., column-wise) c-transform of B.*

Theorem 3.3.7 [5, 7] *Let A be an $s \times t$ binary array.*

(1) *Let B be the c-shear of A (row-wise or column-wise). Then A is perfect if and only if B is perfect;*
(2) *Let c satisfy $ct \equiv s(\text{mod} 2s)$ (resp., $cs \equiv t(\text{mod} 2t)$). Then A is row-wise (resp., column-wise) quasi-perfect if and only if the row-wise (resp., column-wise) c-transform of $\begin{bmatrix} A & A \\ -A & -A \end{bmatrix}$ $\left(\text{resp., } \begin{bmatrix} A & -A \\ A & -A \end{bmatrix}\right)$ is doubly quasi-perfect;*
(3) *Let c satisfy $ct \equiv 0(\text{mod} 2s)$ (resp., $cs \equiv 0(\text{mod} 2t)$). Then A is row-wise (resp., column-wise) quasi-perfect if and only if the row-wise (resp., column-wise) c-transform of $\begin{bmatrix} A & A \\ -A & -A \end{bmatrix}$ $\left(\text{resp., } \begin{bmatrix} A & -A \\ A & -A \end{bmatrix}\right)$ is row-wise (resp., column-wise) quasi-perfect.*
Furthermore A is doubly quasi-perfect if and only if the row-wise (resp. column-wise) c-transform of $\begin{bmatrix} A & -A \\ -A & A \end{bmatrix}$ is doubly quasi-perfect.

Proof The first statement follows immediately from Corollary 3.3.3 with $r = 1$. To prove the second statement, suppose $ct \equiv s(\text{mod} 2s)$. Let $B = \begin{bmatrix} A & A \\ -A & -A \end{bmatrix}$ and let B' be the row-wise c-shear of B. By Definition 3.3.6, we know that

$$\begin{aligned} B'(i,j) &= B(i - cj, j), \\ B'(i, j+t) &= B(i - cj - ct, j+t), \\ B'(i+s, j) &= B(i + s - cj, j), \\ B'(i+s, j+t) &= B(i + s - cj - ct, j+t) \end{aligned}$$

for all $0 \leqslant i \leqslant s-1$, $0 \leqslant j \leqslant t-1$. From the given form of B,

$$B(i,j) = B(i, j+t) = -B(i+s, j) = -B(i+s, j+t)$$

for all $0 \leqslant i \leqslant s-1$, $0 \leqslant j \leqslant t-1$. Hence, using $ct \equiv s (\text{mod} 2s)$,

$$B'(i,j) = -B'(i,j+t) = -B'(i+s,j) = B'(i+s,j+t)$$

for all $0 \leqslant i \leqslant s-1$, $0 \leqslant j \leqslant t-1$. Therefore $B' = \begin{bmatrix} A' & -A' \\ -A' & A' \end{bmatrix}$, where A' is the row-wise c-transform of B. By Corollary 3.3.3 with $r=2$, A is row-wise quasi-perfect if and only if A' is doubly quasi-perfect.

Similarly, we may show, given $cs \equiv t$ (mod$2t$), that A is column-wise quasi-perfect if and only if the column-wise c-transform of $\begin{bmatrix} A & -A \\ A & -A \end{bmatrix}$ is doubly quasi-perfect.

The third statement follows from simple modifications of the above arguments used in the proof of the second statement. □

Thus we have now established the equivalence between quasi-perfect and doubly quasi-perfect binary arrays.

Corollary 3.3.4 [5, 8] *If $t/\gcd(s,t)$ is odd, then there exists a* DQP BA(s,t) *if and only if there exista a* RQPBA(s,t). *If $s/\gcd(s,t)$ is odd, then there exists a* DQPBA(s,t) *if and only if there exists a* CQPBA(s,t).

Proof We note that $ct \equiv s(\text{mod} 2s)$ (resp., $cs \equiv t(\text{mod} 2t)$) if and only if $t/\gcd(s,t)$ (resp., $s/\gcd(s,t)$) is odd and c is an odd multiple of $s/\gcd(s,t)$ (resp., $t/\gcd(s,t)$). The corollary follows from the second statement of Theorem 3.3.7. □

Suppose that $t/\gcd(s,t)$ is odd and A is a RQPBA(s,t). The above proof gives a procedure for obtaining a DQPBA(s,t) A'. Put

$$B = \begin{bmatrix} A & A \\ -A & -A \end{bmatrix}$$

and $c = s/\gcd(s,t)$. Form B', the row-wise c-shear of B, by cycling column j of B by cj places for $j = 0, \cdots, 2t-1$. Then the first s rows and t columns of B' are A', the row-wise c-transform of B.

3.3.3 3-Dimensional Hadamard Matrices Based on PBA$(2^m, 2^m)$ and PBA$(3.2^m, 3.2^m)$

This subsection recursively applies the construction theorems to yield infinite families of 2-dimensional PBAs, and thus 3-dimensional Hadamard matrices are produced by Theorem 3.2.2.

Theorem 3.3.8 [5, 7] *If there exist a* PBA(s,t) *and a* DQPBA(s,t), *then there exist a* PBA$(2s,2t)$ *and a* DQPBA$(2s,2t)$. *If $t/\gcd(s,t)$ is odd there also exist a* PBA$(4s,t)$ *and a* RQPBA$(s,4t)$. *If $s/\gcd(s,t)$ is odd there also exist a* PBA$(s,4t)$ *and a* CQPBA$(4s,t)$.

Proof This theorem is the restatement of Theorems 3.3.4 and 3.3.6, by using the equivalence between quasi-perfect binary arrays and doubly quasi-perfect binary arrays described in Corollary 3.3.4. □

Corollary 3.3.5 [5, 8] *If there exist a* PBA(s,t) *and a* DQPBA(s,t), *then there exist* PBA$(2^y s, 2^y t)$ *and* DQPBA$(2^y s, 2^y t)$ *for each* $y \geqslant 0$. *If* $t/\gcd(s,t)$ *is odd there also exist a* PBA$(2^{y+2}s, 2^y t)$ *and a* RQPBA$(2^y s,\ 2^{y+2}t)$ *for each* $y \geqslant 0$. *If* $s/\gcd(s,t)$ *is odd there also exist a* PBA$(2^y s,\ 2^{y+2}t)$ *and a* CQPBA$(2^{y+2}s, 2^y t)$ *for each* $y \geqslant 0$.

Proof It can be proved by repeating Theorem 3.3.8. □

Corollary 3.3.6 [5, 7] *There exist the following infinite families of* 2*-dimensional* PBAs:

F1.1. PBA$(2^y, 2^y)$, $y \geqslant 1$;

F1.2. PBA$(2^{y+1}, 2^{y-1})$, $y \geqslant 1$;

F1.3. PBA$(2^y.3, 2^y.3)$, $y \geqslant 1$;

F1.4. PBA$(2^{y+1} \cdot 3, 2^{y-1} \cdot 3)$, $y \geqslant 1$.

There exist the following infinite families of doubly quasi-perfect and row-wise quasi-perfect binary arrays:

F2.1. DQPBA$(2^y, 2^y)$, $y \geqslant 1$;

F2.2. DQPBA$(2^y.3, 2^y.3)$, $y \geqslant 1$;

F2.3. RQPBA$(2^{y-1}, 2^{y+1})$, $y \geqslant 1$;

F2.4. RQBA$(2^{y-1}.3, 2^{y+1}.3)$, $y \geqslant 1$.

Proof The proof of this corollary is finished by applying Corollary 3.3.5 to the following:

$$\text{PBA}(2,2): \begin{bmatrix} + & + \\ + & - \end{bmatrix};$$

$$\text{DQPBA}(2,2): \begin{bmatrix} + & + \\ + & + \end{bmatrix},$$

PBA$(6,6)$ stated in the last section;

$$\text{DQPBA}(6,6): \begin{bmatrix} - & + & - & - & + & + \\ + & - & + & - & + & + \\ - & + & - & - & - & + \\ - & - & - & + & - & + \\ + & + & - & - & - & + \\ + & + & + & + & + & + \end{bmatrix};$$

$$\text{PBA}(3,12): \begin{bmatrix} + & + & - & + & + & + & + & - & + & - & - & - \\ - & + & - & + & - & - & - & - & - & + & + & - \\ - & - & + & + & - & + & - & - & - & + & - & - \end{bmatrix}.$$

Thus by the second property following Definition 3.3.3, the existence of PBA$(3,12)$ implies the existence of a RQPBA$(12,3)$. □

Applying Theorem 3.2.2 to the families F1.1 and F1.3 in Corollary 3.3.6, we have finally constructed 3-dimensional Hadamard matrices of orders 2^m and $2^m \cdot 3$. It should be noted that 3-dimensional Hadamard matrices of order $2^{m+n} \cdot 3^m$ can also be simply constructed by the direct multiplication of 3-dimensional Hadamard matrices of orders 2 and 6 which are known to exist.

Theorem 3.3.9 [5, 7] *If there exists a* DQPBA$(2t, t)$, *then there exist a* DQPBA$(4t, 2t)$, *a* RQPBA$(2t, 4t)$ *and a* RQPBA$(2t, 16t)$.

Proof Given a DQPBA$(2t, t)$, by Corollary 3.3.4 there exists a RQPBA$(2t, t)$. Then by Theorem 3.3.6 there exists a RQPBA$(4t, 2t)$ and a PQPBA$(2t, 4t)$. Therefore from Corollary 3.3.4 there exists a DQPBA$(4t, 2t)$. Transposing gives a DQPBA$(2t, 4t)$, which we combine with the RQPBA$(2t, 4t)$ to give a RQPBA$(2t, 16t)$ by Theorem 3.3.6. □

Corollary 3.3.7 [5, 7] *If there exists a* DQPBA$(2t, t)$, *then for each* $y \geqslant 0$ *there exists a* DQPBA$(2^{y+1}t, 2^y t)$, *a* RQPBA$(2^{y+1}t, 2^{y+2}t)$, *and a* RQPBA$(2^{y+1}t, 2^{y+4}t)$.

Corollary 3.3.8 [5, 8] *There exist the following infinite families of doubly quasi-perfect and row-wise quasi-perfect binary arrays*:

(1) DQPBA$(2^y, 2^{y-1})$, $y \geqslant 1$;

(2) RQPBA$(2^y, 2^{y+1})$, $y \geqslant 1$;

(3) RQPBA$(2^y, 2^{y+3})$, $y \geqslant 1$.

Proof $\left[\begin{smallmatrix} + \\ + \end{smallmatrix}\right]$ is a DQPBA$(2, 1)$. □

When we recursively applying Theorems 3.3.4 and 3.3.6 and Corollary 3.3.4 to the trivial array $[+]$, we can obtain a family of perfect, row-wise quasi-perfect and doubly quasi-perfect binary arrays of size $2^m \times 2^m$, $m \geqslant 1$. Specifically, we have the following:

Theorem 3.3.10 [5, 7] *Let* $t = 2^y$, $y \geqslant 0$. *There exist arrays* $A = [A(i,j)]$, $B = [B(i,j)]$, *and* $C = [C(i,j)]$, *which are respectively a* PBA(t, t), *a* RQPBA(t, t), *and a* DQPBA(t, t), *for which the following properties are satisfied for all* $0 \leqslant i, j \leqslant t-1$:

(1) $A(i,j) = A(t-i, t-j)$;

(2) $B(0,j) = B(0, t-j-1)$;

(3) $B(i,j) = -B(t-i, t-j-1)$, $i \neq 0$;

(4) $B(i,j) = B(t-i, j)$, $i \neq 0$;

(5) $C(i,j) = C(i, t-j-1)$;

(6) $C(i,j) = C(t-i-1, j)$.

Proof It can be proved by using induction on y. The case $y = 0$ is trivial by choosing $A = B = C = [+]$. Assume that arrays A, B, and C with the desired properties exist for some $y \geqslant 0$ and let $t = 2^y$. Assume also that $D' = [D'(i,j)] = \begin{bmatrix} C & -C \\ -C & C \end{bmatrix}$ is the row-wise 1-shear of $D = [D(i,j)] = \begin{bmatrix} B & B \\ -B & -B \end{bmatrix}$. From the proof of Theorems 3.3.4 and 3.3.6, $A' = [A'(i,j)] = ic\left(\begin{bmatrix} A \\ A \end{bmatrix}, \begin{bmatrix} B \\ -B \end{bmatrix}\right)$ is a PBA$(2t, 2t)$ and $B' = [B'(i,j)] = ir\left([B\ B], [C\ -C]\right)$ is a RQPBA$(2t, 2t)$. From the proof of Corollary 3.3.4, $C' = [C'(i,j)]$

is a DQPBA$(2t, 2t)$, where $E' = [E'(i,j)] = \begin{bmatrix} C' & -C' \\ -C' & C' \end{bmatrix}$ is the row-wise 1-shear of $E = [E(i,j)] = \begin{bmatrix} B' & B' \\ -B' & -B' \end{bmatrix}$. By the construction of A', for all $0 \leqslant i,j \leqslant t-1$,

$$A'(i+t, 2j) = A'(i, 2j) = A(i,j),$$
$$A'(i+t, 2j+1) = -A'(i, 2j+1) = -B(i,j).$$

To establish the first property for A' we must show that for all $0 \leqslant i,j \leqslant t-1$,

$$A'(i, 2j) = A'(2t-i, 2t-2j),$$
$$A'(i, 2j+1) = A'(2t-i, 2t-2j-1),$$

which is equivalent to

$$A(i,j) = A(t-i, t-j),$$
$$B(0,j) = B(0, t-j-1),$$
$$B(i,j) = -B(t-i, t-j-1), \quad i \neq 0.$$

These relations are given by the first three properties. Similarly, the second, third and fourth properties for B' are given respectively by the second, by the third, fifth and sixth, and by the fourth and sixth properties.

By the definition of E', for all $0 \leqslant i,j \leqslant 2t-1$,

$$E'(i+2t, j+2t) = -E'(i+2t, j) = -E'(i, j+2t) = E'(i,j).$$

To establish the fifth property for C' we must therefore show that for all $0 \leqslant i \leqslant 4t-1$, $0 \leqslant j \leqslant 2t-1$,

$$E'(i,j) = -E'(i, 4t-j-1).$$

Now by Definition 3.3.6, $E'(i,j) = E(i-j, j)$ and so this is equivalent to

$$E(i-j, j) = -E(i+j+1, 4t-j-1).$$

Replacing $i-j$ by $2j$ and then by $2i+1$, we require that for all $0 \leqslant i,j \leqslant 2t-1$,

$$E(2i, j) = -E(2i+2j+1, 4t-j-1),$$
$$E(2i+1, j) = -E(2i+2j+2, 4t-j-1).$$

But by the construction of B', $E = ir\,([DD], [D'D'])$ and so this is equivalent to

$$D(i,j) = -D'(i+j, 2t-j-1),$$
$$D'(i,j) = -D(i+j+1, 2t-j-1).$$

Both of these hold provided

$$D'(i,j) = -D'(i, 2t-j-1)$$

for all $0 \leqslant i,j \leqslant 2t-1$, since $D(i,j) = D'(i+j, j)$, by Definition 3.3.6. By the definition of D' this relation is given by the fifth property. A similar argument gives the sixth property for C'. □

3.4 3-Dimensional Walsh Matrices

3-dimensional Hadamard matrices of order 2^n have been constructed in the last section, whilst this section will continue to investigate a special class of 3-dimensional Hadamard matrices of order 2^n which are called 3-dimensional Walsh matrices. Their definitions, constructions, and analytic representations will be presented.

Recall that every integer i, $0 \leqslant i \leqslant 2^n - 1$, can be uniquely expended as $i = \sum_{k=0}^{n-1} i_k 2^k$, which is uniquely determined by an n-dimensional binary (0 or 1) vector $(i_0, i_1, \cdots, i_{n-1})$. In this section, we use the same symbol i to represent both the integer i and its corresponding vector $(i_0, i_1, \cdots, i_{n-1})$.

3.4.1 Generalized 2-Dimensional Walsh Matrices

Definition 3.4.1 *A (± 1)-valued matrix $H = [h(i,j)]$, $0 \leqslant i, j \leqslant 2^n - 1$, of order 2^n is called a generalized 2-dimensional Walsh matrix if and only if there exist two permutation matrices, say A and B, such that AHB is a regular Walsh matrix.*

Clearly the generalized Walsh matrices are also orthogonal, i.e., $2^n H^{-1} = H$. In addition, their rows are closed under the operation of bit-wise multiplication, i.e., for any two rows $(h(i,0), h(i,1), \cdots, h(i,n-1))$ and $(h(j,0), h(j,1), \cdots, h(j,n-1))$ there exists exactly one row $(h(k,0),\ h(k,1),\ \cdots,\ h(k,n-1))$ such that for each $0 \leqslant s \leqslant n-1$, $h(k,s) = h(i,s)h(j,s)$. The columns of a generalized Walsh matrix are also closed under the operation of bit-wise multiplication. The regular Walsh matrices are special cases of the generalized matrices corresponding to $A = B = I_n$, the unit matrix.

Lemma 3.4.1 *Let*

$$a = (\overbrace{1 \cdots 1}^{2k} \overbrace{-1 \cdots -1}^{2k}),$$

and

$$b = (b_0, b_1, \cdots, b_{4k-1}), \qquad b_i = \pm 1, \qquad \sum_{i=0}^{4k-1} b_i = 0$$

are two vectors of length $4k$. If the vector $c = (c_0, \cdots, c_{4k-1})$ formed by the bit-wise multiplication of the vectors a and b satisfies $\sum_{i=0}^{4k-1} c_i = 0$, then the vector b satisfies

$$\sum_{i=0}^{2k-1} b_i = \sum_{i=0}^{2k-1} b_{2k+i} = 0.$$

Proof The equation $\sum_{i=0}^{4k-1} c_i = 0$ implies

$$\sum_{i=0}^{2k-1} b_i - \sum_{i=0}^{2k-1} b_{2k+i} = 0.$$

On the other hand, we have assumed that

$$\sum_{i=0}^{2k-1} b_i + \sum_{i=0}^{2k-1} b_{2k+i} = \sum_{i=0}^{4k-1} b_i = 0.$$

The lemma follows. □

With the help of Lemma 3.4.1 we can prove the following very important and basic theorem about the generalized Walsh matrix.

Theorem 3.4.1 *A (± 1)-valued matrix $H = [h(i,j)]$, $0 \leqslant i,j \leqslant 2^n - 1$, of order 2^n is a generalized 2-dimensional Walsh matrix if and only if the following conditions are satisfied:*

(1) *The rows of H are closed under the operation of bit-wise multiplication;*

(2) *Except the all one row $(1, 1, \cdots, 1)$, in each other row there are 2^{n-1} 1s and 2^{n-1} −1s.*

Proof $\Longrightarrow$ Permutations among rows and columns keep the properties of: (1) closed bit-wise multiplication and (2) the orthogonality of the original Walsh matrix. The balance between 1s and −1s in each non-all-one row is owed to the orthogonality between that row and the all one row.

$\Longleftarrow$ Because of the property of closed bit-wise multiplication, we know the existence of the all-one row. In fact, the bit-wise multiplication of each row by itself produces the all one row.

Without loss of the generality, we assume that

$$H = \begin{pmatrix} 1 & 1 & \cdots & 1 \\ h(1,0) & h(1,1) & \cdots & h(1,2^n-1) \\ h(2,0) & h(2,1) & \cdots & h(2,2^n-1) \\ \vdots & \vdots & & \vdots \\ h(2^n-1,0) & h(2^n-1,1) & \cdots & h(2^n-1,2^n-1) \end{pmatrix}$$

$$=: \begin{pmatrix} h(0) \\ h(1) \\ \vdots \\ h(2^n-1) \end{pmatrix},$$

where $\sum_{j=0}^{2^n-1} h(i,j) = 0$ for each i, $1 \leqslant i \leqslant 2^n - 1$.

This matrix H can be changed, by column-permutations, to the following form (for simplicity we still use the symbol H for the renewed matrix).

$$H = \begin{pmatrix} 1 & \cdots & 1 & 1 & \cdots & 1 \\ 1 & \cdots & 1 & -1 & \cdots & -1 \\ h(2,0) & \cdots & \cdots & \cdots & \cdots & h(2,2^n-1) \\ h(3,0) & \cdots & \cdots & \cdots & \cdots & h(3,2^n-1) \\ \vdots & & & & & \vdots \\ h(2^n-1,0) & \cdots & \cdots & \cdots & \cdots & h(2^n-1,2^n-1) \end{pmatrix}$$

$$=: \begin{pmatrix} h(0) \\ h(1) \\ \vdots \\ h(2^n-1) \end{pmatrix}.$$

Because of Lemma 3.4.1 and because the rows are closed under bit-wise multiplication, we know that for each j, $j \geqslant 2$, both the first half and the second half of $h(j)$ are (± 1)-balanced. Hence the matrix H can be transformed, still by column permutations, to the following form:

$$H = \begin{pmatrix} 1 & \cdots & 1 & 1 & \cdots & 1 & 1 & \cdots & 1 & 1 & \cdots & 1 \\ 1 & \cdots & 1 & 1 & \cdots & 1 & -1 & \cdots & -1 & -1 & \cdots & -1 \\ 1 & \cdots & 1 & -1 & \cdots & -1 & 1 & \cdots & 1 & -1 & \cdots & -1 \\ h(3,0) & \cdots & \cdots & \cdots & \cdots & \cdots & \cdots & \cdots & \cdots & \cdots & \cdots & h(3,2^n-1) \\ h(4,0) & \cdots & \cdots & \cdots & \cdots & \cdots & \cdots & \cdots & \cdots & \cdots & \cdots & h(4,2^n-1) \\ \vdots & & & & & & & & & & & \vdots \\ h(2^n-1,0) & \cdots & \cdots & \cdots & \cdots & \cdots & \cdots & \cdots & \cdots & \cdots & \cdots & h(2^n-1,2^n-1) \end{pmatrix}$$

$$:= \begin{pmatrix} h(0) \\ h(1) \\ \vdots \\ h(2^n-1) \end{pmatrix}.$$

Similarly, because of Lemma 3.4.1 and because the rows are closed under bit-wise multiplication, we know that for each j, $j \geqslant 3$, the first quarter, the second quarter, the third quarter, and the fourth quarter of $h(j)$ are all (± 1)-balanced. Hence the matrix H can be transformed, by column-permutations, to the following form:

$$H = \begin{pmatrix} 1 & \cdots & 1 & 1 & \cdots & 1 & 1 & \cdots & 1 & 1 & \cdots & 1 \\ 1 & \cdots & 1 & 1 & \cdots & 1 & -1 & \cdots & -1 & -1 & \cdots & -1 \\ 1 & \cdots & 1 & -1 & \cdots & -1 & -1 & \cdots & -1 & 1 & \cdots & 1 \\ 1 & \cdots & 1 & -1 & \cdots & -1 & 1 & \cdots & 1 & -1 & \cdots & -1 \\ h(4,0) & \cdots & \cdots & \cdots & \cdots & \cdots & \cdots & \cdots & \cdots & \cdots & \cdots & h(4,2^n-1) \\ h(5,0) & \cdots & \cdots & \cdots & \cdots & \cdots & \cdots & \cdots & \cdots & \cdots & \cdots & h(5,2^n-1) \\ \vdots & & & & & & & & & & & \vdots \\ h(2^n-1,0) & \cdots & \cdots & \cdots & \cdots & \cdots & \cdots & \cdots & \cdots & \cdots & \cdots & h(2^n-1,2^n-1) \end{pmatrix}$$

$$:= \begin{pmatrix} h(0) \\ h(1) \\ \vdots \\ h(2^n-1) \end{pmatrix}.$$

Thus for $0 \leqslant i \leqslant 3$, $h(i) = \mathrm{Wal}_{i-1}(x)$, the i-th Walsh function. Repeating the above process, we can finally transform the matrix H to the Walsh matrix. □

Theorem 3.4.2 *The following two conditions are equivalent to each other:*

C1. $H = [h(i,j)]$, $0 \leqslant i,j \leqslant 2^n - 1$, *is a generalized* 2*-dimensional Walsh matrix satisfying*

$$h(i_1,j) \cdot h(i_2,j) = h(i_1 \oplus i_2, j) \text{ and } h(i,j_1) \cdot h(i,j_2) = h(i, j_1 \oplus j_2),$$

where $u \oplus v$ *stands for the dyadic summation between the two vectors* $u = (u_0, \cdots, u_{n-1})$ *and* $v = (v_0, \cdots, v_{n-1})$, *i.e.,* $u \oplus v = (u_0 + v_0)\text{mod}2, \cdots, (u_{n-1} + v_{n-1})\text{mod}2)$.

C2. *There exists, in* GF(2), *a non-singular matrix* R *of size* $n \times n$ *such that*

$$h(i,j) = (-1)^{iRj'},$$

where the symbols i *and* j *in* $(-1)^{iRj'}$ *stand for the expended vectors of the integers* i *and* j, *and* j' *the transpose of the vector* j.

Proof C2 $\Rightarrow$ C1. The identities

$$h(i_1, j) \cdot h(i_2, j) = h(i_1 \oplus i_2, j) \text{ and } h(i, j_1) \cdot h(i, j_2) = h(i, j_1 \oplus j_2)$$

are straightforward proof of the assumption $h(i,j) = (-1)^{iRj'}$.

From Theorem 3.4.1 and the equation:

$$\sum_{j=0}^{2^n-1} h(i,j) = \sum_{j=0}^{2^n-1} (-1)^{iRj'} = 0, \quad \text{for } i \neq 0$$

we know that H is indeed a generalized Walsh matrix.

C1 $\Rightarrow$ C2. At first $h(i,j)$ can be formulated by

$$h(i,j) = (-1)^{B(i,j)}, \quad 0 \leqslant i, j \leqslant 2^n - 1,$$

where $B(i,j)$ is a $(0,1)$-valued function of i and j.

Since

$$h(i_1, j) \cdot h(i_2, j) = h(i_1 \oplus i_2, j) \text{ and } h(i, j_1) \cdot h(i, j_2) = h(i, j_1 \oplus j_2),$$

we have

$$B(u,v) \oplus B(w,v) = B(u \oplus w, v) \text{ and } B(u,v) \oplus B(u,w) = B(u, v \oplus w).$$

Hence $B(0,v) = B(v,0) = 0$, and for each $c = 0$ or 1, holds

$$B(cu, v) = B(u, cv) = cB(u,v).$$

Up to now, we have proved that $B(u,v)$ is a bi-linear function. Thus there exists a matrix of size $n \times n$ such that $B(u,v) = uRv'$. It is now sufficient to prove that the matrix R is non-singular.

From the orthogonality we know that

$$\sum_{v=0}^{2^n-1} h(u,v)h(w,v) = \sum_{v=0}^{2^n-1} h(u \oplus w, v) = 2^n \delta_{w,u},$$

or, equivalently,

$$\sum_{v=0}^{2^n-1} (-1)^{uRv'} = \begin{cases} 2^n, & \text{if } u = 0, \\ 0, & \text{otherwise.} \end{cases}$$

In other words, we have $uR \neq (0, \cdots, 0)$ for each non-zero u. Thus the matrix R is indeed non-singular. □

The previous Theorem 3.4.2 states that the Walsh functions in universal orders are special cases of the generalized Walsh matrices.

Theorem 3.4.3 *The direct multiplication of two 2-dimensional generalized Walsh matrices is also a 2-dimensional generalized Walsh matrix.*

Proof Let A and B be two 2-dimensional generalized Walsh matrices.

If $B = W_2 = \begin{pmatrix} 1 & 1 \\ 1 & -1 \end{pmatrix}$ then $B \otimes A = \begin{pmatrix} A & A \\ A & -A \end{pmatrix}$, which is clearly a 2-dimensional generalized Walsh matrix.

If $B = W_{2^n} = W_2 \otimes W_2 \otimes \cdots \otimes W_2$, by induction on n, it can be proved that $B \otimes A$ is also a 2-dimensional generalized Walsh matrix.

In general, let $B = EW_{2^n}D$, where E and D are two permutation matrices of size $2^n \times 2^n$. Then

$$B \otimes A = (EW_{2^n}D) \otimes A = (E \otimes I_{2^n})(W_{2^n} \otimes A)(D \otimes I_{2^n}).$$

Hence their direct multiplication is indeed a 2-dimensional generalized Walsh matrix. □

3.4.2 3-Dimensional Walsh Matrices

The 3-dimensional Walsh matrices studied in this subsection are special cases of 3-dimensional Hadamard matrices of size $2^n \times 2^n$ which are defined by the following definition.

Definition 3.4.2 *A 3-dimensional Hadamard matrix of size $2^n \times 2^n$ is called a 3-dimensional Walsh matrix if all 2-dimensional layers in at least one axis-normal orientation are 2-dimensional generalized Walsh matrices. A 3-dimensional Walsh matrix is said to be proper in the x- (resp., y- and z-) direction if all layers in the x- (resp., y- and z-) direction are 2-dimensional generalized Walsh matrices. A 3-dimensional Walsh matrix is said to be absolutely proper if all of the possible 2-dimensional layers are generalized Walsh matrices, otherwise it is said to be improper.*

All of the absolutely proper 3-dimensional Walsh matrices of size 2×2 are described by

$$A_1(i,j,0) = \begin{bmatrix} 1 & -1 \\ 1 & 1 \end{bmatrix} \text{ and } A_1(i,j,1) = \begin{bmatrix} 1 & 1 \\ -1 & 1 \end{bmatrix},$$

$$A_2(i,j,0) = \begin{bmatrix} 1 & 1 \\ 1 & -1 \end{bmatrix} \text{ and } A_2(i,j,1) = \begin{bmatrix} -1 & 1 \\ 1 & 1 \end{bmatrix},$$

$$A_3(i,j,0) = \begin{bmatrix} 1 & 1 \\ -1 & 1 \end{bmatrix} \text{ and } A_3(i,j,1) = \begin{bmatrix} 1 & -1 \\ 1 & 1 \end{bmatrix}$$

and

$$A_4(i,j,0) = \begin{bmatrix} -1 & 1 \\ 1 & 1 \end{bmatrix} \text{ and } A_4(i,j,1) = \begin{bmatrix} 1 & 1 \\ 1 & -1 \end{bmatrix}.$$

All of the improper 3-dimensional Walsh matrices of size 2×2 that are proper in two directions are described by

$$B_1(i,j,0) = \begin{bmatrix} 1 & 1 \\ -1 & 1 \end{bmatrix} \text{ and } B_1(i,j,1) = \begin{bmatrix} 1 & 1 \\ 1 & -1 \end{bmatrix},$$

$$B_2(i,j,0) = \begin{bmatrix} 1 & 1 \\ 1 & -1 \end{bmatrix} \text{ and } B_2(i,j,1) = \begin{bmatrix} 1 & 1 \\ -1 & 1 \end{bmatrix},$$

$$B_3(i,j,0) = \begin{bmatrix} 1 & 1 \\ -1 & 1 \end{bmatrix} \text{ and } B_3(i,j,1) = \begin{bmatrix} -1 & 1 \\ 1 & 1 \end{bmatrix}$$

and

$$B_4(i,j,0) = \begin{bmatrix} -1 & 1 \\ 1 & 1 \end{bmatrix} \text{ and } B_4(i,j,1) = \begin{bmatrix} 1 & -1 \\ 1 & 1 \end{bmatrix}.$$

Theorem 3.4.4 *The direct multiplication of two 3-dimensional Walsh matrices is also a 3-dimensional Walsh matrix. Furthermore, the product matrix is proper in those directions in which both the parent matrices are proper.*

Proof Let A and B be the two parent 3-dimensional Walsh matrices. Every layer in the x- (resp., y- and z-) direction of $A \otimes B$ is the direct multiplication of two layers of their parent matrices also in the x- (resp., y- and z-) direction. The theorem follows from Definition 3.4.2 and Theorem 3.4.3. □

For example, the direct multiplication $C_1 = A_1 \otimes B_1$ is absolutely proper, which is described

$$C_1(i,j,0) = \begin{bmatrix} 1 & -1 & 1 & -1 \\ 1 & 1 & 1 & 1 \\ 1 & -1 & -1 & 1 \\ 1 & 1 & -1 & -1 \end{bmatrix}; \quad C_1(i,j,1) = \begin{bmatrix} 1 & 1 & 1 & 1 \\ -1 & 1 & -1 & 1 \\ 1 & 1 & -1 & -1 \\ -1 & 1 & 1 & -1 \end{bmatrix}$$

and

$$C_1(i,j,2) = \begin{bmatrix} -1 & 1 & 1 & -1 \\ -1 & -1 & 1 & 1 \\ 1 & -1 & 1 & -1 \\ 1 & 1 & 1 & 1 \end{bmatrix}; \quad C_1(i,j,3) = \begin{bmatrix} -1 & -1 & 1 & 1 \\ 1 & -1 & -1 & 1 \\ 1 & 1 & 1 & 1 \\ -1 & 1 & -1 & 1 \end{bmatrix}.$$

The direct multiplication $C_2 = A_1 \otimes B_2$ is proper in two directions, which is described

$$C_2(i,j,0) = \begin{bmatrix} 1 & 1 & 1 & 1 \\ -1 & 1 & -1 & 1 \\ 1 & 1 & -1 & -1 \\ -1 & 1 & 1 & -1 \end{bmatrix}; \quad C_2(i,j,1) = \begin{bmatrix} 1 & 1 & 1 & 1 \\ 1 & -1 & 1 & -1 \\ 1 & 1 & -1 & -1 \\ 1 & -1 & -1 & 1 \end{bmatrix}$$

and

$$C_2(i,j,2) = \begin{bmatrix} 1 & 1 & 1 & 1 \\ 1 & 1 & 1 & 1 \\ -1 & -1 & 1 & 1 \\ -1 & -1 & 1 & 1 \end{bmatrix}; \quad C_2(i,j,3) = \begin{bmatrix} -1 & -1 & 1 & 1 \\ -1 & -1 & 1 & 1 \\ 1 & 1 & 1 & 1 \\ 1 & 1 & 1 & 1 \end{bmatrix}.$$

The following D is a 3-dimensional Walsh matrix of size 4×4, which is proper in the z-direction but improper in both the x- and y-directions.

$$D(i,j,0) = \begin{bmatrix} 1 & 1 & 1 & 1 \\ -1 & 1 & -1 & 1 \\ -1 & -1 & 1 & 1 \\ 1 & -1 & -1 & 1 \end{bmatrix}; \quad D(i,j,1) = \begin{bmatrix} 1 & 1 & 1 & 1 \\ -1 & 1 & -1 & 1 \\ 1 & 1 & -1 & -1 \\ -1 & 1 & 1 & -1 \end{bmatrix}$$

and

$$D(i,j,2) = \begin{bmatrix} -1 & 1 & -1 & 1 \\ 1 & 1 & 1 & 1 \\ 1 & -1 & -1 & 1 \\ -1 & -1 & 1 & 1 \end{bmatrix}; \quad D(i,j,3) = \begin{bmatrix} -1 & 1 & -1 & 1 \\ 1 & 1 & 1 & 1 \\ -1 & 1 & 1 & -1 \\ 1 & 1 & -1 & -1 \end{bmatrix}.$$

Lemma 3.4.2

$$\sum_{i=0}^{2^n-1} (-1)^{i.j'} = 2^n\delta(j) = \begin{cases} 2^n, & j = 0, \\ 0, & j \neq 0, \end{cases}$$

where $i \cdot j'$ refers to the dot production of the vectors i and j.

This lemma can be easily proved.

Theorem 3.4.5 *Let R_1 and R_2 be two non-singular $n \times n$ matrices in* GF(2). *Then the matrix $H = [h(i,j,k)]$ defined by*

$$h(i,j,k) = (-1)^{iR_1j'+kR_2j'}, \quad 0 \leqslant i,j,k \leqslant 2^n - 1$$

is a 3-dimensional Walsh matrix which is proper in the x- and z-directions.

Proof At first we prove that H is a 3-dimensional Hadamard matrix. In fact,

$$\begin{aligned} \sum_{i=0}^{2^n-1}\sum_{j=0}^{2^n-1} h(i,j,a)h(i,j,b) &= \sum_{i=0}^{2^n-1}\sum_{j=0}^{2^n-1} (-1)^{(a\oplus b)R_2j'} \\ &= 2^n \sum_{j=0}^{2^n-1} (-1)^{(a\oplus b)R_2j'} \\ &= 4^n\delta(a,b). \end{aligned}$$

The last equation is due to Lemma 3.4.2 and the non-singular property of R_2.

Similarly, it can be proved that

$$\sum_{j=0}^{2^n-1}\sum_{k=0}^{2^n-1} h(a,j,k)h(b,j,k) = 4^n\delta(a,b).$$

Furthermore,

$$\begin{aligned} \sum_{i=0}^{2^n-1}\sum_{k=0}^{2^n-1} h(i,a,k)h(i,b,k) &= \sum_{i=0}^{2^n-1}\sum_{k=0}^{2^n-1} (-1)^{iR_1(a\oplus b)'kR_2(a\oplus b)'} \\ &= \left[\sum_{i=0}^{2^n-1} (-1)^{iR_1(a\oplus b)'}\right]\left[\sum_{k=0}^{2^n-1} (-1)^{kR_2(a\oplus b)'}\right] \\ &= 2^n\delta(a,b)2^n\delta(a,b) \\ &= 4^n\delta(a,b). \end{aligned}$$

Therefore we have proved that H is a 3-dimensional Hadamard matrix.

Let $A = [a(j,k)] = [h(i_0,j,k)]$, $0 \leqslant i_0 \leqslant 2^n - 1$, be the i_0-th layer in the x-direction of the matrix H. Then

$$a(j_1,k)a(j_2,k) = a(j_1 \oplus j_2, k).$$

In other words, the first condition in Theorem 3.4.1 is satisfied.

Since

$$\sum_{k=0}^{2^n-1} a(j,k) = (-1)^{i_0 R_1 j'} \sum_{k=0}^{2^n-1} (-1)^{kR_2 j'} = 2^n \delta(j),$$

the second condition in Theorem 3.4.1 is also satisfied. Thus the layer $[a(j,k)]$ is a generalized 2-dimensional Walsh matrix.

By the same way, it can be proved that each layer in the z-direction is also a generalized 2-dimensional Walsh matrix.

Since there is an all-one layer in the y-direction, H is improper in the y-direction. In a word, the matrix H is a 3-dimensional Walsh matrix that is proper in the x- and z-direction. The theorem follows. □

Theorem 3.4.6 *$H = [h(i,j,k)]$ is a matrix defined by*

$$h(i,j,k) = (-1)^{iR_1 j' + kR_2 j'}, \quad 0 \leqslant i,j,k \leqslant 2^n - 1,$$

where R_1 and R_2 are two non-singular $n \times n$ matrices in GF(2), *if and only if H is a 3-dimensional Hadamard matrix satisfying*

$$\begin{aligned} h(i,j_1,k)h(i,j_2,k) &= h(i,j_1 \oplus j_2,k) \text{ and } h(i_1,j,k_1)h(i_2,j,k_2) \\ &= h(i_1 \oplus i_2, j, k_1 \oplus k_2), \end{aligned}$$

$$\textit{for all } 0 \leqslant i,i_1,i_2,j,j_1,j_2,k,k_1,k_2 \leqslant 2^n - 1.$$

Proof $\Longrightarrow$ By direct verification.

$\Longleftarrow$ Let

$$h(i,j,k) = (-1)^{B(i,j,k)}, \quad 0 \leqslant i,j,k \leqslant 2^n - 1,$$

where $B(i,j,k)$ is a $(0,1)$-valued function of i, j, and k.

$$h(i,j,k) = h(0,j,k)h(i,j,0)$$

implies

$$B(i,j,k) = B(0,j,k) \oplus B(i,j,0);$$

$$h(0,j_1,k)h(0,j_2,k) = h(0,j_1 \oplus j_2,k)$$

implies

$$B(0,j_1 \oplus j_2,k) = B(0,j_1,k) \oplus B(0,j_2,k)$$

and

$$h(0,j,k_1)h(0,j,k_2) = h(0,j,k_1 \oplus k_2)$$

implies

$$B(0,j,k_1 \oplus k_2) = B(0,j,k_1) \oplus B(0,j,k_2).$$

Thus $h(0,0,k) = 1$ and $h(0,j,0) = 1$ imply

$$B(0,cj,k) = cB(0,j,k) \text{ and } B(0,j,ck) = cB(0,j,k)$$

for each $c \in \mathrm{GF}(2)$. Therefore $B(0,j,k)$ is bilinear in both j and k.

Similarly, it can be proved that $B(i,j,0)$ is bilinear in both i and j.

In the same way as was used in the proof of Theorem 3.4.2, it can be proved that there exist two $n \times n$ matrices, say R_1 and R_2, in GF(2), such that

$$B(0,j,k) = kR_1 j' \text{ and } B(i,j,0) = iR_2 j'$$

Thus we have proved that

$$h(i,j,k) = (-1)^{iR_1 j' + kR_2 j'}, \quad 0 \leqslant i,j,k \leqslant 2^n - 1.$$

Finally, it is sufficient to prove that both R_1 and R_2 are non-singular.

Because H is a 3-dimensional Hadamard matrix,

$$2^n \sum_{j=0}^{2^n-1} h(0,j,a) = \sum_{i=0}^{2^n-1}\sum_{j=0}^{2^n-1} h(i,j,0)h(i,j,a) = 4^n \delta(a),$$

thus $\sum_{j=0}^{2^n-1} h(0,j,a) = 2^n \delta(a)$, or equivalently,

$$\sum_{j=0}^{2^n-1} (-1)^{aR_2 j'} = 2^n \delta(a).$$

Hence R_2 is non-singular. Similarly, it can be proved that R_1 is also non-singular.

□

Theorem 3.4.7 *Let $D = [D(i,j)]$, $0 \leqslant i,j \leqslant 2^n - 1$, be a 2-dimensional generalized Walsh matrix satisfying*

$$D(i_1,j)D(i_2,j) = D(i_1 \oplus i_2, j) \text{ and } D(i,j_1)D(i,j_2) = D(i, j_1 \oplus j_2),$$

for all $0 \leqslant i,j,i_1,i_2,j_1,j_2 \leqslant 2^n - 1$. and let r be an integer such that $\gcd(r, 2^n) = 1$. *Then the following matrix $H = [h(i,j,k)]$, $0 \leqslant i,j,k \leqslant 2^n - 1$, is a 3-dimensional Walsh matrix that is proper in both the y- and z-directions, where*

$$H(i,j,k) = D(i, (j + kr) \bmod 2^n), \quad 0 \leqslant i,j,k \leqslant 2^n - 1.$$

Proof All layers of H in the y- and z-directions are clearly the generalized Walsh matrices, because they are the shift forms of D. Since there is an all-one layer in the x-direction of H, it is sufficient to prove that H is a 3-dimensional Hadamard matrix.

Let $A = [a(i,j)]$ and $B = [b(i,j)]$ be two layers in the z-direction. Then for each $0 \leqslant i \leqslant 2^n - 1$, the bit-wise multiplication between the rows $(a(i,0), a(i,1), \cdots, a(i, 2^n - 1))$ and $(b(i,0), b(i,1), \cdots, b(i, 2^n - 1))$ is the vector

$$(a(i,0)b(i,0), a(i,1)b(i,1), \cdots, a(i,2^n - 1)b(i, 2^n - 1)),$$

which is a non-zero row of D, because of the closed bit-wise multiplication property of D. Therefore

$$\sum_{j=0}^{2^n-1} a(i,j)b(i,j) = 0, \quad 0 \leqslant i \leqslant 2^n - 1,$$

or equivalently,

$$\sum_{i=0}^{2^n-1}\sum_{j=0}^{2^n-1} h(i,j,a)h(i,j,b) = 4^n\delta(a,b).$$

Similarly, it can be proved that

$$\sum_{i=0}^{2^n-1}\sum_{k=0}^{2^n-1} h(i,a,k)h(i,b,k) = 4^n\delta(a,b).$$

Finally, the equation

$$\sum_{j=0}^{2^n-1}\sum_{k=0}^{2^n-1} h(a,j,k)h(b,j,k) = 4^n\delta(a,b)$$

is the result of the facts: (1) the rows of D are closed under bit-wise multiplication; and (2) the sum of each non-all-ones row is zero. Therefore the matrix H is a 3-dimensional Hadamard matrix. □

For example, if $r = 1$ and

$$D = \begin{bmatrix} 1 & 1 & 1 & 1 \\ 1 & -1 & 1 & -1 \\ 1 & 1 & -1 & -1 \\ 1 & -1 & -1 & 1 \end{bmatrix},$$

then the matrix $H = [h(i,j,k)]$, $0 \leqslant i,j,k \leqslant 2^n - 1$, produced by Theorem 3.4.7 is described by

$$H(i,j,0) = \begin{bmatrix} 1 & 1 & 1 & 1 \\ 1 & -1 & 1 & -1 \\ 1 & 1 & -1 & -1 \\ 1 & -1 & -1 & 1 \end{bmatrix}, \quad H(i,j,1) = \begin{bmatrix} 1 & 1 & 1 & 1 \\ -1 & 1 & -1 & 1 \\ -1 & 1 & 1 & -1 \\ 1 & 1 & -1 & -1 \end{bmatrix},$$

and

$$H(i,j,2) = \begin{bmatrix} 1 & 1 & 1 & 1 \\ 1 & -1 & 1 & -1 \\ -1 & -1 & 1 & 1 \\ -1 & 1 & 1 & -1 \end{bmatrix}, \quad H(i,j,3) = \begin{bmatrix} 1 & 1 & 1 & 1 \\ -1 & 1 & -1 & 1 \\ 1 & -1 & -1 & 1 \\ -1 & -1 & 1 & 1 \end{bmatrix}.$$

This matrix is also the special case of Theorem 3.4.5 for $R_1 = R_2$.

3.4.3 3-Dimensional Pan-Walsh Matrices

In the definition of the 3-dimensional Walsh matrix, it is required that each layer in some direction is a 2-dimensional generalized Walsh matrix. This subsection will study a generalization of the 3-dimensional Walsh matrix by allowing the layers to be a 2-dimensional generalized Walsh matrix H or its minus form $-H$.

Lemma 3.4.3 *Let A and B be 2-dimensional generalized Walsh matrices. Then $(-A)\otimes B$ and $A \otimes (-B)$ are minus 2-dimensional generalized Walsh matrices, and $(-A)\otimes(-B)$ is a 2-dimensional generalized Walsh matrix.*

The proof of this lemma is straightforward and is omitted here.

Definition 3.4.3 *A 3-dimensional Hadamard matrix of size $2^n \times 2^n$ is called a 3-dimensional Pan-Walsh matrix if all 2-dimensional layers in at least one axis-normal orientation are 2-dimensional generalized Walsh matrices or their minus forms. A 3-dimensional Pan-Walsh matrix is said to be proper in the x- (resp., y- and z-) direction if all layers in the x- (resp., y- and z-) direction are 2-dimensional generalized Walsh matrices or their minus forms. A 3-dimensional Pan-Walsh matrix is said to be absolutely proper if all of the possible 2-dimensional layers are generalized Walsh matrices or their minus forms, otherwise it is said to be improper.*

In the same way as that used in the proof of Theorem 3.4.4, the following theorem is proved to be true.

Theorem 3.4.8 *The direct multiplication of two 3-dimensional Pan-Walsh matrices is also a 3-dimensional Pan-Walsh matrix. Furthermore, the product matrix is proper in those directions in which both the parent matrices are proper.*

There are 16 3-dimensional Pan-Walsh matrices of size $2 \times 2 \times 2$ that are not the regular 3-dimensional Walsh matrices. They are

$$A_0(i,j,0) = \begin{bmatrix} 1 & -1 \\ 1 & 1 \end{bmatrix} \text{ and } A_0(i,j,1) = \begin{bmatrix} -1 & -1 \\ 1 & -1 \end{bmatrix},$$

$$A_1(i,j,0) = \begin{bmatrix} -1 & 1 \\ 1 & 1 \end{bmatrix} \text{ and } A_1(i,j,1) = \begin{bmatrix} -1 & -1 \\ -1 & 1 \end{bmatrix},$$

$$A_2(i,j,0) = \begin{bmatrix} 1 & 1 \\ -1 & 1 \end{bmatrix} \text{ and } A_2(i,j,1) = \begin{bmatrix} -1 & 1 \\ -1 & -1 \end{bmatrix},$$

$$A_3(i,j,0) = \begin{bmatrix} 1 & 1 \\ 1 & -1 \end{bmatrix} \text{ and } A_3(i,j,1) = \begin{bmatrix} 1 & -1 \\ -1 & -1 \end{bmatrix},$$

$$A_4(i,j,0) = \begin{bmatrix} 1 & -1 \\ -1 & -1 \end{bmatrix} \text{ and } A_4(i,j,1) = \begin{bmatrix} 1 & 1 \\ 1 & -1 \end{bmatrix},$$

$$A_5(i,j,0) = \begin{bmatrix} -1 & -1 \\ 1 & -1 \end{bmatrix} \text{ and } A_5(i,j,1) = \begin{bmatrix} 1 & -1 \\ 1 & 1 \end{bmatrix},$$

$$A_6(i,j,0) = \begin{bmatrix} -1 & -1 \\ -1 & 1 \end{bmatrix} \text{ and } A_6(i,j,1) = \begin{bmatrix} -1 & 1 \\ 1 & 1 \end{bmatrix},$$

$$A_7(i,j,0) = \begin{bmatrix} -1 & 1 \\ -1 & -1 \end{bmatrix} \text{ and } A_7(i,j,1) = \begin{bmatrix} 1 & 1 \\ -1 & 1 \end{bmatrix},$$

$$A_8(i,j,0) = \begin{bmatrix} 1 & 1 \\ -1 & 1 \end{bmatrix} \text{ and } A_8(i,j,1) = \begin{bmatrix} -1 & -1 \\ -1 & 1 \end{bmatrix},$$

$$A_9(i,j,0) = \begin{bmatrix} 1 & 1 \\ 1 & -1 \end{bmatrix} \text{ and } A_9(i,j,1) = \begin{bmatrix} -1 & 1 \\ -1 & -1 \end{bmatrix},$$

$$A_{10}(i,j,0) = \begin{bmatrix} 1 & -1 \\ 1 & 1 \end{bmatrix} \text{ and } A_{10}(i,j,1) = \begin{bmatrix} 1 & -1 \\ -1 & -1 \end{bmatrix},$$

$$A_{11}(i,j,0) = \begin{bmatrix} -1 & 1 \\ 1 & 1 \end{bmatrix} \text{ and } A_{11}(i,j,1) = \begin{bmatrix} -1 & -1 \\ 1 & -1 \end{bmatrix},$$

$$A_{12}(i,j,0) = \begin{bmatrix} -1 & 1 \\ -1 & -1 \end{bmatrix} \text{ and } A_{12}(i,j,1) = \begin{bmatrix} -1 & 1 \\ 1 & 1 \end{bmatrix},$$

$$A_{13}(i,j,0) = \begin{bmatrix} -1 & -1 \\ 1 & -1 \end{bmatrix} \text{ and } A_{13}(i,j,1) = \begin{bmatrix} 1 & 1 \\ 1 & -1 \end{bmatrix},$$

$$A_{14}(i,j,0) = \begin{bmatrix} -1 & 1 \\ -1 & 1 \end{bmatrix} \text{ and } A_{14}(i,j,1) = \begin{bmatrix} 1 & 1 \\ -1 & -1 \end{bmatrix}$$

and

$$A_{15}(i,j,0) = \begin{bmatrix} -1 & -1 \\ 1 & 1 \end{bmatrix} \text{ and } A_{15}(i,j,1) = \begin{bmatrix} -1 & 1 \\ -1 & 1 \end{bmatrix}.$$

The direct multiplication of the above A_0 and A_1 is the following $B = [B(i,j,k)] = A_0 \otimes A_1$, which is described by

$$B(i,j,0) = \begin{bmatrix} 1 & 1 & 1 & 1 \\ 1 & -1 & 1 & -1 \\ -1 & -1 & 1 & 1 \\ -1 & 1 & 1 & -1 \end{bmatrix}, \quad B(i,j,1) = \begin{bmatrix} 1 & -1 & 1 & -1 \\ -1 & -1 & -1 & -1 \\ -1 & 1 & 1 & -1 \\ 1 & 1 & -1 & -1 \end{bmatrix}$$

and

$$B(i,j,2) = \begin{bmatrix} -1 & -1 & 1 & 1 \\ -1 & 1 & 1 & -1 \\ -1 & -1 & -1 & -1 \\ -1 & 1 & -1 & 1 \end{bmatrix}, \quad B(i,j,3) = \begin{bmatrix} -1 & 1 & 1 & -1 \\ 1 & 1 & -1 & -1 \\ -1 & 1 & -1 & 1 \\ 1 & 1 & 1 & 1 \end{bmatrix}.$$

It is not difficult to see that a minus 3-dimensional Pan-Walsh matrix is also a 3-dimensional Pan-Walsh matrix, and the matrices produced by subtracting some layers in the z-direction of the matrices in Theorem 3.4.7 are also 3-dimensional Pan-Walsh matrices.

3.4.4 Analytic Representations

Sometimes the analytic formulas are much more convenient than others (e.g., the representation based on layers) for the study and application of 3-dimensional Walsh matrices. Thus this subsection concentrates on the list of analytic formulas.

Theorem 3.4.9 *Let M be an $n \times n$ non-singular matrix in* GF(2), *r an integer satisfying* $\gcd(r, 2^n) = 1$, *and*

$$R = [R(i,j)] = [(-1)^{iMj'}], \quad 0 \leqslant i,j \leqslant 2^n - 1.$$

Then the following matrix $H = [h(i,j,k)]$, $0 \leqslant i,j,k \leqslant 2^n - 1$, is a 3-dimensional Walsh matrix that is proper in both the y- and z-directions, where

$$H(i,j,k) = R(i, (j + kr) \bmod 2^n), \quad 0 \leqslant i,j,k \leqslant 2^n - 1.$$

This theorem is a direct corollary of Theorem 3.4.7. In general, we have the following theorem.

Theorem 3.4.10 *Let $f(x,y)$ be a mapping from the set $A = \{(x,y) : 0 \leqslant x, y \leqslant 2^n - 1\}$ to the set $B = \{z : 0 \leqslant z \leqslant 2^n - 1\}$ such that for each prefixed $x_0 \in B$ (resp. $y_0 \in B$), the $f(x_0, y)$ (resp., $f(x, y_0)$) is a one-to-one mapping for B to B. Let $g(x)$ be another one-to-one mapping for B to B, and R an $n \times n$ non-singular matrix in GF(2). Then the matrix $H = [h(i,j,k)]$ defined by*

$$h(i,j,k) = (-1)^{f(i,k)Rg(j)'}, \quad 0 \leqslant i,j,k \leqslant 2^n - 1$$

is a 3-dimensional Walsh matrix that is proper in two directions.

This theorem can be proved by the same way as that used in the proof of Theorem 3.4.5. In fact, Theorem 3.4.5, Theorem 3.4.7, and Theorem 3.4.9 are special cases of this theorem.

Theorem 3.4.11 *Let $H_1 = [H_1(i,j,k)]$, $0 \leqslant i,j,k \leqslant 1$, be a 3-dimensional matrix of size $2 \times 2 \times 2$ defined by*

$$H_1(i,j,k) = (-1)^{1+i+j+k+ij+ik+jk}, \quad 0 \leqslant i,j,k \leqslant 1.$$

Then the matrix

$$H_n =: \overbrace{H_1 \otimes H_1 \otimes \cdots \otimes H_1}^{n\text{-}times}, \ n \geqslant 1,$$

is an absolutely proper 3-dimensional Walsh matrix of size $2^n \times 2^n \times 2^n$ represented by

$$H_n(i,j,k) = (-1)^{n+ii'+jj'+kk'+ij'+ik'+jk'}, \quad 0 \leqslant i,j,k \leqslant 2^n - 1. \tag{3.21}$$

Here and afterwards we use the symbol xy' to simplify the inner product, $x \cdot y'$, between the two vectors x and y.

Proof At first, it is easy to verify that the matrix H_1 itself is an absolutely proper 3-dimensional Walsh matrix of size $2 \times 2 \times 2$. Thus, with Theorem 3.4.4, it is sufficient to prove the analytic formula in Equation 3.21. This will be proven by induction on n as follows.

The case of $n = 1$ is trivial.

Assume that Equation 3.21 works for $n = m$.

Now, consider the case of $n = m + 1$. Let

$$i = (i_m, i_{m-1}, \cdots, i_0), \quad j = (j_m, j_{m-1}, \cdots, j_0)$$

and

$$k = (k_m, k_{m-1}, \cdots, k_0)$$

are the expanded vectors of the integers i, j, and k, respectively.

Let

$$I =: (0, i_{m-1}, \cdots, i_0), \quad J =: (0, j_{m-1}, \cdots, j_0)$$

and

$$K =: (0, k_{m-1}, \cdots, k_0).$$

Because of $H_{m+1} = H_m \otimes H_1$, we have

$$\begin{aligned}H_{m+1}(i,j,k) &= H_m(I,J,K)H_1(i_m,j_m,k_m)\\&= (-1)^{m+II'+JJ'+KK'+IJ'+IK'+JK'}\\&\quad \cdot (-1)^{1+i_m+j_m+k_m+i_m j_m+i_m k_m+j_m k_m}\\&= (-1)^{(m+1)+ii'+jj'+kk'+ij'+ik'+jk'}.\end{aligned}$$

Thus the theorem is also true for the case of $n = m + 1$. □

Corollary 3.4.1 *Let R be an $n \times n$ non-singular matrix in* GF(2), *and $B = RR'$. Then the following matrix $A = [a(i,j,k)]$ is an absolutely proper 3-dimensional Walsh matrix of size $2^n \times 2^n \times 2^n$, where for each $0 \leqslant i,j,k \leqslant 2^n - 1$,*

$$a(i,j,k) = (-1)^{n+iBi'+jBj'+kBk'+iBj'+iBk'+jBk'}.$$

At the end of this chapter we present here the analytic formulas of some basic 3-dimensional Walsh and Pan-Walsh matrices of size $2 \times 2 \times 2$ and their direct multiplications of size $2^n \times 2^n \times 2^n$.

(1) The first matrix is

$$H_1 = [h(i,j,k)] = [(-1)^{1+i+j+k+ij+ik+jk}], \quad 0 \leqslant i,j,k \leqslant 1.$$

Its direct multiplication $H_n = H_1 \otimes H_1 \otimes \cdots \otimes H_1$ is represented by

$$H_n(i,j,k) = (-1)^{n+ii'+jj'+kk'+ij'+ik'+jk'}, \quad 0 \leqslant i,j,k \leqslant 2^n - 1.$$

This matrix has been stated in Theorem 3.4.11.

(2) The second matrix is

$$H_1 = [h(i,j,k)] = [(-1)^{k+ij+ik+jk}], \quad 0 \leqslant i,j,k \leqslant 1.$$

Its direct multiplication $H_n = H_1 \otimes H_1 \otimes \cdots \otimes H_1$ is represented by

$$H_n(i,j,k) = (-1)^{kk'+ij'+ik'+jk'}, \quad 0 \leqslant i,j,k \leqslant 2^n - 1.$$

(3) The third matrix is

$$H_1 = [h(i,j,k)] = [(-1)^{i+ij+ik+jk}], \quad 0 \leqslant i,j,k \leqslant 1.$$

Its direct multiplication $H_n = H_1 \otimes H_1 \otimes \cdots \otimes H_1$ is represented by

$$H_n(i,j,k) = (-1)^{ii'+ij'+ik'+jk'}, \quad 0 \leqslant i,j,k \leqslant 2^n - 1.$$

(4) The fourth matrix is

$$H_1 = [h(i,j,k)] = [(-1)^{j+ij+ik+jk}], \quad 0 \leqslant i,j,k \leqslant 1.$$

Its direct multiplication $H_n = H_1 \otimes H_1 \otimes \cdots \otimes H_1$ is represented by

$$H_n(i,j,k) = (-1)^{jj'+ij'+ik'+jk'}, \quad 0 \leqslant i,j,k \leqslant 2^n - 1.$$

(5) The fifth matrix is

$$H_1 = [h(i,j,k)] = [(-1)^{i+ij+ik}], \quad 0 \leqslant i,j,k \leqslant 1.$$

Its direct multiplication $H_n = H_1 \otimes H_1 \otimes \cdots \otimes H_1$ is represented by

$$H_n(i,j,k) = (-1)^{ii'+ij'+ik'}, \quad 0 \leqslant i,j,k \leqslant 2^n - 1.$$

(6) The sixth matrix is

$$H_1 = [h(i,j,k)] = [(-1)^{ij+ik}], \quad 0 \leqslant i,j,k \leqslant 1.$$

Its direct multiplication $H_n = H_1 \otimes H_1 \otimes \cdots \otimes H_1$ is represented by

$$H_n(i,j,k) = (-1)^{ij'+ik'}, \quad 0 \leqslant i,j,k \leqslant 2^n - 1.$$

(7) The seventh matrix is

$$H_1 = [h(i,j,k)] = [(-1)^{j+k+ij+ik}], \quad 0 \leqslant i,j,k \leqslant 1.$$

Its direct multiplication $H_n = H_1 \otimes H_1 \otimes \cdots \otimes H_1$ is represented by

$$H_n(i,j,k) = (-1)^{jj'+kk'+ij'+ik'}, \quad 0 \leqslant i,j,k \leqslant 2^n - 1.$$

(8) The eighth matrix is

$$H_1 = [h(i,j,k)] = [(-1)^{1+i+j+k+ij+ik}], \quad 0 \leqslant i,j,k \leqslant 1.$$

Its direct multiplication $H_n = H_1 \otimes H_1 \otimes \cdots \otimes H_1$ is represented by

$$H_n(i,j,k) = (-1)^{n+ii'+jj'+kk'+ij'+ik'}, \quad 0 \leqslant i,j,k \leqslant 2^n - 1.$$

(9) The ninth matrix is

$$H_1 = [h(i,j,k)] = [(-1)^{j+k+ij+ik+jk}], \quad 0 \leqslant i,j,k \leqslant 1.$$

Its direct multiplication $H_n = H_1 \otimes H_1 \otimes \cdots \otimes H_1$ is represented by

$$H_n(i,j,k) = (-1)^{jj'+kk'+ij'+ik'+jk'}, \quad 0 \leqslant i,j,k \leqslant 2^n - 1.$$

(10) The tenth matrix is

$$H_1 = [h(i,j,k)] = [(-1)^{1+i+j+ij+ik+jk}], \quad 0 \leqslant i,j,k \leqslant 1.$$

Its direct multiplication $H_n = H_1 \otimes H_1 \otimes \cdots \otimes H_1$ is represented by

$$H_n(i,j,k) = (-1)^{n+ii'+jj'+ij'+ik'+jk'}, \quad 0 \leqslant i,j,k \leqslant 2^n - 1.$$

(11) The eleventh matrix is

$$H_1 = [h(i,j,k)] = [(-1)^{i+k+ij+ik+jk}], \quad 0 \leqslant i,j,k \leqslant 1.$$

Its direct multiplication $H_n = H_1 \otimes H_1 \otimes \cdots \otimes H_1$ is represented by

$$H_n(i,j,k) = (-1)^{ii'+kk'+ij'+ik'+jk'}, \quad 0 \leqslant i,j,k \leqslant 2^n - 1.$$

(12) The twelfth matrix is

$$H_1 = [h(i,j,k)] = [(-1)^{ij+ik+jk}], \quad 0 \leqslant i,j,k \leqslant 1.$$

Its direct multiplication $H_n = H_1 \otimes H_1 \otimes \cdots \otimes H_1$ is represented by

$$H_n(i,j,k) = (-1)^{ij'+ik'+jk'}, \quad 0 \leqslant i,j,k \leqslant 2^n - 1.$$

(13) The thirteenth matrix is

$$H_1 = [h(i,j,k)] = [(-1)^{i+j+ij+ik+jk}], \quad 0 \leqslant i,j,k \leqslant 1.$$

Its direct multiplication $H_n = H_1 \otimes H_1 \otimes \cdots \otimes H_1$ is represented by

$$H_n(i,j,k) = (-1)^{ii'+jj'+ij'+ik'+jk'}, \quad 0 \leqslant i,j,k \leqslant 2^n - 1.$$

(14) The fourteenth matrix is

$$H_1 = [h(i,j,k)] = [(-1)^{1+i+k+ij+ik+jk}], \quad 0 \leqslant i,j,k \leqslant 1.$$

Its direct multiplication $H_n = H_1 \otimes H_1 \otimes \cdots \otimes H_1$ is represented by

$$H_n(i,j,k) = (-1)^{n+ii'+kk'+ij'+ik'+jk'}, \quad 0 \leqslant i,j,k \leqslant 2^n - 1.$$

(15) The fifteenth matrix is

$$H_1 = [h(i,j,k)] = [(-1)^{1+ij+ik+jk}], \quad 0 \leqslant i,j,k \leqslant 1.$$

Its direct multiplication $H_n = H_1 \otimes H_1 \otimes \cdots \otimes H_1$ is represented by

$$H_n(i,j,k) = (-1)^{n+ij'+ik'+jk'}, \quad 0 \leqslant i,j,k \leqslant 2^n - 1.$$

(16) The sixteenth matrix is

$$H_1 = [h(i,j,k)] = [(-1)^{j+k+ij+ik+jk}], \quad 0 \leqslant i,j,k \leqslant 1.$$

Its direct multiplication $H_n = H_1 \otimes H_1 \otimes \cdots \otimes H_1$ is represented by

$$H_n(i,j,k) = (-1)^{jj'+kk'+ij'+ik'+jk'}, \quad 0 \leqslant i,j,k \leqslant 2^n - 1.$$

(17) The seventeenth matrix is

$$H_1 = [h(i,j,k)] = [(-1)^{i+k+ij+ik}], \quad 0 \leqslant i,j,k \leqslant 1.$$

Its direct multiplication $H_n = H_1 \otimes H_1 \otimes \cdots \otimes H_1$ is represented by

$$H_n(i,j,k) = (-1)^{ii'+kk'+ij'+ik'}, \quad 0 \leqslant i,j,k \leqslant 2^n - 1.$$

(18) The eighteenth matrix is

$$H_1 = [h(i,j,k)] = [(-1)^{k+ij+jk}], \quad 0 \leqslant i,j,k \leqslant 1.$$

Its direct multiplication $H_n = H_1 \otimes H_1 \otimes \cdots \otimes H_1$ is represented by

$$H_n(i,j,k) = (-1)^{kk'+ij'+jk'}, \quad 0 \leqslant i,j,k \leqslant 2^n - 1.$$

(19) The ninteenth matrix is

$$H_1 = [h(i,j,k)] = [(-1)^{j+ij+ik}], \quad 0 \leqslant i,j,k \leqslant 1.$$

Its direct multiplication $H_n = H_1 \otimes H_1 \otimes \cdots \otimes H_1$ is represented by

$$H_n(i,j,k) = (-1)^{jj'+ij'+ik'}, \quad 0 \leqslant i,j,k \leqslant 2^n - 1.$$

(20) The twentieth matrix is

$$H_1 = [h(i,j,k)] = [(-1)^{1+i+j+ij+jk}], \quad 0 \leqslant i,j,k \leqslant 1.$$

Its direct multiplication $H_n = H_1 \otimes H_1 \otimes \cdots \otimes H_1$ is represented by

$$H_n(i,j,k) = (-1)^{n+ii'+jj'+ij'+jk'}, \quad 0 \leqslant i,j,k \leqslant 2^n - 1.$$

(21) The twenty-first matrix is

$$H_1 = [h(i,j,k)] = [(-1)^{1+j+ij+ik}], \quad 0 \leqslant i,j,k \leqslant 1.$$

Its direct multiplication $H_n = H_1 \otimes H_1 \otimes \cdots \otimes H_1$ is represented by

$$H_n(i,j,k) = (-1)^{n+jj'+ij'+ik'}, \quad 0 \leqslant i,j,k \leqslant 2^n - 1.$$

(22) The twenty-second matrix is

$$H_1 = [h(i,j,k)] = [(-1)^{1+i+k+ij+ik}], \quad 0 \leqslant i,j,k \leqslant 1.$$

Its direct multiplication $H_n = H_1 \otimes H_1 \otimes \cdots \otimes H_1$ is represented by

$$H_n(i,j,k) = (-1)^{n+ii'+kk'+ij'+ik'}, \quad 0 \leqslant i,j,k \leqslant 2^n - 1.$$

(23) The twenty-third matrix is

$$H_1 = [h(i,j,k)] = [(-1)^{1+j+k+ik+jk}], \quad 0 \leqslant i,j,k \leqslant 1.$$

Its direct multiplication $H_n = H_1 \otimes H_1 \otimes \cdots \otimes H_1$ is represented by

$$H_n(i,j,k) = (-1)^{n+jj'+kk'+ik'+jk'}, \quad 0 \leqslant i,j,k \leqslant 2^n - 1.$$

(24) The twenty-fourth matrix is

$$H_1 = [h(i,j,k)] = [(-1)^{1+i+ik+jk}], \quad 0 \leqslant i,j,k \leqslant 1.$$

Its direct multiplication $H_n = H_1 \otimes H_1 \otimes \cdots \otimes H_1$ is represented by

$$H_n(i,j,k) = (-1)^{n+ii'+ik'+jk'}, \quad 0 \leqslant i,j,k \leqslant 2^n - 1.$$

For more constructions and related backgrounds the readers are recommended to see the papers [13–18].

Bibliography

[1] Shlichta P J. Higher-dimensional Hadamard matrices. *IEEE Trans. On Inform. Theory*, 1979, IT-25(5): 566-572.

[2] Shlichta P J. Three- and four-dimensional Hadamard matrices. *Bull. Amer. Phys. Soc. Ser.*, 1971, 11(16): 825-826.

[3] Yang Y X. Higher-dimensional Walsh-Hadamard transforms. *J. Beijing Univ. of Posts and Telecomm.*, 1988, 11(2): 22-30.

[4] Yang Y X. On the perfect binary arrays. *J. of Electronics(China)*, 1990, 7(2): 175-181.

[5] Jedwab J. Perfect arrays, barker arrays and difference sets. *PhD Thesis.* University of London, 1991.

[6] Kopilovich. On perfect binary arrays. *Electron. Lett.*, 1988, 24(9): 566-567.

[7] Jedwab J, Mitchell C J. Infinite families of quasiperfect and doubly quasiperfect binary arrays. *Electron. Lett.*, 1990, 26(5): 294-295.

[8] Jedwab J. Mitchell C J. Constructing new perfect binary arrays. *Electron. Lett.*, 1988, 24(11): 650-652.

[9] Bomer L, Antweiler M. Perfect binary arrays with 36 elements. *Electron. Lett.*, 1987, 23(9): 730-732.

[10] Bomer L, Antweiler M. Two-dimensional perfect binary arrays with 64 elements. *IEEE Trans. On Inform. Theory*, 36(2): 411-414.

[11] Luke H D, Bomer L and Antweiler M. Perfect binary arrays. *Signal Proc.*, 1989, 17: 69-80.

[12] Wild P. Infinite families of perfect binary arrays. *Electron. Lett.*, 1988, 24(14): 845-847.

[13] Calabro D, Wolf J K. On the synthesis of two-dimensional arrays with desirable correlation properties. *Information and Control*, 1968, 11: 537-560.

[14] Yang Y X, On the construction of 3-dimensional Hadamard matrices. *Proc. of 5-th National Conference on Pattern Recognition and Machine Intelligence.* Xi′an, China, 1986.

[15] Chan W K, Siu M K. Summary of perfect $s \times t$ arrays, $1 \leqslant s \leqslant t \leqslant 100$. *Electron. Lett.*, 1991, 27(9): 709-710.

[16] Chan W K, Siu M K and Tong P. Two-dimensional binary arrays with good autocorrelation. *Inform. and Control*, 1979, 42: 125-130.

[17] Davis J A, Jedwab J. A summary of Menon difference sets. *Congressus Numerantium*, 1993, 93: 203-207.

[18] Davis J A, Jedwab J. A unifying construction for difference sets. *HP Laboratories Technical Report*, 1996.

Chapter 4

Multi-Dimensional Walsh-Hadamard Transforms

We know from Part I of this book that orthogonal transforms based on 2-dimensional Walsh and Hadamard matrices are very useful in (1-dimensional) signal processing. In engineering practice, sometimes we have to process higher-dimensional digital signals (e.g., the 2-dimensional image signals and 3-dimensional seismic waves, etc.). Thus this chapter concentrates on the Walsh-Hadamard transforms used for higher-dimensional digital signals.

4.1 Conventional 2-Dimensional Walsh-Hadamard Transforms

4.1.1 2-Dimensional Walsh-Hadamard Transforms

For more details of this subsection the readers are recommended to see [1].

An image signal can be represented by a 2-dimensional light intensity function $f(x, y)$, where x and y denote spatial coordinates and the value $f(x, y)$ at any point (x, y) is proportional to the brightness (or grey level) of the image at that point. In order to be in a form suitable for computer processing, an image function $f(x, y)$ must be digitized both spatially and in amplitude. Digitization of the spatial coordinates (x, y) is called image sampling, while amplitude digitization is called grey-level quantization. After digitization, a continuous image $f(x, y)$ is approximated by equally spaced samples arranged in the form of an $N \times N$ matrix:

$$f(x,y) \simeq \begin{bmatrix} f(0,0) & f(0,1) & \cdots & f(0,N-1) \\ f(1,0) & f(1,1) & \cdots & f(1,N-1) \\ \vdots & \vdots & & \vdots \\ f(N-1,0) & f(N-1,1) & \cdots & f(N-1,N-1) \end{bmatrix}, \tag{4.1}$$

where each element of this matrix is a discrete quantity, which is called an image element, picture element, pixel, or pel. The right hand side of this equation represents what is commonly called a digital image. The conventional 2-dimensional Walsh-Hadamard transforms are used for the digital image processing.

Let $N = 2^n$, and W_N be the Walsh-Hadamard matrix of size $N \times N$ defined by $W_1 = [1]$ and $W_{2^{n+1}} = \begin{bmatrix} W_{2^n} & W_{2^n} \\ W_{2^n} & -W_{2^n} \end{bmatrix}$. The conventional 2-dimensional Walsh-Hadamard forward transform of the digital image $f = [f(i, j)]$, $0 \leqslant i, j \leqslant 2^n - 1$, is the matrix $F = [F(u, v)]$,

$0 \leqslant u, v \leqslant 2^n - 1$, of the same size defined by[1]

$$F = W_N f W_N. \tag{4.2}$$

Because of the known identity $W_N^{-1} = 1/NW_N$, it is easy to see, from Equation (4.2), that the inverse transform is of the form

$$f = \frac{1}{N^2} W_N F W_N. \tag{4.3}$$

For example, the conventional 2-dimensional Walsh-Hadamard forward transform of the basic digital image

$$f = [f(i,j)] = \begin{bmatrix} + & + & + & + \\ - & - & + & + \\ - & - & + & + \\ + & + & - & - \end{bmatrix}$$

is

$$\begin{aligned} F = [F(u,v)] &= W_4 f W_4 \\ &= \begin{bmatrix} + & + & + & + \\ + & + & - & - \\ + & - & - & + \\ + & - & + & - \end{bmatrix} \begin{bmatrix} + & + & + & + \\ - & - & + & + \\ - & - & + & + \\ + & + & - & - \end{bmatrix} \begin{bmatrix} + & + & + & + \\ + & + & - & - \\ + & - & - & + \\ + & - & + & - \end{bmatrix} \\ &= \begin{bmatrix} 0 & 0 & 0 & 0 \\ 0 & 0 & 0 & 0 \\ 0 & 16 & 0 & 0 \\ 0 & 0 & 0 & 0 \end{bmatrix}. \end{aligned}$$

Recall that another equivalent definition of the Walsh-Hadamard matrix $W_{2^n} = [W(i,j)]$, $0 \leqslant i, j \leqslant 2^n - 1$ is

$$W(i,j) = (-1)^{ij'}, \tag{4.4}$$

where $ij' =: \sum_{k=0}^{N-1} i_k j_k$ is the inner product of the two vectors $i = (i_0, \cdots, i_{n-1})$ and $j = (j_0, \cdots, j_{n-1})$, which are the binary expended vectors of the integers $i = \sum_{k=0}^{n-1} i_k 2^k$ and $j = \sum_{k=0}^{n-1} j_k 2^k$. Thus by Equations (4.2)–(4.4), the conventional 2-dimensional Walsh-Hadamard forward transform and its inverse are equivalently defined by

$$F(u,v) = \sum_{i=0}^{n-1}\sum_{j=0}^{n-1} f(i,j)(-1)^{iu'+jv'} \tag{4.5}$$

and

$$f(i,j) = \frac{1}{N^2}\sum_{u=0}^{n-1}\sum_{v=0}^{n-1} F(u,v)(-1)^{iu'+jv'}, \tag{4.6}$$

where $0 \leqslant i, j, u, v \leqslant 2^n - 1$.

The conventional 2-dimensional Walsh-Hadamard transforms keep most of the properties satisfied by the 1-dimensional cases, e.g., the following Parseval theorem is true:

$$\sum_{u=0}^{n-1}\sum_{v=0}^{n-1} [F(u,v)]^2 = \sum_{i=0}^{n-1}\sum_{j=0}^{n-1} [f(i,j)]^2. \tag{4.7}$$

The fast algorithm for the conventional 2-dimensional Walsh-Hadamard transforms can be completed by using the fast ones for some suitable 1-dimensional signals. In fact, Equation (4.5) is equivalent to

$$F(u,v)=\sum_{i=0}^{n-1}(-1)^{iu'}\sum_{j=0}^{n-1}f(i,j)(-1)^{jv'}. \tag{4.8}$$

The inner summation in Equation (4.8) is

$$\begin{aligned}&\sum_{j=0}^{2^n-1}f(i,j)(-1)^{jv'}\\&=f(i,0)(-1)^{0v'}+f(i,1)(-1)^{1v'}+\cdots+f(i,N-1)(-1)^{(N-1)v'}\\&=:B(i,v),\end{aligned} \tag{4.9}$$

which is clearly the known 1-dimensional Walsh-Hadamard transform of the vector $(f(i,0), f(i,1),\cdots,f(i,2^n-1)$, the i-th row of the image $f=[f(i,j)]$. Thus the $B(i,v)$ in Equation (4.9) can be calculated by using the fast Walsh-Hadamard transforms introduced in Part I of the book. After N times of 1-dimensional fast Walsh-Hadamard algorithms, we get the following $N\times N$ matrix:

$$B=[B(i,v)]=\begin{bmatrix}B(0,0)&B(0,1)&\cdots&B(0,2^n-1)\\B(1,0)&B(1,1)&\cdots&B(1,2^n-1)\\\vdots&\vdots&&\vdots\\B(2^n-1,0)&B(2^n-1,1)&\cdots&B(2^n-1,2^n-1)\end{bmatrix}. \tag{4.10}$$

Substituting of Equation (4.9)into Equation (4.8), we have

$$F(u,v)=\sum_{i=0}^{2^n-1}B(i,v)(-1)^{iu'}, \tag{4.11}$$

which is the 1-dimensional Walsh-Hadamard transform for the vector $(B(0,v),B(1,v),\cdots, B(2^n-1,v))$, the v-th column of the matrix $B=[B(i,v)]$ defined in Equation (4.10).

Hence we have seen that a fast algorithm for the conventional 2-dimensional Walsh-Hadamard transform can be completed by using N^2 fast algorithms for the 1-dimensional cases.

4.1.2 Definitions of 4-Dimensional Hadamard Matrices[2]

It will be clear that the conventional 2-dimensional Walsh-Hadamard transforms introduced in the last subsection are, in fact, special cases of Walsh-Hadamard transforms based on 4-dimensional Hadamard matrices $H=[H(i,j,k,l)]=[(-1)^{ij'+kl'}]$. Thus in this preliminary subsection we state the definitions of 4- and higher-dimensional Hadamard matrices.

A 2-dimensional Hadamard matrix is defined by a binary matrix such that all of its parallel $(2-1)$-dimensional layers are orthogonal to each other. A 3-dimensional Hadamard matrix is defined by a binary matrix such that all of its parallel $(3-1)$-dimensional layers are orthogonal to each other. Thus it is natural to define a 4-dimensional Hadamard matrix as such a binary matrix that all of its parallel $(4-1)$-dimensional layers are mutually orthogonal[2, 3]. In other words, $H = [H(i,j,k,l)], 0 \leqslant i,j,k,l \leqslant m-1$, is a four-dimensional Hadamard matrix iff $H(i,j,k,l) = \pm 1$ and

$$\begin{aligned}
&\sum_{p=0}^{m-1}\sum_{q=0}^{m-1}\sum_{r=0}^{m-1} H(p,q,r,a)H(p,q,r,b) \\
&= \sum_{p=0}^{m-1}\sum_{q=0}^{m-1}\sum_{s=0}^{m-1} H(p,q,a,s)H(p,q,b,s) \\
&= \sum_{p=0}^{m-1}\sum_{r=0}^{m-1}\sum_{s=0}^{m-1} H(p,a,r,s)H(p,b,r,s) \\
&= \sum_{q=0}^{m-1}\sum_{r=0}^{m-1}\sum_{s=0}^{m-1} H(a,q,r,s)H(b,q,r,s) \\
&= m^3\delta_{ab}.
\end{aligned} \tag{4.12}$$

A four-dimensional Hadamard matrix is called "absolutely proper," if all of its two-dimensional sections, in all possible directions (and therefore all three-dimensional sections), are themselves two-dimensional Hadamard matrices. Besides the extreme case "absolutely proper," there are several other kinds of intermediate propriety. For example, a four-dimensional Hadamard matrix is called "three-dimensionally proper," if all of its three-dimensional sections are proper or improper Hadamard matrices. A four-dimensional Hadamard matrix is called "proper in two directions," if all of its three-dimensional sections in some two directions are proper or improper Hadamard matrices[2].

The definition of direct multiplications can also be extended to four-dimensional cases. In fact, it will be proved that[2, 3]

(1) The direct multiplication of two four-dimensional Hadamard matrices is also a four-dimensional Hadamard matrix. Thus a four-dimensional Hadamard matrix of order 2^t can be generated by direct multiplication of the ones of order 2.

(2) A four-dimensional Hadamard matrix is called "absolutely improper" if none of its two- or three-dimensional sections is a Hadamard matrix. A four-dimensional absolutely improper Hadamard matrix of order m^2 can be generated by the successive direct multiplication of four two-dimensional Hadamard matrices, of order m, which are in a set of mutually perpendicular orientations in which each axis appears twice (e.g., the wx, xy, yz, and zw planes).

(3) A four-dimensional matrix is Hadamardian whenever

- all three-dimensional sections in one direction are three-dimensional Hadamard matrices which are mutually orthogonal, or
- all three-dimensional sections in two directions are three-dimensional Hadamard matrices.

The first requirement can be satisfied by constructing an m^4 matrix from a set of m^3 Hadamard matrices formed by the successive cyclic m^2-layer permutations of a single m^3

Hadamard matrix. Since the latter can be formed, as shown in the last chapter, from a single m^2 Hadamard matrix, it follows that every two-dimensional Hadamard matrix implies a three-dimensional and a four-dimensional one with at least partial propriety.

The four-dimensional Hadamard matrices can be generalized further to the following general higher-dimensional cases[2, 3]:

An n-dimensional Hadamard matrix $H = [H(i_1, i_2, \cdots, i_n)]$ of order m is a (± 1)-valued matrix in which all parallel $(n-1)$-dimensional sections are mutually orthogonal, that is

$$\sum_{i_1=0}^{m-1} \cdots \sum_{i_{n-1}=0}^{m-1} H(i_1, i_2, \cdots, i_{n-1}, a) H(i_1, i_2, \cdots, i_{n-1}, b) = 2^{(n-1)} \delta_{ab},$$

$$\sum_{i_1=0}^{m-1} \cdots \sum_{i_{n-2}=0}^{1} \sum_{i_n=0}^{1} H(i_1, i_2, \cdots, i_{n-2}, a, i_n) H(i_1, i_2, \cdots, i_{n-2}, b, i_n) = 2^{(n-1)} \delta_{ab},$$

$$\cdots\cdots\cdots$$

$$\sum_{i_2=0}^{m-1} \cdots \sum_{i_n=0}^{m-1} H(a, i_2, i_3, \cdots, i_n) H(b, i_2, i_3, \cdots, i_n) = 2^{(n-1)} \delta_{ab}.$$

4.2 Algebraic Theory of Higher-Dimensional Matrices

An n-dimensional matrix of size $m_1 \times \cdots \times m_n$ can be denoted by $A = [A(i_1, \cdots, i_n)]$, where each $A(i_1, \cdots, i_n)$, $0 \leqslant i_1 \leqslant m_1 - 1, \cdots, 0 \leqslant i_n \leqslant m_n - 1$, is called an element of this matrix. Let r be a number. Then rA refers to the matrix

$$rA := [rA(i_1, \cdots, i_n)],$$

i.e., the product of each element with the number r.

Two matrices A and B are called "equal to each other," denoted by $A = B$, if they have the same size and

$$A(i_1, \cdots, i_n) = B(i_1, \cdots, i_n)$$

for all possible $0 \leqslant i_1 \leqslant m_1 - 1, \cdots, 0 \leqslant i_n \leqslant m_n - 1$.

Let A and B be two n-dimensional matrices having the same size. Then their summation $C = A + B$ is another matrix defined by

$$C =: A + B =: [A(i_1, \cdots, i_n) + B(i_1, \cdots, i_n)],$$

i.e., the matrix formed by the element-wise summation of their parent matrices A and B. Similarly, $A - B$ is defined as the matrix formed by element-wise minuses of their parent matrices.

One of the most useful definitions for the higher-dimensional matrices is, possibly, the following operation, called the multiplication.

Definition 4.2.1 *Let $A = [A(i_1, \cdots, i_n)]$ and $B = [B(j_1, \cdots, j_n)]$ be two n-dimensional matrices of sizes $a_1 \times a_2 \times a_n$ and $b_1 \times b_2 \times b_n$, respectively.*

(1) *If $n = 2s$ is even and $(a_{s+1}, \cdots, a_n) = (b_1, \cdots, b_s)$, then the multiplication between A and B is the following n-dimensional matrix $C = [C(k_1, \cdots, k_n)] =: AB$ of size*

$a_1 \times \cdots \times a_s \times b_{s+1} \times \cdots \times b_n$ defined by

$$C(k_1, \cdots, k_n) = \sum_{e(1)=0}^{b_1-1} \sum_{e(2)=0}^{b_2-1} \cdots \sum_{e(s)=0}^{b_s-1} A(k_1, \cdots, k_s, e(1), \cdots, e(s))$$

$$\cdot B(e(1), \cdots, e(s), k_{s+1}, \cdots, k_n).$$

Clearly, this definition is the natural generalization of the regular multiplication between two 2-dimensional matrices.

(2) *If $n = 2s + 1$ is odd and $(a_{s+1}, \cdots, a_{2s}) = (b_1, \cdots, b_s)$ and $a_n = b_n$, then the multiplication between A and B is the following n-dimensional matrix $D = [D(d_1, \cdots, d_n)] =: AB$ of size $a_1 \times \cdots \times a_s \times b_{s+1} \times \cdots \times b_n$ defined by*

$$D(d_1, \cdots, d_n) = \sum_{e(1)=0}^{b_1-1} \sum_{e(2)=0}^{b_2-1} \cdots \sum_{e(s)=0}^{b_s-1} A(d_1, \cdots, d_s, e(1), \cdots, e(s), d_n)$$

$$\cdot B(e(1), \cdots, e(s), d_{s+1}, \cdots, d_n). \tag{4.13}$$

With these definitions in mind, it is easy to verify the following theorem.

Theorem 4.2.1 *Let k and s be two integers. And let A, B, and C be three n-dimensional matrices such that the following operations are well defined. Then*

$$\begin{aligned} A(B + C) &= AB + AC; \\ (B + C)A &= BA + CA; \\ (k + s)A &= kA + sA; \\ k(A + B) &= kA + kB; \\ k(sA) &= (ks)A; \\ 1A &= A; \\ k(AB) &= (kA)B = A(kB). \end{aligned}$$

Theorem 4.2.2 *Let A, B, and C be three n-dimensional matrices of suitable sizes such that the following multiplication are well defined. Then*

$$A(BC) = (AB)C.$$

Proof We prove only the case of even $n = 2s$, because the odd n case can be proved in the same way.

Let $A = [A(i_1, \cdots, i_n)]$, $B = [B(j_1, \cdots, j_n)]$, and $C = [C(k_1, \cdots, k_n)]$.

Assume that

$$V =: A(BC) = [V(v_1, \cdots, v_n)] \text{ and } U =: (AB)C = [U(u_1, \cdots, u_n)].$$

When the above multiplications have been well defined, the sizes of U and V are clearly equivalent to each other. By Definition 4.2.1, we have, on the one hand,

$$V(v_1, \cdots, v_n) = \sum_{e(1),\cdots,e(s)} A(v_1, \cdots, v_s, e(1), \cdots, e(s)) \Bigg\{ \sum_{f(1),\cdots,f(s)} B(e(1), \cdots,$$

$$e(s), f(1), \cdots, f(s))C(f(1), \cdots, f(s), v_{s+1}, \cdots, v_n)\Big\}$$

$$= \sum_{e(1),\cdots,e(s),f(1),\cdots,f(s)} A(v_1, \cdots, v_s, e(1), \cdots, e(s))$$

$$\cdot B(e(1), \cdots, e(s), f(1), \cdots, f(s))$$

$$\cdot C(f(1), \cdots, f(s), v_{s+1}, \cdots, v_n). \tag{4.14}$$

On the other hand,

$$U(u_1, \cdots, u_n) = \sum_{f(1),\cdots,f(s)} \Big\{ \sum_{e(1),\cdots,e(s)} A(u_1, \cdots, u_s, e(1), \cdots, e(s)) \cdot B(e(1), \cdots,$$

$$e(s), f(1), \cdots, f(s))\Big\} \cdot C(f(1), \cdots, f(s), u_{s+1}, \cdots, u_n)$$

$$= \sum_{e(1),\cdots,e(s),f(1),\cdots,f(s)} A(u_1, \cdots, u_s, e(1), \cdots, e(s))$$

$$\cdot B(e(1), \cdots, e(s), f(1), \cdots, f(s))$$

$$\cdot C(f(1), \cdots, f(s), u_{s+1}, \cdots, u_n). \tag{4.15}$$

The proof is finished by comparing the right hand sides of Equation (4.14) and Equation (4.15). □

Definition 4.2.2 *Let $A = [A(i_1, \cdots, i_n)]$ be an n-dimensional matrix.*

(1) *If $n = 2s$ is an even integer, then the matrix*

$$A' =: [A'(j_1, \cdots, j_n)] = [A(j_{s+1}, \cdots, j_n, j_1, \cdots, j_s)]$$

is called the transpose matrix of the mother matrix A;

(2) *If $n = 2s + 1$ is an odd integer, then the matrix*

$$A' =: [A'(j_1, \cdots, j_n)] = [A(j_{s+1}, \cdots, j_{2s}, j_1, \cdots, j_s, j_n)]$$

is called the transpose matrix of the mother matrix A.

Theorem 4.2.3 *Let A and B be two n-dimensional matrices, and k a number. Then the transpose operations defined by Definition 4.2.2 satisfy the following identities:*

$$(A + B)' = A' + B',$$
$$(A')' = A,$$
$$(AB)' = B'A',$$
$$(kA)' = kA',$$

where it has been assumed that all of these operations are well-defined.

Proof We prove only the third identity $(AB)' = B'A'$, because the other identities are trivial.

Without loss of the generality, we consider the case of $n = 2s$ even (the case $n = 2s + 1$ odd can be proved by the same way).

Let $A = [A(i_1, \cdots, i_n)]$ and $B = [B(j_1, \cdots, j_n)]$ be of the sizes $a_1 \times \cdots \times a_n$ and $b_1 \times \cdots \times b_n$, respectively. First, it is clear that the two matrices:

$$(AB)' =: U = [U(u_1, \cdots, u_n)] \text{ and } B'A' =: V = [V(v_1, \cdots, v_n)]$$

have the same size of $b_{s+1} \times \cdots \times b_{2s} \times a_1 \times \cdots \times a_s$.

In addition, by Definition 4.2.1, we have

$$U(u_1, \cdots, u_n) = \sum_{e(1),\cdots,e(s)} A(u_{s+1}, \cdots, u_n, e(1), \cdots, e(s)) \cdot B(e(1), \cdots, e(s), u_1, \cdots, u_s) \tag{4.16}$$

and

$$V(v_1, \cdots, v_n) = \sum_{e(1),\cdots,e(s)} B(e(1), \cdots, e(s), v_1, \cdots, v_s) \cdot A(v_{s+1}, \cdots, v_n, e(1), \cdots, e(s)). \tag{4.17}$$

The proof is finished by comparing Equation (4.16) with Equation (4.16). □

Definition 4.2.3 *An n-dimensional matrix I of size $I_1 \times \cdots \times I_n$ is called a unit matrix iff for every n-dimensional matrix A, the following identity is satisfied*

$$IA = AI = A$$

provided that AI and IA are well defined.

Remark 4.2.1 *The requirement $AI = IA$ in Definition 4.2.3 implies that:*

(1) *If $n = 2s$ is even, then the size of I should satisfy $(I_1, \cdots, I_s) = (I_{s+1}, \cdots, I_n)$;*

(2) *If $n = 2s + 1$ is odd, then the size of I should satisfy $(I_1, \cdots, I_s) = (I_{s+1}, \cdots, I_{n-1})$.*

Theorem 4.2.4 *For every given integer n and the size $I_1 \times \cdots \times I_n$ satisfying the conditions stated in the previous remark, there exists one and only one unit matrix. In fact,*

(1) *If $n = 2s$ is even, then the unit matrix $I = [I(i_1, \cdots, i_n)]$ of size $I_1 \times \cdots \times I_s \times I_1 \times \cdots \times I_s$ is defined by*

$$I(i_1, \cdots, i_n) = \begin{cases} 1, & if\ (i_1, \cdots, i_s) = (i_{s+1}, \cdots, i_n), \\ 0, & otherwise; \end{cases}$$

(2) *If $n = 2s + 1$ is odd, then the unit matrix $J = [J(j_1, \cdots, j_n)]$ of size $J_1 \times \cdots \times J_s \times J_1 \times \cdots \times J_s \times J_n$ is defined by*

$$J(j_1, \cdots, j_n) = \begin{cases} 1, & if\ (j_1, \cdots, j_s) = (j_{s+1}, \cdots, j_{n-1}), \\ 0, & otherwise. \end{cases}$$

This theorem can be proved by direct verification.

For example, the matrix

$$I = [I(i,j)] = \begin{bmatrix} 1 & 0 \\ 0 & 1 \end{bmatrix}$$

is the 2-dimensional unit matrix of size 2×2. The 3-dimensional unit matrix $I = [I(i,j,k)]$ of size $2 \times 2 \times 2$ is described by

$$[I(i,j,0)] = \begin{bmatrix} 1 & 0 \\ 0 & 1 \end{bmatrix} \text{ and } [I(i,j,1)] = \begin{bmatrix} 1 & 0 \\ 0 & 1 \end{bmatrix}.$$

The 3-dimensional unit matrix $I = [I(i,j,k)]$ of size $2 \times 2 \times 3$ is described by

$$[I(i,j,0)] = \begin{bmatrix} 1 & 0 \\ 0 & 1 \end{bmatrix}, \ [I(i,j,1)] = \begin{bmatrix} 1 & 0 \\ 0 & 1 \end{bmatrix} \text{ and } [I(i,j,2)] = \begin{bmatrix} 1 & 0 \\ 0 & 1 \end{bmatrix}.$$

Definition 4.2.4 *A matrix B is called the inverse matrix of A, denoted by A^{-1}, if $BA = AB = I$, the unit matrix. An invertible matrix is called non-singular.*

It is not difficult to prove the following statements:

(1) If the matrix A is non-singular, then it has a unique inverse matrix;

(2) If both A and B are non-singular, then so is their multiplication matrix AB. Moreover, $(AB)^{-1} = B^{-1}A^{-1}$ (of course, it has been assumed that these operations are well defined).

The following theorem makes it reasonable to concentrate on the even-dimensional matrices.

Theorem 4.2.5 *Let $n = 2s+1$ be odd, and $A = [A(i_1, \cdots, i_n)]$ a matrix of size $a_1 \times \cdots \times a_n$. Then A is non-singular if and only if its $(n-1)$-dimensional section $[B(j_1, \cdots, j_{n-1})] =: [A(j_1, \cdots, j_{n-1}, i_n)]$ is non-singular for each i_n, $0 \leqslant i_n \leqslant a_n - 1$.*

Proof $\Longrightarrow$ Let $C = [C(i_1, \cdots, i_n)] = A^{-1}$. Then it is easy to verify that for each $0 \leqslant k \leqslant a_n - 1$,

$$[A(i_1, \cdots, i_{n-1}, k)]^{-1} = [C(i_1, \cdots, i_{n-1}, k)].$$

$\Longleftarrow$ Let

$$V^{(k)} = [V^{(k)}(i_1, \cdots, i_{2s})] =: [A(i_1, \cdots, i_{2s}, k)]^{-1}, \quad 0 \leqslant k \leqslant a_n - 1.$$

Then it can be verified that $A^{-1} = V^{(a_n - 1)}$. □

Theorem 4.2.6 *Let $n = 2s > 2$ be even. The n-dimensional matrix $A = [A(i_1, \cdots, i_s, \cdots, i_n)]$ of size $a_1 \times \cdots \times a_s \times a_1 \times \cdots \times a_s$ is non-singular if the following two conditions are satisfied:*

(1) If $i_s \neq i_n$, then $A(i_1, \cdots, i_s, \cdots, i_{2s}) = 0$;

(2) If $i_s = i_n = k$, then the following $(n-2)$-dimensional section B is non-singular, where

$$B = [B(i_1, \cdots, i_{s-1}, i_{s+1}, \cdots, i_{n-1})]$$
$$:= [A(i_1, \cdots, i_{s-1}, k, i_{s+1}, \cdots, i_{n-1}, k)].$$

This theorem can be generalized as the following theorem. In fact, Theorem 4.2.6 is the special case of the following Theorem 4.2.7 by letting $m = 2$ and V be the 2-dimensional unit matrix.

Theorem 4.2.7 *Let $n = 2s$ and $m = 2r$ be even, $U = [U(i_1, \cdots, i_n)]$ and $V = [V(j_1, \cdots, j_m)]$ two non-singular matrices of sizes $a_1 \times \cdots \times a_s \times a_1 \times \cdots \times a_s$ and $b_1 \times \cdots \times b_r \times b_1 \times \cdots \times b_r$, respectively. Then the following $(m+n)$-dimensional matrix $H = [H(h_1, \cdots, h_{m+n})]$ of size*

$$a_1 \times \cdots \times a_s \times b_1 \times \cdots \times b_r \times a_1 \times \cdots \times a_s \times b_1 \times \cdots \times b_r$$

is non-singular too, where

$$H(h_1, \cdots, h_{m+n}) = U(h_1, \cdots, h_s, h_{s+r+1}, \cdots, h_{2s+r}) \cdot V(h_{s+1}, \cdots, h_{s+r}, h_{2s+r+1}, \cdots, h_{m+n}).$$

Proof Let $A = [A(i_1, \cdots, i_n)] = U^{-1}$ and $B = [B(j_1, \cdots, j_m)] = V^{-1}$ be the inverse matrices of the matrices U and V, respectively. And let $F = [F(h_1, \cdots, h_{m+n})]$ be the $(m+n)$-dimensional matrix defined by

$$F(h_1, \cdots, h_{m+n}) = A(h_1, \cdots, h_s, h_{s+r+1}, \cdots, h_{2s+r}) \cdot B(h_{s+1}, \cdots, h_{s+r}, h_{2s+r+1}, \cdots, h_{m+n}).$$

Assume $HF = C = [C(c_1, \cdots, c_{m+n})]$. Then by Definition 4.2.1, we have

$$\begin{aligned}
C(c_1, \cdots, c_{m+n}) = & \sum_{e(1), \cdots, e(r+s)} H(c_1, \cdots, c_{r+s}, e(1), \cdots, e(r+s)) \\
& \cdot F(e(1), \cdots, e(r+s), c_{r+s+1}, \cdots, c_{m+n}) \\
= & \sum_{e(1), \cdots, e(r+s)} U(c_1, \cdots, c_s, e(1), \cdots, e(s)) \\
& \cdot V(c_{s+1}, \cdots, c_{s+r}, e(s+1), \cdots, e(s+r)) \\
& \cdot A(e(1), \cdots, e(s), c_{r+s+1}, \cdots, c_{r+2s}) \\
& \cdot B(e(s+1), \cdots, e(s+r), c_{2s+r+1}, \cdots, c_{2s+2r}) \\
= & \left[\sum_{e(1), \cdots, e(s)} U(c_1, \cdots, c_s, e(1), \cdots, e(s)) \right. \\
& \left. \cdot A(e(1), \cdots, e(s), c_{r+s+1}, \cdots, c_{r+2s}) \right] \\
& \cdot \left[\sum_{e(s+1), \cdots, e(r+s)} V(c_{s+1}, \cdots, c_{s+r}, e(s+1), \cdots, e(s+r)) \right. \\
& \left. \cdot B(e(s+1), \cdots, e(s+r), c_{2s+r+1}, \cdots, c_{2s+2r}) \right] \\
= & \delta[(c_1, \cdots, c_s) - (c_{r+s+1}, \cdots, c_{r+2s})] \\
& \cdot \delta[(c_{s+1}, \cdots, c_{s+r}) - (c_{r+2s+1}, \cdots, c_{2r+2s})] \\
= & \delta[(c_1, \cdots, c_{r+s}) - (c_{r+s+1}, \cdots, c_{2r+2s})]
\end{aligned}$$

$$= \begin{cases} 1, & \text{if } (c_1, \cdots, c_{r+s}) = (c_{r+s+1}, \cdots, c_{2r+2s}), \\ 0, & \text{otherwise.} \end{cases}$$

Thus we have proved that $HF = I$, the unit matrix.

In the same way, it can be proved that $FH = I$, the unit matrix. Thus H is non-singular and its inverse matrix is F. □

The following useful corollary follows from Theorem 4.2.7:

Corollary 4.2.1 *Let $n = 2s$ be even, and let $A_1 = [A_1(i,j)]$, $\cdots$, $A_s = [A_s(i,j)]$ be 2-dimensional non-singular matrices of sizes $n_1 \times n_1$, $\cdots$, $n_s \times n_s$, respectively. Then the following n-dimensional matrix $H = [H(i_1, \cdots, i_n)]$ of size $n_1 \times \cdots \times n_s \times n_1 \times \cdots \times n_s$ is also non-singular, where*

$$H(i_1, \cdots, i_n) =: A_1(i_1, i_{s+1}) A_2(i_2, i_{s+2}) \cdots A_s(i_s, i_n).$$

Proof Let $B_1 = [B_1(i,j)]$, $\cdots$, $B_s = [B_s(i,j)]$ be the inverse matrices of $A_1 = [A_1(i,j)]$, $\cdots$, $A_s = [A_s(i,j)]$, respectively. And let

$$F(i_1, \cdots, i_n) =: B_1(i_1, i_{s+1}) B_2(i_2, i_{s+2}) \cdots B_s(i_s, i_n). \tag{4.18}$$

In the same way as that used in the proof of Theorem 4.2.7 it can be proved that the matrix $F = [F(i_1, \cdots, i_n)]$ defined by Equation (4.18)is the inverse matrix of H. □

The definitions of direct multiplications for 2- and 3-dimensional matrices have been shown to be very useful. Now we generalize this concept to the general higher-dimensional cases.

Definition 4.2.5 *Let $A = [A(i_1, \cdots, i_n)]$, $0 \leqslant i_s \leqslant a_s - 1$, $1 \leqslant s \leqslant n$, and $B = [B(j_1, \cdots, j_n)]$, $0 \leqslant j_s \leqslant b_s - 1$, $1 \leqslant s \leqslant n$, be two n-dimensional matrices of sizes $a_1 \times \cdots \times a_n$ and $b_1 \times \cdots \times b_n$, respectively. The direct multiplication of A and B, denoted by $A \otimes B$, is an n-dimensional matrix, $C = [C(k_1, \cdots, k_n)]$, $0 \leqslant k_s \leqslant (a_s b_s) - 1$, $1 \leqslant s \leqslant n$, of size $(a_1 b_1) \times \cdots \times (a_n b_n)$, defined by*

$$C(k_1, \cdots, k_n) = A\left(\left\lfloor \frac{k_1}{a_1} \right\rfloor, \cdots, \left\lfloor \frac{k_n}{a_n} \right\rfloor\right) B([k_1]_{a_1}, \cdots, [k_n]_{a_n}),$$

where $\lfloor x \rfloor$ stands for the largest integer upper-bounded by x, and $[x]_m \equiv x \bmod (m)$.

Theorem 4.2.8 *Let $A = [A(i_1, \cdots, i_n]$ and $B = [B(j_1, \cdots, j_n]$ be two n-dimensional matrices of sizes $a_1 \times \cdots \times a_n$ and $b_1 \times \cdots \times b_n$, respectively.*

(1) *If both A and B are non-singular, then so is their direct multiplication. Precisely, $(B \otimes A)^{-1} = B^{-1} \otimes A^{-1}$;*

(2) *$(B \otimes A)' = B' \otimes A'$.*

Proof We prove only the first statement.

If $n = 2s$ is even, then the sizes of A^{-1} and B^{-1} are $a_{s+1} \times \cdots \times a_n \times a_1 \times \cdots \times a_s$ and $b_{s+1} \times \cdots \times b_n \times b_1 \times \cdots \times b_s$, respectively. Write

$$A^{-1} = [A^{-1}(i_{s+1}, \cdots, i_n, i_1, \cdots, i_s)]$$

and

$$B^{-1} = [B^{-1}(j_{s+1}, \cdots, j_n, j_1, \cdots, j_s)].$$

Thus $B^{-1} \otimes A^{-1} =: [D(i_{s+1}, \cdots, i_n, i_1, \cdots, i_s)]$ is determined by

$$\begin{aligned} D(i_{s+1}, \cdots, i_n, i_1, \cdots, i_s) = & A^{-1}\left([i_{s+1}]_{a_{s+1}}, \cdots, [i_n]_{a_n}, [i_1]_{a_1}, \cdots, [i_s]_{a_s}\right) \\ & \cdot B^{-1}\left(\left\lfloor \frac{i_{s+1}}{a_{s+1}} \right\rfloor, \cdots, \left\lfloor \frac{i_n}{a_n} \right\rfloor, \left\lfloor \frac{i_1}{a_1} \right\rfloor, \cdots, \left\lfloor \frac{i_s}{a_s} \right\rfloor\right). \end{aligned}$$

By Definition 4.2.5, we have

$$B \otimes A =: [C(k_1, \cdots, k_n)] = \left[A\left([k_1]_{a_1}, \cdots, [k_n]_{a_n}\right) B\left(\left\lfloor \frac{k_1}{a_1} \right\rfloor, \cdots, \left\lfloor \frac{k_n}{a_n} \right\rfloor\right)\right].$$

It can be proved that the matrix $E = [E(e_1, \cdots, e_n)] = (B \otimes A)\left(B^{-1} \otimes A^{-1}\right)$ is a unit matrix. In fact,

$$\begin{aligned} E(e_1, \cdots, e_n) = & \sum_{c(s+1), \cdots, c(n)} C(e_1, \cdots, e_s, c(s+1), \cdots, c(n)) \\ & \cdot D(c(s+1), \cdots, c(n), e_{s+1}, \cdots, e_n) \\ = & \sum_{c(s+1), \cdots, c(n)} A\left([e_1]_{a_1}, \cdots, [e_s]_{a_s}, [c(s+1)]_{a_{s+1}}, \cdots, [c(n)]_{a_n}\right) \\ & \cdot B\left(\left\lfloor \frac{e_1}{a_1} \right\rfloor, \cdots, \left\lfloor \frac{e_s}{a_s} \right\rfloor, \left\lfloor \frac{c(s+1)}{a_{s+1}} \right\rfloor, \cdots, \left\lfloor \frac{c(n)}{a_n} \right\rfloor\right) \\ & \cdot A^{-1}\left([c(s+1)]_{a_{s+1}}, \cdots, [c(n)]_{a_n}, [e_{s+1}]_{a_1}, \cdots, [e_n]_{a_s}\right) \\ & \cdot B^{-1}\left(\left\lfloor \frac{c(s+1)}{a_{s+1}} \right\rfloor, \cdots, \left\lfloor \frac{c(n)}{a_n} \right\rfloor, \left\lfloor \frac{e_{s+1}}{a_1} \right\rfloor, \cdots, \left\lfloor \frac{e_n}{a_s} \right\rfloor\right) \\ = & \left\{ \sum_{a(s+1), \cdots, a(n)} A\left([e_1]_{a_1}, \cdots, [e_s]_{a_s}, a(s+1), \cdots, a(n)\right) \right. \\ & \left. \cdot A^{-1}\left(a(s+1), \cdots, a(n), [e_{s+1}]_{a_1}, \cdots, [e_n]_{a_s}\right) \right\} \\ & \cdot \left\{ \sum_{b(s+1), \cdots, b(n)} B\left(\left\lfloor \frac{e_1}{a_1} \right\rfloor, \cdots, \left\lfloor \frac{e_s}{a_s} \right\rfloor, b(s+1), \cdots, b(n)\right) \right. \\ & \left. \cdot B^{-1}\left(b(s+1), \cdots, b(n), \left\lfloor \frac{e_{s+1}}{a_1} \right\rfloor, \cdots, \left\lfloor \frac{e_n}{a_s} \right\rfloor\right) \right\} \\ = & \delta\left\{\left([e_1]_{a_1}, \cdots, [e_s]_{a_s}\right) - \left([e_{s+1}]_{a_1}, \cdots, [e_n]_{a_s}\right)\right\} \\ & \cdot \delta\left\{\left(\left\lfloor \frac{e_1}{a_1} \right\rfloor, \cdots, \left\lfloor \frac{e_s}{a_s} \right\rfloor\right) - \left(\left\lfloor \frac{e_{s+1}}{a_1} \right\rfloor, \cdots, \left\lfloor \frac{e_n}{a_s} \right\rfloor\right)\right\} \\ = & \delta\left((e_1, \cdots, e_s) - (e_{s+1}, \cdots, e_n)\right) \\ = & \begin{cases} 1, & \text{if } (e_1, \cdots, e_s) = (e_{s+1}, \cdots, e_n), \\ 0, & \text{otherwise}. \end{cases} \end{aligned} \tag{4.19}$$

Therefore the matrix E is a unit matrix.

In the same way, it can be proved that the matrix $\left(B^{-1} \otimes A^{-1}\right)(B \otimes A)$ is also a unit matrix. Hence it is proved that $B^{-1} \otimes A^{-1} = (B \otimes A)^{-1}$.

Similarly, it can be proved that if $n = 2s + 1$ is odd, then the equation $B^{-1} \otimes A^{-1} = (B \otimes A)^{-1}$ is also satisfied. □

Theorem 4.2.9 *The following equations are true for the direct multiplication operation:*

(1) $(A \otimes B) \otimes C = A \otimes (B \otimes C)$;

(2) $(A + B) \otimes (C + D) = A \otimes C + A \otimes D + B \otimes C + B \otimes D$;

(3) $(A \otimes B)(C \otimes D) = (AC) \otimes (BD)$;

(4) $(A_1 B_1) \otimes (A_2 B_2) \otimes \cdots \otimes (A_n B_n) = (A_1 \otimes \cdots \otimes A_n)(B_1 \otimes \cdots \otimes B_n)$;

(5) $(AB)^{(k)} = A^{(k)} B^{(k)}$, *where* $X^{(k)} =: \overbrace{X \otimes \cdots \otimes X}^{k}$.

Proof We prove only the third identity, because the fourth and fifth identities can be implied by the third one, and the first two identities can be proved by direct verification.

Let

$$A = [A(a(1), \cdots, a(n))], \quad 0 \leqslant a(i) \leqslant a_i - 1,\ 1 \leqslant i \leqslant n,$$

$$B = [B(b(1), \cdots, b(n))], \quad 0 \leqslant b(i) \leqslant b_i - 1,\ 1 \leqslant i \leqslant n,$$

$$C = [C(c(1), \cdots, c(n))], \quad 0 \leqslant c(i) \leqslant c_i - 1,\ 1 \leqslant i \leqslant n$$

and

$$D = [D(d(1), \cdots, d(n))], \quad 0 \leqslant d(i) \leqslant d_i - 1,\ 1 \leqslant i \leqslant n.$$

If $n = 2s$ is even, then AC and BD are well defined only if $(a_{s+1}, \cdots, a_n) = (c_1, \cdots, c_s)$ and $(b_{s+1}, \cdots, b_n) = (d_1, \cdots, d_s)$. For $0 \leqslant e(i) \leqslant a_i b_i - 1$, $0 \leqslant f(i) \leqslant c_i d_i - 1$, $1 \leqslant i \leqslant n$. Let

$$\begin{aligned} A \otimes B = E &= [E(e(1), \cdots, e(n))] \\ &= \left[B\left([e(1)]_{b_1}, \cdots, [e(n)]_{b_n}\right) A\left(\left\lfloor \frac{e(1)}{b_1} \right\rfloor, \cdots, \left\lfloor \frac{e(n)}{b_n} \right\rfloor\right)\right] \end{aligned}$$

and

$$\begin{aligned} C \otimes D = F &= [F(f(1), \cdots, f(n))] \\ &= \left[D\left([f(1)]_{d_1}, \cdots, [f(n)]_{d_n}\right) C\left(\left\lfloor \frac{f(1)}{d_1} \right\rfloor, \cdots, \left\lfloor \frac{f(n)}{d_n} \right\rfloor\right)\right]. \end{aligned}$$

Hence the matrix $(A \otimes B)(C \otimes D) =: G = [G(g(1), \cdots, g(n))]$ has the size $(a_1 b_1) \times \cdots \times (a_s b_s) \times (c_{s+1} d_{s+1}) \times \cdots \times (c_n d_n)$ and its general term represented by

$$\begin{aligned} G(g(1), \cdots, g(n)) = &\sum_{0 \leqslant f(i) \leqslant c_i d_i - 1, 1 \leqslant i \leqslant s} E(g(1), \cdots, g(s), f(1), \cdots, f(s)) \\ &\cdot F(f(1), \cdots, f(s), g(s+1), \cdots, f(n)) \\ = &\sum_{0 \leqslant f(i) \leqslant c_i d_i - 1, 1 \leqslant i \leqslant s} B\left([g(1)]_{b_1}, \cdots, [g(s)]_{b_s}, [f(1)]_{b_{s+1}}, \cdots,\right. \\ &\left.[f(s)]_{b_n}\right) A\left(\left\lfloor \frac{g(1)}{b_1} \right\rfloor, \cdots, \left\lfloor \frac{g(s)}{b_s} \right\rfloor, \left\lfloor \frac{f(1)}{b_{s+1}} \right\rfloor, \cdots, \left\lfloor \frac{f(s)}{b_n} \right\rfloor\right) \end{aligned}$$

$$
\begin{aligned}
&\cdot D\left([f(1)]_{d_1},\cdots,[f(s)]_{d_s},[g(s+1)]_{d_{s+1}},\cdots,[g(n)]_{d_n}\right)\\
&\cdot C\left(\left\lfloor\frac{f(1)}{d_1}\right\rfloor,\cdots,\left\lfloor\frac{f(s)}{d_s}\right\rfloor,\left\lfloor\frac{g(s+1)}{d_{s+1}}\right\rfloor,\cdots,\left\lfloor\frac{g(n)}{d_n}\right\rfloor\right)\\
=&\left\{\sum_{0\leqslant c(i)\leqslant c_i-1,1\leqslant i\leqslant s} A\left(\left\lfloor\frac{g(1)}{b_1}\right\rfloor,\cdots,\left\lfloor\frac{g(s)}{b_s}\right\rfloor,c(1),\cdots,c(s)\right)\right.\\
&\left.\cdot C\left(c(1),\cdots,c(s),\left\lfloor\frac{g(s+1)}{d_{s+1}}\right\rfloor,\cdots,\left\lfloor\frac{g(n)}{d_n}\right\rfloor\right)\right\}\\
&\cdot\left\{\sum_{0\leqslant d(i)\leqslant d_i-1,\ 1\leqslant i\leqslant s} B\left([g(1)]_{b_1},\cdots,[g(s)]_{b_s},d(1),\cdots,d(s)\right)\right.\\
&\left.\cdot D\left(d(1),\cdots,d(s),[g(s+1)]_{d_{s+1}},\cdots,[g(n)]_{d_n}\right)\right\}
\end{aligned}
$$

Because the size of the matrix BD is $b_1\times\cdots\times b_s\times d_{s+1}\times\cdots\times d_n$, and then by the definition of direct multiplication we have proved that $(AC)\otimes(BD)=[G(g(1),\cdots,g(n))]$. In other word, the equation $(AC)\otimes(BD)=(A\otimes B)(C\otimes D)$ has been proved.

By the same way it can be proved that the equation $(AC)\otimes(BD)=(A\otimes B)(C\otimes D)$ is also true when $n=2s+1$ is odd. □

Theorem 4.2.10 *Let I and J be two unit matrices. Then $I\otimes J$ is also a unit matrix.*

Corollary 4.2.2 *$I_m\otimes I_n=I_{mn}$, where I_k stands for the n-dimensional unit matrix of size $k\times\cdots\times k$.*

The following three theorems are very useful for the fast algorithms of higher-dimensional Walsh-Hadamard transforms.

Theorem 4.2.11 *Let A be an n-dimensional matrix of size $p\times\cdots\times p$, and*

$$B=A^{(m)}=:\overbrace{A\otimes\cdots\otimes A}^{m}.$$

Then the matrix B can be decomposed as the multiplication of the following m sparse matrices, precisely,

$$B=\prod_{s=0}^{m-1}\left(I_{p^s}\otimes A\otimes I_{p^{m-1-s}}\right),$$

where I_k refers to the n-dimensional unit matrix of size $k\times\cdots\times k$.

Proof It can be proved by using induction on the integer m.

If $m=2$, the theorem follows by applying the third identity of Theorem 4.2.9 for $B=C=I_p$ and $A=D$.

Assume that the theorem is true for $m=M-1$.

Now it is sufficient to prove that the theorem is true for $m=M$.

$$B=\left(\overbrace{A\otimes\cdots\otimes A}^{M-1}\right)\otimes A$$

$$= \left[\prod_{s=0}^{M-2} \left(I_{p^s} \otimes A \otimes I_{p^{M-2-s}} \right) \right] \otimes A$$

(by the assumption on the case of $n = M - 1$)

$$= \left\{ \left[\prod_{s=0}^{M-2} \left(I_{p^s} \otimes A \otimes I_{p^{M-2-s}} \right) \right] \otimes I_p \right\} \left(I_{p^{M-1}} \otimes A \right)$$

(by the third identity of Theorem 4.2.9)

$$= \left[\prod_{s=0}^{M-2} \left(I_{p^s} \otimes A \otimes I_{p^{M-2-s}} \otimes I_p \right) \right] \left[I_{p^{M-1}} \otimes A \right]$$

(by the equation$(AB) \otimes I = (A \otimes I)(B \otimes I)$)

$$= \prod_{s=0}^{M-1} \left(I_{p^s} \otimes A \otimes I_{p^{M-1-s}} \right) \text{ (by Corollary 4.2.2).}$$

Thus the theorem works for the case of $m = M$. □

Theorem 4.2.12 *Let* $H = [H(i,j,k,l)] = \left[(-1)^{B(i,j,k)+C(j,k,l)} \right]$, $0 \leqslant i, j, k, l \leqslant 1$, *be a 4-dimensional* (± 1)*-valued matrix of size* $2 \times 2 \times 2 \times 2$. *Then the* H *can be decomposed as the multiplication of two sparse matrices, i.e.,* $H = UV$, *where* $U = [U(i,j,k,l)] = \left[\delta(j,l)(-1)^{B(i,j,k)} \right]$ *and* $V = [V(i,j,k,l)] = \left[\delta(i,k)(-1)^{C(j,k,l)} \right]$.

Theorem 4.2.13 *The following decompositions hold:*

(1) *Let* $H = [H(i,j,k,l,m,n)] = \left[(-1)^{B(i,j,k,l)+C(j,k,l,m)+D(k,l,m,n)} \right]$, $0 \leqslant i, j, k, l, m, n \leqslant 1$, *be a 6-dimensional* (± 1)*-valued matrix of size* $2 \times 2 \times 2 \times 2 \times 2 \times 2$. *Then this matrix can be decomposed into the multiplication of three sparse matrices, i.e.,* $H = UVW$, *where*

$$U = [U(i,j,k,l,m,n)] = \left[\delta((j,k)-(m,n))(-1)^{B(i,j,k,l)} \right],$$
$$V = [V(i,j,k,l,m,n)] = \left[\delta((i,k)-(l,n))(-1)^{C(j,k,l,m)} \right],$$
$$W = [W(i,j,k,l,m,n)] = \left[\delta((i,j)-(l,m))(-1)^{D(k,l,m,n)} \right].$$

(2) *Let* $H = [H(i,j,k,l,m,n)] = \left[(-1)^{B(i,j,k,l,m)+C(k,l,m,n)} \right]$, $0 \leqslant i, j, k, l, m, n \leqslant 1$, *be a 6-dimensional* (± 1)*-valued matrix of size* $2 \times 2 \times 2 \times 2 \times 2 \times 2$. *Then this matrix can be decomposed into the multiplication of two sparse matrices, i.e.,* $H = UV$, *where*

$$U = [U(i,j,k,l,m,n)] = \left[\delta(k,n)(-1)^{B(i,j,k,l,m)} \right],$$
$$V = [V(i,j,k,l,m,n)] = \left[\delta((i,j)-(l,m))(-1)^{C(k,l,m,n)} \right].$$

The proofs for Theorems 4.2.12 and 4.2.13 can be finished by direct verifications. Now we are ready for the introduction of higher-dimensional Walsh-Hadamard transforms and their fast algorithms.

4.3 Multi-Dimensional Walsh-Hadamard Transforms

This section will introduce many new higher-dimensional orthogonal transforms used for the processing of 2-dimensional (e.g., images) and/or 3-dimensional (e.g., seismic waves) digital signals. These new transforms share the following advantages:

(1) Their transform matrices are higher-dimensional (± 1)-valued matrices. Thus only plus and minus are required. In other words, it is easier to implement these new transforms than some other known orthogonal transforms that employ productions;

(2) The forward transforms are almost the same as their inverse transforms. Thus the hardware implementation is also much easier;

(3) Fast algorithms have been found;

(4) The known 2-dimensional Walsh-Hadamard transforms introduced in the first section of this chapter are special cases of these new transforms;

(5) The Parseval theorems and some other properties are also retained.

4.3.1 Transforms Based on 3-Dimensional Hadamard Matrices

Let $W = [W(i,j,k)]$, $0 \leqslant i,j,k \leqslant 2^n - 1$, be a 3-dimensional ($\pm 1$)-valued matrix of size $2^n \times 2^n \times 2^n$ such that

$$W^{-1} = \frac{1}{2^n} W, \tag{4.20}$$

or equivalently, $WW = 2^n I$. A 2-dimensional digital signal $f = [f(i,j)]$, $0 \leqslant i,j \leqslant 2^n - 1$, can be treated as a 3-dimensional matrix of size $2^n \times 1 \times 2^n$. Thus

$$F = Wf \tag{4.21}$$

and

$$f = \frac{1}{2^n} WF \tag{4.22}$$

are a pair of orthogonal forward and inverse transforms.

Equation (4.20) can be satisfied by the following matrices:

$$W = [W(i,j,k)] = [(-1)^{ij'+ik'+jk'+an+b(ii'+jj')+ckk'}], \tag{4.23}$$

where $0 \leqslant i,j,k \leqslant 2^n - 1$, $a,b,c \in \{0,1\}$. Here and henceforth, we use ij' to stand for the dot product of the vectors $i = (i_0, \cdots, i_{n-1})$ and $j = (j_0, \cdots, j_{n-1})$, which are the binary expanded vectors of the integers $i = \sum_{k=0}^{n-1} i_k 2^k$ and $j = \sum_{k=0}^{n-1} j_k 2^k$, respectively.

Theorem 4.3.1 *Let $W = [W(i,j,k)]$ be the matrix in Equation 4.23, $f = [f(i,j)]$, $0 \leqslant i,j \leqslant 2^n - 1$, an image signal. Then the transforms defined by Equations* (4.21) *and* (4.21) *satisfy*

(1) $F = [F(i,j)] = Wf$ *and* $f = \frac{1}{2^n} WF$.

(2) *The fast algorithm is*

$$F = Wf = \prod_{s=0}^{n-1} (I_{2^s} \otimes A \otimes I_{2^{n-1-s}}) f,$$

where $A = [A(p,q,r)]$, $0 \leqslant p,q,r \leqslant 1$, *is a 3-dimensional matrix of size* $2 \times 2 \times 2$ *defined by*

$$A(p,q,r) = (-1)^{pq+pr+qr+a+b(p+q)+cr}.$$

This fast algorithm can be finished by $n \times 4^n$ *plus and/or minus operations.*

(3) (*Parseval's Theorem*) *If* $F = [F(i,j)] = Wf$, *then*

$$\sum_{i,k=0}^{2^n-1} F(i,k)^2 = 2^n \sum_{i,k=0}^{2^n-1} f(i,k)^2.$$

(4) *If all elements in* $f = [f(i,j)]$ *are integers then the elements in* $F = [F(i,j)]$ *are also integers, and the elements in each column of* F *have the same parity.*

4.3.2 Transforms Based on 4-Dimensional Hadamard Matrices

Let $W = [W(i,j,k,l)]$, $0 \leqslant i,j,k,l \leqslant 2^n - 1$, be a 4-dimensional ($\pm$)-valued matrix of size $2^n \times 2^n \times 2^n \times 2^n$ such that

$$W^{-1} = \frac{1}{4^n} W, \tag{4.24}$$

or equivalently, $WW = 4^n I$. A 2-dimensional digital signal $f = [f(i,j)]$, $0 \leqslant i,j \leqslant 2^n - 1$, can be treated as a 4-dimensional matrix of size $2^n \times 2^n \times 1 \times 1$. Thus $F = Wf$ and $f = (1/4^n)WF$ are a pair of orthogonal forward and inverse transforms. For different transform matrices W we have the following transforms and their fast algorithms:

Class 1. The transform matrix $W = [W(i,j,k,l)]$, $0 \leqslant i,j,k,l \leqslant 2^n - 1$ is defined by

$$W(i,j,k,l) = (-1)^{ik'+jl'+an+b(ii'+kk')+c(jj'+ll')},$$

where $a,b,c \in \{0,1\}$;

The forward and inverse transforms are $F = Wf$ and $f = \dfrac{1}{4^n}WF$, respectively.

The fast transform is

$$F = \prod_{s=0}^{n-1} (I_{2^s} \otimes A_1 \otimes I_{2^{n-1-s}})(I_{2^s} \otimes B_1 \otimes I_{2^{n-1-s}}) f,$$

where $A_1 = [A_1(p,q,r,s)]$, $B_1 = [B_1(p,q,r,s)]$, $0 \leqslant p,q,r,s \leqslant 1$ are defined by

$$A_1(p,q,r,s) = \delta(q,s)(-1)^{pr+a+b(p+r)}$$

and

$$B_1(p,q,r,s) = \delta(p,r)(-1)^{qs+c(q+s)}.$$

This fast algorithm needs $n \cdot 2^{2n+1}$ plus and/or minus operations.

Remark 4.3.1 *If* $a = b = c = 0$, *then this transform reduces to the conventional 2-dimensional Walsh-Hadamard transform introduced in the first section of this chapter.*

Class 2. The transform matrix $W = [W(i,j,k,l)]$, $0 \leqslant i,j,k,l \leqslant 2^n - 1$, is defined by

$$W(i,j,k,l) = (-1)^{ij'+ik'+jl'+kl'+an+b(ii'+kk')+c(jj'+ll')},$$

where $a,b,c \in \{0,1\}$.

The forward and inverse transforms are $F = Wf$ and $f = \frac{1}{4^n}WF$, respectively.

The fast transform is

$$F = \prod_{s=0}^{n-1}(I_{2^s} \otimes A_2 \otimes I_{2^{n-1-s}})(I_{2^s} \otimes B_2 \otimes I_{2^{n-1-s}})f,$$

where $A_2 = [A_2(p,q,r,s)]$, $B_2 = [B_2(p,q,r,s)]$, $0 \leqslant p,q,r,s \leqslant 1$ are defined by

$$A_2(p,q,r,s) = \delta(q,s)(-1)^{pq+pr+a+b(p+r)}$$

and

$$B_2(p,q,r,s) = \delta(p,r)(-1)^{rs+qs+c(q+s)}.$$

This fast algorithm needs $n \cdot 2^{2n+1}$ plus and/or minus operations.

Class 3. The transform matrix $W = [W(i,j,k,l)]$, $0 \leqslant i,j,k,l \leqslant 2^n - 1$, is defined by

$$W(i,j,k,l) = (-1)^{il'+kj'+an+b(ii'+kk')+c(jj'+ll')},$$

where $a,b,c \in \{0,1\}$.

The forward and inverse transforms are $F = Wf$ and $f = \frac{1}{4^n}WF$, respectively.

The fast transform is

$$F = \prod_{s=0}^{n-1}(I_{2^s} \otimes A_3 \otimes I_{2^{n-1-s}})f,$$

where $A_3 = [A_3(p,q,r,s)]$, $0 \leqslant p,q,r,s \leqslant 1$, is defined by

$$A_3(p,q,r,s) = (-1)^{ps+qr+a+b(p+r)+c(q+s)}.$$

This fast algorithm needs $3n \cdot 2^{2n}$ plus and/or minus operations.

Class 4. The transform matrix $W = [W(i,j,k,l)]$, $0 \leqslant i,j,k,l \leqslant 2^n - 1$, is defined by

$$W(i,j,k,l) = (-1)^{ij'+il'+kj'+kl'+an+b(ii'+kk')+c(jj'+ll')},$$

where $a,b,c \in \{0,1\}$.

The forward and inverse transforms are $F = Wf$ and $f = \frac{1}{4^n}WF$, respectively.

The fast transform is

$$F = \prod_{s=0}^{n-1}(I_{2^s} \otimes A_4 \otimes I_{2^{n-1-s}}), f,$$

where $A_4 = [A_4(p,q,r,s)]$, $0 \leqslant p,q,r,s \leqslant 1$ is defined by

$$A_4(p,q,r,s) = (-1)^{pq+ps+qr+qs+a+b(p+r)+c(q+s)}.$$

This fast algorithm needs $3n \cdot 2^{2n}$ plus and/or minus operations.

Class 5. The transform matrix $W = [W(i,j,k,l)]$, $0 \leqslant i,j,k,l \leqslant 2^n - 1$ is defined by

$$W(i,j,k,l) = (-1)^{ik'+il'+kj'+an+b(ii'+kk')+c(jj'+ll')},$$

where $a,b,c \in \{0,1\}$.

The forward and inverse transforms are $F = Wf$ and $f = \dfrac{1}{4^n}WF$, respectively.

The fast transform is

$$F = \prod_{s=0}^{n-1} (I_{2^s} \otimes A_5 \otimes I_{2^{n-1-s}})f,$$

where $A_5 = [A_5(p,q,r,s)]$, $0 \leqslant p,q,r,s \leqslant 1$, is defined by

$$A_5(p,q,r,s) = (-1)^{pr+ps+qr+a+b(p+r)+c(q+s)}.$$

This fast algorithm needs $3n \cdot 2^{2n}$ plus and/or minus operations.

Class 6. The transform matrix $W = [W(i,j,k,l)]$, $0 \leqslant i,j,k,l \leqslant 2^n - 1$ is defined by

$$W(i,j,k,l) = (-1)^{ij'+ik'+il'+jk'+kl'+an+b(ii'+kk')+c(jj'+ll')},$$

where $a, b, c \in \{0, 1\}$.

The forward and inverse transforms are $F = Wf$ and $f = \dfrac{1}{4^n}WF$, respectively.

The fast transform is

$$F = \prod_{s=0}^{n-1} (I_{2^s} \otimes A_6 \otimes I_{2^{n-1-s}})f,$$

where $A_6 = [A_6(p,q,r,s)]$, $0 \leqslant p,q,r,s \leqslant 1$, is defined by

$$A_6(p,q,r,s) = (-1)^{pq+pr+ps+qr+qs+a+b(p+r)+c(q+s)}.$$

This fast algorithm needs $3n \cdot 2^{2n}$ plus and/or minus operations.

Class 7. The transform matrix $W = [W(i,j,k,l)]$, $0 \leqslant i,j,k,l \leqslant 2^n - 1$ is defined by

$$W(i,j,k,l) = (-1)^{il'+kj'+jl'+an+b(ii'+kk')+c(jj'+ll')},$$

where $a, b, c \in \{0, 1\}$.

The forward and inverse transforms are $F = Wf$ and $f = \dfrac{1}{4^n}WF$, respectively.

The fast transform is

$$F = \prod_{s=0}^{n-1} (I_{2^s} \otimes A_7 \otimes I_{2^{n-1-s}})f,$$

where $A_7 = [A_7(p,q,r,s)]$, $0 \leqslant p,q,r,s \leqslant 1$ is defined by

$$A_7(p,q,r,s) = (-1)^{ps+qr+qs+a+b(p+r)+c(q+s)}.$$

This fast algorithm needs $3n \cdot 2^{2n}$ plus and/or minus operations.

Class 8. The transform matrix $W = [W(i,j,k,l)]$, $0 \leqslant i,j,k,l \leqslant 2^n - 1$, is defined by

$$W(i,j,k,l) = (-1)^{ij'+il'+kj'+jl'+kl'+an+b(ii'+kk')+c(jj'+ll')},$$

where $a, b, c \in \{0, 1\}$.

The forward and inverse transforms are $F = Wf$ and $f = \dfrac{1}{4^n}WF$, respectively.

The fast transform is

$$F = \prod_{s=0}^{n-1} (I_{2^s} \otimes A_8 \otimes I_{2^{n-1-s}})f,$$

where $A_8 = [A_8(p, q, r, s)]$, $0 \leqslant p, q, r, s \leqslant 1$, is defined by

$$A_8(p, q, r, s) = (-1)^{pq+ps+qr+qs+rs+a+b(p+r)+c(q+s)}.$$

This fast algorithm needs $3n \cdot 2^{2n}$ plus and/or minus operations.

Theorem 4.3.2 *All of the transforms described in this subsection satisfy the following properties:*

(1) (*Parseval's Theorem*) *If* $F = [F(i, j)] = Wf$, *then*

$$\sum_{i,j=0}^{2^n-1} F(i, j)^2 = 4^n \sum_{i,k=0}^{2^n-1} f(i, k)^2.$$

(2) *If all elements in* $f = [f(i, j)]$ *are integers, then the elements in* $F = [F(i, j)]$ *have the same parity.*

4.3.3 Transforms Based on 6-Dimensional Hadamard Matrices

Let $W = [W(i, j, k, p, q, r)]$, $0 \leqslant i, j, k, p, q, r \leqslant 2^n - 1$, be a 6-dimensional Hadamard matrix of size $2^n \times 2^n \times 2^n \times 2^n \times 2^n \times 2^n$ such that

$$W^{-1} = \frac{1}{8^n} W, \tag{4.25}$$

or equivalently, $WW = 8^n I$. A 3-dimensional digital signal $f = [f(i, j, k)]$, $0 \leqslant i, j, k \leqslant 2^n - 1$, can be treated as a 6-dimensional matrix of size $2^n \times 2^n \times 2^n \times 1 \times 1 \times 1$. Thus $F = Wf$ and $f = \frac{1}{8^n} WF$ produce a pair of orthogonal forward and inverse transforms. For different transform matrices W, we have the following transforms and their fast algorithms:

First Class. The transform matrix $W = [W(i, j, k, p, q, r)]$, $0 \leqslant i, j, k, p, q, r \leqslant 2^n - 1$ is defined by

$$W(i, j, k, p, q, r) = (-1)^{g(i,j,k,p,q,r)},$$

where $g(i, j, k, p, q, r) = a_1 ij' + ip' + a_2 jk' + jq' + kr' a_1 pq' + a_2 qr' + a_3 n + a_4(ii' + pp') + a_5(jj' + qq') + a_6(kk' + rr')$, $a_1, a_2, a_3, a_4, a_5, a_6 \in \{0, 1\}$.

The forward and inverse transforms are: $F = Wf$ and $f = \frac{1}{8^n} WF$, respectively.

The fast transform is

$$F = \prod_{s=0}^{n-1} (I_{2^s} \otimes A_1 \otimes I_{2^{n-1-s}})(I_{2^s} \otimes B_1 \otimes I_{2^{n-1-s}})(I_{2^s} \otimes C_1 \otimes I_{2^{n-1-s}}) f,$$

where $A_1 = [A_1(i, j, k, p, q, r)]$, $B_1 = [B_1(i, j, k, p, q, r)]$, $C_1 = [C_1(i, j, k, p, q, r)]$, $0 \leqslant i, j, k, p, q, r \leqslant 1$, are defined by

$$A_1(i, j, k, p, q, r) = \delta((j, k) - (q, r))(-1)^{a_1 ij + ip + a_3 + a_4(i+p)},$$

$$B_1(i, j, k, p, q, r) = \delta((i, k) - (p, r))(-1)^{a_2 jk + jq + a_5(j+q)}$$

and

$$C_1(i, j, k, p, q, r) = \delta((i, j) - (p, q))(-1)^{kr + a_1 pq + a_2 qr + a_5(k+r)}.$$

This fast algorithm needs $3n \cdot 2^{3n}$ plus and/or minus operations.

Second Class. The transform matrix $W = [W(i,j,k,p,q,r)]$, $0 \leqslant i,j,k,p,q,r \leqslant 2^n - 1$ is defined by

$$W(i,j,k,p,q,r) = (-1)^{g(i,j,k,p,q,r)},$$

where $g(i,j,k,p,q,r) = a_1 ij' + a_2 ik' + iq' + a_3 jk' + jp' + kr' + a_1 pq' + a_2 pr' + a_3 qr' + a_4 n + a_5(ii' + pp') + a_6(jj' + qq') + a_7(kk' + rr')$, $a_1, a_2, a_3, a_4, a_5, a_6, a_7 \in \{0,1\}$.

The forward and inverse transforms are $F = Wf$ and $f = \dfrac{1}{8^n} WF$, respectively.

The fast transform is

$$F = \prod_{s=0}^{n-1} (I_{2^s} \otimes A_2 \otimes I_{2^{n-1-s}})(I_{2^s} \otimes B_2 \otimes I_{2^{n-1-s}}) f,$$

where $A_2 = [A_2(i,j,k,p,q,r)]$, $B_2 = [B_2(i,j,k,p,q,r)]$, $0 \leqslant i,j,k,p,q,r \leqslant 1$, are defined by

$$A_2(i,j,k,p,q,r) = \delta(k,r)(-1)^{a_1 ij + a_2 ik + iq + a_3 jk + jp + a_4 + a_5(i+p) + a_6(j+q)}$$

and

$$B_2(i,j,k,p,q,r) = \delta((i,j) - (p,q))(-1)^{kr + a_1 pq + a_2 pr + a_3 qr + a_7(k+r)}.$$

This fast algorithm needs $4n \cdot 2^{3n}$ plus and/or minus operations.

Third Class. The transform matrix $W = [W(i,j,k,p,q,r)]$, $0 \leqslant i,j,k,p,q,r \leqslant 2^n - 1$, is defined by

$$W(i,j,k,p,q,r) = (-1)^{g(i,j,k,p,q,r)},$$

where $g(i,j,k,p,q,r) = a_1 ij' + a_2 ik + ip' + iq' + a_3 jk' + jp' + kr' + a_1 pq' + a_2 pr' + a_3 qr + a_4 n + a_5(ii' + pp') + a_6(jj' + qq') + a_7(kk' + rr')$, $a_1, a_2, a_3, a_4, a_5, a_6, a_7 \in \{0,1\}$.

The forward and inverse transforms are: $F = Wf$ and $f = \frac{1}{8^n} WF$, respectively.

The fast transform is

$$F = \prod_{s=0}^{n-1} (I_{2^s} \otimes A_3 \otimes I_{2^{n-1-s}})(I_{2^s} \otimes B_3 \otimes I_{2^{n-1-s}}) f,$$

where $A_3 = [A_3(i,j,k,p,q,r)]$ and $B_3 = [B_3(i,j,k,p,q,r)]$ are defined by

$$A_3(i,j,k,p,q,r) = \delta(k,r)(-1)^{a_1 ij + a_2 ik + ip + iq + a_3 jk + jp + a_4 + a_5(i+p) + a_6(j+q)}$$

and

$$B_3(i,j,k,p,q,r) = \delta((i,j) - (p,q))(-1)^{kr + a_1 pq + a_2 pr + a_3 qr + a_7(k+r)}.$$

This fast algorithm needs $4n \cdot 2^{3n}$ plus and/or minus operations.

Fourth Class. The transform matrix $W = [W(i,j,k,p,q,r)]$, $0 \leqslant i,j,k,p,q,r \leqslant 2^n - 1$, is defined by

$$W(i,j,k,p,q,r) = (-1)^{g(i,j,k,p,q,r)},$$

where $g(i,j,k,p,q,r)$ is one of the following 17 functions:

(1) $g(i,j,k,p,q,r) = a_1 ij' + a_2 ik' + ir' + a_3 jk' + jq' + kp' + a_1 pq' + a_2 pr' + a_3 qr' + a_4 n + a_5(ii' + pp') + a_6(jj' + qq') + a_7(kk' + rr')$;

$$(2)\ g(i,j,k,p,q,r) = a_1ij' + a_2ik' + ip' + a_3jk' + jr' + kq' + a_1pq' + a_2pr' + a_3qr' + a_4n + a_5(ii' + pp') + a_6(jj' + qq') + a_7(kk' + rr');$$

$$(3)\ g(i,j,k,p,q,r) = a_1ij' + a_2ik' + ip' + iq' + a_3jk' + jp' + jq' + jr' + kq' + a_1pq' + a_2pr' + a_3qr' + a_4n + a_5(ii' + pp') + a_6(jj' + qq') + a_7(kk' + rr');$$

$$(4)\ g(i,j,k,p,q,r) = a_1ij' + a_2ik' + ip' + iq' + a_3jk' + jp' + jr' + kq' + a_1pq' + a_2pr' + a_3qr' + a_4n + a_5(ii' + pp') + a_6(jj' + qq') + a_7(kk' + rr');$$

$$(5)\ g(i,j,k,p,q,r) = a_1ij' + a_2ik' + ip' + iq' + a_3jk' + jp' + jq' + jr' + kq' + kr' + a_1pq' + a_2pr' + a_3qr' + a_4n + a_5(ii' + pp') + a_6(jj' + qq') + a_7(kk' + rr');$$

$$(6)\ g(i,j,k,p,q,r) = a_1ij' + a_2ik' + iq' + ir' + a_3jk' + jp' + jq' + jr' + kp' + kq' + a_1pq' + a_2pr' + a_3qr' + a_4n + a_5(ii' + pp') + a_6(jj' + qq') + a_7(kk' + rr');$$

$$(7)\ g(i,j,k,p,q,r) = a_1ij' + a_2ik' + iq' + ir' + a_3jk' + jp' + jq' + kp' + a_1pq' + a_2pr' + a_3qr' + a_4n + a_5(ii' + pp') + a_6(jj' + qq') + a_7(kk' + rr');$$

$$(8)\ g(i,j,k,p,q,r) = a_1ij' + a_2ik' + iq' + ir' + a_3jk' + jp' + jr' + kp' + kq' + kr' + a_1pq' + a_2pr' + a_3qr' + a_4n + a_5(ii' + pp') + a_6(jj' + qq') + a_7(kk' + rr');$$

$$(9)\ g(i,j,k,p,q,r) = a_1ij' + a_2ik' + iq' + ir' + a_3jk' + jp' + kp' + kr' + a_1pq' + a_2pr' + a_3qr' + a_4n + a_5(ii' + pp') + a_6(jj' + qq') + a_7(kk' + rr');$$

$$(10)\ g(i,j,k,p,q,r) = a_1ij' + a_2ik' + ip' + ir' + a_3jk' + jq' + jr' + kp' + kq' + kr' + a_1pq' + a_2pr' + a_3qr' + a_4n + a_5(ii' + pp') + a_6(jj' + qq') + a_7(kk' + rr');$$

$$(11)\ g(i,j,k,p,q,r) = a_1ij' + a_2ik' + ip' + ir' + a_3jk' + jq' + kq' + a_1pq' + a_2pr' + a_3qr' + a_4n + a_5(ii' + pp') + a_6(jj' + qq') + a_7(kk' + rr');$$

$$(12)\ g(i,j,k,p,q,r) = a_1ij' + a_2ik' + ip' + ir' + a_3jk' + jr' + kp' + kq' + kr' + a_1pq' + a_2pr' + a_3qr' + a_4n + a_5(ii' + pp') + a_6(jj' + qq') + a_7(kk' + rr');$$

$$(13)\ g(i,j,k,p,q,r) = a_1ij' + a_2ik' + ip' + ir' + a_3jk' + jr' + kp' + kq' + a_1pq' + a_2pr' + a_3qr' + a_4n + a_5(ii' + pp') + a_6(jj' + qq') + a_7(kk' + rr');$$

$$(14)\ g(i,j,k,p,q,r) = a_1ij' + a_2ik' + ip' + iq' + ir' + a_3jk' + jp' + jq' + kp' + kr' + a_1pq' + a_2pr' + a_3qr' + a_4n + a_5(ii' + pp') + a_6(jj' + qq') + a_7(kk' + rr');$$

(15) $g(i,j,k,p,q,r) = a_1ij' + a_2ik' + ip' + iq' + ir' + a_3jk' + jp' + jq' + kp' + a_1pq'$
$+ a_2pr' + a_3qr' + a_4n + a_5(ii' + pp') + a_6(jj' + qq')$
$+ a_7(kk' + rr')$;

(16) $g(i,j,k,p,q,r) = a_1ij' + a_2ik' + ip' + iq' + ir' + a_3jk' + jp' + jr' + kp' + kq'$
$+ a_1pq' + a_2pr' + a_3qr' + a_4n + a_5(ii' + pp')$
$+ a_6(jj' + qq') + a_7(kk' + rr')$;

(17) $g(i,j,k,p,q,r) = a_1ij' + a_2ik' + ip' + iq' + ir' + a_3jk' + jp' + kp' + kr'$
$+ a_1pq' + a_2pr' + a_3qr' + a_4n + a_5(ii' + pp') + a_6(jj' + qq')$
$+ a_7(kk' + rr')$.

The forward and inverse transforms are $F = Wf$ and $f = \frac{1}{8^n}WF$, respectively. The fast transform is

$$F = \prod_{s=0}^{n-1}(I_{2^s} \otimes A \otimes I_{2^{n-1-s}})f,$$

where the matrix A is the corresponding transform matrix of size $2 \times 2 \times 2 \times 2 \times 2 \times 2$. This fast algorithm needs $7n \cdot 2^{3n}$ plus and/or minus operations.

For more details of three-dimensional Hadamard matrices the readers are referred to the papers [4–8].

Bibliography

[1] Ahmed N and Rao K R. *Orthogonal Transforms for Digital Signal Processing.* Springer-Verlag, 1975.

[2] Shlichta P J. Higher-dimensional Hadamard matrices. *IEEE Trans. On Inform. Theory,* 1979, IT-25(5): 566-572.

[3] Shlichta P J. Three- and four-dimensional Hadamard matrices. *Bull. Amer. Phys. Soc. Ser.,* 1971. 11(16): 825-826.

[4] Yang Y X. Operations and applications of higher dimensional matrices. *J. of Chengdu Institute of Radio Engineering,* 1987, 16(2): 191-199.

[5] Yang Y X. Higher-dimensional Walsh-Hadamard transforms. *J. of Beijing Univ. of Posts and Telecomm.,* 1988, 11(2): 22-30.

[6] Elliott D F and Rao K R. *Fast Transforms Algorithms, Analyses, Applications.* New York: Academic Press, 1982.

[7] Harmuth H F. *Transmission of Information by orthogonal functions. 2nd ed.* New York: Springer-Verlag, 1972.

[8] Gonzalez R C and Wintz P. *Digital Image Processing (Second Edition).* Addison-Wesley Publishing Company, 1987.

Part III

General Higher-Dimensional Cases

Chapter 5

n-Dimensional Hadamard Matrices of Order 2

The simplest higher-dimensional Hadamard matrices are those of order 2. Informally, an n-dimensional Hadamard matrix of order 2 (in short, a 2^n Hadamard matrix) is a binary n-cube of order 2 in which all parallel $(n-1)$-dimensional sections are mutually orthogonal. This chapter will concentrate on the constructions, enumeration and applications of 2^n Hadamard matrices.

5.1 Constructions of 2^n Hadamard Matrices

The number of n-dimensional Hadamard matrices of order 2 is probably a substantial fraction of the number of possible binary 2^n matrices, but the latter are so numerous that for $n > 4$, exhaustive search routines are impractical. Thus the first question to be answered should be: how can one construct as many 2^n Hadamard matrices as possible?

Here are four intuitive constructions of 2^n Hadamard matrices[1]:

Minimal[1] If the population of black positions has the minimum value of $1/4$, then the only requirement for a Hadamard matrix is that black-black correlation be absent. For a 2^n matrix this means merely that black-black nearest neighbors must be avoided. Here and henceforth we treat a binary 2^n matrix as an n-cube of order 2 with the element "1" being replaced by the black position and "-1" by the white position.

Petrie Polygon[1] These are n-cubes in which the Petrie polygon is colored black and the rest of the matrix white, and all of the latter are the nearest neighbors to at least one black position. This method may be extendable to higher-dimensions by using a Petrie-polygon-colored matrix as a trial solution for a computer search.

Antipodal $(n-2)$-Dimensional Sections[1] This construction can be finished by the following two steps:

(1) A pair of antipodal $(n-2)$-dimensional sections are selected; one is colored black and the other white.

(2) The remaining positions are regarded as chains between the two $(n-2)$-dimensional sections; alternate ones are colored black and white.

Double Proximity Shells[1] A position in a 2^n matrix may be considered to be surrounded by a shell of nearest neighbors (having one intervening tie line), which in turn is surrounded by successive proximity shells until the ultimate shell, a single antipodal position, is reached (the population of these shells is given by the binomial coefficients of order n). Colors are assigned by giving the next two proximity shells the opposite color, and so on. This scheme results in one coloring pattern when n is even and two coloring patterns when n is odd. All of these are completely proper 2^n Hadamard matrices. Moreover, all of the lower-dimensional sections of these matrices are members of the same family.

Clearly the above four constructions cannot exhaust all 2^n Hadamard matrices. In order to produce more 2^n Hadamard matrices, we introduce, in the following subsection, the concept of H-Boolean function, which is one of the equivalent forms of the 2^n Hadamard matrix.

5.1.1 Equivalence between 2^n Hadamard Matrices and H-Boolean Functions

Definition 5.1.1 *A Boolean function $f(x_1, x_2, \cdots, x_n)$ of n variables is called an* H-*Boolean function if and only if the Hamming weights of the following n Boolean functions $g_1(\cdot), g_2(\cdot), \cdots, g_n(\cdot)$ of $(n-1)$ variables are 2^{n-2}, where*

$$g_1(x_1, x_2, \cdots, x_{n-1}) = f(x_1, x_2, \cdots, x_{n-1}, 0) + f(x_1, x_2, \cdots, x_{n-1}, 1),$$
$$g_2(x_1, \cdots, x_{n-2}, x_n) = f(x_1, \cdots, x_{n-2}, 0, x_n) + f(x_1, \cdots, x_{n-2}, 1, x_n),$$
$$\cdots\cdots\cdots$$
$$g_n(x_2, x_3, \cdots, x_n) = f(0, x_2, \cdots, x_n) + f(1, x_2, \cdots, x_n).$$

Remark 5.1.1 *The Hamming weight of a Boolean function means the number of* 1s *in the truth table of this Boolean function. The Boolean functions $g_1(\cdot), g_2(\cdot), \cdots, g_n(\cdot)$ in Definition* 5.1.1 *must be treated as Boolean functions of $(n-1)$ but not of n variables.*

On the other hand, the 2^n Hadamard matrix is defined by

Definition 5.1.2 *An n-dimensional Hadamard matrix $H = [H(i_1, i_2, \cdots, i_n)]$ of order* 2 *is a binary matrix in which all parallel $(n-1)$-dimensional sections are mutually orthogonal; that is, all $H(i_1, i_2, \cdots, i_n) = -1$ or 1, $i_1, i_2, \cdots, i_n = 0$ or 1, and*

$$\sum_{i_1=0}^{1}\sum_{i_2=0}^{1}\cdots\sum_{i_{n-1}=0}^{1} H(i_1, i_2, \cdots, i_{n-1}, a)H(i_1, i_2, \cdots, i_{n-1}, b) = 2^{(n-1)}\delta_{ab},$$
$$\sum_{i_1=0}^{1}\cdots\sum_{i_{n-2}=0}^{1}\sum_{i_n=0}^{1} H(i_1, i_2, \cdots, i_{n-2}, a, i_n)H(i_1, i_2, \cdots, i_{n-2}, b, i_n) = 2^{(n-1)}\delta_{ab},$$
$$\cdots\cdots\cdots$$
$$\sum_{i_2=0}^{1}\sum_{i_3=0}^{1}\cdots\sum_{i_n=0}^{1} H(a, i_2, i_3, \cdots, i_n)H(b, i_2, i_3, \cdots, i_n) = 2^{(n-1)}\delta_{ab}.$$

Comparing the above Definition 5.1.1 and Definition 5.1.2, the equivalent relationship between H-Boolean functions and 2^n Hadamard matrices becomes straightforward, i.e.,

Theorem 5.1.1 *A Boolean function $f(x_1, x_2, \cdots, x_n)$ is an* H-*Boolean function of n variables if and only if the n-dimensional binary matrix $[H(i_1, i_2, \cdots, i_n)]$, $i_1, i_2, \cdots, i_n = 0$ or 1, defined by*

$$H(i_1, i_2, \cdots, i_n) = (-1)^{f(i_1, i_2, \cdots, i_n)}$$

is a 2^n Hadamard matrix.

It is Theorem 5.1.1 that motivates the study of H-Boolean functions in the following subsections.

5.1.2 Existence of H-Boolean Functions

This subsection states some basic properties and necessary and/or sufficient conditions of H-Boolean functions.

At first, it is not difficult to verify that the following Boolean functions are all H-Boolean:

$$f_1(x_1, \cdots, x_n) = \sum_{1 \leqslant i < j \leqslant n} x_i x_j,$$

$$f_2(x_1, \cdots, x_n) = x_1 \sum_{k=2}^{n} x_k + \prod_{k=m}^{n} x_k, \quad 2 \leqslant m \leqslant n,$$

$$f_3(x_1, \cdots, x_n) = x_1 \sum_{k=2}^{n} x_k + x_1 x_2 \cdots x_{n-1} + x_2 x_3 \cdots x_n,$$

$$f_4(x_1, \cdots, x_n) = \sum_{1 \leqslant i < j \leqslant n} x_i x_j + x_2 x_3 \cdots x_n,$$

$$f_5(x_1, \cdots, x_n) = x_1(x_2 + \cdots + x_n) + x_2(x_3 + \cdots + x_n) + x_1 x_3 x_4 \cdots x_n,$$

$$f_6(x_1, \cdots, x_n) = x_1(x_2 + \cdots + x_n) + x_3 \cdots x_n + x_2 \cdots x_n + x_1 x_3 x_4 \cdots x_n.$$

These examples imply that for each n, there exists at least one n-dimensional Hadamard matrix of order 2.

Definition 5.1.3 *Two Boolean functions $f(x_1, x_2, \cdots, x_n)$ and $g(x_1, x_2, \cdots, x_n)$ are said to be equivalent to each other if and only if there is a permutation $\tau(\cdot)$ of the set $\{1, 2, \cdots, n\}$ and a binary vector $(a_1, a_2, \cdots, a_n)$, $a_i = 0$ or 1 for $1 \leqslant i \leqslant n$, such that*

$$f(x_1, x_2, \cdots, x_n) = g(\tau(x_1 + a_1, x_2 + a_2, \cdots, x_n + a_n)),$$

where $\tau(y_1, y_2, \cdots, y_n) = (y_{\tau(1)}, y_{\tau(2)}, \cdots, y_{\tau(n)})$.

From the definition of H-Boolean functions and Definition 5.1.3, one can prove the following theorem.

Theorem 5.1.2 *The Boolean function equivalent to an* H-*Boolean function is also an* H-*Boolean function.*

Proof Equivalent Boolean functions have the same Hamming weight. □

Lemma 5.1.1 *Let $f(x_1, x_2, \cdots, x_n)$ and $g(x_1, x_2, \cdots, x_n)$ be two Boolean functions. Then*

$$w(f(\cdot) + g(\cdot)) = w(f(\cdot)) + w(g(\cdot)) - 2w(f(\cdot)g(\cdot)),$$

where $w(h(\cdot))$ refers to the Hamming weight of the Boolean function $h(x_1, x_2, \cdots, x_n)$.

Proof This lemma is, in fact, a direct corollary of the well known identity $(a + b)\text{mod}2 = a + b - 2ab$, for $a, b = 0$ or 1. □

Lemma 5.1.2 *The following are true:*

(1) *When all of the following Boolean functions are treated as of* n *variables, we have*

$$\begin{aligned} w(x_1) &= 2^{n-1}; \\ w(x_1x_2) &= 2^{n-2}; \\ &\cdots\cdots\cdots \\ w(x_1x_2\cdots x_k) &= 2^{n-k}; \\ &\cdots\cdots\cdots \\ w(x_1x_2\cdots x_n) &= 1. \end{aligned}$$

(2) *If* $f(x_1, x_2, \cdots, x_n)$ *is a Boolean function of odd Hamming weight, then there is a Boolean function* $g(x_1, x_2, \cdots, x_n)$ *of even Hamming weight such that*

$$f(x_1, x_2, \cdots, x_n) = g(x_1, x_2, \cdots, x_n) + x_1x_2\cdots x_n.$$

This lemma can be directly verified by the definition of Hamming weight.

Theorem 5.1.3 *The Hamming weight of every* H-*Boolean function* $f(x_1, x_2, \cdots, x_n)$, $n \geqslant 3$, *is even.*

Proof If, on the contrary, $f(x_1, x_2, \cdots, x_n)$ is an H-Boolean function of odd Hamming weight, then there is a Boolean function $g(x_1, x_2, \cdots, x_n)$ of even Hamming weight such that

$$f(x_1, x_2, \cdots, x_n) = g(x_1, x_2, \cdots, x_n) + x_1x_2\cdots x_n.$$

Thus

$$\begin{aligned} g_1(x_1, x_2, \cdots, x_{n-1}) &= f(x_1, x_2, \cdots, x_{n-1}, 0) + f(x_1, x_2, \cdots, x_{n-1}, 1) \\ &= [g(x_1, x_2, \cdots, x_{n-1}, 0) + g(x_1, x_2, \cdots, x_{n-1}, 1)] \\ &\quad + x_1x_2\cdots x_{n-1} \\ &= r(x_1, x_2, \cdots, x_{n-1}) + x_1x_2\cdots x_{n-1}. \end{aligned}$$

By Lemma 5.1.2, the Hamming weight of this Boolean function of $(n-1)$ variables must be odd, which is different from 2^{n-2}, $n \geqslant 3$. In other words, $f(x_1, x_2, \cdots, x_n)$ cannot be H-Boolean. □

Every Boolean function is a linear combination of the functions x_1; x_1x_2; $\cdots$; $x_1x_2\cdots x_k$; $\cdots$; $x_1x_2\cdots x_n$ and their equivalent forms. Formally, every Boolean function $f(x_1, x_2, \cdots, x_n)$ can be denoted by its polynomial form as

$$\begin{aligned} f(x_1, x_2, \cdots, x_n) &= a + \sum_{i=1}^{n} b(i)x_i + \sum_{1\leqslant i<j\leqslant n} c(i,j)x_ix_j \\ &\quad + \sum_{1\leqslant i<j<k\leqslant n} d(i,j,k)x_ix_jx_k + \cdots, \end{aligned}$$

in which every term "x_i" is said to be of degree one, "x_ix_j" of degree two, and "$x_ix_jx_k$" of degree three, and so on, where $b(i)$, $c(i,j)$, and $d(i,j,k)$, etc., are zero or one.

Lemma 5.1.3 *Let $f(x_1, x_2, \cdots, x_n)$ be a Boolean function consisting of the sum of K terms of degree $n-1$. Then its Hamming weight $w(f)$ is*

$$w(f) = 2\lfloor (K+1)/2 \rfloor,$$

where $\lfloor x \rfloor$ refers to the floor function, i.e., the largest integer up to x, e.g., $\lfloor 1/2 \rfloor = 0$ and $\lfloor 3/2 \rfloor = 1$.

Proof It can be proved by using induction on the integer K.

Case $K = 1$. The Hamming weight of the single term

$$f(x_1, x_2, \cdots, x_n) = x_1 x_2 \cdots x_{n-1}$$

is equal to 2, i.e., the lemma is right if $K = 1$.

Case $K = 2$. Let

$$f(x_1, x_2, \cdots, x_n) = g(x_1, x_2, \cdots, x_n) + h(x_1, x_2, \cdots, x_n)$$

be the sum of two terms of degree $n-1$. Then by Lemma 5.1.1 and Lemma 5.1.2, the Hamming weight of $f(\cdot)$ is $w(f) = w(g) + w(h) - 2w(g \cdot h) = 2 + 2 - 2 \times 1 = 2$. In other words, the lemma is right if $K = 2$.

In general, suppose that the lemma is correct for the cases of $K = 2m-1$ and $K = 2m$. It is sufficient to prove the correctness of the cases of $K = 2m+1$ and $K = 2m+2$.

For $K = 2m+1$ let

$$\begin{aligned} f(x_1, x_2, \cdots, x_n) = & g_1(x_1, x_2, \cdots, x_n) + g_2(x_1, x_2, \cdots, x_n) + \cdots \\ & + g_{2m}(x_1, x_2, \cdots, x_n) + g_{2m+1}(x_1, x_2, \cdots, x_n), \end{aligned}$$

where $g_i(x_1, x_2, \cdots, x_n)$, $1 \leqslant i \leqslant 2m+1$, are different single-terms of degree $n-1$. Hence, for each $i \neq j$,

$$g_i(x_1, x_2, \cdots, x_n) g_j(x_1, x_2, \cdots, x_n) = x_1 x_2 \cdots x_n.$$

Therefore,

$$\begin{aligned} w(f) = & w([g_1 + \cdots + g_{2m}] + g_{2m+1}) \\ = & w(g_1 + \cdots + g_{2m}) + w(g_{2m+1}) - 2w([g_1 + \cdots + g_{2m}] g_{2m+1}) \\ = & 2\lfloor (2m+1)/2 \rfloor + 2 \\ = & 2m + 2 \\ = & 2\lfloor [(2m+1)+1]/2 \rfloor, \end{aligned}$$

where the second equation is due to the facts of $[g_1 + \cdots + g_{2m}] g_{2m+1} = 0$ and the assumption about the case of $K = 2m$. Up to now, it is clear that the lemma is correct for the case of $K = 2m+1$.

For $K = 2m+2$, let

$$\begin{aligned} f(x_1, x_2, \cdots, x_n) = & g_1(x_1, x_2, \cdots, x_n) + g_2(x_1, x_2, \cdots, x_n) + \cdots \\ & + g_{2m+1}(x_1, x_2, \cdots, x_n) + g_{2m+2}(x_1, x_2, \cdots, x_n), \end{aligned}$$

where $g_i(x_1, x_2, \cdots, x_n)$, $1 \leqslant i \leqslant 2m+2$, are different single-terms of degree $n-1$. Therefore,

$$\begin{aligned} w(f) = & w([g_1 + \cdots + g_{2m+1}] + g_{2m+2}) \\ = & w(g_1 + \cdots + g_{2m+1}) + w(g_{2m+2}) - 2w([g_1 + \cdots + g_{2m+1}] g_{2m+2}) \\ = & 2(m+1) + 2 - 2 \\ = & 2(m+1) \\ = & 2\lfloor [(2m+2)+1]/2 \rfloor, \end{aligned}$$

where the second equation is owed to the property $[g_1 + \cdots + g_{2m+1}]g_{2m+2} = x_1 x_2 \cdots x_n$ and the assumption on the case of $K = 2m+1$. It is clear that the lemma is also correct for the case of $K = 2m+2$. □

One can easily verify that $f(x_1, x_2, x_3) = x_1x_2 + x_1x_3$ is an H-Boolean function of 3-variables consisting of 2 terms of degree 2, and that

$$g(x_1, x_2, x_3, x_4) = x_1x_2x_3 + x_1x_2x_4 + x_1x_3x_4 + x_2x_3x_4$$

is an H-Boolean function of 4-variables consisting of 3 terms of degree 3. Motivated by these two examples, it is natural to ask that, except for these two examples, are there any other H-Boolean functions of n-variable ($n > 4$) consisting of terms of degree just $n-1$? The following theorem gives us a negative answer to this problem.

Theorem 5.1.4 *No* H-*Boolean functions of* n-*variable consist of terms of degree* $n-1$ *if* $n > 4$.

Proof On the contrary, if $f(x_1, x_2, \cdots, x_n)$ is an H-Boolean function consisting of k terms of degree $(n-1)$, then the following Boolean function

$$g(x_1, x_2, \cdots, x_{n-1}) = f(x_1, x_2, \cdots, x_{n-1}, 0) + f(x_1, x_2, \cdots, x_{n-1}, 1)$$

of $(n-1)$-variable consists of no more than k terms of degree $(n-2)$. Hence, by Lemma 5.1.3 the Hamming weight of $g(x_1, x_2, \cdots, x_{n-1})$ is bounded above by

$$w(g) = 2\lfloor (k+1)/2 \rfloor \leqslant 2\lfloor (n+1)/2 \rfloor < 2^{n-2}, \quad \text{if } n > 4.$$

Therefore, by Definition 5.1.1, $f(x_1, x_2, \cdots, x_n)$ is not an H-Boolean function. □

In order to introduce more non-existence results, we prove the following popular lemma.

Lemma 5.1.4 *If* $f(x_1, x_2, \cdots, x_n)$ *and* $g(x_1, x_2, \cdots, x_n)$ *are two different Boolean functions, then*

$$\begin{aligned} w(f(x_1, x_2, \cdots, x_n) + g(x_1, x_2, \cdots, x_n) \leqslant & w(f(x_1, x_2, \cdots, x_n)) \\ & + w(g(x_1, x_2, \cdots, x_n)). \end{aligned}$$

The equation is satisfied if and only if

$$f(x_1, x_2, \cdots, x_n) g(x_1, x_2, \cdots, x_n) = 0.$$

Theorem 5.1.5 *For* $n \geqslant 9$ *no* H-*Boolean functions of* n-*variable consist of terms of degrees* $(n-1)$ *or* $(n-2)$.

Proof Let $f(x_1, x_2, \cdots, x_n)$ be a Boolean function consisting of terms of degrees $(n-1)$ or $(n-2)$. Consider the following Boolean function:

$$g(x_1, x_2, \cdots, x_{n-1}) = f(x_1, x_2, \cdots, x_{n-1}, 0) + f(x_1, x_2, \cdots, x_{n-1}, 1),$$

which consists of terms of degrees $(n-2)$ or $(n-3)$.

Because of that, among the set of $(n-1)$-variable's single-terms, there are at most $(n-1)$ terms of degree $(n-2)$ and $(n-1)(n-2)/2$ terms of degree $(n-3)$. On the other hand, the Hamming weights of each $(n-1)$-variable's single-term of degree $(n-1)$ and $(n-2)$ are 2 and 4, respectively. Therefore, in Lemma 5.1.4 the Hamming weight of $g(x_1, x_2, \cdots, x_{n-1})$ is upper-bounded by

$$4 \times (n-1)(n-2)/2 + 2 \times (n-1) < 2^{n-2}, \quad \text{for} \quad n \geqslant 9.$$

Thus by Definition 5.1.1, the function $f(x_1, x_2, \cdots, x_n)$ is not an H-Boolean function. □

By the same approach as that used in Theorem 5.1.5, we have the following theorems (their proofs are all omitted to save space):

Theorem 5.1.6 *For $n \geqslant 13$, no H-Boolean functions of n-variable consist of terms of degrees just $(n-1)$, $(n-2)$, or $(n-3)$.*

Theorem 5.1.7 *For $n \geqslant 18$, no H-Boolean functions of n-variable consist of terms of degrees just $(n-1)$, $(n-2)$, $(n-3)$ or $(n-4)$.*

Definition 5.1.4 *The minimum degree of a Boolean function is defined by the minimum value of the degrees of the single-terms contained in this Boolean function. For example, the minimum degree of*

$$f(x_1, x_2, \cdots, x_n) = x_1 + x_1 x_2$$

is 1, while the minimum degree of

$$f(x_1, x_2, \cdots, x_n) = x_1 x_3 + x_2 x_4 + x_5 x_6 x_7$$

is 2.

The previous Theorems 5.1.4–5.1.7 can be generalized as:

Theorem 5.1.8 *For every given integer k, there exists an integer N such that no H-Boolean functions of n-variable are of minimum degree $(n-k)$ for $n \geqslant N$.*

Proof From Lemma 5.1.4, we know that the Hamming weights of n-variable's Boolean functions of minimum degree $(n-k)$ are upper bounded by $\sum_{i=1}^{k} 2^i n!/[i!(n-i)!]$, whilst, on the other hand,

$$\sum_{i=1}^{k} 2^i n!/[i!(n-i)!] \leqslant 2^{n-1}$$

for sufficiently large n. The theorem follows from Definition 5.1.1. □

Definition 5.1.5 *A Boolean function $f(x_1, x_2, \cdots, x_n)$ is said to be independent of some variable, say x_i, if and only if the equation*

$$f(x_1, \cdots, x_{i-1}, 0, x_{i+1}, \cdots, x_n) = f(x_1, \cdots, x_{i-1}, 1, x_{i+1}, \cdots, x_n)$$

is satisfied by all $x_1, \cdots, x_{i-1}, x_{i+1}, \cdots, x_n = 0$ or 1, otherwise the function is said to be dependent on x_i.

A Boolean function $f(x_1, x_2, \cdots, x_n)$ is said to be linear in some variable, say x_i, if and only if the function $f(x_1, x_2, \cdots, x_n) + x_i$ is independent of x_i.

Theorem 5.1.9 *The following two results are true:*

(1) *H-Boolean functions of n-variable are dependent on every variable.*

(2) *No* H-*Boolean functions are linear in any variable.*

Proof If the Boolean function $f(x_1, x_2, \cdots, x_n)$ is independent of x_i, then

$$f(x_1, x_2, \cdots, x_{i-1}, 0, x_{i+1}, \cdots, x_n) + f(x_1, x_2, \cdots, x_{i-1}, 1, x_{i+1}, \cdots, x_n) = 0.$$

i.e., the function is not H-Boolean.

The second statement can be proved in the same way. □

The other equivalent definition of the H-Boolean function is stated by

Theorem 5.1.10 *The necessary and sufficient condition for $f(x_1, x_2, \cdots, x_n)$ being an* H-*Boolean function of n-variable is that*

$$\begin{aligned}&2^{n-2} + 2w(f(x_1, x_2, \cdots, x_{i-1}, 0, x_{i+1}, \cdots, x_n) f(x_1, x_2, \cdots, x_{i-1}, 1, x_{i+1}, \cdots, x_n))\\ &= w(f(x_1, x_2, \cdots, x_n)),\end{aligned}$$

for each $1 \leqslant i \leqslant n$.

Proof The proof is finished by the following equivalent equations:

$$\begin{aligned}&w(f(x_1, x_2, \cdots, x_{i-1}, 0, x_{i+1}, \cdots, x_n)\\ &+f(x_1, x_2, \cdots, x_{i-1}, 1, x_{i+1}, \cdots, x_n)) = 2^{n-2}\end{aligned}$$

is equivalent to

$$\begin{aligned}2^{n-2} =& w(f(x_1, x_2, \cdots, x_{i-1}, 0, x_{i+1}, \cdots, x_n))\\ &+w(f(x_1, x_2, \cdots, x_{i-1}, 1, x_{i+1}, \cdots, x_n))\\ &-2w(f(x_1, x_2, \cdots, x_{i-1}, 0, x_{i+1}, \cdots, x_n)\\ &\times f(x_1, x_2, \cdots, x_{i-1}, 1, x_{i+1}, \cdots, x_n))\\ =& w(f(x_1, x_2, \cdots, x_n))\\ &-2w(f(x_1, x_2, \cdots, x_{i-1}, 0, x_{i+1}, \cdots, x_n)\\ &\times f(x_1, x_2, \cdots, x_{i-1}, 1, x_{i+1}, \cdots, x_n)).\end{aligned}$$

The last equation is due to the identity

$$\begin{aligned}&w(f(x_1, x_2, \cdots, x_{i-1}, 0, x_{i+1}, \cdots, x_n))\\ &+w(f(x_1, x_2, \cdots, x_{i-1}, 1, x_{i+1}, \cdots, x_n)) = w(f(x_1, x_2, \cdots, x_n)).\end{aligned}$$ □

Theorem 5.1.11 *Let*

$$f(x_1, x_2, \cdots, x_n) = x_{i_1} x_{i_2} \cdots x_{i_r} + x_{j_1} x_{j_2} \cdots x_{j_s}$$

be a Boolean function consisting of two terms. Then the following statements are true:

- *$f(\cdot)$ is not* H-*Boolean if $n = 1$ or $n = 2$; item For $n = 3$, $f(\cdot)$ is* H-*Boolean if and only if $f(x_1, x_2, x_3)$ is equivalent to $x_1x_2 + x_1x_3$;*
- *For $n = 4$, $f(\cdot)$ is* H-*Boolean if and only if $f(x_1, x_2, x_3, x_4)$ is equivalent to $x_1x_2 + x_3$ x_4;*
- *For $n \geqslant 5$, $f(\cdot)$ cannot be* H-*Boolean.*

Proof The statements (1), (2) and (3) are from Theorems 5.1.2, 5.1.9, and 5.1.16. The proof for the fourth statement is divided into the following three steps:

Step 1. First, we prove that a necessary condition of $f(\cdot)$ being H-Boolean is $r \geqslant 3$.

In fact, if $r = 1$, from Theorems 5.1.2 and 5.1.9 it is reasonable to assume that

$$f(x_1, x_2, \cdots, x_n) = x_1 + x_1^a x_2 x_3 \cdots x_n, \tag{5.1}$$

where $a = 0$ or 1 and $x_1^1 =: x_1$ and $x_1^0 =: 1$.

From Theorem 5.1.9, the function $f(\cdot)$ in Equation (5.1) is not H-Boolean for $a = 0$. Moreover, from Theorem 5.1.16, this $f(\cdot)$ is not H-Boolean for $a = 1$ and $n \geqslant 5$ too. Thus $r > 1$.

If $r = 2$,

$$f(x_1, x_2, \cdots, x_n) = x_1 x_2 + x_1^a x_2^b x_3 \cdots x_n. \tag{5.2}$$

In the same way as the case of $r = 1$, it can be proved that the $f(\cdot)$ in Equation (5.2) is not H-Boolean for $r = 2$.

Therefore we have $r \geqslant 3$.

Step 2. Then, in the same way as in Step 1, it can be proved that $s \geqslant 3$.

Step 3. Finally, we prove that $f(\cdot)$ is not an H-Boolean function even if $r \geqslant 3$ and $s \geqslant 3$.

In fact, from Theorem 5.1.9, we obtain the following necessary condition for $f(\cdot)$ being H-Boolean

$$x_{i_1} x_{i_2} \cdots x_{i_r} x_{j_1} x_{j_2} \cdots x_{j_s} = x_1 x_2 \cdots x_n.$$

Thus the Hamming weight of $f(\cdot)$ is

$$w(f(\cdot)) = 2^{n-r} + 2^{n-s} - 2.$$

From Theorem 5.1.10, another necessary condition for $f(\cdot)$ being H-Boolean is that

$$2w(f(0, x_2, \cdots, x_n) f(1, x_2, \cdots, x_n)) = 2^{n-r} + 2^{n-s} - 2 - 2^{n-2}.$$

Because of the conditions $r \geqslant 3$ and $s \geqslant 3$, we have

$$2^{n-r} + 2^{n-s} - 2 - 2^{n-2} < 0$$

or equivalently

$$w(f(0, x_2, \cdots, x_n) f(1, x_2, \cdots, x_n)) < 0,$$

which is clearly impossible.

In a word, for $n \geqslant 5$, the function $f(\cdot)$ is not H-Boolean. □

An n-dimensional Hadamard matrix of order m is called absolutely improper if and only if none of its subsections is an Hadamard matrix of lower dimension. In the following chapter many absolutely improper Hadamard matrices will be constructed for different n and m. The following theorem proves that no 2^n-Hadamard matrix is absolutely improper when $n \geqslant 3$.

Theorem 5.1.12 *There exists no absolutely improper n-dimensional Hadamard matrix of order 2, if $n \geqslant 3$.*

Proof Every binary n-dimensional matrix of order 2 can be denoted by $A = [A(x_1, x_2, \cdots, x_n)]$, where

$$A(x_1, x_2, \cdots, x_n) = (-1)^{f(x_1, x_2, \cdots, x_n)}, \quad 0 \leqslant x_1, x_2, \cdots, x_n \leqslant 1.$$

Case 1. If the function $f(x_1, x_2, \cdots, x_n)$ is linear, then, from Theorem 5.1.9, the matrix A is not Hadamard.

Case 2. If the function $f(\cdot)$ is not linear, without loss of generality,

$$x_1 x_2 x_{i_1} \cdots x_{i_k},$$

say, is a term of this function, then

$$f(x_1, x_2, d_3, d_4, \cdots, d_n) = x_1 x_2 + a x_1 + b x_2 + c, \tag{5.3}$$

where $a, b, c = 0$ or 1, and $d_j = 1$ iff $j \in \{i_1, i_2, \cdots, i_k\}$, otherwise $d_j = 0$.

The function in Equation (5.3) is clearly an H-Boolean function of 2-variables, i.e., at least one 2-dimensional subsection of the matrix A is Hadamard, which is equivalent to saying that A is not an absolutely improper n-dimensional Hadamard matrix of order 2.

The proof is completed by the above two cases. □

Opposite to the concept of absolutely improper, an n-dimensional Hadamard matrix of order m is called absolutely proper if and only if all of its subsections are Hadamard matrices of lower-dimensional ones. The following theorem lists all of the possible n-dimensional absolutely proper Hadamard matrices of order 2.

Theorem 5.1.13 *Let $A = [A(x_1, x_2, \cdots, x_n)]$, $0 \leqslant x_i \leqslant 1$, $A(x_1, x_2, \cdots, x_n) = 1$ or -1. Then the matrix A is an n-dimensional absolutely proper Hadamard matrix of order 2 if and only if there is a Boolean function $f(x_1, x_2, \cdots, x_n)$ of the following form*

$$f(x_1, x_2, \cdots, x_n) = \sum_{1 \leqslant i < j \leqslant n} x_i x_j + a_0 + \sum_{i=1}^{n} a_i x_i$$

satisfying

$$A(x_1, x_2, \cdots, x_n) = (-1)^{f(x_1, x_2, \cdots, x_n)},$$

where $a_i = 0$ or 1.

Proof $\Longleftarrow$ The matrix defined by

$$A(x_1, x_2, \cdots, x_n) = (-1)^{f(x_1, x_2, \cdots, x_n)}$$

with

$$f(x_1, x_2, \cdots, x_n) = \sum_{1 \leqslant i < j \leqslant n} x_i x_j + a_0 + \sum_{i=1}^{n} a_i x_i$$

is clearly an n-dimensional absolutely proper Hadamard matrix of order 2. In fact, each of its lower subsections is of the form $[(-1)^{h(x_1, x_2, \cdots, x_m)}]$, $1 \leqslant m \leqslant n$, with

$$h(x_1, x_2, \cdots, x_m) = \sum_{1 \leqslant i < j \leqslant m} x_i x_j + b_0 + \sum_{i=1}^{m} b_i x_i,$$

which is clearly an H-Boolean function.

$\Longrightarrow$ Assume that the Boolean function $g(x_1, x_2, \cdots, x_n)$ satisfies

$$A(x_1, x_2, \cdots, x_n) = (-1)^{g(x_1, x_2, \cdots, x_n)}.$$

Now we try to prove that if $[A]$ is an n-dimensional absolutely proper Hadamard matrix of order 2, then the function $g(\cdot)$ must be in the form of

$$g(x_1, x_2, \cdots, x_n) = \sum_{1 \leqslant i < j \leqslant n} x_i x_j + a_0 + \sum_{i=1}^{n} a_i x_i.$$

For any prefixed $1 \leqslant i < j \leqslant n$ and x_k, $(k \neq i,\ k \neq j)$, the function $g(\cdot)$ can be divided into

$$\begin{aligned} g(x_1, x_2, \cdots, x_n) = & x_i E(x_1, \cdots, x_{i-1}, x_{i+1}, \cdots, x_{j-1}, x_{j+1}, \cdots, x_n) \\ & + x_j B(x_1, \cdots, x_{i-1}, x_{i+1}, \cdots, x_{j-1}, x_{j+1}, \cdots, x_n) \\ & + x_i x_j C(x_1, \cdots, x_{i-1}, x_{i+1}, \cdots, x_{j-1}, x_{j+1}, \cdots, x_n) \\ & + D(x_1, \cdots, x_{i-1}, x_{i+1}, \cdots, x_{j-1}, x_{j+1}, \cdots, x_n), \end{aligned}$$

where $E(\cdot)$, $B(\cdot)$, $C(\cdot)$ and $D(\cdot)$ are Boolean functions of $(n-2)$-variable independent of x_i and x_j.

Let $r(x_i, x_j) = g(x_1, \cdots, x_i, \cdots, x_j, \cdots, x_n)$, which is a Boolean function of 2-variable x_i and x_j.

Because the Hadamard matrix A is absolutely proper, the matrix

$$[F(x_i, x_j)] = [(-1)^{r(x_i, x_j)}]$$

should be a 2-dimensional Hadamard matrix of order 2. Thus

$$\sum_{x_j=0}^{1} (-1)^{r(0, x_j) + r(1, x_j)} = 0,$$

i.e., $r(0,0) + r(0,1) + r(1,0) + r(1,1) \equiv 1 \bmod 2$. Hence

$$C(x_1, \cdots, x_{i-1}, x_{i+1}, \cdots, x_{j-1}, x_{j+1}, \cdots, x_n) = 1.$$

Therefore $g(\cdot)$ is in the form

$$\begin{aligned} g(x_1, x_2, \cdots, x_n) = & x_i E(x_1, \cdots, x_{i-1}, x_{i+1}, \cdots, x_{j-1}, x_{j+1}, \cdots, x_n) \\ & + x_j B(x_1, \cdots, x_{i-1}, x_{i+1}, \cdots, x_{j-1}, x_{j+1}, \cdots, x_n) \\ & + x_i x_j + D(x_1, \cdots, x_{i-1}, x_{i+1}, \cdots, x_{j-1}, x_{j+1}, \cdots, x_n). \end{aligned}$$

Because of that the above equation is true for any $1 \leqslant i < j \leqslant n$, the function $g(\cdot)$ must be in the form of

$$g(x_1, x_2, \cdots, x_n) = \sum_{1 \leqslant i < j \leqslant n} x_i x_j + a_0 + \sum_{i=1}^{n} a_i x_i. \qquad \square$$

5.1.3 Constructions of H-Boolean Functions

We have known that the construction of H-Boolean functions is in fact the construction of 2^n Hadamard matrices. This subsection shows some powerful such constructions.

First, it is easy to verify that the following result is true:

Theorem 5.1.14 *If $f(x_1, x_2, \cdots, x_n)$ is an* H*-Boolean function, then so is*

$$f(x_1, x_2, \cdots, x_n) + a_0 + \sum_{i=1}^{n} a_i x_i.$$

Because of this theorem and Theorem 5.1.4, we have:

Corollary 5.1.1 *No* H*-Boolean functions are of the form*

$$f(x_1, x_2, \cdots, x_n) = a + \sum_{i=1}^{n} a_i x_i + \sum_{k=1}^{n} b_k x_1 x_2 \cdots x_{k-1} x_{k+1} \cdots x_n$$

if $n > 4$.

The following construction makes it possible to produce many H-Boolean functions of large m-variable from those of small m-variable ones, which can be found by computer search.

Theorem 5.1.15 *If $f(x_1, x_2, \cdots, x_n)$ and $g(y_1, y_2, \cdots, y_m)$ are* H*-Boolean functions of variables n and m, respectively, then*

$$r(x_1, x_2, \cdots, x_n, y_1, y_2, \cdots, y_m) = f(x_1, x_2, \cdots, x_n) + g(y_1, y_2, \cdots, y_m)$$

is an H*-Boolean function of $(m + n)$ variables.*

Proof It can be proved directly by using the definition of H-Boolean functions.

□

Theorem 5.1.16 *Let $f(x_1, x_2, \cdots, x_n, y_1, y_2, \cdots, y_m) = h(x_1, x_2, \cdots, x_n) \cdot r(y_1, y_2, \cdots, y_m)$, where $h(\cdot)$ and $r(\cdot)$ are Boolean functions of n- and m-variable, respectively. Then $f(\cdot)$ is* H*-Boolean function of $(m + n)$-variable if and only if $h(x_1, x_2, \cdots, x_n) = a + x_1 + x_2 + \cdots + x_n$ and $r(y_1, y_2, \cdots, y_m) = b + y_1 + y_2 + \cdots + y_m$.*

Proof $\Longleftarrow$ It is easy to verify that the function

$$[a + x_1 + x_2 + \cdots + x_n][b + y_1 + y_2 + \cdots + y_m]$$

is indeed H-Boolean.

$\Longrightarrow$ At first we prove that both $h(\cdot)$ and $r(\cdot)$ should be linear functions, if $f(\cdot)$ is H-Boolean.

In fact, one necessary condition for $f(\cdot)$ being H-Boolean is that

$$\begin{aligned}2^{m+n-2} &= w(f(0, x_2, \cdots, x_n, y_1, y_2, \cdots, y_m) \\ &\quad + f(1, x_2, \cdots, x_n, y_1, y_2, \cdots, y_m)) \\ &= w(r(\cdot))w(h(0, x_2, \cdots, x_n) + h(1, x_2, \cdots, x_n)).\end{aligned}$$

Therefore there should exist integers, say p and q, such that

$$w(r(y_1, y_2, \cdots, y_m)) = 2^p$$

and

$$w(h(0, x_2, \cdots, x_n) + h(1, x_2, \cdots, x_n)) = 2^q.$$

Because no H-Boolean functions are linear, we have $p < m$. And because that H-Boolean functions are dependent on each of their variables, we have $q < n$. Hence the identity $m+n-2 = p+q$ implies $q = n-1$, in other words, $h(0, x_2, \cdots, x_n) + h(1, x_2, \cdots, x_n) \equiv 1$, i.e., the function $h(\cdot)$ is linear in variable x_1.

In the same way it can be proved that $h(\cdot)$ (and $r(\cdot)$) should be linear in every variable.

Then because H-Boolean functions should be dependent on every variable, $h(\cdot)$ and $r(\cdot)$ are of the forms:

$$h(x_1, x_2, \cdots, x_n) = a + x_1 + x_2 + \cdots + x_n$$

and

$$r(y_1, y_2, \cdots, y_m) = b + y_1 + y_2 + \cdots + y_m. \qquad \square$$

This theorem provides us numerous H-Boolean functions consisting of terms of degree 2.

Theorem 5.1.17[1−3] *A Boolean function $f(x_1, x_2, \cdots, x_n)$ is H-Boolean if the following three conditions are satisfied:*

C1. $f(x_1, x_2, \cdots, x_{n-2}, 1, 1) = 1$ *and* $f(x_1, x_2, \cdots, x_{n-2}, 0, 0) = 0$;

C2. *The Hamming weights of*

$$h_1(x_1, x_2, \cdots, x_{n-2}) := f(x_1, x_2, \cdots, x_{n-2}, 0, 1)$$

and

$$h_2(x_1, x_2, \cdots, x_{n-2}) := f(x_1, x_2, \cdots, x_{n-2}, 1, 0)$$

are 2^{n-3}.

C3. *For every* $(a_1, a_2, \cdots, a_{n-2})$, *if* $h_i(a_1, a_2, \cdots, a_{n-2}) = 1$ *(or resp.* 0*),* $i = 1$ *or* 2*, then*

$$\begin{aligned} h_i(1-a_1, a_2, \cdots, a_{n-2}) &= h_i(a_1, 1-a_2, \cdots, a_{n-2}) \\ &\cdots\cdots\cdots \\ &= h_i(a_1, a_2, \cdots, 1-a_{n-2}) \\ &= 0 \quad (\textit{or, resp.}, 1). \end{aligned}$$

Proof The proof is finished by the following steps:

Step 1. At first we prove that the Hamming weight of

$$A(x_2, x_3, \cdots, x_n) =: f(0, x_2, x_3, \cdots, x_n) + f(1, x_2, x_3, \cdots, x_n)$$

is 2^{n-2}. In fact, the known conditions imply that

$$\begin{aligned} A(x_2, \cdots, x_{n-2}, 1, 1) &= f(0, x_2, \cdots, x_{n-2}, 1, 1) + f(1, x_2, \cdots, x_{n-2}, 1, 1) \\ &= 0 + 0 = 0, \\ A(x_2, \cdots, x_{n-2}, 0, 0) &= f(0, x_2, \cdots, x_{n-2}, 0, 0) + f(1, x_2, \cdots, x_{n-2}, 0, 0) \\ &= 1 + 1 = 0, \\ A(x_2, \cdots, x_{n-2}, 0, 1) &= f(0, x_2, \cdots, x_{n-2}, 0, 1) + f(1, x_2, \cdots, x_{n-2}, 0, 1) \\ &= h_1(0, x_2, \cdots, x_{n-2}) + h_1(1, x_2, \cdots, x_{n-2}) \\ &= 0 + 1 \text{ (or 1+0)} \end{aligned}$$

$$
\begin{aligned}
&= 1, \\
A(x_2, \cdots, x_{n-2}, 1, 0) &= f(0, x_2, \cdots, x_{n-2}, 1, 0) + f(1, x_2, \cdots, x_{n-2}, 1, 0) \\
&= h_2(0, x_2, \cdots, x_{n-2}) + h_2(1, x_2, \cdots, x_{n-2}) \\
&= 0 + 1 \text{ (or } 1{+}0) \\
&= 1.
\end{aligned}
$$

The above four equations imply that the Hamming weight of

$$A(x_2, x_3, \cdots, x_n)$$

is indeed 2^{n-2}. In fact, $A(x_2, x_3, \cdots, x_n) = 1$ is satisfied by the points of the form $(x_2, x_3, \cdots, x_{n-2}, 0, 1)$ or $(x_2, x_3, \cdots, x_{n-2}, 1, 0)$.

Because the variables $x_1, x_2, \cdots, x_{n-2}$ are replaceable by each other, the function $A(x_2, x_3, \cdots, x_n)$ has the same Hamming weight as those of the following $n-2$ Boolean functions of the $(n-1)$-variable:

$$
\begin{gathered}
f(x_1, 0, x_3, \cdots, x_n) + f(x_1, 1, x_3, \cdots, x_n), \\
f(x_1, x_2, 0, x_4, \cdots, x_n) + f(x_1, x_2, 1, x_4, \cdots, x_n), \\
\cdots\cdots\cdots \\
f(x_1, x_2, \cdots, x_{n-3}, 0, x_{n-1}, x_n) + f(x_1, x_2, \cdots, x_{n-3}, 1, x_{n-1}, x_n).
\end{gathered}
$$

Step 2. Then, we try to prove that the Hamming weight of

$$B(x_1, \cdots, x_{n-2}, x_n) =: f(x_1, \cdots, x_{n-2}, 0, x_n) + f(x_1, \cdots, x_{n-2}, 1, x_n)$$

is also 2^{n-2}.

In fact, for the Hamming weight of $h_2(x_1, x_2, \cdots, x_{n-2})$ is 2^{n-3}, the function

$$
\begin{aligned}
B(x_1, \cdots, x_{n-2}, 0) &= f(x_1, \cdots, x_{n-2}, 0, 0) + f(x_1, \cdots, x_{n-2}, 1, 0) \\
&= 1 + h_2(x_1, x_2, \cdots, x_{n-2})
\end{aligned}
$$

is equal to 1 at 2^{n-3} points of the form $(x_1, \cdots, x_{n-2}, 0)$.

Similarly, because the Hamming weight of $h_1(x_1, x_2, \cdots, x_{n-2})$ is 2^{n-3}, the function

$$
\begin{aligned}
B(x_1, \cdots, x_{n-2}, 1) &= f(x_1, \cdots, x_{n-2}, 0, 1) + f(x_1, \cdots, x_{n-2}, 1, 1) \\
&= h_1(x_1, x_2, \cdots, x_{n-2}) + 0 \\
&= h_1(x_1, x_2, \cdots, x_{n-2})
\end{aligned}
$$

is equal to 1 at 2^{n-3} points of the form $(x_1, \cdots, x_{n-2}, 1)$.

Therefore, the Hamming weight of $B(x_1, \cdots, x_{n-2}, x_n)$ is proved to be $2^{n-3} + 2^{n-3} = 2^{n-2}$.

Step 3. In the same way as in Step 2, it can be proved that the Hamming weight of

$$C(x_1, \cdots, x_{n-1}) =: f(x_1, \cdots, x_{n-1}, 0) + f(x_1, \cdots, x_{n-1}, 1)$$

is also 2^{n-2}.

By the definition of H-Boolean functions and the above three steps, it is clear that $f(x_1, x_2, \cdots, x_n)$ is an H-Boolean function of n-variables. □

Remark 5.1.2 *The conditions in Theorem* 5.1.17 *can be stated in many equivalent forms. For example, they can be replaced by the following three new conditions*:

C′1. *$f(1,0,x_3,x_4,\cdots,x_n)=1$ and $f(1,1,x_3,x_4,\cdots,x_n)=0$;*

C′2. *The Hamming weights of*

$$R_1(x_3,x_4,\cdots,x_n)=:f(0,0,x_3,x_4,\cdots,x_n)$$

and

$$R_2(x_3,x_4,\cdots,x_n)=:f(0,1,x_3,x_4,\cdots,x_n)$$

are 2^{n-3};

C′3. *For every $(a_3,a_4,\cdots,a_n)$, if $R_i(a_3,a_4,\cdots,a_n)=1$ (or, resp., 0), $i=1$ or 2, then*

$$\begin{aligned}R_i(1-a_3,a_4,\cdots,a_n)&=R_i(a_3,1-a_4,\cdots,a_n)\\&\cdots\cdots\cdots\\&=R_i(a_3,a_4,\cdots,1-a_n)\\&=0\quad(\text{or, resp., }1).\end{aligned}$$

Theorem 5.1.18[1−3] *A Boolean function $f(x_1,x_2,\cdots,x_n)$ is* H*-Boolean if the following conditions are satisfied:*

CC1. *The Hamming weight of $f(x_1,x_2,\cdots,x_n)$ itself is 2^{n-2};*

CC2. *For every $(a_1,a_2,\cdots,a_n)$ if $f(a_1,a_2,\cdots,a_n)=1$, then*

$$\begin{aligned}f(1-a_1,a_2,\cdots,a_n)&=f(a_1,1-a_2,\cdots,a_n)\\&\cdots\cdots\cdots\\&=f(a_1,a_2,\cdots,1-a_n).\end{aligned}$$

Proof It is sufficient to prove that the Hamming weight of

$$g(x_1,x_2,\cdots,x_{n-1})=:f(x_1,x_2,\cdots,x_{n-1},0)+f(x_1,x_2,\cdots,x_{n-1},1)$$

is 2^{n-2}.

In fact, group the binary vectors of length n into 2^n groups, say X_1, X_2, $\cdots$, $X_{2^{n-1}}$, such that each X_i consists of just two vectors and $(a_1,a_2,\cdots,a_{n-1},a_n)$ and $(b_1,b_2,\cdots,b_{n-1},b_n)$ belong to the same group if and only if $(a_1,a_2,\cdots,a_{n-1})=(b_1,b_2,\cdots,b_{n-1})$.

From the condition CC2, at most one of the two points in each X_i satisfies $f(\cdot)=1$. While, on the other hand, the condition CC1 ensures that just one of the two points in each X_i satisfies $f(\cdot)=1$.

If $f(a_1,a_2,\cdots,a_n)=1$, then, by the condition CC2, $f(a_1,a_2,\cdots,1-a_n)=0$. Thus

$$\begin{aligned}g(a_1,a_2,\cdots,a_{n-1})&=f(a_1,a_2,\cdots,1-a_n)+f(a_1,a_2,\cdots,a_n)\\&=1+0\\&=1.\end{aligned}$$

In other words, the Hamming weight of $g(x_1,x_2,\cdots,x_{n-1})$ is larger than or equal to that of $f(x_1,x_2,\cdots,x_n)$.

If $g(a_1,a_2,\cdots,a_{n-1})=1$, i.e.,

$$f(a_1,\cdots,a_{n-1},1)+f(a_1,\cdots,a_{n-1},0)=1,$$

then one and only one of $f(a_1,\cdots,a_{n-1},1)$ and $f(a_1,\cdots,a_{n-1},0)$ is "1". Therefore, the Hamming weight of $g(x_1,\cdots,x_{n-1})$ is less than or equal to that of $f(x_1,\cdots,x_n)$.

It has been proved that the Hamming weight of $g(x_1,x_2,\cdots,x_{n-1})$ is equal to 2^{n-2}, the Hamming weight of $f(x_1,x_2,\cdots,x_n)$. □

5.2 Enumeration of 2^n Hadamard Matrices

The enumeration for n-dimensional Hadamard matrices of order 2 is a very difficult problem, which is still open up to now. This section will show only some enumerations about small n and special H-Boolean functions.

5.2.1 Classification of 2^4 Hadamard Matrices

It is easy to see that the numbers of 2^2 and 2^3 Hadamard matrices are 8 and 64, respectively.

According to the definition of H-Boolean function, a 4-variable's Boolean function $B(i, j, k, l)$ is H-Boolean if and only if the following four conditions are simultaneously satisfied:

C1. $w(B(0, j, k, l) + B(1, j, k, l)) = 4$;

C2. $w(B(i, 0, k, l) + B(i, 1, k, l)) = 4$;

C3. $w(B(i, j, 0, l) + B(i, j, 1, l)) = 4$;

C4. $w(B(i, j, k, 0) + B(i, j, k, 1)) = 4$,

where $w(g(\cdot))$ refers to the Hamming weight of $g(\cdot)$.

In order to enumerate the 2^4 Hadamard matrices we divide the Boolean functions of the form

$$f(i, j, k, l) = b_1ij + b_2ik + b_3il + b_4kl + b_5jk + b_6jl, \quad b_h = 0 \text{ or } 1,$$

into ten equivalence classes with their representative functions being:

(1) $f_1(i, j, k, l) = ij$;
(2) $f_2(i, j, k, l) = ij + ik$;
(3) $f_3(i, j, k, l) = ij + kl$;
(4) $f_4(i, j, k, l) = ij + ik + il$;
(5) $f_5(i, j, k, l) = ij + ik + kj$;
(6) $f_6(i, j, k, l) = ij + ik + jl$;
(7) $f_7(i, j, k, l) = ij + ik + il + jk$;
(8) $f_8(i, j, k, l) = ij + ik + jl + lk + ik$;
(9) $f_9(i, j, k, l) = ij + ik + il + jk + jl$;
(10) $f_{10}(i, j, k, l) = ij + ik + jl + jk + il + kl$;

respectively, where two functions $f(i, j, k, l)$ and $g(i, j, k, l)$ belong to the same equivalence class if and only if there exists a permutation of 4 elements, say $\tau(\cdot)$, such that $f(i, j, k, l) = g(\tau(i, j, k, l))$.

The general form of 4-variable's Boolean functions without linear terms is

$$\begin{aligned} B(i, j, k, l) = &b_1ij + b_2ik + b_3il + b_4jk + b_5jl + b_6kl \\ &+ c_1ijk + c_2jkl + c_3ikl + c_4jkl + d_1ijkl. \end{aligned}$$

If the number of H-Boolean functions of this form is N, then there are 2^5N H-Boolean functions of 4-variables, because there are 2^5 4-variable's linear terms in all and $B(i, j, k, l)$ is H-Boolean iff so is $B(i, j, k, l)+a_0 + a_1i + a_2j + a_3k + a_4l$.

Based on the above representative functions of equivalence classes, we find that every 4-variable's H-Boolean function without linear terms is equivalent to one of the following 11 types:

Type 1. $B_1(i, j, k, l) = c_1ijk + c_2jkl + c_3ikl + c_4jkl + d_1ijkl$;

Type 2. $B_2(i, j, k, l) = f_1(i, j, k, l) + B_1(i, j, k, l)$;

Type 3. $B_3(i,j,k,l) = f_2(i,j,k,l) + B_1(i,j,k,l)$;

Type 4. $B_4(i,j,k,l) = f_3(i,j,k,l) + B_1(i,j,k,l)$;

Type 5. $B_5(i,j,k,l) = f_4(i,j,k,l) + B_1(i,j,k,l)$;

Type 6. $B_6(i,j,k,l) = f_5(i,j,k,l) + B_1(i,j,k,l)$;

Type 7. $B_7(i,j,k,l) = f_6(i,j,k,l) + B_1(i,j,k,l)$;

Type 8. $B_8(i,j,k,l) = f_7(i,j,k,l) + B_1(i,j,k,l)$;

Type 9. $B_9(i,j,k,l) = f_8(i,j,k,l) + B_1(i,j,k,l)$;

Type 10. $B_{10}(i,j,k,l) = f_9(i,j,k,l) + B_1(i,j,k,l)$;

Type 11. $B_{11}(i,j,k,l) = f_{11}(i,j,k,l) + B_1(i,j,k,l)$.

Now we try to list those non-equivalent H-Boolean functions of different types.

About Type 1. A Boolean function of Type 1 is H-Boolean iff

$$\begin{cases} w(c_3kl + c_1jk + c_2jl + d_1jkl) = 4, & \text{(A)} \\ w(c_4kl + c_1ik + c_2il + d_1ikl) = 4, & \text{(B)} \\ w(c_1ij + c_3il + c_4jl + d_1ijl) = 4, & \text{(C)} \\ w(c_2ij + c_3ik + c_4jk + d_1ijk) = 4. & \text{(D)} \end{cases}$$

The left side of the equation (A) is

$$2c_3 + w(c_1jk + c_2jl + d_1jkl) - 2c_3(c_1 + c_2 + d_1)\text{mod}2.$$

The left side of the equation (B) is

$$2c_4 + w(c_1ik + c_2il + d_1ikl) - 2c_4(c_1 + c_2 + d_1)\text{mod}2.$$

(A)−(B) implies

$$(c_3 - c_4)[1 - (c_1 + c_2 + d_1)\text{mod}2] = 0 \tag{E}$$

and similarly, (C)−(D) implies

$$(c_1 - c_2)[1 - (c_3 + c_4 + d_1)\text{mod}2] = 0 \tag{F}$$

The solution of equations (A) to (F) is $c_1 = c_2 = c_3 = c_4 = 1$ and d_1=0. Hence there is only one H-Boolean function in Type 1 which is denoted by

$$A_1(i,j,k,l) = ijk + ijl + jkl + ikl.$$

To simplify the mathematical expressions, we introduce the following notations:

$$\begin{aligned} {}_1B_m(j,k,l) &= B_m(0,j,k,l) + B_m(1,j,k,l); \\ {}_2B_m(i,k,l) &= B_m(i,0,k,l) + B_m(i,1,k,l); \\ {}_3B_m(i,j,l) &= B_m(i,j,0,l) + B_m(i,j,1,l); \\ {}_4B_m(i,j,k) &= B_m(i,j,k,0) + B_m(i,j,k,1). \end{aligned}$$

About Type 2. Because the simultaneous equations

$$\begin{cases} w[{}_1B_2(j,k,l)]=4,\\ w[{}_2B_2(i,k,l)]=4,\\ w[{}_3B_2(i,j,l)]=4,\\ w[{}_4B_2(i,j,k)]=4 \end{cases}$$

have no solutions, no H-Boolean functions are of the form of Type 2.

About Type 3. Because the simultaneous equations

$$\begin{cases} w[{}_3B_3(i,j,l)]=4,\\ w[{}_1B_3(j,k,l)]=4,\\ w[{}_2B_3(i,k,l)]=4,\\ w[{}_4B_3(i,j,k)]=4 \end{cases}$$

have no solutions, H-Boolean functions are of the form of Type 3.

About Type 4. Because the simultaneous equations

$$\begin{cases} w[{}_1B_4(j,k,l)]=4,\\ w[{}_2B_4(i,k,l)]=4,\\ w[{}_3B_4(i,j,l)]=4,\\ w[{}_4B_4(i,j,k)]=4 \end{cases}$$

have only one solution $c_1 = c_2 = c_3 = c_4 = d_1 = 0$, there is only one H-Boolean function in Type 4. This H-Boolean is

$$A_2(i,j,k,l) = ij + kl.$$

About Type 5. Because the simultaneous equations

$$\begin{cases} w[{}_1B_5(j,k,l)]=4,\\ w[{}_2B_5(i,k,l)]=4,\\ w[{}_3B_5(i,j,l)]=4,\\ w[{}_4B_5(i,j,k)]=4 \end{cases}$$

have five solutions

(1) $c_1 = c_2 = c_3 = c_4 = d_1 = 0$;

(2) $c_1 = c_2 = c_3 = 0,\ c_4 = 1,\ d_1 = 0$;

(3) $c_1 = 1,\ c_2 = c_3 = 0,\ c_4 = 1,\ d_1 = 0$;

(4) $c_1 = 0,\ c_2 = 1,\ c_3 = 0,\ c_4 = 1,\ d_1 = 0$;

(5) $c_1 = c_2 = 0,\ c_3 = c_4 = 1,\ d_1 = 0$,

while the H-Boolean functions produced by the last three solutions are equivalent to each other, there are three non-equivalent H-Boolean functions in Type 5. They are

$$\begin{aligned} A_3(i,j,k,l) &= ij + ik + il,\\ A_4(i,j,k,l) &= ij + ik + il + jkl,\\ A_5(i,j,k,l) &= ij + ik + il + ijk + jkl. \end{aligned}$$

About Type 6. Since the simultaneous equations

$$\begin{cases} w[{}_1B_6(j,k,l)]=4,\\ w[{}_2B_6(i,k,l)]=4,\\ w[{}_3B_6(i,j,l)]=4,\\ w[{}_4B_6(i,j,k)]=4 \end{cases}$$

have one solution $c_1 = c_2 = c_3 = c_4 = 1$ and $d_1 = 0$, there is one H-Boolean function of Type 6 which is

$$A_6(i,j,k,l) = ij + ik + kj + ijk + ijl + ikl + jkl.$$

About Type 7. The simultaneous equations

$$\begin{cases} w[{}_1B_7(j,k,l)] = 4, \\ w[{}_2B_7(i,k,l)] = 4, \\ w[{}_3B_7(i,j,l)] = 4, \\ w[{}_4B_7(i,j,k)] = 4 \end{cases}$$

have two solutions: (1) $c_1 = c_2 = c_3 = c_4 = d_1 = 0$; (2) $c_1 = c_2 = 0$, $c_3 = c_4 = 1$, $d_1 = 0$. They correspond to the following two non-equivalent H-Boolean functions:

$$A_7(i,j,k,l) = ij + ik + jl,$$
$$A_8(i,j,k,l) = ij + ik + jl + ikl + jkl.$$

About Type 8. The simultaneous equations

$$\begin{cases} w[{}_1B_8(j,k,l)] = 4, \\ w[{}_2B_8(i,k,l)] = 4, \\ w[{}_3B_8(i,j,l)] = 4, \\ w[{}_4B_8(i,j,k)] = 4 \end{cases}$$

have three solutions:

(1) $c_1 = 1$, $c_2 = c_3 = 0$, $c_4 = 1$, and $d_1 = 0$;

(2) $c_1 = c_2 = c_3 = c_4 = d_1 = 0$;

(3) $c_1 = c_2 = c_3 = 0$, $c_4 = 1$, $d_1 = 0$,

which correspond to the following three non-equivalent H-Boolean functions:

$$A_9(i,j,k,l) = ij + ik + il + jk + jkl,$$
$$A_{10}(i,j,k,l) = ij + ik + il + jk + ijk + jkl,$$
$$A_{11}(i,j,k,l) = ij + ik + il + ik.$$

About Type 9. The simultaneous equations

$$\begin{cases} w[{}_1B_9(j,k,l)] = 4, \\ w[{}_2B_9(i,k,l)] = 4, \\ w[{}_3B_9(i,j,l)] = 4, \\ w[{}_4B_9(i,j,k)] = 4 \end{cases}$$

have six solutions:

(1) $c_1 = c_2 = c_3 = c_4 = d_1 = 0$;

(2) $c_1 = c_2 = c_3 = c_4 = 1$, $d_1 = 0$;

(3) $c_1 = 0$, $c_2 = 1$, $c_3 = 0$, $c_4 = 1$, $d_1 = 0$;

(4) $c_1 = 1$, $c_2 = 0$, $c_3 = 1$, $c_4 = d_1 = 0$;

(5) $c_1 = c_2 = 0$, $c_3 = c_4 = 1$, $d_1 = 0$;

(6) $c_1 = c_2 = 1$, $c_3 = c_4 = 0$, $d_1 = 0$,

while the Boolean functions corresponding to the last four solutions are equivalent to each other. There are three non-equivalent H-Boolean functions in Type 9. They are

$$A_{12}(i,j,k,l) = ij + ik + jl + kl,$$
$$A_{13}(i,j,k,l) = ij + ik + jl + kl + ijk + ijl + jkl + ikl,$$
$$A_{14}(i,j,k,l) = ij + ik + jl + kl + ijl + jkl.$$

About Type 10. Since the simultaneous equations

$$\begin{cases} w[{}_1B_{10}(j,k,l)] = 4, \\ w[{}_2B_{10}(i,k,l)] = 4, \\ w[{}_3B_{10}(i,j,l)] = 4, \\ w[{}_4B_{10}(i,j,k)] = 4 \end{cases}$$

have three solutions:

(1) $c_1 = c_2 = c_3 = c_4 = d_1 = 0$;

(2) $c_1 = c_2 = c_3 = 0,\ c_4 = 1,\ d_1 = 0$;

(3) $c_1 = c_2 = 0,\ c_3 = 1,\ c_4 = d_1 = 0$,

the Boolean functions corresponding to the last two solutions are equivalent to each other, and there are two non-equivalent H-Boolean functions in Type 10. They are

$$A_{15}(i,j,k,l) = ij + ik + il + jl + jk,$$
$$A_{16}(i,j,k,l) = ij + ik + il + jk + jl + ikl.$$

About Type 11. Since the simultaneous equations

$$\begin{cases} w[{}_1B_{11}(j,k,l)] = 4, \\ w[{}_2B_{11}(i,k,l)] = 4, \\ w[{}_3B_{11}(i,j,l)] = 4, \\ w[{}_4B_{11}(i,j,k)] = 4 \end{cases}$$

have five solutions:

(1) $c_1 = c_2 = c_3 = c_4 = d_1 = 0$;

(2) $c_1 = c_2 = c_3 = 0,\ c_4 = 1,\ d_1 = 0$;

(3) $c_1 = c_2 = 0,\ c_3 = 1,\ c_4 = d_1 = 0$;

(4) $c_1 = 0,\ c_2 = 1,\ c_3 = c_4 = d_1 = 0$;

(5) $c_1 = 1,\ c_2 = c_3 = c_4 = d_1 = 0$,

the Boolean functions corresponding to the last four solutions are equivalent to each other, and there are two non-equivalent H-Boolean functions in Type 11. They are

$$A_{17}(i,j,k,l) = ij + ik + il + jl + jk + kl,$$
$$A_{18}(i,j,k,l) = ij + ik + il + jk + jl + kl + jkl.$$

In a word, there are in total 18 non-equivalent H-Boolean functions from Type 1 to Type 11 denoted by $A_m(i,j,k,l)$, $(1 \leqslant m \leqslant 18)$. Therefore all the H-Boolean functions of 4-variable without linear terms can be divided into 18 equivalent classes $X_1, X_2, \cdots, X_{18}$ such that X_m, $1 \leqslant m \leqslant 18$, consists of all functions equivalent to $A_m(i,j,k,l)$.

When the linear terms are considered, all 4-variable's H-Boolean functions are divided into 18 classes, say Y_1, Y_2, $\cdots$, Y_{18}, with

$$Y_m = \{f(i,j,k,l) + g(i,j,k,l) : f(\cdot) \in X_m,\ g(\cdot) \text{ linear}\}.$$

The construction of 2^4 Hadamard matrices is clear: every 2^4 Hadamard matrix corresponds to a 4-variable's H-Boolean function belonging to one of Y_m ($1 \leqslant m \leqslant 18$).

Finally, we turn to enumerating the 2^4 Hadamard matrices. Let N be the number of different Boolean functions produced by all $A_m(i,j,k,l)$, $1 \leqslant m \leqslant 18$, produced by permutations of $\{1,2,3,4\}$. Then the number of 2^4 Hadamard matrices should be $2^5N = 32N$.

The number N is found by the following 18 steps:

Step 1. $A_1(i,j,k,l)$ keeps unchanged under any permutations among the variables i, j, k, and l.

Step 2. After permuting the four variables i, j, k, and l, $A_2(i,j,k,l)$ results in three different H-Boolean functions: $ij + kl$; $ik + jl$; and $il + jk$.

Step 3. After permuting the four variables i, j, k, and l, $A_3(i,j,k,l)$ results in four different H-Boolean functions:

$$\begin{array}{ll} ij + ik + il; & ij + jk + kl, \\ ik + jk + kl; & il + jl + kl. \end{array}$$

Step 4. After permuting the four variables i, j, k, and l, $A_4(i,j,k,l)$ results in four different H-Boolean functions:

$$\begin{array}{ll} ij + ik + il + jkl; & ij + jk + jl + ikl, \\ ik + jk + kl + ijl; & il + jl + kl + ijk. \end{array}$$

Step 5. After permuting the four variables i, j, k, and l, $A_5(i,j,k,l)$ results in 12 different H-Boolean functions:

$$\begin{array}{ll} ij + ik + il + ijk + jkl; & ij + ik + il + ijl + jkl, \\ ij + ik + il + ikl + jkl; & ij + jk + jl + ijk + ikl, \\ ij + jk + jl + ijl + ikl; & ij + jk + jl + jkl + ikl, \\ ik + jk + kl + ijk + ijl; & ik + jk + kl + ikl + jkl, \\ ik + jk + kl + jkl + ijl; & il + jl + kl + ijl + ijk, \\ il + jl + kl + ikl + ijk; & il + jl + kl + jkl + ijk. \end{array}$$

Step 6. After permuting the four variables i, j, k, and l, $A_6(i,j,k,l)$ results in 4 different H-Boolean functions:

$$\begin{array}{ll} ij + ik + jk + ijk + ijl + ikl + jkl; & ij + il + jl + ijk + ijl + ikl + jkl, \\ ik + il + kl + ijk + ijl + ikl + jkl; & jk + jl + kl + ijk + ijl + ikl + jkl. \end{array}$$

Step 7. After permuting the four variables i, j, k, and l, $A_7(i,j,k,l)$ results in 12 different H-Boolean functions:

$$\begin{array}{llll} ij + ik + jl; & ij + il + jk; & ik + il + jk; & ij + ik + kl, \\ il + ik + jl; & il + ij + lk; & jk + ji + kl; & jk + jl + ik, \\ ij + jl + kl; & jl + jk + il; & kl + ik + jl; & kl + kj + jl. \end{array}$$

Step 8. After permuting the four variables i, j, k, and l, $A_8(i,j,k,l)$ results in 12 different H-Boolean functions:

$$\begin{array}{ll}
ij+ik+jl+ikl+jkl; & ij+il+jk+ikl+jkl, \\
ik+il+jk+ijl+jkl; & ij+ik+kl+ijl+jkl, \\
il+ik+jl+ijk+jkl; & il+ij+lk+ijk+jkl, \\
jk+ji+kl+ijl+ikl; & jk+jl+ik+jkl+ikl, \\
ij+jl+kl+ijk+ikl; & jl+jk+il+ijk+ikl, \\
kl+ik+jl+ijk+ijl; & kl+kj+il+ijk+ijl.
\end{array}$$

Step 9. After permuting the four variables i, j, k, and l, $A_9(i,j,k,l)$ results in 12 different H-Boolean functions:

$$\begin{array}{lll}
ij+ik+il+jk; & ij+ik+il+jl; & ij+ik+il+kl, \\
ij+jk+jl+ik; & ij+jk+jl+il; & ij+jk+jl+kl, \\
ik+jk+kl+ij; & ik+jk+kl+il; & ik+jk+kl+jl, \\
il+jl+kl+ik; & il+jl+kl+ij; & il+jl+kl+jk.
\end{array}$$

Step 10. After permuting the four variables i, j, k, and l, $A_{10}(i,j,k,l)$ results in 12 different H-Boolean functions:

$$\begin{array}{ll}
ij+ik+il+jk+ijk+jkl; & ij+ik+il+jl+ijl+jkl, \\
ij+ik+il+kl+ikl+jkl; & ij+jk+jl+ik+ijk+ikl, \\
ij+jk+jl+il+ijl+ikl; & ij+jk+jl+kl+jkl+ikl, \\
ik+jk+kl+ij+ijk+ijl; & ik+jk+kl+il+ikl+ijl, \\
ik+jk+kl+jl+jkl+ijl; & il+jl+kl+ik+ikl+ijk, \\
il+jl+kl+ij+ijl+ijk; & il+jl+kl+jk+ijk+jkl.
\end{array}$$

Step 11. After permuting the four variables i, j, k, and l, $A_{11}(i,j,k,l)$ results in 12 different H-Boolean functions:

$$\begin{array}{lll}
ij+ik+il+jk+jkl; & ij+ik+il+jl+jkl; & ij+ik+il+kl+jkl, \\
ij+jk+jl+ik+ikl; & ij+jk+jl+il+ikl; & ij+jk+jl+kl+ikl, \\
ik+jk+kl+ij+ijl; & ik+jk+kl+il+ijl; & ik+jk+kl+jl+ijl, \\
il+jl+kl+ik+ijk; & il+jl+kl+ij+ijk; & il+jl+kl+jk+ijk.
\end{array}$$

Step 12. After permuting the four variables i, j, k, and l, $A_{12}(i,j,k,l)$ results in 3 different H-Boolean functions: $ij+ik+jl+lk$; $ij+il+jk+lk$; and $ik+il+jk+jl$.

Step 13. After permuting the four variables i, j, k, and l, $A_{13}(i,j,k,l)$ results in 3 different H-Boolean functions:

$$ij+ik+jl+lk+ijk+ijl+ikl+jkl,$$

$$ij+il+jk+kl+ijk+ijl+ikl+jkl,$$

$$ik+il+jk+jl+ijk+ijl+ikl+jkl.$$

Step 14. After permuting the four variables i, j, k, and l, $A_{14}(i,j,k,l)$ results in 12 different H-Boolean functions:

$$\begin{array}{ll}
ij+ik+jl+lk+ijl+jkl; & ij+il+jk+kl+ijk+jkl, \\
ik+il+jk+jl+ijk+jkl; & ik+ij+kl+jl+ikl+jkl, \\
il+ik+jl+jk+ijl+jkl; & il+ij+kl+kj+ikl+jkl, \\
ij+jk+il+kl+ijl+ikl; & ij+jl+ki+kl+ijk+ikl, \\
jk+jl+ik+il+ijk+ikl; & jl+jk+il+ik+ijl+ikl, \\
ik+kl+ij+jl+ijk+ijl; & jk+kl+ij+il+ijk+ijl.
\end{array}$$

Step 15. After permuting the four variables i, j, k, and l, $A_{15}(i,j,k,l)$ results in 6 different H-Boolean functions:

$$\begin{array}{ll} ij+ik+il+jk+jl; & ij+ik+il+kl+jk, \\ ij+ik+il+lk+jl; & jk+ij+jl+ik+kl, \\ jl+ij+jk+il+kl; & kl+ki+kj+il+jl. \end{array}$$

Step 16. After permuting the four variables i, j, k, and l, $A_{16}(i,j,k,l)$ results in 12 different H-Booleanr functions:

$$\begin{array}{ll} ij+ik+il+jk+jl+ikl; & ij+ik+il+kl+jk+ijl, \\ ij+ik+il+kl+lj+ijk; & ij+jk+jl+ik+il+jkl, \\ ij+jk+jl+ik+kl+ijl; & ij+jk+jl+li+kl+ijk, \\ ik+jk+kl+ij+il+jkl; & ik+jk+kl+li+lj+ijk, \\ il+jl+lk+ij+ik+jkl; & il+jl+kl+jk+ij+ikl, \\ il+jl+kl+kj+ik+ijl; & ik+jk+kl+ij+jl+ikl. \end{array}$$

Step 17. The $A_{17}(i,j,k,l)$ keeps unchanged under any permutation of the four variables i, j, k, and l.

Step 18. After permuting the four variables i, j, k, and l, $A_{18}(i,j,k,l)$ results in 4 different H-Boolean functions:

$$\begin{array}{ll} A_{17}(i,j,k,l)+jkl; & A_{17}(i,j,k,l)+ikl, \\ A_{17}(i,j,k,l)+ijl; & A_{17}(i,j,k,l)+ijk. \end{array}$$

From the above 18 steps, we find that

$$N = 1+3+4+4+12+4+12+12+12+12+12+3+3+12+6+12+1+4 = 129.$$

With the above we have finished the proof of the following theorem.

Theorem 5.2.1 *There are* $32 \times 129 = 4128$ 4*-dimensional Hadamard matrices of order* 2.

5.2.2 Enumeration of 2^5 Hadamard Matrices

Based on the classification approach used in the last subsection and with the help of a computer search, we obtain the following theorem.

Theorem 5.2.2 *There are* 12,086,336 5*-dimensional Hadamard matrices of order* 2.

To save space we omit its proof here.

Besides Theorem 5.2.2, the following partial results are also true.

(1) The number of 5-variable's H-Boolean functions of the form

$$a_0 + \sum_{i=1}^{5} a_i x_i + \sum_{1 \leqslant i<j \leqslant 5} b_{ij} x_i x_j, \quad a_i, b_{ij} = 0 \text{ or } 1$$

is 49152;

(2) There exists no 5-variable's H-Boolean functions of the form

$$\sum_{1 \leqslant i<j<k<l \leqslant 5} c_{ijkl} x_i x_j x_k x_l, c_{ijkl} = 0 \text{ or } 1;$$

(3) The number of 5-variable's H-Boolean functions of the form

$$a_0+\sum_{i=1}^{5}a_ix_i+\sum_{1\leqslant i<j\leqslant 5}b_{ij}x_ix_j+\sum_{1\leqslant i<j<k<l\leqslant 5}c_{ijkl}x_ix_jx_kx_l$$

is 60416;

(4) The number of 5-variable's H-Boolean functions of the form

$$a_0+\sum_{i=1}^{5}a_ix_i+\sum_{1\leqslant i<j<k\leqslant 5}b_{ijk}x_ix_jx_k,\quad a_i,b_{ijk}=0\text{ or }1$$

is 640;

(5) The number of 5-variable's H-Boolean functions of the form

$$a_0+\sum_{i=1}^{5}a_ix_i+\sum_{1\leqslant i<j<k\leqslant 5}b_{ijk}x_ix_jx_k+\sum_{1\leqslant i<j<k<l\leqslant 5}c_{ijkl}x_ix_jx_kx_l$$

is 6720.

5.2.3 Enumeration of General 2^n Hadamard Matrices

The problem of enumerating the general 2^n Hadamard matrices, or equivalently the H-Boolean functions of n-variable, is still open. This subsection will provide some partial enumeration results.

Theorem 5.2.3 *The number of m-variable's* H-*Boolean functions of the form*

$$a_0+\sum_{i=1}^{m}a_ix_i+\sum_{1\leqslant i<j\leqslant m}b_{ij}x_ix_j$$

is equal to

$$2^{m+1}\left[2^{m(m-1)/2}-\sum_{k=1}^{m}(-1)^k\binom{m}{k}2^{m(m-1)/2-k(2m-k-1)/2}\right].$$

Proof It is sufficient to prove that the number of m-variable's H-Boolean functions of the form

$$g(x_1,x_2,\cdots,x_m)=\sum_{1\leqslant i<j\leqslant m}b_{ij}x_ix_j$$

is equal to

$$2^{m(m-1)/2}-\sum_{k=1}^{m}(-1)^k\binom{m}{k}2^{m(m-1)/2-k(2m-k-1)/2}.$$

The Hamming weight of every non-constant linear Boolean function of $(m-1)$-variable is 2^{m-2}. Thus the necessary condition for

$$\begin{aligned}w[g(x_1,\cdots,x_{k-1},0,x_{k+1},\cdots,x_m)+g(x_1,\cdots,x_{k-1},1,x_{k+1},\cdots,x_m)]\\ =w\left[\sum_{i=1}^{k-1}b_{ik}x_i+\sum_{j=k+1}^{m}b_{kj}x_j\right]\\ =2^{m-2}\qquad(1\leqslant k\leqslant m)\end{aligned}$$

is that $(b_{1,k}, \cdots, b_{k-1,k}, b_{k,k+1}, \cdots, b_{k,m})$ is not the zero-vector for each $1 \leqslant k \leqslant m$.

Therefore the enumeration of the above $g(x_1, x_2, \cdots, x_m)$ is equal to the number of binary symmetric matrices

$$A = \begin{bmatrix} 0 & b_{12} & b_{13} & b_{14} & \cdots & b_{1m} \\ b_{12} & 0 & b_{23} & b_{24} & \cdots & b_{2m} \\ b_{13} & b_{23} & 0 & b_{34} & \cdots & b_{3m} \\ \vdots & \vdots & \vdots & \vdots & & \vdots \\ b_{1m} & b_{2m} & b_{3m} & b_{4m} & \cdots & 0 \end{bmatrix}$$

satisfying that none of its rows are all-zero.

Let $A_k, 1 \leqslant k \leqslant m$, be the set of binary symmetric matrices with its k-th row the zero-vector; and $N(A_k)$ be the number of matrices contained in A_k.

Using the well-known Polya's enumeration identity, we have

$$\begin{aligned} N(A_1 \bigcup \cdots \bigcup A_m) = & \sum_{i=1}^{m} N(A_i) - \sum_{1 \leqslant i < j \leqslant m} N(A_i \bigcap A_j) \\ & + \sum_{1 \leqslant i < j < k \leqslant m} N(A_i \bigcap A_j \bigcap A_k) + \cdots \\ & + (-1)^{m-1} N(A_1 \bigcap \cdots \bigcap A_m). \end{aligned}$$

Because of the identity

$$N(A_{i_1} \bigcap \cdots \bigcap A_{i_k}) = 2^{m(m-1)/2 - k(2m-k-1)/2},$$

we have

$$N(A_1 \bigcup \cdots \bigcup A_m) = \sum_{k=1}^{m} (-1)^{k-1} \binom{m}{k} 2^{m(m-1)/2 - k(2m-k-1)/2}.$$

The enumeration of the above Boolean functions $g(x_1, \cdots, x_m)$ is

$$N\left(\overline{A_1 \bigcup \cdots \bigcup A_m}\right) = 2^{m(m-1)/2} - N\left(A_1 \bigcup \cdots \bigcup A_m\right).$$

The theorem follows. □

Theorem 5.2.4 *The number of m-variable's* H-*Boolean functions of the form*

$$a_0 + \sum_{i=1}^{m} a_i x_i + \sum_{1 \leqslant i < j \leqslant m} b_{ij} x_i x_j + x_{i_1} x_{i_2} \cdots x_{i_k},$$

where $1 \leqslant i_1 < i_2 < \cdots < i_k \leqslant m$ *and* $3 \leqslant k \leqslant m-1$, *is equal to*

$$\begin{aligned} & 2^{m+1} \times \binom{m}{k} \times 2^{k(k-1)/2} \times \{2^{m(m-1)/2 - k(k-1)/2} \\ & - \sum_{s=1}^{m} (-1)^{s-1} \sum_{r=0}^{k} \binom{k}{r} \binom{m-k}{s-r} \\ & \times 2^{(m-k)(k-r)} \times 2^{(m-k)(m-k-1)/2 - [(m-k-1)(s-r) - (s-r)(s-r-1)/2]}\}. \end{aligned}$$

Proof It is sufficient to prove that the number of H-Boolean functions of the form

$$\sum_{1\leqslant i<j\leqslant m} b_{ij}x_ix_j + x_1x_2\cdots x_k$$

is equal to

$$2^{k(k-1)/2}\times\{2^{m(m-1)/2-k(k-1)/2}-\sum_{s=1}^{m}(-1)^{s-1}\sum_{r=0}^{k}\binom{k}{r}\binom{m-k}{s-r}$$
$$\times 2^{(m-k)(k-r)}\times 2^{(m-k)(m-k-1)/2-[(m-k-1)(s-r)-(s-r)(s-r-1)/2]}\}.$$

By the definition of H-Boolean function, we know that the function

$$\sum_{1\leqslant i<j\leqslant m} b_{ij}x_ix_j + x_1x_2\cdots x_k$$

is an H-Boolean function of m-variable if and only if the following two identities are satisfied:

$$\begin{cases} w\left(\sum_{j=1}^{i-1} b_{ji}x_j+\sum_{j=i+1}^{m} b_{ij}x_j + x_1\cdots x_{i-1}x_{i+1}\cdots x_k\right)=2^{m-2}, & \text{if } 1\leqslant i\leqslant k\\ w\left(\sum_{j=1}^{i-1} b_{ji}x_j+\sum_{j=i+1}^{m} b_{ij}x_j\right)=2^{m-2}, & \text{if } k+1\leqslant i\leqslant m. \end{cases}$$

The second identity is equivalent to the condition: at least one of b_{1i}, b_{2i}, $\cdots$, b_{i-1i}, b_{ii+1}, $\cdots$, b_{im} is 1, for each i, $k+1\leqslant i\leqslant m$.

And the first identity is equivalent to the condition: at least one of b_{ik+1}, b_{ik+2}, $\cdots$, b_{im} is 1, for each i, $1\leqslant i\leqslant k$. In fact, if b_{ik+1}, b_{ik+2}, $\cdots$, b_{im} are all zero, then

$$\begin{aligned} & w\Big(\sum_{j=1}^{i-1} b_{ji}x_j+\sum_{j=i+1}^{k} b_{ij}x_j + x_1\cdots x_{i-1}x_{i+1}\cdots x_k\Big)\\ =& w\left(\sum_{j=1}^{i-1} b_{ji}x_j+\sum_{j=i+1}^{k} b_{ij}x_j\right)+w(x_1\cdots x_{i-1}x_{i+1}\cdots x_k)\\ & -2w\left[\left(\sum_{j=1}^{i-1} b_{ji}x_j+\sum_{j=i+1}^{k} b_{ij}x_j\right)x_1\cdots x_{i-1}x_{i+1}\cdots x_k\right]\\ =& 2^{m-2}+2^{m-k}-2^{m-k+1}\left[\left(\sum_{j=1}^{i-1} b_{ji}+\sum_{j=i+1}^{k} b_{ij}\right)\bmod 2\right]\\ \neq & 2^{m-2}, \end{aligned}$$

where the last to the second equation is due to the treated Hamming weights being of Boolean functions of $(m-1)$ variables.

Because of the above two equivalent identities, it is sufficient to prove that there are

$$2^{m(m-1)/2-k(k-1)/2}-\left\{\sum_{s=1}^{m}(-1)^{s-1}\sum_{r=0}^{k}\binom{k}{r}\binom{m-k}{s-r}\right.$$
$$\left.\times 2^{(m-k)(k-r)}\times 2^{(m-k)(m-k-1)/2-[(m-k-1)(s-r)-(s-r)(s-r-1)/2]}\right\}$$

binary (0 or 1) matrices of the form

$$\begin{bmatrix} 0 & 0 & \cdots & 0 & b_{1k+1} & \cdots & \cdots & b_{1m} \\ 0 & 0 & \cdots & 0 & b_{2k+1} & \cdots & \cdots & b_{2m} \\ \vdots & \vdots & & \vdots & \vdots & & & \vdots \\ 0 & 0 & \cdots & 0 & b_{kk+1} & \cdots & \cdots & b_{km} \\ b_{1k+1} & b_{2k+1} & \cdots & b_{kk+1} & 0 & b_{k+1k+2} & \cdots & b_{k+1m} \\ \vdots & \vdots & & \vdots & \vdots & \vdots & & \vdots \\ b_{1m} & b_{2m} & \cdots & b_{km} & b_{k+1m} & \cdots & b_{m-1m} & 0 \end{bmatrix}$$

without all-zero rows. This enumeration can be finished in the same way as that of Theorem 5.2.3. □

Similarly, it has been proved that

Theorem 5.2.5 *The number of* H*-Boolean functions of the form*

$$a_0 + \sum_{i=1}^{m} a_i x_i + \sum_{1 \leqslant i < j \leqslant m} b_{ij} x_i x_j + \sum_{1 \leqslant i < j < k \leqslant m} c_{ijk} x_i x_j x_k$$

is bounded above by

$$2^{n+X} \times \left\{ 2^{n+Y} - \sum_{i \neq 2^{n-2}} A_i \right\},$$

where

$$A_{2^{n-2} \pm 2^{n-2-h}} = 2^{h(h+1)} \times \frac{(2^{n-1}-1)(2^{n-2}-1)\cdots(2^{n-2h}-1)}{(2^{2k}-1)(2^{2k-2}-1)\cdots(2^2-1)}$$

and the other A_is are all zero, and where $1 \leqslant h \leqslant \left\lfloor \frac{n-1}{2} \right\rfloor$, and

$$X =: \binom{n-1}{2} + \binom{n-1}{3}, \quad and \quad Y = \binom{n-1}{2}.$$

In order to introduce more enumerations about H-Boolean function, we need the following definitions.

Definition 5.2.1 *An n-dimensional vector $X = (x_1, x_2, \cdots, x_n)$ is called a characteristic vector of a Boolean function $f(\cdot)$, iff $f(X) = 1$. The matrix produced by arranging all characteristic vectors of $f(\cdot)$ in the dictionary order is called the ordered characteristic matrix of this Boolean function.*

Clearly, Boolean functions and their ordered characteristic matrices are uniquely determined by each other.

Definition 5.2.2 *A binary $(0$ or $1)$ matrix of size $2k \times n$ is said to be a matrix of type A_r, $0 \leqslant r \leqslant k$, if and only if*

(1) *The rows of this matrix are different from each other;*

(2) *The sub-matrices consisting of any $n-1$ columns have and only have r same-row pairs, while the other rows are different from each other.*

For example, $f(x_1, x_2, x_3, x_4) = x_1x_2 + x_3x_4$ is an H-Boolean function of 4-variable, and its characteristic matrix is

$$\begin{bmatrix} 0 & 0 & 1 & 1 \\ 0 & 1 & 1 & 1 \\ 1 & 0 & 1 & 1 \\ 1 & 1 & 0 & 0 \\ 1 & 1 & 0 & 1 \\ 1 & 1 & 1 & 0 \end{bmatrix}.$$

This matrix is of type A_1, e.g., its sub-matrix consisting of the first 3 columns has just one pair of same-row (the 4-th and 5-th rows are the same vector (1,1,0)). In fact, this example can be generalized as the following theorem, which indicates another equivalent definition of H-Boolean function.

Theorem 5.2.6 *An n-variable Boolean function of Hamming weight 2k is* H-*Boolean if and only if its ordered characteristic matrix is of type* $A_{k-2^{n-3}}$.

Proof If $w(f(\cdot)) = 2k$ then the equation

$$w(f(x_1, \cdots, x_{i-1}, 0, \cdots, x_n) + f(x_1, \cdots, x_{i-1}, 1, \cdots, x_n)) = 2^{n-2}$$

is equivalent to

$$w(f(x_1, \cdots, x_{i-1}, 0, \cdots, x_n)f(x_1, \cdots, x_{i-1}, 1, \cdots, x_n)) = k - 2^{n-3},$$

which infers that the sub-matrix produced by canceling the i-th column of its ordered characteristic matrix has and only has $k - 2^{n-3}$ same-row pairs. The proof is finished by applying $1 \leqslant i \leqslant n$. □

A straightforward corollary of Definition 5.2.2 is

Lemma 5.2.1 *A binary matrix is of type* A_0 *if and only if the Hamming distances between any two rows are larger than or equal to* 2.

Theorem 5.2.7 *There are at least*

$$2\sum_{s=0}^{2^{n-3}-1}\binom{2^{n-1}}{s}\binom{2^{n-1}-ns}{2^{n-2}-s}+\binom{2^{n-1}}{2^{n-3}}\binom{2^{n-1}-n2^{n-3}}{2^{n-3}}$$

n-variable H-*Boolean functions of Hamming weight* 2^{n-2}.

Proof A matrix of type A_0 can be constructed by: (1) choose any s n-dimensional vectors of even Hamming weights as s rows of the matrix; (2) thus there are at least $2^{n-1} - ns$ n-dimensional vectors of odd Hamming weights that have Hamming distances larger than 1 apart from the above chosen s even weight's vectors. Choose the other $2^{n-2} - s$ rows of the matrix from these odd weight ones. □

Theorem 5.2.8 *There are at least*

$$2^{n-1}\sum_{s=0}^{2^{n-3}-\lfloor(n-1)/2\rfloor}\binom{2^{n-1}-n^2-n}{s}\binom{2^{n-1}-n^2-n-ns}{2^{n-2}-n+1}$$

n-variable H-*Boolean functions of Hamming weight* $2^{n-2}+2$.

Proof A matrix of type A_1 can be constructed by: (1) choose $(0, 0, \cdots, 0)$ and $e_i =: (0, \cdots, 1, 0, \cdots, 0)$, $1 \leqslant i \leqslant n$, as $n + 1$ rows of the matrix; (2) there are at least $2^n - n^2$ n-dimensional vectors that have Hamming distances larger than 1 apart from the e_i, $1 \leqslant i \leqslant n$. Choose the other $2^{n-2} - n + 1$ rows of the matrix from these vectors. □

5.3 Applications

n-dimensional Hadamard matrices of order 2, or equivalently, the H-Boolean functions, can be widely used in modern cryptography, error-correcting codes, and signal processing. This section concentrates on the close relationships between H-Boolean functions and the strict avalanche criterion, propagation characteristics, Bent functions, and Reed-Muller codes, respectively.

5.3.1 Strict Avalanche Criterion and H-Boolean Functions[4–8]

The symmetric or private-key cryptosystem is one of the most important kinds of cryptosystems. This kind of cryptosystem, e.g., the famous DES, is constructed by the combination of substitutions and permutations. For the development of a symmetric cryptosystem, a significant portion of time has been spent on design or on analysis of the substitution boxes (called S-boxes in short). Because the remainder of the cryptosystem algorithm is linear, severe weaknesses in the S-boxes can therefore lead to an insecure cryptosystem.

The Strict Avalanche Criterion (SAC) was introduced by Webster and Tavares[4] in order to combine the ideas of completeness and the avalanche effect of the design of an S-box. A cryptographic transformation is said to be complete if each output bit depends on all of the input bits, and it exhibits the avalanche effect if an average of one-half of the output bits changes whenever a single input bit is changed. Forre[7] extended the notion of SAC by defining higher-order Strict Avalanche Criteria. This subsection will prove that the SAC is, in fact, another equivalent form of H-Boolean. Thus all results about H-Boolean functions can be applied to the study of SAC, and vice versa.

The cryptographic significance of the SAC is highlighted by considering the situation where a cryptographer needs some "complex" mapping f of n bits onto one bit, although there is no precise mathematical definition for the expression "complex." In order to make a more intuitively pleasing meaning of "complex," we present here an information-theoretical statement: If the conditional entropy

$$H[f(x_1, x_2, \cdots, \overline{x_i}, \cdots, x_n) \mid f(x_1, x_2, \cdots, x_n)]$$

is maximized for all i, $1 \leqslant i \leqslant n$, then the Boolean $f(x_1, x_2, \cdots, x_n)$ is said to satisy the SAC. In other words, little information of $f(x_1, x_2, \cdots, \overline{x_i}, \cdots, x_n)$ will be exposed by a Boolean $f(x_1, x_2, \cdots, x_i, \cdots, x_n)$ satisfying the SAC. The higher-order SAC goes even further, by keeping one or more input bits of $f(\cdot)$ constant, and making the obtained "sub-functions" complex as well. It is worthwhile pointing out that any function $g(\cdot)$ of $n - 1$ variables will be a relatively bad approximation of $f(\cdot)$ if $f(\cdot)$ satisfies the SAC. Indeed, the output of the best possible $g(\cdot)$ will differ from the output of f with a probability of $1/4$. This lack of accuracy of lower-dimensional approximations is a desired property of cryptosystems, because the existence of some lower-dimensional approximation of an encryption could reduce the amount of work for an exhaustive decryption according to the dimension of the domain of the approximation.

The mathematical definition of the Strict Avalanche Criterion (SAC) is[4]:

Definition 5.3.1 *A Boolean function $f(x_1, x_2, \cdots, x_n)$ is called "satisfying SAC" if and only if for each unit vector $e_i = (0, \cdots, 0, 1, 0, \cdots, 0)$, the following equation is held*

$$w[f(x) + f(x + e_i)] = 2^{n-1},$$

where $x = (x_1, x_2, \cdots, x_n)$.

Comparing this definition with the definition of an H-Boolean function, i.e., Definition 5.1.1, it is easy to see that they are the same thing. Hence we have:

Theorem 5.3.1 *A Boolean function $f(x_1, x_2, \cdots, x_n)$ satisfies the SAC if and only if it is an H-Boolean function of n variables.*

Definition 5.3.2[8] *A Boolean function $f(x_1, x_2, \cdots, x_n)$ is said to satisfy a SAC to order k, $0 \leqslant k \leqslant n-2$, if and only if whenever k variables are fixed arbitrarily, the resulting function of $(n-k)$ variables satisfies the SAC.*

It is easy to see that a SAC of order 0 is the SAC defined in Definition 5.3.1. In addition, if a function satisfies the SAC of order $k > 0$, then it also satisfies the SAC of order j for any $j = 0, 1, \cdots, k-1$. Thus if $f(x_1, \cdots, x_n)$ is a SAC of order k, then $f(\cdot)$ generates a higher-dimensional Hadamard matrix such that all of its $m(\geqslant n-k)$-dimensional sections are Hadamard matrices of order 2. In fact, this higher-dimensional Hadamard matrix is defined by $[(-1)^{f(x_1, \cdots, x_n)}]$.

One of the hot topics in SAC criteria is to count the Boolean functions satisfying the SAC criterion. We use the abbreviation SAC(k) for the strict avalanche criterion of order k. Here are a few fundamental results on Boolean functions satisfying SAC(k).

Lemma 5.3.1[7,8] *If $f(x_1, x_2, \cdots, x_n)$ satisfies SAC(k) for some k, $0 \leqslant k \leqslant n-2$, then so does*

$$f(x_1, x_2, \cdots, x_n) + a + \sum_{i=1}^{n} b_i x_i.$$

Lemma 5.3.2 [5,8] *If $f(x_1, x_2, \cdots, x_n)$ satisfies SAC(k) for some k, $0 \leqslant k \leqslant n-3$, then $f(\cdot)$ is a Boolean function of degree up to $n-k-1$. If $f(x_1, x_2, \cdots, x_n)$ satisfies SAC($n-2$), then $f(\cdot)$ is a Boolean function of degree 2.*

Proof Recall that every H-Boolean function of n, $n \geqslant 3$, variables is of the degree less than or equal to $(n-1)$ (see Theorem 5.1.3). If $f(x_1, x_2, \cdots, x_n)$ is a Boolean function of degree larger than $n-k-1$, $0 \leqslant k \leqslant n-3$, say it contains the term $x_1 x_2 \cdots x_d$, $d \geqslant n-k$, then by fixing $x_{n-k+1} = \cdots = x_n = 0$ we get a function $f(x_1, \cdots, x_{n-k}, 0, \cdots, 0)$, which cannot be an H-Boolean of $(n-k)$ variables, because it is of degree $n-k$. Thus this function $f(x_1, x_2, \cdots, x_n)$ does not satisfy SAC(k).

The fact that a function satisfying SAC($n-2$) is of degree 2 can be proved by the following observations. On one hand, linear function is not H-Boolean (see Theorem 5.1.9) and does not satisfy SAC($n-2$). On the other hand, a function satisfying SAC($n-2$) satisfies SAC($n-3$) too. Thus, by the first assertion of this lemma, the function has a degree upper bounded by $n-(n-3)-1 = 2$. □

Lemma 5.3.3[6,8] *A quadratic Boolean function of the form*

$$f(x_1, x_2, \cdots, x_n) = \sum_{1 \leqslant i < j \leqslant n} a_{ij} x_i x_j, \quad n \geqslant 2$$

satisfies SAC(k), $0 \leqslant k \leqslant n-2$, if and only if every variable x_i occurs at least $k+1$ times.

Proof At first it is easy to prove that a quadratic Boolean function is H-Boolean if and only if this function is dependent on each of its variables. On the one hand, if x_1 occurs no more than $k+1$ times, say,

$$f(x_1, x_2, \cdots, x_n) = x_1(x_2 + x_3 + \cdots + x_d) + \sum_{2 \leqslant i < j \leqslant n} a_{ij} x_i x_j, \quad d \leqslant k+1,$$

so the function $f(x_1, 0, \cdots, 0, x_{k+2}, \cdots, x_n)$ is independent of x_1 and thus not an H-Boolean function. Hence this function does not satisfies SAC(k).

On the other hand, if every variable occurs at least $k+1$ times, then the function produced by fixing any k variables is a quadratic function dependent on each of its variables. Hence it is an H-Boolean function. □

Theorem 5.3.2 [5,8] *There are 2^{n+1} Boolean funcions of $n \geqslant 2$ variables which satisfy* SAC$(n-2)$, *they are exactly the functions of the form*

$$a + \sum_{i=1}^{n} b_i x_i + \sum_{1 \leqslant i < j \leqslant n} x_i x_j.$$

Proof Using Lemma 5.3.1 it is sufficient to prove that a quadratic Boolean function of the form $\sum_{1 \leqslant i < j \leqslant n} a_{ij} x_i x_j$ satisfies SAC$(n-2)$ if and only if $a_{ij} = 1$ for all $1 \leqslant i < j \leqslant n$ which is, in fact, a direct corollary of Lemma 5.3.3. □

Lemma 5.3.1, Lemma 5.3.2, and Lemma 5.3.3 can also be used to enumerate the Boolean functions satisfying SAC$(n-3)$.

Theorem 5.3.3 [5,8] *Define a sequence $\{W_i\}$ of integers by $W_1 = 1$, $W_2 = 2$, and*

$$W_n = W_{n-1} + (n-1)W_{n-2}, \quad \textit{for } n \geqslant 3.$$

Then there are $2^{n+1} W_n$ Boolean functions of $n \geqslant 3$ variables which satisfy SAC$(n-3)$.

Proof Using Lemma 5.3.1, it suffices to show that there are W_n Boolean functions of the form $\sum_{1 \leqslant i < j \leqslant n} a_{ij} x_i x_j$ satisfying SAC$(n-3)$. Using Lemma 5.3.2 and Lemma 5.3.3, any such function is obtained by deleting zero or more terms from the sum $\sum_{1 \leqslant i < j \leqslant n} x_i x_j$ in such a way that the remaining sum has the property that every variable x_i occurs in at least $n-2$ terms. Thus S is a set of terms which we are allowed to delete if and only if no subscript i occurs in a term $x_i x_j$ in S more than once. It is easy to find a recursion for the number W_n of such sets S. Obviously $W_1 = 1$ (the empty set) and $W_2 = 2$. Clearly, any set of terms $T = \{x_i x_j\}$ which is counted in W_{n-1} is also a set which must be counted in W_n, and this includes all sets which do not contain any term $x_i x_n$. If we have any set T which includes less than or equal to $n-2$ variables from $x_1, x_2, \cdots, x_{n-1}$, we may add a term $x_k x_n$ to T and get a set to be counted in W_n if and only if x_k does not already occur in a term in T. There are W_{n-2} such sets of T, by our definitions, so we count W_{n-2} sets for each k, $1 \leqslant k \leqslant n-1$. Hence $W_n = W_{n-1} + (n-1)W_{n-2}$ and the theorem follows. □

The number W_n appearing in Theorem 5.3.3 is the number of permutations in the symmetric group S_n whose square is the identity. In particular, the following asymptotic formula for W_n has been proved:

$$W_n \overset{n \to \infty}{\longrightarrow} (e^{1/4}\sqrt{2})^{-1} e^{\sqrt{n}} (n/e)^{n/2}.$$

The problem of counting the Boolean functions satisfying the SAC of order $k \leqslant n-4$ is still open. The updated bounds for the number, T_n, of Boolean functions satisfying SAC($n-4$) is[8]

$$2^n(n-1)! < T_n < n^5 2^n n!.$$

The cases of $k \leqslant n-4$ are much difficult than the cases of $k=n-2$ and $k=n-3$, because many of such functions are non-quadratic.

Now, we turn to the spectral characterization of Boolean functions satisfying SAC. Every Boolean function $f(x_1, x_2, \cdots, x_n)$ corresponds to a unique function defined by

$$f'(x_1, x_2, \cdots, x_n) =: (-1)^{f(x_1, x_2, \cdots, x_n)},$$

which takes values in the range $\{-1, 1\}$. By this notation, Definition 5.3.1 can be equivalently stated as

Lemma 5.3.4[7,9] *A Boolean function $f(x_1, x_2, \cdots, x_n)$ satisfies* SAC *if and only if its $f'(x_1, x_2, \cdots, x_n)$ satisfies*

$$\sum_{x \in Z_2^n} f'(x) f'(x+e_i) = 0. \tag{5.4}$$

Recall that the Walsh Transform of $f'(x)$, $x =: (x_1, x_2, \cdots, x_n)$, is

$$F'(w) = \sum_{x \in Z_2^n} f'(x)(-1)^{x_1 w_1 + x_2 w_2 + \cdots + x_n w_n}, \tag{5.5}$$

where $w = (w_1, w_2, \cdots, w_n)$. The function $f'(x)$ can be recovered from $F'(w)$ by the inverse Walsh transform:

$$f'(x) = 2^{-n} \sum_{w \in Z_2^n} F'(w)(-1)^{x_1 w_1 + x_2 w_2 + \cdots + x_n w_n}. \tag{5.6}$$

From the well-known convolution theorem, which states that

$$h'(x) = \sum_{y \in Z_2^n} f'(y) \cdot g'(y+x) \text{ if and only if } H'(w) = F'(w) \cdot G'(w), \tag{5.7}$$

we see that the left hand side of Equation (5.4) is also the inverse Walsh transform of $F'(w) \cdot F'(w) = [F'(w)]^2$ and by using Equation (5.6) we obtain:

$$\sum_{x \in Z_2^n} f'(x) \cdot f'(x+e_i) = 2^{-n} \sum_{w \in Z_2^n} [F'(w)]^2 (-1)^{w_i}, \tag{5.8}$$

where w_i is the i-th coordinate of the vector w. This equation proves the following theorem.

Theorem 5.3.4[7,9] *A Boolean function $f(x_1, x_2, \cdots, x_n)$ satisfies* SAC *if and only if its $f'(x)$ has a Walsh transform $F'(w)$ satisfying*

$$\sum_{w \in Z_2^n} [F'(w)]^2 (-1)^{w_i} = 0 \tag{5.9}$$

for all i, $1 \leqslant i \leqslant n$.

A geometrical interpretation of Theorem 5.3.4 can be introduced[7] if we treat the n-tuples $(w_1, w_2, \cdots, w_n)$ as the corners of an n-dimensional cube with edges of length one. Let's attach to each corner $w = (w_1, w_2, \cdots, w_n)$ a weight $m(w)$ equal to $[F'(w)]^2$. The center of gravity of this n-dimensional body has the coordinates $(\overline{w_1}, \overline{w_2}, \cdots, \overline{w_n})$ with

$$\overline{w_i} = \frac{\sum_{w\in Z_2^n} m(w)w_i}{\sum_{w\in Z_2^n} m(w)} = \frac{\sum_{w\in Z_2^n,\ w_i=1}[F'(w)]^2}{\sum_{w\in Z_2^n}[F'(w)]^2} \tag{5.10}$$

for all $1 \leqslant i \leqslant n$. If a Boolean function $f(x)$ satisfies SAC, we know from Theorem 5.3.4 that

$$\sum_{w\in Z_2^n,\ w_i=0} [F'(w)]^2 - \sum_{w\in Z_2^n,\ w_i=1} [F'(w)]^2 = 0. \tag{5.11}$$

Thus

$$\sum_{w\in Z_2^n,\ w_i=0} [F'(w)]^2 = \sum_{w\in Z_2^n,\ w_i=1} [F'(w)]^2. \tag{5.12}$$

And in that case, we have

$$\overline{w_i} = \frac{\sum_{w\in Z_2^n,\ w_i=1}[F'(w)]^2}{\sum_{w\in Z_2^n}[F'(w)]^2} = \frac{\sum_{w\in Z_2^n,\ w_i=0}[F'(w)]^2}{\sum_{w\in Z_2^n}[F'(w)]^2}, \tag{5.13}$$

which shows that the coordinate w_i of the center of gravity of the considered cubic body remains unchanged if all the weights on one "face" of the cube (the face with $w_i = 0$) are moved to the opposite "face" (the face with $w_i = 1$) and conversely. Therefore, we can state that a Boolean function $f(x)$ satisfies SAC if and only if the n-cube with weights equal to $[F'(w)]^2$ attached to its corners has a center of gravity which is equidistant from any two opposite "faces" of the cube, and thus from all the corners of the cube. The center of gravity of the body associated to the Walsh-spectrum of an SAC-fulfilling function therefore has the coordinates $\left(\frac{1}{2}, \frac{1}{2}, \cdots, \frac{1}{2}\right)$.

The idea that now naturally arises is to use this as a construction for new Boolean functions satisfying SAC from known ones. The pitfall is that $F'(w)$ might be taken as $\pm\sqrt{[F'(w)]^2}$ for each one of the 2^n possible $w's$. For the worst case where all 2^n $w's$ are associated to nonzero values of $[F(w)]^2$, this will yield 2^{2^n} possible choices for the mapping $F(w)$, and some of them have no valid functions $f'(x)$ (i.e., (±1)-valued) as inverse Walsh transforms. In fact, a function $f'(x)$ is a (±1)-valued if and only if

$$[f'(x)]^2 = 1, \quad \text{for all possible } x \in Z_2^n. \tag{5.14}$$

By the convolution theorem, we see that this is equivalent to

Theorem 5.3.5[7] *$F'(w)$ is the Walsh transform of a (±1)-valued function $f'(x)$ if and only if*

$$\sum_{w\in Z_2^n} F'(w)\cdot F'(w+s) = 2^n\delta(s) = \begin{cases} 2^n, & \textit{for } s = (0, 0, \cdots, 0), \\ 0, & \textit{otherwise.} \end{cases}$$

5.3.2 Bent Functions and H-Boolean Functions

Bent functions, defined and first analyzed by Rothaus in 1976, have been the subject of some interest in logic synthesis, digital communications (especially spread-spectrum multiple access communications), coding theory, and cryptography. We will show in this subsection that every Bent function is an H-Boolean which corresponds to a higher-dimensional Hadamard matrix of order 2. The concept of propagation characteristics (PC) and the relationships among PC, Bent, SAC, and H-Boolean will be introduced too.

Definition 5.3.3[10] *A Boolean function $f(x_1, x_2, \cdots, x_n)$ is called a Bent function if its Walsh transform coefficients are all $\pm 2^{n/2}$, i.e.,*

$$F(u_1, u_2, \cdots, u_n) = \sum_{x \in Z_2^n} (-1)^{u \cdot x + f(x)} \tag{5.15}$$

is a vector with elements $\pm 2^{n/2}$, where $x = (x_1, \ldots, x_n)$, $u = (u_1, \cdots, u_n)$, and $u \cdot x = \sum_{i=1}^{n} u_i x_i$.

Because the transform coefficients are summations of integers, $\pm 2^{n/2}$ must be an integer, i.e., n must be an even integer for $f(x_1, x_2, \cdots, x_n)$ being Bent.

The following three functions are Bent on 6 variables[10]:

$$\begin{aligned}
f_1 =& x_1x_4 + x_2x_5 + x_3x_6 + x_1x_2x_3, \\
f_2 =& x_1x_2 + x_1x_4 + x_2x_6 + x_3x_5 + x_4x_5 + x_1x_2x_3 + x_2x_4x_5, \\
f_3 =& x_1x_4 + x_2x_6 + x_3x_4 + x_3x_5 + x_3x_6 + x_4x_5 + x_4x_6 \\
&+ x_1x_2x_3 + x_2x_4x_5 + x_3x_4x_6.
\end{aligned}$$

Theorem 5.3.6[2,10] *$f(x_1, x_2, \cdots, x_n)$ is Bent if and only if the $2^n \times 2^n$ matrix H whose (u, v)-th entry is $(1/2^{n/2})F(u + v)$ is a 2-dimensional Hadamard matrix.*

Proof It is sufficient to prove that if $f(\cdot)$ is Bent, then $(1/2^{n/2})F(u + v)$ is a Hadamard matrix. On the one hand, using Definition 5.3.3, this matrix is clearly a (± 1)-matrix. On the other hand,

$$\begin{aligned}
&\sum_{u \in Z_2^n} \frac{1}{2^{n/2}} F(u)(1/2^{n/2})F(u + v) \\
=& \frac{1}{2^n} \sum_{u \in Z_2^n} \sum_{w \in Z_2^n} (-1)^{u \cdot w}(-1)^{f(w)} \sum_{x \in Z_2^n} (-1)^{(u+v) \cdot x}(-1)^{f(x)} \\
=& \frac{1}{2^n} \sum_{w, x \in Z_2^n} (-1)^{v \cdot x}(-1)^{f(w)+f(x)} \sum_{u \in Z_2^n} (-1)^{u \cdot (w+x)} \\
=& \sum_{w \in Z_2^n} (-1)^{v \cdot w}[(-1)^{f(w)}]^2 \\
=& \sum_{w \in Z_2^n} (-1)^{v \cdot w} \\
=& 2^n \delta_{v,0}.
\end{aligned}$$

Thus this matrix is orthogonal. □

Note that if $f(x_1, x_2, \cdots, x_n)$ is Bent, we then may write

$$F(u)/2^{n/2} = (-1)^{g(u)},$$

which defines a Boolean function $g(u)$. It is easy to verify that the Walsh transform coefficients of $g(u)$ are

$$2^{n/2}(-1)^{f(u)} = \pm 2^{n/2}.$$

Therefore $g(u)$ is also Bent. Thus there is a natural pairing $f \longleftrightarrow g$ of Bent functions. The other straight implications of Theorem 5.3.6 are:

Theorem 5.3.7 [2,10] *Bent functions can also be equivalently defined by:*

(1) $f(x_1, x_2, \cdots, x_n)$ *is Bent if and only if the matrix whose* (u, v)*-th entry is* $(-1)^{f(u+v)}$, *for* $u, v \in Z_2^n$, *is a 2-dimensional Hadamard matrix.*

(2) $f(x_1, x_2, \cdots, x_n)$ *is Bent if and only if for all* $v \neq 0$, $v \in Z_2^n$, *the Hamming weight of* $g(x) = f(x + v) + f(x)$ *is* 2^{n-1}.

(3) $f(x_1, x_2, \cdots, x_n)$ *is Bent if and only if the function*

$$f(x_1, x_2, \cdots, x_n) + a + \sum_{i=1}^{n} b_i x_i$$

is Bent.

Theorem 5.3.8 *Every Bent function is* H*-Boolean.*

Proof It can be proved by applying the unit vectors $e_i = (0, \cdots, 0, 1, 0, \cdots, 0)$, $1 \leqslant i \leqslant n$, to the second statement of Theorem 5.3.7. □

Thus all constructions for Bent functions are valid for H-Boolean and thus also for a n-dimensional Hadamard matrix of order 2.

Theorem 5.3.9[10] *For any Boolean funcion* $g(y_1, \cdots, y_n)$, *the function*

$$f(x_1, \cdots, x_n; y_1, \cdots, y_n) = \sum_{i=1}^{m} x_i y_i + g(y_1, \cdots, y_n)$$

is Bent.

Proof Let $a = (a_1, \cdots, a_n)$ and $b = (b_1, \cdots, b_n)$ be two binary vectors.

Case 1. If $b \neq 0$, then

$$\begin{aligned} & f(x_1, \cdots, x_n; y_1, \cdots, y_n) + f(x_1 + a_1, \cdots, x_n + a_n; y_1 + b_1, \cdots, y_n + b_n) \\ = & \sum_{i=1}^{n} b_i x_i + h(y_1, \cdots, y_n), \end{aligned}$$

which is linear in the variables $x_1, x_2, \cdots, x_n$ and thus its Hamming weight is 2^{2n-1}.

Case 2. If $b = 0$, then $a \neq 0$ and

$$\begin{aligned} & f(x_1, \cdots, x_n; y_1, \cdots, y_n) + f(x_1 + a_1, \cdots, x_n + a_n; y_1 + b_1, \cdots, y_n + b_n) \\ = & \sum_{i=1}^{n} a_i y_i, \end{aligned}$$

which is linear in the variables $y_1, y_2, \cdots, y_n$, and thus its Hamming weight is also 2^{2n-1}.

The proof is finished by the second statement of Theorem 5.3.7. □

A Bent function $f(x_1, x_2, \cdots, x_{2n})$ can also be equivalently treated as a (± 1)-valued vector of length 4^n. We call such a vector the Bent sequence[12]. In other words, a (± 1)-sequence $y = (y_1, y_2, \cdots, y_N)$, $N = 4^n$, is a Bent sequence if and only if its normalized Walsh transform $Y = (1/2^n)H_{2n}y$ is also a (± 1)-sequence, where H_m is the Walsh matrix of order 2^m defined in the first chapter of this book.

It is easy to check that there are eight Bent sequences of length 4, and they are

$$\begin{aligned} B_4 = \{&111-1;\ 11-11;\ 1-111;\ -1111;\\ &-1-1-11;\ -1-11-1;\ -11-1-1;\ 1-1-1-1\}. \end{aligned}$$

Theorem 5.3.10 [12,13] *Let m, n be positive even integers and let $y(1), y(2), \cdots, y(2^m)$ be Bent sequences of length 2^n. Furthermore, let z be the concatenation of the normalized Walsh transforms of these Bent sequences, that is*

$$z^{\mathrm{T}} = (Y^{\mathrm{T}}(1)Y^{\mathrm{T}}(2)\cdots Y^{\mathrm{T}}(2^m)).$$

The sequence z is Bent if and only if the sequence $(y(1)_i, y(2)_i, \cdots, y(2^m)_i)$ is Bent for all i, where $y(j)_i$ is the i-th bit of the sequence $y(j)$.

Proof To show that z is Bent, we show that its normalized Walsh transform Z is a (± 1)-sequence. We write z as a $2^n \times 2^m$ matrix: $a = (Y(1)Y(2)\cdots Y(2^m))$. It is easy to check that

$$Z = \frac{1}{2^{(m+n)/2}} H_{m+n} z \quad \Longrightarrow \quad A = \frac{1}{2^{m/2}} \frac{1}{2^{n/2}} H_n a H_m$$

(i.e., that A is the 2-dimensional Walsh transform of a). Thus,

$$\begin{aligned} A &= \frac{1}{2^{m/2}} \frac{1}{2^{n/2}} H_n (Y(1)Y(2)\cdots Y(2^m)) H_m \\ &= \frac{1}{2^{m/2}} (y(1)y(2)\cdots y(2^m)) H_m \\ &= \frac{1}{2^{m/2}} B H_m, \end{aligned}$$

where $B = (y(1)y(2)\cdots y(2^m))$ (note that the entries of B are all ± 1). The rows of A are the normalized Walsh transforms of the rows of B, and are therefore (± 1)-sequences of length 2^m if and only if $(y(1)_i, y(2)_i, \cdots, y(2^m)_i)$ is Bent for all i. □

By using Theorem 5.3.10 we can construct long Bent sequences from short ones. For example, if x is a Bent sequence of length 4^k, then the following four sequences are Bent of length 4^{k+1}[12, 13]:

$$\begin{aligned} &(X^{\mathrm{T}}X^{\mathrm{T}}X^{\mathrm{T}} - X^{\mathrm{T}}),\quad (X^{\mathrm{T}}X^{\mathrm{T}} - X^{\mathrm{T}}X^{\mathrm{T}}),\\ &(X^{\mathrm{T}} - X^{\mathrm{T}}X^{\mathrm{T}}X^{\mathrm{T}}),\quad (-X^{\mathrm{T}}X^{\mathrm{T}}X^{\mathrm{T}}X^{\mathrm{T}}). \end{aligned}$$

If x and y are two Bent sequences of length 4^k, then the following six sequences are Bent of length 4^{k+1}[12, 13]:

$$\begin{aligned} &(X_1^{\mathrm{T}}X_1^{\mathrm{T}}X_2^{\mathrm{T}} - X_2^{\mathrm{T}}),\quad (X_1^{\mathrm{T}}X_2^{\mathrm{T}}X_1^{\mathrm{T}} - X_2^{\mathrm{T}}),\quad (X_1^{\mathrm{T}}X_2^{\mathrm{T}} - X_2^{\mathrm{T}}X_1^{\mathrm{T}}),\\ &(X_2^{\mathrm{T}}X_1^{\mathrm{T}}X_1^{\mathrm{T}} - X_2^{\mathrm{T}}),\quad (X_2^{\mathrm{T}}X_1^{\mathrm{T}} - X_2^{\mathrm{T}}X_1^{\mathrm{T}}),\quad (X_2^{\mathrm{T}} - X_2^{\mathrm{T}}X_1^{\mathrm{T}}X_1^{\mathrm{T}}). \end{aligned}$$

By using the same proof as that for Theorem 5.3.10, we have the following theorem:

Theorem 5.3.11[14] *If $x = (x_1, x_2, \cdots, x_{4^k})$ and $y = (y_1, y_2, \cdots, y_{4^k})$ are two Bent sequences of length 4^k, then the following two sequences, u and v, are Bent sequences of length 4^{k+1}, where*

$$\begin{aligned} u = (&x_1, x_2, x_3, x_4, y_1, y_2, y_3, y_4, x_5, x_6, x_7, x_8, y_5, y_6, y_7, y_8, \cdots, y_{4^k}, \\ &x_1, x_2, x_3, x_4, -y_1, -y_2, -y_3, -y_4, x_5, x_6, x_7, x_8, \\ &-y_5, -y_6, -y_7, -y_8, \cdots, -y_{4^k}) \end{aligned}$$

and

$$\begin{aligned} v = (&x_1, x_2, x_3, x_4, y_1, y_2, y_3, y_4, x_5, x_6, x_7, x_8, y_5, y_6, y_7, y_8, \cdots, y_{4^k}, \\ &-x_1, -x_2, -x_3, -x_4, y_1, y_2, y_3, y_4, -x_5, -x_6, -x_7, -x_8, \\ &y_5, y_6, y_7, y_8, \cdots, y_{4^k}). \end{aligned}$$

Besides the above constructions, Bent sequences can also be produced by Kronecker product, dyadic shift, Boolean variable transform, threshold logic synthesis, and so on. (e.g., see [12, 15–18].)

The definitions of SAC and Bent functions have been generalized as the following propagation criterion[9].

Definition 5.3.4 *A Boolean function $f(x_1, \cdots, x_n)$ is said to "satisfy the propagation criterion of degree k"* (PC *of degree k*) *if $f(x_1, \cdots, x_n)$ changes with a probability of one half whenever i, $1 \leqslant i \leqslant k$, bits of $x = (x_1, \cdots, x_n)$ are complemented, i.e.,*

$$w(f(x) + f(x+a)) = 2^{n-1}, \quad if \quad 1 \leqslant w(a) \leqslant k.$$

Thus the SAC (or equivalently H-Boolean functions) is equivalent to the PC of degree 1, and Bent functions satisfy the PC of the maximum degree n. The PC of degree k, $k > 1$, is also a PC of degree $k-1$, and hence a PC of degree 1, the H-Boolean function.

It seems plausible to study what happens if m bits are kept constant in functions that satisfy PC of degree k. This allows for the following more general classification of propagation characteristics of Boolean functions.

Definition 5.3.5[9] *A Boolean function $f(x_1, \cdots, x_n)$ is said to satisfy the propagation criterion of degree k and order m* (PC *of degree k and order m*) *if any function obtained from $f(x_1, \cdots, x_n)$ by keeping m input bits constant satisfies the* PC *of degree k.*

A function $f(x_1, x_2, \cdots, x_n)$ satisfying PC of degree 1 and order m produces an n-dimensional Hadamard matrix $A = [(-1)^{f(x_1, x_2, \cdots, x_n)}]$, in which all of the $(n-m)$-dimensional sections are $(n-m)$-dimensional Hadamard matrices of order 2. Thus all k-dimensional sections of this matrix A are Hadamard matrices of order 2 if $k \geqslant (n-m)$. In particular, functions satisfying PC of degree 1 and order $n-2$ are those H-Boolean functions that produce absolutely proper higher-dimensional Hadamard matrices of order 2. Thus, to Theorem 5.1.13, they are of the form

$$f(x_1, x_2, \cdots, x_n) = \sum_{1 \leqslant i < j \leqslant n} x_i x_j + a + \sum_{i=1}^{n} b_i x_i.$$

Similarly, functions satisfying PC of degree 1 and order $(n-3)$ are those functions of the form

$$f(x_1, x_2, \cdots, x_n) = \sum_{1 \leqslant i < j \leqslant n} x_i x_j + g(x_1, \cdots, x_n),$$

where

$$g(x_1, x_2, \cdots, x_n) = \sum_{1 \leqslant i < j \leqslant n} b_{ij} x_i x_j + a + \sum_{i=1}^{n} a_i x_i,$$

with the condition $\sum_i b_{ij} \leqslant 1$ and $\sum_j b_{ij} \leqslant 1$, i.e., $g(x_1, \cdots, x_n)$ satisfies $\deg(g(\cdot)) \leqslant 2$ such that every variable x_i occurs at most once in the second order terms.

5.3.3 Reed-Muller Codes and H-Boolean Functions[19]

Reed-Muller codes are one of the oldest and best understood families of error-correcting codes. This subsection explores connections between n-dimensional Hadamard matrices of order 2 and Reed-Muller codes[19].

Reed-Muller code can be defined very simply in terms of Boolean functions[10]:

Definition 5.3.6 *The r-th order binary Reed-Muller code $R(r, n)$ of length 2^n, for $0 \leqslant r \leqslant n$, is the set of all truth tables of Boolean functions of degree up to r.*

Let $H(2, r, n)$ denote an n-dimensional Hadamard matrix of order 2 in which all of the r-dimensional sections are r-dimensional Hadamard matrices of order 2. Thus a $H(2, 2, n)$ is an absolutely proper Hadamard matrix of order 2. An n-dimensional Hadamard matrix is clearly an $H(2, n, n)$. In general, an $H(2, r, n)$ is also an $H(2, r + i, n)$ for any $i = 1, 2, \cdots, n - r$. In fact, it is easy to prove that a (± 1)-valued higher-dimensional matrix is a Hadamard matrix if its lower-dimensional layers are Hadamard matrices.

Recall that every n-dimensional matrix of the form

$$X(f) = [(-1)^{f(x_1, x_2, \cdots, x_n)}]$$

corresponds to the Boolean function $f(x)$, where $x = (x_1, x_2, \cdots, x_n)$. Similarly, every 2^n-dimensional binary vector corresponds to a unique Boolean function $f(x)$: the binary vector $c(f) = (f(x))_x$ being the truth table of $f(x)$. Let $C(2, r, n)$ denote the set of vectors $c(f)$ where $X(f)$ is an $H(2, r, n)$.

Lemma 5.3.5[19] *Let $n > r \geqslant 1$ be two integers, and let $f(x)$ be a Boolean function of n variables. Then $c(f) \in R(r - 1, n)$ if and only if every r-dimensional section of $X(f)$ has even Hamming weight, where the Hamming weight of some section X, formulated by $w(X)$, refers to the number of 1s contained in that section.*

Proof First consider the Boolean function $f(x) = x_1 x_2 \cdots x_m$ where $m \leqslant n$. We show that every r-dimensional section of $X(f)$ has even weight if and only if $m < r$. Suppose $m \geqslant r$, and let X be the m-dimensional section of $X(f)$ defined by setting

$$x_{r+1} = x_{r+2} = \cdots = x_m = 1 \quad \text{and} \quad x_{m+1} = x_{m+2} = \cdots = x_n = 0.$$

Then its Hamming weight is $w(X) = 1$. Conversely, suppose $m < r$, and consider any r-dimensional section X of $X(f)$. For some $t > m$, x_t is not fixed. Let Y_0 and Y_1 be the $(r-1)$-dimensional sections of X that are defined by setting $x_t = 0$ and $x_t = 1$, respectively. Then $w(X) = w(Y_0) + w(Y_1)$, and $Y_0 = Y_1$; so $w(X)$ is even.

Now we prove the general result. Without loss of generality we may assume $x_1 x_2 \cdots x_m$ is the highest degree term in $f(x)$. By the argument above, if $m < r$, $X(f)$ is the sum of n-dimensional matrices whose r-dimensional sections all have even Hamming weights. Conversely, if $m \geqslant r$, then the m-dimensional section defined by setting $x_{m+1} = x_{m+2} =$

$\cdots = x_n = 0$ will have an odd Hamming weight, and, since this is the sum of Hamming weights of 2^{m-r} r-dimensional sections parallel to each other, there must be at least one r-dimensional section with odd Hamming weight. Since $c(f) \in R(r-1, n)$ if and only if the degree of $f(x)$ is less than r, the lemma is proved. □

Theorem 5.3.12[19] *Let $n \geqslant 2$. Then $C(2,2,n)$ is a coset of $R(1,n)$, the first-order Reed-Muller code.*

Proof Fix $c(f_1) \in C(2,2,n)$. Let

$$B = \{c(f_1) + c(f) : c(f) \in C(2,2,n)\}.$$

Observe that every 2-dimensional Hadamard matrix of order 2 has odd Hamming weight; so $c(f) \in C(2,2,n)$ if and only if every 2-dimensional section of $X(f)$ has odd Hamming weight, and, hence, $c(f) \in B$ if and only if every 2-dimensional section of $X(f)$ has even Hamming weight. Therefore, using Lemma 5.3.5, $B = R(1,n)$. □

The above Theorem 5.3.12 states that the set of absolutely proper Hadamard matrices of order 2 is equivalent to a coset of the first-order Reed-Muller code. This has the following three implications[19]:

(1) If the entries are taken in a prescribed order, each matrix (and its negation) corresponds to a row of a Hadamard matrix that is equivalent to the Sylvester matrix.

(2) Absolutely proper Hadamard matrices of order 2 may be repaired by any of the methods used to decode first-order Reed-Muller codes (the Yates transform will do this in order $n2^n$ steps).

(3) The values taken by any entry and its n adjacent entries determine the entire matrix. In particular, since each of these adjacent entries lies in separate $(n-1)$-dimensional sections, the absolutely proper n-dimensional Hadamard matrix of order 2 is unique up to the complementation of $(n-1)$-dimensional sections.

Using Theorem 5.1.3,

$$C(2,2,n) = \left\{ \sum_{1 \leqslant i < j \leqslant n} x_i x_j + a_0 + \sum_{i=1}^{n} a_i x_i \right\}.$$

In particular, $f(x) = \sum_{1 \leqslant i<j \leqslant n} x_i x_j$ is an $H(2,2,n)$. Put $m = \lfloor n/2 \rfloor$, and

$$g(x) = \begin{cases} \displaystyle\sum_{i=1}^{m} x_{2i-1}x_{2i}, & \text{for } n \text{ even ,} \\ \displaystyle\sum_{i=1}^{m} x_{2i-1}x_{2i} + x_n \sum_{i=1}^{2m} x_i, & \text{for } n \text{ odd.} \end{cases}$$

Then the Walsh transform of $g'(x) = (-1)^{g(x)}$ is

$$G'(w) = \begin{cases} 2^m \displaystyle\prod_{i=1}^{m} (-1)^{w_{2i}w_{2i-1}}, & \text{for } n \text{ even,} \\ 2^m (1 + (-1)^{m+|w|}) \displaystyle\prod_{i=1}^{m} (-1)^{w_{2i}w_{2i-1}}, & \text{for } n \text{ odd.} \end{cases}$$

Now put $b_i = \lfloor (i-1)/2 \rfloor$, and let L be the linear map such that $y = Lx$, where $y_n = x_n$ and for $i = 1, 2, \cdots, m$,

$$y_{2i-1} = x_{2i-1} + (x_{2i+1} + x_{2i+2} + \cdots + x_{2m}),$$

$$y_{2i} = x_{2i} + (x_{2i+1} + x_{2i+2} + \cdots + x_{2m}).$$

Then

$$f(x) + \sum_{i=1}^{2m} b_i x_i = g(y).$$

Finally, put

$$b = \begin{cases} (b_1, b_2, \cdots, b_{2m}), & \text{for } n \text{ even}, \\ (b_1, b_2, \cdots, b_{2m}, 0), & \text{for } n \text{ odd}. \end{cases}$$

Then the Walsh transform of $f'(x) = (-1)^{f(x)}$ satisfies

$$F'(wL^{\mathrm{T}}+b)=G'(w)=\begin{cases} 2^m \prod_{i=1}^{m} (-1)^{w_{2i}w_{2i-1}}, & \text{for } n \text{ even}, \\ 2^m(1+(-1)^{m+|w|}) \prod_{i=1}^{m} (-1)^{w_{2i}w_{2i-1}}, & \text{for } n \text{ odd}. \end{cases} \tag{5.16}$$

It follows that when $n = 2m$, half of the $H(2,2,n)$ has $2^{m-1}(2^m - 1)$ entries equal to -1, and the remainder has $2^{m-1}(2^m + 1)$ entries equal to -1. When $n = 2m + 1$, half of it has 2^{2m} entries equal to -1, one quarter of it has $2^{m-1}(2^m - 1)$ entries equal to -1, and the rest has $2^{m-1}(2^m + 1)$ entries equal to -1. The Equation 5.16 implies

Theorem 5.3.13[19] *Every $H(2, 2, 2m)$ corresponds to a Bent Function, or equivalently, every absolutely proper $(2m)$-dimensional Hadamard matrix of order 2 corresponds to a Bent function.*

This theorem is also a simple implication of Theorem 5.1.3.

Theorem 5.3.14[19] *If $n \geqslant r \geqslant 3$, then $C(2, r, n) \subset R(r-1, n)$.*

Proof Because of Theorem 5.1.3, the Hamming weight of every $H(2, n, n)$ is even. Now suppose that $X(f)$ is an $H(2, r, n)$, and $X(g)$ is a r-dimensional section of $X(f)$. Then $X(g)$ is Hadamard and the Hamming weight $w(g)$ is even. Indeed, from Lemma 5.3.5, $c(f) \in R(r-1, n)$ as required. To show that $C(2, r, n) \neq R(r-1, n)$, consider $f(x) = x_1 x_2 \cdots x_{r-1}$ and the r-dimensional section of $X(f)$ defined by putting $x_{r+1} = \cdots = x_n = 0$. □

So, for $n \geqslant r \geqslant 3$, if no more than $2^{n-r} - 1$ entries of an $H(2, r, n)$ are corrupted, then the usual methods used for decoding a corrupted codeword of $R(r-1, n)$ can be used to repair the $H(2, r, n)$. Precisely, by the known result of the minimum Hamming distance of Reed-Muller code $R(r-1, n)$, we have the following corollary:

Corollary 5.3.1[19] *If $n \geqslant r \geqslant 2$, then $C(2, r, n)$ has a minimum Hamming distance of at least 2^{n-r+1}.*

Theorem 5.3.15[19] *Let $n-1 \geqslant t \geqslant r \geqslant 1$ be integers. Let $g(\cdot)$ be any Boolean function of $(n-t)$ variables, and let $X(h)$ be an $H(2,r+1,n)$. If*

$$f(x_1,x_2,\cdots,x_n)=g(x_1,x_2,\cdots,x_{n-t})+h(x_1,x_2,\cdots,x_n),$$

then $X(f)$ is an $H(2,n+r-t,n)$.

Proof It is sufficient to show that any two opposed $(n-r-t-1)$-dimensional sections are orthogonal. The proof can be made by two cases according to whether g depends on the opposed index i_m. If $m>n-t$, then g has no bearing on the inner product. If $m\leqslant n-t$, then the result is obtained by fixing the varied indexes among the indexes $i_1,i_2,\cdots,i_{n-t}$, and using the orthogonality properties of $X(h)$. We do this explicitly for $t=1$ and leave the details of the general result to the interested reader. Without loss of generality, we need only show that the pair of $(n-1)$-dimensional sections defined by setting $x_1=0$ and 1 and the pair obtained by setting $x_n=0$ and 1 are orthogonal pairs. Equivalently, we must show that

$$w_1=w(f(0,x_2,\cdots,x_n)+f(1,x_2,\cdots,x_n))=2^{n-2}$$

and

$$w_2=w(f(x_1,\cdots,x_{n-1},0)+f(x_1,\cdots,x_{n-1},1))=2^{n-2}.$$

Now

$$w_2=w(h(x_1,\cdots,x_{n-1},0)+h(x_1,\cdots,x_{n-1},1))=2^{n-2},$$

since $X(h)$ is Hadamard, and

$$\begin{aligned}w_1=&w(g(0,x_2,\cdots,x_{n-r})+g(1,x_2,\cdots,x_{n-r})\\&+h(0,x_2,\cdots,x_n)+h(1,x_2,\cdots,x_n))\\=&\sum_{i=2}^{n-r}(\sum_{a_i=0}^{1}w(g(0,a_2,\cdots,a_{n-r})+g(1,a_2,\cdots,a_{n-r})\\&+h(0,a_2,\cdots,a_{n-r},x_{n-r+1},\cdots,x_n)\\&+h(1,a_2,\cdots,a_{n-r},x_{n-r+1},\cdots,x_n)))\\=&2^{n-r-1}2^{r-1},\end{aligned}$$

since every $(r+1)$-dimensional section of $X(h)$ is Hadamard. □

As a corollary, we obtain an extension of Theorem 5.3.12.

Corollary 5.3.2[19] *If $n\geqslant r\geqslant 2$, then $C(2,r,n)$ is a union of cosets of $R(1,n)$.*

Proof Let $s=r-1$. It is sufficient to prove that $X(f)$, where $f(x)=h(x)+\sum_{i=1}^{n}a_ix_i+a_0$, is an $H(2,s+1,n)$ whenever $X(h)$ is an $H(2,s+1,n)$. To do so, fix $h(x)$ and apply Theorem 5.3.15 (with $r=s$) successively to $g=g_i$, where $g_i=a_ix_i$, $i=1,2,\cdots,n$, and $g_0=a_0$. Since $t=n-1$, the result follows. □

Arguments similar to those used in the proof of Theorem 5.3.15 can be used to prove the following result.

Theorem 5.3.16[19] *Let h and g be Boolean functions of, respectively, t and $n-t$ variables and let*

$$\begin{aligned}f(x_1,x_2,\cdots,x_n)=&g(x_1,x_2,\cdots,x_{n-1})\\&+\sum_{i\leqslant t,\ j>t}x_ix_j+h(x_{n-t+1},\cdots,x_n).\end{aligned}$$

If $X(h)$ is an $H(2,r,t)$, the $X(f)$ is an $H(2,n-t+r,n)$. Moreover, if $X(g)$ is an $H(2,s,n-t)$ and $u = \max(t+s, n-t+r)$, then $X(f)$ is an $H(2,u,n)$.

Corollary 5.3.3[19] *Let $n \geqslant r \geqslant 2$. Then*

$$|C(2,r,n)| \geqslant \sum_{t=n-r+2}^{n} \binom{n}{t} \sum_{i=0}^{n-t} (-1)^i \binom{n-t}{i} 2^{t+i+2^{n-t-i}}.$$

Proof For all $i = 1, 2, \cdots, n$, let $h_i(x) = x_i \sum_{j=1}^{n} x_j$. We will say f "has attribute i" if $f(x) = h_i(x) + g(x) + ax_i$, where $a = 0$ or 1, and $g(x)$ is independent of x_i. Now let $N = \{1, 2, \cdots, n\}$, and, for all $T \subset N$, let $S(T)$ denote the set of $H(2,n,n)$ of the form $X(f)$ where, for all $i \in T$, f has property i. The typical element of $S(T)$ is $X(f)$ where

$$f(x) = g(x) + \sum_{i \in T,\ j \in N-T} x_i x_j + \sum_{i \in T} a_i x_i + \sum_{i<j,\ i,j \in T} x_i x_j,$$

where $g(x)$ is independent of x_i for all $i \in T$. Hence, if $t = | T |$, $| S(T) | = 2^t 2^{2^{n-t}}$. Using Theorem 5.3.16, every element of $S(T)$ is an $H(2, n-t+2, n)$. Now let $E(T)$ be the set of matrices of the form $X(f)$ where f has property i precisely when $i \in T$, then, by the Principle of Inclusion and Exclusion,

$$| E(T) | = \sum_{i=0}^{n-t} (-1)^i \binom{n-t}{i} 2^{t+i+2^{n-t-i}}.$$ □

For more details of higher-dimensional Hadamard matrices of order 2, the readers are recommended to see the papers [20–24].

Bibliography

[1] Shlichta P J. Higher-dimensional Hadamard matrices. *IEEE Trans. On Inform. Theory*, 1979, IT-25(5): 566-572.

[2] Xing Y and Yixian Yang. On the H-Boolean functions (II). *J. of Electronics (China)*, 1989, 19(2): 214-216.

[3] Yang Y X. On the H-Boolean functions. *J. of Beijing Univ. of Posts and Telecomm*, 1988, 11(3): 1-9.

[4] Webster A F and Tavares S E. On the design of S-box//*Advances in Cryptology, Proc. Crypto'85*. Berlin: Springer-Verlag, 1986: 523-534.

[5] Lloyd S A. Characterising and counting functions satisfying the strict avalanche criterion of order $(n-3)$. *Proc. of the Second IMA Conf. on Cryptography and Coding*, 1989. Oxford: Clarendon Press, 1992: 165-172.

[6] Lloyd S A. Counting binary functions with certain cryptographic properties. *J. of Cryptology*, 1992, 5: 107-131.

[7] Forre R. The strict avalanche criterion: spectral properties of Boolean functions and an extended definition//*Advances in Cryptology, Proc. Crypto*'88. Berlin: Springer-Verlag, 1986: 450-468.

[8] Cusick T W. Boolean functions satisfying a higher order strict avalanche criterion. *Proc. of Eurocrypt*'93, 86-95.

[9] Preneel B, Vanleekwijk W etc. Propagation characteristics of Boolean functions. *Eurocrypt*'90, 1991: 161-173.

[10] Macwilliams F J, Sloane N. *The Theory of Error-Correcting Codes*. New York: North-Holland, 1977.

[11] Yixian Yang and Lin X D. *Coding and Cryptography*. Beijing: PPT Press, 1992.

[12] Yarlagadda R and Hershey J E. Analysis and synthesis of Bent sequences. *IEE Proc.*, 1989, 136(2): 112-123.

[13] Adams C M and Tavares S E. Generating Bent sequences. *IEEE Trans. Inform. Theory*, 1990, 36(5): 1170-1173.

[14] Guo B A and Cai C N. Generating and counting a class of binary Bent functions which is neither Bent-based nor linear-based. *Chinese Science Bulletin*, 1992, 37(6): 517-520.

[15] Carlet C. Two new classes of Bent functions. *Proc. of Eurocrypt'93*. Lofthus, Norway, 1993: 75-85.

[16] Yixian Yang and Hu Z M. Dyadic codes with single valued correlations (I). *Chinese J. of Electronics*, 1988, 16(6): 50-55.

[17] Hu Z M and Yixian Yang. Dyadic codes with single valued correlations (II). *J. of China Institute of Communications*, 1989, 10(5): 42-46.

[18] Hu Z M and Yixian Yang. On the dyadic cross-correlations of dyadic codes. *J. of China Institute of Communications*, 1993, 14(1): 15-21.

[19] Launey W. A note on n-dimensional Hadamard matrices of drder 2^t and Reed-Muller codes. *IEEE Trans. on Inform. Theory*, 1991, 27(3): 664-667.

[20] Yang Y X. n-dimensional Hadamard matrices of order two. *J. of Beijing Univ. of Posts and Telecomm*, 1991, 14(4): 1-8.

[21] Yixian Yang and Zhenming Hu. On the classification of 4-dimensional Hadamard matrices of order 2. *J. of Systems Science and Mathematical Sciences*, 1987, 7(1): 40-46.

[22] Xinan Pan and Yixian Yang. On the enumeration of 5-dimensional Hadamard matrices of order 2. *J. of Beijing Univ. of Posts and Telecomm*, 1987, 10(4): 11-19.

[23] Li S Q and Yixian Yang. The final solution of enumerating the 5-dimensional Hadamard matrices of order 2. *J. of Beijing Univ. of Posts and Telecomm*, 1988, 11(2): 17-21.

[24] Hammer J and Seberry J. Higher-dimensional orthogonal designs and applications. *IEEE Trans. Inform. Theory*, 1981, 27(6): 772-779.

Chapter 6

General Higher-Dimensional Hadamard Matrices

When m^n ($n \geqslant 2$ an integer) elements are given they can be arranged in the form of an n-dimensional cube of order m (in short, an n-cube). An n-cube can be mathematically described by the matrix form $A = [A(i_1, i_2, \cdots, i_n)]$, $0 \leqslant i_1, i_2, \cdots, i_n \leqslant m - 1$. The elements which have all the same suffixes, with the exception of i_k and/or i_s, lie in the same two-dimensional layer parallel to a coordinate axis (a plane). Thus an n-cube can be treated as a set of m^{n-2} two-dimensional square matrices $B = [B(x, y)] = [A(a_1, a_2, \cdots, x, \cdots, y, \cdots, a_n)]$, $0 \leqslant x, y \leqslant m - 1$, where the a_is are prefixed integers. The elements which have all the same suffies, except i_k, lie in the same row (line). Thus an n-cube can be treated as a set of m^{n-1} lines of length m. The elements which have only one suffix in common lie in an $(n - 1)$-dimensional layer. Thus an n-cube can also be treated as a set of m $(n - 1)$-dimensional layers. With regard to practical applications, the most obvious advantage of higher-dimensional Hadamard matrices is the presence or absence of property. Some higher-dimensional Hadamard matrices, especially, those proper Hadamard matrices of dimension $n \geqslant 4$ and those n-dimensional Hadamard matrices of order two will prove advantageous in error correcting codes. Certain forms of higher-dimensional Hadamard matrices, e.g., H-Boolean functions, may be of value in security coding because of their resemblance to random matrices. Hadamard matrices are kinds of paradigms of random binary matrices in which the correlation values for any pair of parallel $(n - 1)$-dimensional sections (i.e., zero) are the expected values for such correlations in a random matrix. It therefore seems plausible to regard any sufficiently large random binary matrix as potentially a Hadamard matrix with errors which can be located and corrected. Those Hadamard matrices derived by checking a random binary matrix might usefully combine error correction with immunity from unauthorized decoding, because of the absence of an obvious pattern and their resemblance to random matrices.

This chapter concentrates on the theory of higher-dimensional Hadamard matrices and their generalizations. Definitions, properties, existences, constructions, and the other related topics will be systematically introduced in the coming sections.

6.1 Definitions, Existences and Constructions

The definition of n-dimensional Hadamard matrices has been informally stated in the last few chapters. Now we present, here, its mathematical definition.

Definition 6.1.1[1] *A general n-dimensional Hadamard matrix of order m is an n-dimensional binary matrix*

$$H = [H(i_1, i_2, \cdots, i_n)], \quad 0 \leqslant i_k \leqslant m-1, \quad 1 \leqslant k \leqslant n$$

of size $\overbrace{m \times \cdots \times m}^{n}$ *in which all parallel $(n-1)$-dimensional sections (layers) are mutually orthogonal; that is, all $H(i_1, i_2, \cdots, i_n) = \pm 1$ and*

$$\sum_{i_1=0}^{m-1} \cdots \sum_{i_{n-1}=0}^{m-1} H(i_1, i_2, \cdots, i_{n-1}, a) H(i_1, i_2, \cdots, i_{n-1}, b) = m^{(n-1)} \delta_{ab},$$

$$\sum_{i_1=0}^{m-1} \cdots \sum_{i_{n-2}=0}^{m-1} \sum_{i_n=0}^{m-1} H(i_1, i_2, \cdots, i_{n-2}, a, i_n) H(i_1, i_2, \cdots, i_{n-2}, b, i_n) = m^{(n-1)} \delta_{ab},$$

$$\cdots\cdots\cdots$$

$$\sum_{i_2=0}^{m-1} \cdots \sum_{i_n=0}^{m-1} H(a, i_2, i_3, \cdots, i_n) H(b, i_2, i_3, \cdots, i_n) = m^{(n-1)} \delta_{ab}. \tag{6.1}$$

6.1.1 n-Dimensional Hadamard Matrices of Order $2k$

Recall that the order m of a non-trivial 2-dimensional Hadamard matrix has to be $m = 2$ or $m = 4k$, while the order m of a non-trivial 3-dimensional Hadamard matrix can be $m = 2k$. In this subsection, it will be proved, at first, that the orders of the non-trivial general n-dimensional Hadamard matrices are also necessarily even integers. Then infinite families of $n(\geqslant 4)$-dimensional Hadamard matrices of order $2k$ will be constructed by algebraic approaches.

Theorem 6.1.1 *Let $H = [H(h_1, \cdots, h_n)]$, $0 \leqslant h_k \leqslant m-1$, $1 \leqslant k \leqslant n$, be an $n(\geqslant 3)$-dimensional Hadamard matrix of order m. If $m > 1$, then m is necessarily even.*

Proof From Definition 6.1.1, we have the following

$$\sum_{0 \leqslant h_2, \cdots, h_n \leqslant m-1} H(a, h_2, \cdots, h_n) H(b, h_2, \cdots, h_n) = m^{n-1} \delta(a, b), \tag{6.2}$$

which implies

$$\begin{aligned}
&\sum_{0 \leqslant h_2, \cdots, h_n \leqslant m-1} [H(0, h_2, \cdots, h_n) + H(1, h_2, \cdots, h_n)] H(0, h_2, \cdots, h_n) \\
= &\sum_{0 \leqslant h_2, \cdots, h_n \leqslant m-1} [H(0, h_2, \cdots, h_n) H(0, h_2, \cdots, h_n) \\
&+ H(0, h_2, \cdots, h_n) H(1, h_2, \cdots, h_n)] \\
= &\sum_{0 \leqslant h_2, \cdots, h_n \leqslant m-1} [1 + H(0, h_2, \cdots, h_n) H(1, h_2, \cdots, h_n)] \\
= &\, m^{n-1} \quad \text{(by Equation (6.2))}.
\end{aligned} \tag{6.3}$$

Because $H(h_1, \cdots, h_n) = \pm 1$, $H(0, h_2, \cdots, h_n) + H(1, h_2, \cdots, h_n)$ is even. In other words, the left hand side of Equation (6.3) is the summation of some even integers, and its right hand side, m^{n-1}, is also even. So the order m has to be even. □

Theorem 6.1.2 *Let $1 \leqslant i, j, k, l \leqslant 6$ be integers with their binary expended forms*

$$i = \sum_{s=0}^{2} i_s 2^s, \quad j = \sum_{s=0}^{2} j_s 2^s, \quad k = \sum_{s=0}^{2} k_s 2^s, \quad \text{and} \quad l = \sum_{s=0}^{2} l_s 2^s.$$

Then the matrix $H = [H(i, j, k, l)]$, $1 \leqslant i, j, k, l \leqslant 6$, defined by

$$H(i, j, k, l) = (-1)^{i_0 j_0 + i_1 k_1 + i_2 l_2 + j_1 l_1 + j_2 k_0 + k_2 l_0}$$

is a 4-dimensional Hadamard matrix of order 6.

Proof Let $a = \sum_{s=0}^{2} a_s 2^s$ and $b = \sum_{s=0}^{2} b_s 2^s$ be two different integers where $1 \leqslant a \neq b \leqslant 6$. Then

$$\begin{aligned}
&\sum_{1 \leqslant j,k,l \leqslant 6} H(a, j, k, l) H(b, j, k, l) \\
&= \sum_{1 \leqslant j,k,l \leqslant 6} (-1)^{(a_0+b_0)j_0 + (a_1+b_1)k_1 + (a_2+b_2)l_2} \\
&= \left[\sum_{j=1}^{6} (-1)^{(a_0+b_0)j_0}\right] \left[\sum_{k=1}^{6} (-1)^{(a_1+b_1)k_1}\right] \left[\sum_{l=1}^{6} (-1)^{(a_2+b_2)l_2}\right].
\end{aligned} \tag{6.4}$$

The non-equality $a \neq b$ implies $a_0 + b_0 = 1$, $a_1 + b_1 = 1$, or $a_2 + b_2 = 1$. These three possible cases are separately studied as follows:

Case 1. $a_0 + b_0 = 1$ implies $\sum_{j=1}^{6} (-1)^{(a_0+b_0)j_0} = 0$, and the first term in Equation (6.4) is vanished;

Case 2. $a_1 + b_1 = 1$ implies $\sum_{k=1}^{6} (-1)^{(a_1+b_1)k_1} = 0$, and the second term in Equation (6.4) is vanished;

Case 3. $a_2 + b_2 = 1$ implies $\sum_{l=1}^{6} (-1)^{(a_2+b_2)l_2} = 0$, and the third term in Equation (6.4) is vanished.

Thus the left hand side of Equation (6.4) always vanishes, i.e.,

$$\sum_{1 \leqslant j,k,l \leqslant 6} H(a, j, k, l) H(b, j, k, l) = 6^3 \delta(a, b).$$

Similarly, it can be proved that the following equations are true:

$$\begin{aligned}
\sum_{1 \leqslant i,k,l \leqslant 6} H(i, a, k, l) H(i, b, k, l) &= \sum_{1 \leqslant i,j,l \leqslant 6} H(i, j, a, l) H(i, j, b, l) \\
&= \sum_{1 \leqslant i,j,k \leqslant 6} H(i, j, k, a) H(i, j, k, b) \\
&= 6^3 \delta(a, b).
\end{aligned}$$

Hence the matrix $H = [H(i, j, k, l)]$ is a 4-dimensional Hadamard matrix of order 6. □

In order to construct more higher-dimensional Hadamard matrices of general order $2k$, we first prove the following important lemma.

Lemma 6.1.1 *Let n and k be two integers such that $k \leqslant 2^{n-1}$. Then there exists a 2-dimensional $(0,1)$-valued matrix $A = [A(i,j)]$ of size $(2k) \times n$ which satisfies the following two conditions:*

(1) *The rows of A are different from each other;*

(2) *Each column of A contains k 0s and k 1s.*

A matrix satisfying both of these two conditions is called a column-balanced matrix, because each of its columns is balanced by 1s and 0s.

Proof We show a constructive proof for this lemma.

Let $A_0 = [A_0(i,j)]$, $0 \leqslant i \leqslant 2^n - 1$, $0 \leqslant j \leqslant n-1$, be the $(0,1)$-valued matrix such that its i-th row $(A_0(i,0), \cdots, A_0(i,n-1))$ is the binary expended vector of the integer i, i.e., $i = \sum_{k=0}^{n-1} A_0(i,k)2^k$. It is easy to verify that this matrix A_0 is a column-balanced matrix of size $2^n \times n$.

Two binary vectors $(i_0, \cdots, i_{n-1})$ and $(j_0, \cdots, j_{n-1})$ are said to be a complementary-pair if

$$(i_0, \cdots, i_{n-1}) + (j_0, \cdots, j_{n-1}) = (1, \cdots, 1).$$

The rows of A_0 clearly consist of 2^{n-1} complementary-pairs. For any integer r, $0 \leqslant r \leqslant 2^{n-1} - 1$, a column-balanced matrix of size $(2^n - 2r) \times n$ is obtained by deleting r complementary-pairs from the matrix A_0. □

Theorem 6.1.3 *Let $n = 2s$ be even, $m \leqslant s - 1$, and $k \leqslant 2^{m-1}$. And let $B = [B(i,j)]$, $0 \leqslant i \leqslant 2k-1$, be a column-balanced matrix of size $(2k) \times m$. Then the following matrix $H = [H(h(0), \cdots, h(n-1))]$, $0 \leqslant h(i) \leqslant 2k-1$, $0 \leqslant i \leqslant n-1$, defined by*

$$H(h(0), \cdots, h(n-1)) = (-1)^{\sum_{i=0}^{s-1} \sum_{j=0}^{m-1} B(h(i),j) B(h((i+j) \bmod s + s), j)}$$

is an n-dimensional Hadamard matrix of order $2k$.

Proof For every prefixed integer l, $0 \leqslant l \leqslant n-1$, let

$$\begin{aligned}\alpha(a,b) = & \sum_{0 \leqslant h(0), \cdots, h(l-1), h(l+1), \cdots, h(n-1) \leqslant 2k-1} H(h(0), \cdots, h(l-1), \\ & a, h(l+1), \cdots, h(n-1)) H(h(0), \cdots, h(l-1), \\ & b, h(l+1), \cdots, h(n-1)),\end{aligned} \tag{6.5}$$

where $0 \leqslant a, b \leqslant 2k-1$.

It is sufficient to prove that $\alpha(a,b) = (2k)^{n-1} \delta(a,b)$. The following two cases are separately considered:

Case 1. If $0 \leqslant l \leqslant s-1$, then by the definition of the matrix H, we have

$$\begin{aligned}\alpha(a,b) = & \sum_{0 \leqslant h(0), \cdots, h(l-1), h(l+1), \cdots, h(n-1) \leqslant 2k-1} \\ & (-1)^{\sum_{j=0}^{m-1} [B(a,j) + B(b,j)] B(h((l+j) \bmod s + s), j)} \\ = & \sum_{0 \leqslant h(r) \leqslant 2k-1,\ r \neq l,\ l+s,\ (l+1) \bmod s + s,\ \cdots,\ (l+m-1) \bmod s + s}\end{aligned}$$

$$\left\{\sum_{0\leqslant h(l+s),h((l+1)\bmod s+s),\cdots,h((l+m-1)\bmod s+s)\leqslant 2k-1} (-1)^{\sum_{j=0}^{m-1}[B(a,j)+B(b,j)]B(h((l+j)\bmod s+s),j)}\right\}$$

$$= \sum_{0\leqslant h(r)\leqslant 2k-1,\ r\neq l,\ l+s,\ (l+1)\bmod s+s,\ \cdots,\ (l+m-1)\bmod s+s} \prod_{j=0}^{m-1} \left\{\sum_{0\leqslant h((l+j)\bmod s+s)\leqslant 2k-1} (-1)^{[B(a,j)+B(b,j)]B(h((l+j)\bmod s+s),j)}\right\}.$$

Because the rows of the matrix $B=[B(i,j)]$ are different from each other, $a\neq b$ implies the existence of some integer j_0, $0\leqslant j_0\leqslant m-1$, such that $B(a,j_0)+B(b,j_0)=1$, which implies, further, the following

$$\sum_{0\leqslant h((l+j_0)\bmod s+s)\leqslant 2k-1} (-1)^{[B(a,j_0)+B(b,j_0)]B(h((l+j_0)\bmod s+s),j_0)} = 0,$$

which is owed to the columns of $B=[B(i,j)]$ being balanced by 1s and 0s.

Hence, we have proved that if $0\leqslant l\leqslant s-1$, then $\alpha(a,b)=(2k)^{n-1}\delta(a,b)$.

Case 2. If $s\leqslant l\leqslant n-1$, then in the same way as that used in Case 1, it can be proved that the equation $\alpha(a,b)=(2k)^{n-1}\delta(a,b)$ is also true. The theorem follows. □

Theorem 6.1.4 *Let $A=[A(i,j)]$, $0\leqslant i,j\leqslant(2t)^s-1$, be a 2-dimensional Hadamard matrix of order $(2t)^s$, where $s>1$ and t are positive integers. Then the following matrix $H=[H(x_0,\cdots,x_{s-1},y_0,\cdots,y_{s-1})]$, $0\leqslant x_i,y_i\leqslant 2t-1$, $0\leqslant i\leqslant s-1$, defined by*

$$\begin{aligned}&H(x_0,\cdots,x_{s-1},y_0,\cdots,y_{s-1})\\&=A((2t)^{s-1}x_{s-1}+(2t)^{s-2}x_{s-2}+\cdots+2tx_1\\&\quad+x_0,(2t)^{s-1}y_{s-1}+(2t)^{s-2}y_{s-2}+\cdots+2ty_1+y_0)\end{aligned}$$

is a $(2s)$-dimensional Hadamard matrix of order $2t$.

Proof From the definition of the matrix H, we have

$$\begin{aligned}\alpha(a,b)=&\sum_{0\leqslant x_0,\cdots,x_{s-1},y_0,\cdots,y_{s-1}\leqslant 2t-1} H(a,x_1,\cdots,x_{s-1},y_0,\cdots,y_{s-1})\\&\times H(b,x_1,\cdots,x_{s-1},y_0,\cdots,y_{s-1})\\=&\sum_{0\leqslant x_0,\cdots,x_{s-1}\leqslant 2t-1}\left[\sum_{0\leqslant y_0,\cdots,y_{s-1}\leqslant 2t-1} A((2t)^{s-1}x_{s-1}+\cdots+2tx_1+a,\right.\\&(2t)^{s-1}y_{s-1}+\cdots+2ty_1+y_0)A((2t)^{s-1}x_{s-1}+\cdots+2tx_1+b,\\&\left.(2t)^{s-1}y_{s-1}+\cdots+2ty_1+y_0)\right].\end{aligned} \tag{6.6}$$

If $a\neq b$, then

$$\begin{aligned}&(2t)^{s-1}x_{s-1}+(2t)^{s-2}x_{s-2}+\cdots+2tx_1+a\\&\neq(2t)^{s-1}x_{s-1}+(2t)^{s-2}x_{s-2}+\cdots+2tx_1+b.\end{aligned}$$

Thus the inner summation of Equation (6.6) is

$$\sum_{0\leqslant y_0,\cdots,y_{s-1}\leqslant 2t-1}[A((2t)^{s-1}x_{s-1}+\cdots+2tx_1+a,\ (2t)^{s-1}y_{s-1}+\cdots$$

$$+2ty_1+y_0)A((2t)^{s-1}x_{s-1}+\cdots+2tx_1+b,\ (2t)^{s-1}y_{s-1}+\cdots+2ty_1+y_0)]$$

$$=\sum_{r=0}^{(2t)^s-1}A((2t)^{s-1}x_{s-1}+\cdots+2tx_1+a,r)A((2t)^{s-1}x_{s-1}+\cdots+2tx_1+b,r)$$

$$=0,$$

where the last equation is due to the fact that the matrix $A=[A(i,j)]$ is a 2-dimensional Hadamard matrix. Thus we have

$$\sum_{r=0}^{(2t)^s-1}A(u,r)A(v,r)=0,\quad \text{if } u\neq v.$$

In the same way it can be proved that for each l, $0\leqslant l\leqslant 2s-1$, we have the equation

$$\sum_z H(z_0,\cdots,z_{l-1},a,z_{l+1},\cdots,z_{2s-1})H(z_0,\cdots,z_{l-1},b,z_{l+1},\cdots,z_{2s-1})$$
$$=(2t)^{2s-1}\delta(a,b).$$

Hence the matrix H is a $(2s)$-dimensional Hadamard matrix of order $2t$. □

Lemma 6.1.2 *If $0\leqslant a,b,j\leqslant N-1$ and $a\neq b$, then $(a+j)\mathrm{mod}N\neq(b+j)\bmod N$.*

The following theorem provides us a recursive construction of higher-dimensional Hadamard matrices from lower-dimensional ones.

Theorem 6.1.5 *Let $H=[H(i_1,\cdots,i_n)]$, $0\leqslant i_1,\cdots,i_n\leqslant N-1$, be an n-dimensional Hadamard matrix of order N. Then the matrix $A=[A(i_1,\cdots,i_n,i_{n+1})]$, $0\leqslant i_1,\cdots,i_n,i_{n+1}\leqslant N-1$, defined by*

$$A(i_1,\cdots,i_n,i_{n+1})=H(i_1,\cdots,i_{n-1},(i_n+i_{n+1})\mathrm{mod}N))$$

is an $(n+1)$-dimensional Hadamard matrix of order N.

Proof Let

$$\begin{aligned}\alpha(a,b)&=\sum_{0\leqslant i_2,\cdots,i_{n+1}\leqslant N-1}A(a,i_2,\cdots,i_{n+1})A(b,i_2,\cdots,i_{n+1})\\&=\sum_{0\leqslant i_2,\cdots,i_{n+1}\leqslant N-1}H(a,i_2,\cdots,i_{n-1},(i_n+i_{n+1})\mathrm{mod}N))\\&\quad\times H(b,i_2,\cdots,i_{n-1},(i_n+i_{n+1})\mathrm{mod}N))\\&=\sum_{i_{n+1}=0}^{N-1}\Bigg\{\sum_{0\leqslant i_2,\cdots,i_n\leqslant N-1}H(a,i_2,\cdots,i_{n-1},(i_n+i_{n+1})\mathrm{mod}N))\\&\quad\times H(b,i_2,\cdots,i_{n-1},(i_n+i_{n+1})\mathrm{mod}N))\Bigg\}.\end{aligned}\tag{6.7}$$

For each i_{n+1}, $0 \leqslant i_{n+1} \leqslant N-1$, and $a \neq b$, the inner summation of Equation (6.7) is

$$\begin{aligned}
&\sum_{0\leqslant i_2,\cdots,i_n\leqslant N-1} H(a,i_2,\cdots,i_{n-1},(i_n+i_{n+1})\text{mod}N)) \\
&\times H(b,i_2,\cdots,i_{n-1},(i_n+i_{n+1})\text{mod}N)) \\
=&\sum_{0\leqslant i_2,\cdots,i_n\leqslant N-1} H(a,i_2,\cdots,i_n)H(b,i_2,\cdots,i_n) \\
=&0,
\end{aligned}$$

where the last equation is due to the fact that $H = [H(i_1,\cdots,i_n)]$ is an n-dimensional Hadamard matrix of order N.

Similarly, it can be proved that for each l, $1 \leqslant l \leqslant n$,

$$\begin{aligned}
&\sum_i A(i_1,\cdots,i_{l-1},a,i_{l+1},\cdots,i_{n+1})A(i_1,\cdots,i_{l-1},b,i_{l+1},\cdots,i_{n+1}) \\
=&N^n\delta(a,b).
\end{aligned}$$

Let

$$\begin{aligned}
\beta(a,b)=&\sum_{0\leqslant i_1,\cdots,i_n\leqslant N-1} A(i_1,\cdots,i_n,a)A(i_1,\cdots,i_n,b) \\
=&\sum_{0\leqslant i_1,\cdots,i_n\leqslant N-1} H(i_1,\cdots,i_{n-1},(i_n+a)\ \text{mod}N)) \\
&\times H(i_1,\cdots,i_{n-1},(i_n+b)\ \text{mod}N)) \\
=&\sum_{i_n=0}^{N-1}\Bigg\{\sum_{0\leqslant i_1,\cdots,i_{n-1}\leqslant N-1} H(i_1,\cdots,i_{n-1},(i_n+a)\ \text{mod}N)) \\
&\times H(i_1,\cdots,i_{n-1},(i_n+b)\ \text{mod}N))\Bigg\}.
\end{aligned} \tag{6.8}$$

From Lemma 6.1.2, $a \neq b$ implies $(i_n+b)\text{mod}N \neq (i_n+a)\text{mod}N$. Thus for each i_n, $0 \leqslant i_n \leqslant N-1$, the inner summation of the last Equation (6.8) is

$$\begin{aligned}
&\sum_{0\leqslant i_1,\cdots,i_{n-1}\leqslant N-1} H(i_1,\cdots,i_{n-1},(i_n+a)\text{mod}N)) \\
&\times H(i_1,\cdots,i_{n-1},(i_n+b)\text{mod}N)) = 0.
\end{aligned}$$

Hence $\beta(a,b) = N^n\delta(a,b)$. □

Recall that the famous Hadamard Conjecture states that "for each positive integer k, there exists at least one 2-dimensional Hadamard of order $4k$." Theorems 6.1.5 and 6.1.3 lead to the interesting statement: "If the Hadamard Conjecture had been proved, then for each positive $n(\geqslant 4)$ and $2t$, there exists at least one n-dimensional Hadamard matrix of order $2t$, even though t is odd."

Theorem 6.1.5 can be generalized further. For example, we have

Theorem 6.1.6 *Let $H = [H(i_1,\cdots,i_n)]$, $0 \leqslant i_1,\cdots,i_n \leqslant N-1$ be an n-dimensional Hadamard matrix of order N. And let r be an integer satisfying $\gcd(r,N) = 1$. Then the matrix $A = [A(i_1,\cdots,i_{n+1})]$, $0 \leqslant i_1,\cdots,i_{n+1} \leqslant N-1$, defined by*

$$A(i_1, \cdots, i_{n+1}) = H(i_1, \cdots, i_{n-1}, (i_n + ri_{n+1}) \bmod N)$$

is an $(n+1)$-dimensional Hadamard matrix of order N.

Clearly, this theorem reduces to Theorem 6.1.5 if $r = 1$.

Proof For $0 \leqslant a, b \leqslant N-1$, let

$$\alpha(a, b) = \sum_{0 \leqslant i_2, \cdots, i_{n+1} \leqslant N-1} A(a, i_2, \cdots, i_{n+1}) A(b, i_2, \cdots, i_{n+1}),$$

$$\beta(a, b) = \sum_{0 \leqslant i_1, \cdots, i_{n-1}, i_{n+1} \leqslant N-1} A(i_1, \cdots, i_{n-1}, a, i_{n+1}) A(i_1, \cdots, i_{n-1}, b, i_{n+1})$$

and

$$\gamma(a, b) = \sum_{0 \leqslant i_1, \cdots, i_n \leqslant N-1} A(i_1, \cdots, i_n, a) A(i_1, \cdots, i_n, b).$$

It is sufficient to prove that

$$\alpha(a, b) = \beta(a, b) = \gamma(a, b) = N^n \delta(a, b).$$

The equation $\alpha(a, b) = N^n \delta(a, b)$ can be proved in the same way as that used in the proof of Theorem 6.1.5.

If $a \neq b$, then

$$\begin{aligned}
\beta(a, b) &= \sum_{0 \leqslant i_1, \cdots, i_{n-1}, i_{n+1} \leqslant N-1} H(i_1, \cdots, i_{n-1}, (a + ri_{n+1}) \bmod N) \\
&\quad \times H(i_1, \cdots, i_{n-1}, (b + ri_{n+1}) \bmod N) \\
&= \sum_{i_{n+1}=0}^{N-1} \left\{ \sum_{0 \leqslant i_1, \cdots, i_{n-1} \leqslant N-1} \times H(i_1, \cdots, i_{n-1}, (a + ri_{n+1}) \bmod N) \right. \\
&\quad \left. \times H(i_1, \cdots, i_{n-1}, (b + ri_{n+1}) \bmod N) \right\} \\
&= \sum_{i_{n+1}=0}^{N-1} 0 \\
&= 0,
\end{aligned}$$

where the last two equations are due to the facts (1) H is a Hadamard matrix, and (2) $(b + ri_{n+1}) \bmod N \neq (a + ri_{n+1}) \bmod N$, if $0 \leqslant a \neq b \leqslant N-1$.

Hence we have proved that $\beta(a, b) = N^n \delta(a, b)$.

If $0 \leqslant a \neq b \leqslant N-1$, and $\gcd(r, N) = 1$, then $[(i_n + ra) - (i_n + rb)] \bmod N = [r(a-b)] \bmod N \neq 0$, or equivalently, $(i_n + ra) \bmod N \neq (i_n + rb) \bmod N$. Thus we have

$$\begin{aligned}
\gamma(a, b) &= \sum_{0 \leqslant i_1, \cdots, i_n \leqslant N-1} H(i_1, \cdots, i_{n-1}, (i_n + ra) \bmod N) \\
&\quad \times H(i_1, \cdots, i_{n-1}, (i_n + rb) \bmod N) \\
&= \sum_{i_n=0}^{N-1} \left\{ \sum_{0 \leqslant i_1, \cdots, i_{n-1} \leqslant N-1} H(i_1, \cdots, i_{n-1}, (i_n + ra) \bmod N) \right. \\
&\quad \left. \times H(i_1, \cdots, i_{n-1}, (i_n + rb) \bmod N) \right\}
\end{aligned}$$

$$= \sum_{i_n=0}^{N-1} 0$$
$$= 0,$$

where the last two equations result from H being a Hadamard matrix.

Hence we have proved that $\gamma(a, b) = N^n \delta(a, b)$. The theorem follows. □

Let p be a positive integer. Then every integer a, $0 \leqslant a \leqslant p^m - 1$, can be uniquely decomposed by $a = \sum_{i=0}^{m-1} a_i p^i$, $0 \leqslant a_i \leqslant p - 1$, $0 \leqslant i \leqslant m - 1$. Thus there is a one-to-one mapping from the integer a to the m-dimensional vector $(a_0, \cdots, a_{m-1})$. Similarly to the binary ($p = 2$) case, we use the symbol a to represent both the integer a or its corresponding p-ary vector $(a_0, \cdots, a_{m-1})$. Let $a = (a_0, \cdots, a_{m-1})$ and $b = (b_0, \cdots, b_{m-1})$ be two integers, $0 \leqslant a, b \leqslant p^m - 1$. Then the p-dyadic summation between a and b, denoted by $a \oplus_p b$, is defined by the integer corresponding to $((a_0 + b_0) \bmod p, \cdots, (a_{m-1} + b_{m-1}) \bmod p)$. Thus the regular dyadic summation, bit-wise mod2 summation, is the special case of p-dyadic summation for $p = 2$.

Theorem 6.1.7 *Let $A = [A(i_1, \cdots, i_n)]$, $0 \leqslant i_1, \cdots, i_n \leqslant p^m - 1$, be an n-dimensional Hadamard matrix of order p^m, then the following matrix $H = [H(i_1, \cdots, i_{n+1})]$, $0 \leqslant i_1, \cdots, i_{n+1} \leqslant p^m - 1$, defined by*

$$H(i_1, \cdots, i_{n+1}) = A(i_1, \cdots, i_{n-1}, i_n \oplus_p i_{n+1}),$$

is an $(n+1)$-dimensional Hadamard matrix of order p^m.

The proof of Theorem 6.1.7 can be finished in the same way as that of Theorem 6.1.5.

Let $X = \{0, 1, \cdots, N-1\}$ be an integer-set, and $f(x, y)$ a mapping from $X \times X$ to X such that the following conditions are satisfied:

(1) For any given $a \in X$, both $f(a, y)$ and $f(x, a)$ are one-to-one mapping from X to itself;

(2) If $y_1 \neq y_2$, then $f(x, y_1) \neq f(x, y_2)$ for each $x \in X$. If $x_1 \neq x_2$, then $f(x_1, y) \neq f(x_2, y)$ for each $y \in X$.

A mapping $f(x, y)$ satisfying the above two conditions is called an H-mapping.

Theorem 6.1.8 *Let $A = [A(i_1, \cdots, i_n)]$, $0 \leqslant i_1, \cdots, i_n \leqslant N - 1$ be an n-dimensional Hadamard matrix of order N, and $f(x, y)$ an H-mapping. Then the following matrix $H = [H(i_1, \cdots, i_{n+1})]$, $0 \leqslant i_1, \cdots, i_{n+1} \leqslant N - 1$, defined by*

$$H(i_1, \cdots, i_{n+1}) = A(i_1, \cdots, i_{n-1}, f(i_n, i_{n+1}))$$

is an $(n+1)$-dimensional Hadamard matrix of order N.

It is easy to see that Theorem 6.1.8 reduces to Theorems 6.1.5–6.1.7, if we choose $f(x, y) = (x + y) \bmod N$, $f(x, y) = (x + ry) \bmod N$, $\gcd(r, N) = 1$, and $f(x, y) = x \oplus_p y$, respectively. The proof of Theorem 6.1.8 is almost the same as that of Theorem 6.1.5.

From the last chapter, we know that H-Boolean functions are equivalent to n-dimensional Hadamard matrices of order 2. The following theorem shows us another construction of the general n-dimensional Hadamard matrices by using the H-Boolean functions consisting of terms of degree 2.

Theorem 6.1.9 *Let $f(x_1, \cdots, x_n)$ be an* H*-Boolean function of n variables defined by*

$$f(x_1, \cdots, x_n) = \sum_{i<j,\ (i,j)\in B} x_i x_j,$$

where B is a subset of $\{(x, y) : 1 \leqslant x, y \leqslant n\}$. And let $a = [A(i, j)]$, $1 \leqslant i, j \leqslant N$, be a 2-dimensional Hadamard matrix of order N. Then the following matrix $C = [C(c(1), \cdots, c(n))]$, $1 \leqslant c(i) \leqslant N$, $1 \leqslant i \leqslant n$, defined by

$$C(c(1), \cdots, c(n)) = \prod_{1\leqslant i<j\leqslant n,\ (i,j)\in B} A(c(i), c(j))$$

is an n-dimensional Hadamard matrix of order N.

Proof Let k, $1 \leqslant k \leqslant n$. It is sufficient to prove that

$$\begin{aligned}\alpha(a,b) = & \sum_{1\leqslant c(1),\cdots,c(k-1),c(k+1),\cdots,c(n)\leqslant N} C(c(1), \cdots, c(k-1), a, \\ & c(k+1), \cdots, c(n)) C(c(1), \cdots, c(k-1), b, c(k+1), \cdots, c(n)) \\ = & N^{n-1}\delta(a, b).\end{aligned}$$

Rewrite the Boolean function $f(x_1, \cdots, x_n)$ as

$$\begin{aligned}f(x_1, \cdots, x_n) = & x_k \sum_{j>k,(k,j)\in B} x_j + x_k \sum_{i<k,(i,k)\in B} x_i \\ & + \sum_{i<j,(i,j)\in B,i\neq k,j\neq k} x_i x_j.\end{aligned}$$

Because every H-Boolean function depends on each of their variables, the set

$$\{j > k : (k, j) \in B\} \cup \{i < k : (i, k) \in B\}$$

is non-empty. Thus by employing the identity $A(i, j)^2 = 1$, and $a \neq b$, we have

$$\begin{aligned}\alpha(a,b) = & \sum_{1\leqslant c(1),\cdots,c(k-1),c(k+1),\cdots,c(n)\leqslant N} \left\{ \prod_{j<k,(k,j)\in B} A(a, c(j))A(b, c(j)) \right\} \\ & \times \left\{ \prod_{i<k,(i,k)\in B} A(c(i), a)A(c(i), b) \right\} \\ = & \rho \left[\prod_{j>k,(k,j)\in B} \left(\sum_{c(j)=1}^{N} A(a, c(j))A(b, c(j)) \right) \right] \\ & \times \left[\prod_{i<k,(i,k)\in B} \left(\sum_{c(i)=1}^{N} A(c(i), a)A(c(i), b) \right) \right] \\ = & 0 \quad (\text{for } A = [A(i, j)] \text{ is a Hadamard matrix}),\end{aligned}$$

where ρ is a constant defined by $\rho = N^{r_1+r_2}$, with $r_1 = |\{j : j > k, (k, j) \notin B\}|$, and $r_2 = |\{i : i < k, (i, k) \notin B\}|$.

Thus we have proved that $\alpha(a, b) = N^{r_1+r_2}\delta(a, b)$. The theorem follows. □

The m-sequences, or equivalently the longest linear shift register sequences, are very popular (± 1)-valued sequences because they are the best pseudo-random sequences and have good correlation functions. One of the interesting properties about the m-sequences is[2]:

Lemma 6.1.3 *If $a_0, \cdots, a_{N-1}$, $N = 2^k - 1$, is an m-sequence of length N, then the matrix*

$$A = \begin{bmatrix} 1 & 1 & 1 & \cdots & \cdots & 1 \\ 1 & a_0 & a_1 & \cdots & a_{N-2} & a_{N-1} \\ 1 & a_1 & a_2 & \cdots & a_{N-1} & a_0 \\ 1 & a_2 & a_3 & \cdots & a_0 & a_1 \\ \vdots & \vdots & \vdots & & \vdots & \vdots \\ 1 & a_{N-1} & a_0 & \cdots & a_{N-3} & a_{N-2} \end{bmatrix}$$

is a 2-dimensional Hadamard matrix of order $N + 1$.

This lemma provides us with another construction of higher-dimensional Hadamard matrices:

Theorem 6.1.10 *Let $a_0, \cdots, a_{N-1}$, $N = 2^k - 1$, be an m-sequence of length N. And let r, $1 \leqslant r \leqslant n$, be an integer. Then the matrix $A = [A(i_1, \cdots, i_n)]$, $0 \leqslant i_1, \cdots, i_n \leqslant N$, defined by*

$$A(i_1, \cdots, i_n) = (a_{(i_1+\cdots+i_n-2)\bmod N})^{\delta((i_1+\cdots+i_r)(i_{r+1}+\cdots+i_n))}$$

is an n-dimensional Hadamard matrix of order $N + 1$, where $\delta(x) = 1$, iff $x = 0$.

Proof From Lemma 6.1.3, we know that the matrix

$$B = [B(i, j)] = [(a_{(i+j-2)\bmod N})^{\delta(ij)}]$$

is a 2-dimensional Hadamard matrix of order $N + 1$. The proof is finished by recursively applying Theorem 6.1.5 to the matrix B. □

Remark 6.1.1 *The m-sequences used in Theorem 6.1.10 can be replaced by every (± 1)-valued sequence provided that its periodic out-of-phase autocorrelation is the constant -1.*

Theorem 6.1.11 *Let A and B be two n-dimensional Hadamard matrices. Then their direct multiplication $A \otimes B$ is also a n-dimensional Hadamard matrix.*

A more general result will be proved later, so we omitted here the proof of Theorem 6.1.11.

Theorem 6.1.12 *Let $B = [B(i_1, \cdots, i_n)]$ be an n-dimensional Hadamard matrix of order N. Then the matrix $A = [A(i_1, \cdots, i_n)]$, $0 \leqslant i_1, \cdots, i_n \leqslant N - 1$, defined by*

$$A(i_1, \cdots, i_n) = B(i_1, \cdots, i_n)(-1)^{\sum_{u=0}^{M-1} \sum_{v=0}^{n-1} a_{uv} i(u,v)}$$

is an n-dimensional Hadamard matrix of order N, where $a_{uv} = 0$ or 1, $N \leqslant 2^M$, and $i_k = (i(M-1, k), i(M-2, k), \cdots, i(0, k))$, the binary expended vector of the integer i_k.

In the case of $n = 2$, Theorem 6.1.12 implies that every Hadamard matrix is transformed to another Hadamard matrix if some rows and/or columns are minused.

Theorem 6.1.13 *Let $\tau(.)$ be a permutation of the set $\{1,2,\cdots,n\}$, and let $A=[A(a(1),\cdots,a(n))]$ be an n-dimensional Hadamard matrix. Then the matrix $B=[B(a(1),\cdots,a(n))]$ defined by*

$$B(a(1),\cdots,a(n))=A(a(\tau(1)),\cdots,a(\tau(n)))$$

is also an n-dimensional Hadamard matrix.

In the case of $n=2$, Theorem 6.1.13 implies that the transpose of a Hadamard matrix is also a Hadamard matrix.

Theorem 6.1.14 *Let $f_1(\cdot),\cdots,f_n(\cdot)$ be one-to-one mappings of the set $\{0,1,\cdots,m-1\}$. And let $A=[A(i_1,\cdots,i_n)]$, $0\leqslant i_k\leqslant m-1$, $1\leqslant k\leqslant n$, be an n-dimensional Hadamard matrix of order m. Then the matrix $B=[B(i_1,\cdots,i_n)]$ defined by*

$$B(i_1,\cdots,i_n)=A(f_1(i_1),\cdots,f_n(i_n))$$

is also an n-dimensional Hadamard matrix of order m.

In the case of $n=2$, Theorem 6.1.14 implies that every Hadamard matrix is transformed to another Hadamard matrix if the rows and/or columns are permutated.

Theorem 6.1.15 *Let $A=[A(i_1,\cdots,i_n)]$ and $B=[B(j_1,\cdots,j_m)]$ be two Hadamard matrices of the same order N and dimensions n and m, respectively. Then the matrix $C=[C(i_1,\cdots,i_n,j_1,\cdots,j_m)]$ defined by*

$$C(i_1,\cdots,i_n,j_1,\cdots,j_m)=A(i_1,\cdots,i_n)\times B(j_1,\cdots,j_m)$$

is a $(m+n)$-dimensional Hadamard matrix of order N.

The proofs of Theorems 6.1.12 to 6.1.15 are trivial, so we have omitted the details here.

Theorem 6.1.16 *Let $B=[B(b(1),\cdots,b(n))]$, $0\leqslant b(i)\leqslant m-1$, $1\leqslant i\leqslant n$, be an n-dimensional matrix of order m. And let $A=[A(a(1),\cdots,a(n))]$, $0\leqslant a(i)\leqslant m^n-1$, $1\leqslant i\leqslant n$, be an n-dimensional matrix of order m^n which is defined by*

$$A(a(1),\cdots,a(n))=\prod_{i=1}^{n}B(a(1,i),a(2,i),\cdots,a(n,i)),$$

where $(a(1,i),a(2,i),\cdots,a(n,i))$ is the m-ary expended vector of the integer $a(i)$, i.e., $a(i)=\sum_{j=1}^{n}a(j,i)m^{j-1}$. Then A is a Hadamard matrix if and only if B is a Hadamard matrix.

Proof $\Leftarrow$ Let $0\leqslant a\neq b\leqslant m^n-1$, and let $(a_1,\cdots,a_n)$ and $(b_1,\cdots,b_n)$ be the m-ary expended vectors of the integers a and b, respectively. Then there is at least one k, $1\leqslant k\leqslant n$, such that $a_k\neq b_k$. Hence

$$\sum_{a(2),\cdots,a(n)}A(a,a(2),\cdots,a(n))A(b,a(2),\cdots,a(n))$$
$$=\sum_{0\leqslant a(i,j)\leqslant m-1,\ 2\leqslant i\leqslant n,\ 1\leqslant j\leqslant n}\prod_{p=1}^{n}B(a_p,a(2,p),\cdots,a(n,p))$$

$$
\begin{aligned}
&\times \prod_{p=1}^{n} B(b_p, a(2,p), \cdots, a(n,p)) \\
=& \sum_{0\leqslant a(i,j)\leqslant m-1,\ 2\leqslant i\leqslant n,\ 1\leqslant j\leqslant n,\ j\neq k} \rho\Bigg[\sum_{0\leqslant a(i,k)\leqslant m-1,\ 2\leqslant i\leqslant n} B(a_k, a(2,k), \cdots, \\
& a(n,k))B(b_k, a(2,k), \cdots, a(n,k))\Bigg] \\
=& \sum \rho \times 0, \quad (\text{for } B \text{ is a Hadamard matrix}) \\
=& 0,
\end{aligned}
$$

where ρ is a constant that is independent of $a(i,k)$.

Similarly, it can be proved that for each r, $1 \leqslant r \leqslant n$,

$$
\begin{aligned}
&\sum_{a(\cdot)} A(a(1), \cdots, a(r-1), a, a(r+1), \cdots, a(n)) \\
&\times A(a(1), \cdots, a(r-1), b, a(r+1), \cdots, a(n)) = (m^n)^{n-1}\delta(a,b).
\end{aligned}
$$

Thus B is an n-dimensional Hadamard matrix of order m^n.

$\Longrightarrow$ If B is not a Hadamard matrix, then there exists a pair a and b, $0 \leqslant a \neq b \leqslant m-1$, such that

$$
\sum_{b(2),\cdots,b(n)} B(a, b(2), \cdots, b(n))B(b, b(2), \cdots, b(n)) = r \neq 0.
$$

Let u and v be two integers satisfying $u = \sum_{i=0}^{n-1} am^i$ and $v = \sum_{i=0}^{n-1} bm^i$, respectively. Thus $0 \leqslant u \neq v \leqslant m^n - 1$, and

$$
\begin{aligned}
&\sum_{a(2),\cdots,a(n)} A(u, a(2), \cdots, a(n))A(v, a(2), \cdots, a(n)) \\
=& \sum_{0\leqslant a(i,j)\leqslant m-1,\ 2\leqslant i\leqslant n,\ 1\leqslant j\leqslant n} \prod_{k=1}^{n} \{B(a, a(2,k), \cdots, a(n,k)) \\
& \times B(b, a(2,k), \cdots, a(n,k))\} \\
=& \prod_{k=1}^{n} \Bigg\{ \sum_{0\leqslant a(2,k),\cdots,a(n,k)\leqslant m-1} B(a, a(2,k), \cdots, a(n,k)) \\
& \times B(b, a(2,k), \cdots, a(n,k)) \Bigg\} \\
=& \prod_{k=1}^{n} r = r^n \neq 0.
\end{aligned}
$$

Therefore A is not a Hadamard matrix. The theorem follows. □

Theorem 6.1.17 *The following are true:*

(1) *Let $A = [A(a(1), \cdots, a(2n))]$, $1 \leqslant a(i) \leqslant m$, $1 \leqslant i \leqslant 2n$, be a $(2n)$-dimensional (± 1)-valued matrix of order m. If $AA' = m^n I$. Then A is a $(2n)$-dimensional Hadamard matrix, where A' refers to the transpose of the matrix A, and I the $(2n)$-dimensional unit matrix of order m (Remark: It has been known that $AA' = m^2 I$ is both a necessary and sufficient condition of A being a 2-dimensional Hadamard matrix. However, it should be pointed out that if $n > 2$, there does exist a $(2n)$-dimensional Hadamard matrix A satisfying $AA' \neq m^n I$).*

(2) *Let* $B = [B(b(1), \cdots, b(2n+1))]$, $1 \leqslant b(i) \leqslant m$, $1 \leqslant i \leqslant 2n+1$, *be a* $(2n+1)$*-dimensional* (± 1)*-valued matrix of order* m. *If* B *simultaneously satisfies the following two equations:*

$$BB' = m^n I$$

and, for $1 \leqslant a, b \leqslant m$,

$$\sum_{1\leqslant b(1),\cdots,b(2n)\leqslant m} B(b(1), \cdots, b(2n), a)B(b(1), \cdots, b(2n), b) = m^{2n}\delta(a, b).$$

Then B *is a* $(2n+1)$*-dimensional Hadamard matrix, where* B' *refers to the transpose of the matrix* B, *and* I *the* $(2n+1)$*-dimensional unit matrix of order* m.

Proof The proofs for these two statements are almost the same, so we prove the first one only.

$$\begin{aligned}
\alpha(a, b) &= \sum_{1\leqslant a(2),\cdots,a(2n)\leqslant m} A(a, a(2), \cdots, a(2n))A(b, a(2), \cdots, a(2n)) \\
&= \sum_{1\leqslant a(2),\cdots,a(2n)\leqslant m} A(a, a(2), \cdots, a(n), a(n+1), \cdots, a(2n)) \\
&\quad \times A'(a(n+1), \cdots, a(2n), b, a(2), \cdots, a(n)) \\
&\qquad \text{(by the definition of } A'\text{)} \\
&= \sum_{1\leqslant a(2),\cdots,a(n)\leqslant m} \left\{ \sum_{1\leqslant a(n+1),\cdots,a(2n)\leqslant m} A(a, a(2), \cdots, a(n), \right. \\
&\quad \left. a(n+1), \cdots, a(2n))A'(a(n+1), \cdots, a(2n), b, a(2), \cdots, a(n)) \right\} \\
&= \sum_{1\leqslant a(2),\cdots,a(n)\leqslant m} m^n\delta(a, b) \text{ (because of } AA' = m^n I\text{)} \\
&= m^{2n-1}\delta(a, b).
\end{aligned}$$

Now let

$$\begin{aligned}
\beta(a, b) = &\sum_{1\leqslant a(1),\cdots,a(n),a(n+2),\cdots,a(2n)\leqslant m} A(a(1), \cdots, a(n), a, a(n+2), \cdots, \\
&a(2n))A(a(1), \cdots, a(n), b, a(n+2), \cdots, a(2n)).
\end{aligned}$$

$AA' = m^n I$ implies $A' = m^n A^{-1}$. Thus $A'A = m^n A^{-1}A = m^n I$, where A^{-1} refers to the inverse matrix of A. Hence the equation $\beta(a, b) = m^{2n-1}\delta(a, b)$ can be proved by the same way as the $\alpha(a, b) = m^{2n-1}\delta(a, b)$.

Similarly, we can prove that, for each $1 \leqslant r \leqslant 2n$,

$$\begin{aligned}
&\sum_{a(\cdot)} A(a(1), \cdots, a(r-1), a, a(r+1), \cdots, a(2n)) \\
&\times A(a(1), \cdots, a(r-1), b, a(r+1), \cdots, a(2n)) = m^{2n-1}\delta(a, b).
\end{aligned}$$

Thus A is a $(2n)$-dimensional Hadamard matrix of order m. □

6.1.2 Proper and Improper n-Dimensional Hadamard Matrices

Absolutely proper and improper higher-dimensional Hadamard matrices are two extreme special cases of the general Hadamard matrices that are defined by:

Definition 6.1.2[1] *An absolutely proper n-dimensional Hadamard matrix is a (± 1)-valued matrix in which all two-dimensional sections in all possible axis-normal orientations are Hadamard matrices.*

Similar to the consequences of Definition 6.1.2, it is easy to see that[1]

(1) All intermediate-dimensional sections of an absolutely proper n-dimensional Hadamard matrix are also absolutely proper Hadamard matrices;

(2) Absolutely proper Hadamard matrices are themselves Hadamard matrices. In fact, an n-dimensional Hadamard matrix is specified if either all $(n-1)$-dimensional sections in one direction are Hadamard matrices and also are mutually orthogonal or if all $(n-1)$-dimensional sections in two directions are Hadamard matrices.

It has been proved, in the previous chapter, that

Theorem 6.1.18 *An n-dimensional (± 1)-valued matrix of order 2 is an absolutely proper Hadamard matrix if and only if this matrix is defined by $A = [A(x_1, \cdots, x_n)]$, $0 \leqslant x_1, \cdots, x_n \leqslant 1$,*

$$A(x_1, \cdots, x_n) = (-1)^{f(x_1, \cdots, x_n)}$$

for some Boolean function of the form

$$f(x_1, \cdots, x_n) = \sum_{1 \leqslant i,j \leqslant n} x_i x_j + a + \sum_{i=1}^{n} b_i x_i,$$

where $a, b_1, \cdots, b_n \in \{0, 1\}$.

Besides this theorem, another important result about the absolutely proper Hadamard matrix is:

Theorem 6.1.19 *Let $A = [A(i,j)]$, $0 \leqslant i,j \leqslant m-1$, be a 2-dimensional Hadamard matrix of order m. Then the following matrix $C = [C(c(1), \cdots, c(n))]$, $0 \leqslant c(k) \leqslant m-1$, $1 \leqslant k \leqslant n$, defined by*

$$C(c(1), \cdots, c(n)) = \prod_{1 \leqslant i < j \leqslant n} A(c(i), c(j)), \tag{6.9}$$

is an n-dimensional absolutely proper Hadamard matrix of order m.

Proof It is sufficient to prove that all 2-dimensional sections, in all possible axis-normal orientations, are Hadamard matrices. In other words, for any prefixed $1 \leqslant i, j \leqslant n$, and $c(k)$, $k \neq i$, $k \neq j$, we have to prove that the 2-dimensional section

$$\begin{aligned} D &= [D(c(i), c(j))] \\ &= [C(c(1), \cdots, c(i), \cdots, c(j), \cdots, c(n))], \ 0 \leqslant c(i), c(j) \leqslant m-1, \end{aligned} \tag{6.10}$$

is a Hadamard matrix of order m.

From Equations (6.9) and (6.10) , we have

$$D(c(i),c(j))=A(c(i),c(j))\times\left\{\prod_{i<k\leqslant n,\ k\neq j}A(c(i),c(k))\right\}$$
$$\times\left\{\prod_{1\leqslant k<i}A(c(k),c(i))\right\}\times\left\{\prod_{j<k\leqslant n}A(c(j),c(k))\right\}$$
$$\times\left\{\prod_{1\leqslant k<j,\ k\neq i}A(c(k),c(j))\right\}$$
$$\times\left\{\prod_{1\leqslant r<s\leqslant n,\ r\neq i,j\ s\neq i,j}A(c(r),c(s))\right\}.$$

Thus because $A(i,j)^2=1$, we know that, for $0\leqslant a\neq b\leqslant m-1$,

$$\sum_{c(j)=0}^{m-1}D(a,c(j))D(b,c(j))$$
$$=\sum_{c(j)=0}^{m-1}\left\{\prod_{i<k\leqslant n}A(a,c(k))A(b,c(k))\right\}\times\left\{\prod_{1\leqslant k<i}A(c(k),a)A(c(k),b)\right\}$$
$$=\left\{\prod_{i<k\leqslant n,\ k\neq j}A(a,c(k))A(b,c(k))\right\}\times\left\{\prod_{1\leqslant k<i}A(c(k),a)A(c(k),b)\right\}$$
$$\times\sum_{c(j)=0}^{m-1}A(a,c(j))A(b,c(j))$$
$$=\left\{\prod_{i<k\leqslant n,\ k\neq j}A(a,c(k))A(b,c(k))\right\}\times\left\{\prod_{1\leqslant k<i}A(c(k),a)A(c(k),b)\right\}\times 0$$
(for A is a Hadamard matrix)
$$=0.$$

Hence $\sum_{c(j)=0}^{m-1}D(a,c(j))D(b,c(j))=m\delta(a,b)$.

Similarly, it can be proved that

$$\sum_{c(i)=0}^{m-1}D(c(i),a)D(c(i),b)=m\delta(a,b).$$

Thus D is indeed a Hadamard matrix. The theorem follows. □

Let A be an n-dimensional absolutely proper Hadamard matrix of order $m>2$. The requirement of "A's 2-dimensional sections are Hadamard matrices" implies that $m=4k$, for some integer k. On the other hand, Theorem 6.1.19 confirms that every 2-dimensional Hadamard matrix produces an n-dimensional absolutely proper Hadamard matrix of the same order. Hence if the Hadamard conjecture were proved, there would exist at least one n-dimensional absolutely proper Hadamard matrix of order $4k$, for each $n\geqslant 2$ and $4k$.

The other concept opposite to the absolutely proper Hadamard matrix is the following absolutely improper one.

Definition 6.1.3[1] *An absolutely improper n-dimensional Hadamard matrix is a Hadamard matrix in which no lower-dimensional section is a Hadamard matrix.*

An absolutely improper n-dimensional Hadamard matrix of order m^2 may be formed by $n-1$ successive direct multiplications of n two-dimensional Hadamard matrices of order m in appropriately different orientations, e.g., the i_1i_2, i_2i_3, $\cdots$, $i_{n-1}i_n$, and i_ni_1 planes. To be precise, we have the following theorem.

Theorem 6.1.20 *Let $A = [A(i,j)]$, $0 \leqslant i,j \leqslant m-1$, be a 2-dimensional Hadamard matrix. And let A_1, A_2, $\cdots$, and A_n be the n-dimensional matrices produced by treating A as the n-dimensional ones of orders $m\times m\times 1\times\cdots\times 1$, $1\times m\times m\times 1\times\cdots\times 1$, $\cdots$, $1\times 1\times\cdots\times m\times m$, and $m\times 1\times\cdots\times 1\times m$, respectively. Then the following direct multiplication matrix B, defined by*

$$B = [B(b(1),\cdots,b(n))] =: A_1 \otimes A_2 \otimes \cdots \otimes A_n,$$

is an n-dimensional absolutely imporper Hadamard matrix of order m^2, where $0 \leqslant b(1),\cdots,$ $b(n) \leqslant m^2-1$.

Proof At first we prove that B is a Hadamard matrix. In fact, from the definition of direct multiplication, we know that

$$\begin{aligned}B(b(1),\cdots,b(n)) = &A([b(1)]_m,[b(2)]_m)\times A\left(\left\lfloor\frac{b(2)}{m}\right\rfloor,[b(3)]_m\right)\\&\times A\left(\left\lfloor\frac{b(3)}{m}\right\rfloor,[b(4)]_m\right)\times\cdots\times A\left(\left\lfloor\frac{b(n-1)}{m}\right\rfloor,[b(n)]_m\right)\\&\times A\left(\left\lfloor\frac{b(1)}{m}\right\rfloor,\left\lfloor\frac{b(n)}{m}\right\rfloor\right).\end{aligned} \tag{6.11}$$

Because of the identity $A(i,j)^2=1$,

$$\begin{aligned}&\sum_{0\leqslant b(2),\cdots,b(n)\leqslant m^2-1} B(a,b(2),\cdots,b(n))B(b,b(2),\cdots,b(n))\\=&\sum_{0\leqslant b(2),\cdots,b(n)\leqslant m^2-1} A([a]_m,[b(2)]_m)\\&\times A\left(\left\lfloor\frac{a}{m}\right\rfloor,\left\lfloor\frac{b(n)}{m}\right\rfloor\right)\times A([b]_m,[b(2)]_m)\times A\left(\left\lfloor\frac{b}{m}\right\rfloor,\left\lfloor\frac{b(n)}{m}\right\rfloor\right)\\=&\sum_{0\leqslant b(2),\cdots,b(n-1)\leqslant m^2-1}\left[\sum_{b(2)=0}^{m^2-1} A([a]_m,[b(2)]_m)\times A([b]_m,[b(2)]_m)\right]\\&\times\left[\sum_{b(2)=0}^{m^2-1} A\left(\left\lfloor\frac{a}{m}\right\rfloor,\left\lfloor\frac{b(n)}{m}\right\rfloor\right)A\left(\left\lfloor\frac{b}{m}\right\rfloor,\left\lfloor\frac{b(n)}{m}\right\rfloor\right)\right].\end{aligned} \tag{6.12}$$

If $a\neq b$, then either $[a]_m \neq [b]_m$ or $\left\lfloor\frac{a}{m}\right\rfloor \neq \left\lfloor\frac{b}{m}\right\rfloor$. In the first case, the summation in the first bracket of Equation (6.12) vanishes, otherwise the summation in the second bracket of Equation (6.12) vanishes.

In the same way it can be proved that, for each $1 \leqslant r \leqslant n$,

$$\sum_{B(\cdot)} B(b(1), \cdots, b(r-1), a, b(r+1), \cdots, b(n))$$
$$\times B(b(1), \cdots, b(r-1), b, b(r+1), \cdots, b(n)) = m^{2(n-1)}\delta(a, b).$$

Thus B is an n-dimensional Hadamard matrix of order m^2.

Then we prove that no lower section of B is a Hadamard matrix.

Let $E = [E(e(1), \cdots, e(k))]$, $k < n$, be a k-dimensional section of B. Without loss of the generality, we assume that E is obtained by fixing the coordinate $b(i_0)$, $1 \leqslant i_0 \leqslant n$, but not $b(i_0 + 1)$. Thus, we have

$$E(e(1), \cdots, e(k)) = A(\{b(i_0)\}, \{e(j)\}) \times A(\{e(j)\}, (\cdot)) \times (:),$$

where $\{x\}$ stands for either $[x]_m$ or $\lfloor x/m \rfloor$; $(\cdot)$ is a constant if $b(i_0+2)$ has been fixed during the construction of E, otherwise $(\cdot) = \lfloor e(j+1) \rfloor$; and $(:)$ is a term independent of $e(j)$.

Then

$$\begin{aligned}\rho(a,b) = & \sum_{0 \leqslant e(1), \cdots, e(j-1), e(j+1), \cdots, e(k) \leqslant m^2-1} E(e(1), \cdots, e(j-1), a, \\ & e(j+1), \cdots, e(k)) E(e(1), \cdots, e(j-1), b, e(j+1), \cdots, e(k)) \\ = & \sum_{0 \leqslant e(1), \cdots, e(j-1), e(j+1), \cdots, e(k) \leqslant m^2-1} A(\{b(i_0)\}, \{a\}) \\ & \times A(\{b(i_0)\}, \{b\}) \times A(\{a\}, \{\cdot\}) A(\{b\}, \{\cdot\}).\end{aligned}$$

If $(\cdot)$ is a constant, then

$$|\rho(a, b)| = m^{2(k-1)} \neq m^{2(k-1)}\delta(a, b).$$

If $(\cdot) = \{e(j+1)\}$, then

$$\begin{aligned}\rho(a,b) = & A(\{b(i_0)\}, \{a\}) \times A(\{b(i_0)\}, \{b\}) \\ & \times \left[\sum_{0 \leqslant e(1), \cdots, e(j-1), e(j+2), \cdots, e(k) \leqslant m^2-1} \left(\sum_{e(j+1)=0}^{m^2-1} A(\{a\}, \{e(j+1)\}) \right.\right. \\ & \left.\left. \times A(\{b\}, \{e(j+1)\}) \right) \right].\end{aligned} \tag{6.13}$$

Because we can always find two integers a_0 and b_0 such that $a_0 \neq b_0$ and $\{a_0\} = \{b_0\}$, from Equation (6.13), we have

$$\rho(a_0, b_0) = m^{2(k-1)} \neq 0 = m^{2(k-1)}\delta(a_0, b_0).$$

Hence we have proved that the k-dimensional section E is not a Hadamard matrix. The theorem follows. □

A more general version of Theorem 6.1.20 is the following.

Theorem 6.1.21 *Let $A = [A(a(1), \cdots, a(k))]$ be a k-dimensional Hadamard matrix of order m. And let A_1, A_2, $\cdots$, and A_n, $n > k$, be the n-dimensional matrices produced by treating A as the n-dimensional ones of orders $m \times \cdots \times m \times 1 \times \cdots \times 1$, $1 \times m \times \cdots \times m \times 1 \times \cdots \times$*

$1, \cdots, 1\times 1\times\cdots\times 1\times m\times\cdots\times m, m\times 1\times 1\times\cdots\times 1\times m\times\cdots\times m, \cdots, m\times\cdots\times m\times 1\times\cdots\times 1\times m,$ *respectively. Then the following direct multiplication matrix B, defined by*

$$B = [B(b(1), \cdots, b(n))] =: A_n \otimes A_{n-1} \otimes \cdots \otimes A_1,$$

is an n-dimensional absolutely imporper Hadamard matrix of order m^k.

If $k = 2$, this theorem clearly reduces to Theorem 6.1.20.

Between the two extreme cases of absolutely proper and absolutely improper Hadamard matrices, there exist a wide variety of intermediate degrees of property. For example, an n-dimensional Hadamard matrix A is called "proper in some direction," if all of its $(n-1)$-dimensional sections in that direction are Hadamard matrices, otherwise A is called improper in that direction. These partially proper n-dimensional Hadamard matrices can be constructed from a lower-dimensional one, by some "bootstrap" sequence of cyclic section-permutations.

6.1.3 Generalized Higher-Dimensional Hadamard Matrices

Let $A = [A(i_1, \cdots, i_n)]$, $0 \leqslant i_k \leqslant a_k - 1$, $1 \leqslant k \leqslant n$, be an n-dimensional matrix of size $a_1 \times \cdots \times a_n$. Recall that a necessary condition for this A to be Hadamard matrix is that its size satisfying $a_1 = \cdots = a_n$, i.e., the length of every side is the same constant. While in this subsection, we will generalize the concept of Hadamard matrices as those which have different side lengths.

Definition 6.1.4 *An n-dimensional (± 1)-valued matrix $A = [A(i_1, \cdots, i_n)]$, $0 \leqslant i_k \leqslant a_k - 1$, $1 \leqslant k \leqslant n$, of size $a_1 \times \cdots \times a_n$ is called a generalized Hadamard matrix if the following two conditions are simultaneously satisfied:*

(1) *At least two of $a_1, a_2, \cdots, a_n$ are larger than 1;*

(2) *If $a_i > 1$, then*

$$\sum_{a(1),\cdots,a(i-1),a(i+1),\cdots,a(n)} A(a(1), \cdots, a(i-1), a, a(i+1), \cdots, a(n)) \times A(a(1), \cdots, a(i-1), b, a(i+1), \cdots, a(n)) = \left(\prod_{j=1, j\neq i}^{n} a_j\right) \delta(a, b).$$

Clearly, a generalized Hadamard matrix of size $m \times \cdots \times m$ is in fact a regular higher-dimensional Hadamard matrix as studied in last subsections. In addition, an n-dimensional Hadamard matrix of order m can be treated as an $(n+k)$-dimensional Hadamard matrix of size $1 \times \cdots \times m \cdots \times 1 \cdots \times m \times \cdots \times 1$. Particularly, the matrices $A_1, \cdots, A_n$ used in Theorems 6.1.21 and 6.1.20 are all generalized Hadamard matrices. In other words, these two theorems assert that the direct multiplication of some generalized Hadamard matrices may produce other Hadamard matrices. The following theorem proves the rightness of this assertation in general.

Theorem 6.1.22 *Let $A = [A(i_1, \cdots, i_n)]$, $0 \leqslant i_k \leqslant a_k - 1$, $1 \leqslant k \leqslant n$, and $B = [B(j_1, \cdots, j_n)]$, $0 \leqslant j_k \leqslant b_k - 1$, $1 \leqslant k \leqslant n$, be n-dimensional generalized Hadamard matrix of size $a_1 \times \cdots \times a_n$, and $b_1 \times \cdots \times b_n$, respectively. Then their direct multiplication $C = B \otimes A$ is an n-dimensional generalized Hadamard matrix of size $(a_1 b_1) \times \cdots \times (a_n b_n)$.*

Proof Let $C = [C(c(1), \cdots, c(n))]$, $0 \leqslant c(i) \leqslant a_i b_i - 1$, $1 \leqslant i \leqslant n$. From the definition of direct multiplication, we have

$$C(c(1), \cdots, c(n)) = A([c(1)]_{a_1}, \cdots, [c(n)]_{a_n}) \times B\left(\left\lfloor \frac{c(1)}{a_1} \right\rfloor, \cdots, \left\lfloor \frac{c(n)}{a_n} \right\rfloor\right).$$

If $a_i b_i > 1$, and $0 \leqslant a \neq b \leqslant a_i b_i - 1$, then

$$\begin{aligned}
\rho(a,b) = & \sum_{c(1),\cdots,c(i-1),c(i+1),\cdots,c(n)} C(c(1), \cdots, c(i-1), a, c(i+1), \cdots, c(n)) \\
& \times C(c(1), \cdots, c(i-1), b, c(i+1), \cdots, c(n)) \\
= & \sum_{c(1),\cdots,c(i-1),c(i+1),\cdots,c(n)} B\left(\left\lfloor \frac{c(1)}{a_1} \right\rfloor, \cdots, \left\lfloor \frac{c(i-1)}{a_{i-1}} \right\rfloor, \left\lfloor \frac{a}{a_i} \right\rfloor, \right. \\
& \left. \left\lfloor \frac{c(i+1)}{a_{i+1}} \right\rfloor, \cdots, \left\lfloor \frac{c(n)}{a_n} \right\rfloor\right) \\
& \times B\left(\left\lfloor \frac{c(1)}{a_1} \right\rfloor, \cdots, \left\lfloor \frac{c(i-1)}{a_{i-1}} \right\rfloor, \left\lfloor \frac{b}{a_i} \right\rfloor, \left\lfloor \frac{c(i+1)}{a_{i+1}} \right\rfloor, \cdots, \left\lfloor \frac{c(n)}{a_n} \right\rfloor\right) \\
& \times A([c(1)]_{a_1}, \cdots, [c(i-1)]_{a_{i-1}}, [a]_{a_i}, [c(i+1)]_{a_{i+1}}, \cdots, [c(n)]_{a_n}) \\
& \times A([c(1)]_{a_1}, \cdots, [c(i-1)]_{a_{i-1}}, [b]_{a_i}, [c(i+1)]_{a_{i+1}}, \cdots, [c(n)]_{a_n}).
\end{aligned}$$

Now we prove $\rho(a, b) = 0$ in the following two separate cases:

Case 1. $a_i = 1$. Thus $b_i > 1$.

Because of $[a]_{a_i} = [b]_{a_i} = 0$, $\lfloor \frac{a}{a_i} \rfloor = a$, $\lfloor \frac{b}{a_i} \rfloor = b$, and $A(a(1), \cdots, a(n))^2 = 1$, we have

$$\begin{aligned}
\rho(a,b) = & \sum_{c(1),\cdots c(i-1),c(i+1),\cdots,c(n)} B\left(\left\lfloor \frac{c(1)}{a_1} \right\rfloor, \cdots, \left\lfloor \frac{c(i-1)}{a_{i-1}} \right\rfloor, \left\lfloor \frac{a}{a_i} \right\rfloor, \right. \\
& \left. \left\lfloor \frac{c(i+1)}{a_{i+1}} \right\rfloor, \cdots, \left\lfloor \frac{c(n)}{a_n} \right\rfloor\right) \\
& \times B\left(\left\lfloor \frac{c(1)}{a_1} \right\rfloor, \cdots, \left\lfloor \frac{c(i-1)}{a_{i-1}} \right\rfloor, \left\lfloor \frac{b}{a_i} \right\rfloor, \left\lfloor \frac{c(i+1)}{a_{i+1}} \right\rfloor, \cdots, \left\lfloor \frac{c(n)}{a_n} \right\rfloor\right) \\
= & b_1 \times \cdots \times b_{i-1} \times b_{i+1} \times \cdots \times b_n \\
& \times \sum_{b(1),\cdots,b(i-1),b(i+1),\cdots,b(n)} B(b(1), \cdots, b(i-1), a, b(i+1), \cdots, b(n)) \\
& \times B(b(1), \cdots, b(i-1), b, b(i+1), \cdots, b(n)) \\
= & 0,
\end{aligned}$$

where the last equation is due to the fact that B is an n-dimensional generalized Hadamard matrix.

Case 2. $a_i > 1$. Thus

$$\rho(a,b)=\left[\sum_{0\leqslant k_1\leqslant b_1-1,\cdots,0\leqslant k_{i-1}\leqslant b_{i-1}-1,0\leqslant k_{i+1}\leqslant b_{i+1}-1,\cdots,0\leqslant k_n\leqslant b_n-1} B\left(k_1,\cdots,\right.\right.$$

$$\left.\left. k_{i-1},\left\lfloor\frac{a}{a_i}\right\rfloor,k_{i+1},\cdots,k_n\right)\times B\left(k_1,\cdots,k_{i-1},\left\lfloor\frac{b}{a_i}\right\rfloor,k_{i+1},\cdots,k_n\right)\right]$$

$$\times\left[\sum_{0\leqslant l_1\leqslant a_1-1,\cdots,0\leqslant l_{i-1}\leqslant a_{i-1}-1,0\leqslant l_{i+1}\leqslant a_{i+1}-1,\cdots,0\leqslant l_n\leqslant a_n-1} A\left(l_1,\cdots,\right.\right.$$

$$\left.\left. l_{i-1},[a]_{a_i},l_{i+1},\cdots,l_n\right)\times A(l_1,\cdots,l_{i-1},[b]_{a_i},l_{i+1},\cdots,l_n)\right]. \tag{6.14}$$

$a \neq b$ implies either $[a]_{a_i} \neq [b]_{a_i}$ or $\lfloor a/a_i\rfloor \neq \lfloor b/a_i\rfloor$.

If $[a]_{a_i} \neq [b]_{a_i}$, then the summation in the second bracket of Equation 6.14 is zero. If $\lfloor a/a_i\rfloor \neq \lfloor b/a_i\rfloor$, then the summation in the first bracket of Equation (6.14) is zero.

Thus we have proved that $\rho(a,b)=0$, if $a\neq b$. The proof is finished. □

A trivial corollary of Theorem 6.1.22 is

Corollary 6.1.1 *The direct multiplication of two n-dimensional Hadamard matrices is also a Hadamard matrix, which is proper in some direction if its parent matrices are proper in that direction.*

The other application of Theorem 6.1.22 is that it provides us a recursive construction of higher-dimensional Hadamard matrices from lower-dimensional ones. For example, we have the following two theorems.

Theorem 6.1.23 *Let A and B be two n-dimensional generalized Hadamard matrices of sizes $a_1\times\cdots\times a_n$ and $b_1\times\cdots\times b_n$, respectively. If $a_1b_1=\cdots=a_nb_n=m$, then $A\otimes B$ is an n-dimensional regular Hadamard matrix of order m.*

Lemma 6.1.4 *Let $A=[A(a(1),\cdots,a(n))]$ be an n-dimensional generalized Hadamard matrix of size $a_1\times\cdots\times a_n$. Then,*

(1) *A is also an $(n+k)$-dimensional generalized Hadamard matrix of size $1\times\cdots\times a_1\times\cdots\times 1\cdots\times a_n\times\cdots\times 1$;*

(2) *If $\tau(\cdot)$ is a permutation of the set $\{1,2,\cdots,n\}$, then the matrix $B=[A(a(\tau(1)),\cdots,a(\tau(n))]$ is an n-dimensional generalized Hadamard matrix of size $a_{\tau(1)}\times\cdots\times a_{\tau(n)}$.*

Theorem 6.1.24 *Every n-dimensional generalized Hadamard matrix of size $a_1\times\cdots\times a_n$ produces an n-dimensional regular Hadamard matrix of order $\prod_{i=1}^n a_i$.*

Proof Let A be an n-dimensional generalized Hadamard matrix of size $a_1\times\cdots\times a_n$. And let $\tau(x)=x+1$, if $1\leqslant x\leqslant n-1$; $\tau(n)=1$. This $\tau(\cdot)$ is clearly a permutation of the set $\{1,2,\cdots,n\}$. Thus, from Theorem 6.1.22 and Lemma 6.1.4, the direct multiplication of the matrices $A\otimes A(\tau)\otimes\cdots\otimes A(\tau^k)\otimes\cdots\otimes A(\tau^{n-1})$ is the required n-dimensional regular Hadamard matrix of order $\prod_{i=1}^n a_i$, where $A(\tau^k)=[A(a(\tau^k(1)),\cdots,a(\tau^k(n)))]$. □

6.2 Higher-Dimensional Hadamard Matrices Based on Perfect Binary Arrays

6.2.1 n-Dimensional Hadamard Matrices Based on PBAs

The definition of 2-dimensional perfect binary arrays (PBA) can be generalized for higher-dimensional cases which will be used to construct the general higher-dimensional Hadamard matrices.

Definition 6.2.1[3] *Let* $A = [A(a(1), \cdots, a(n))]$, $0 \leqslant a(j) \leqslant a_j - 1$, $1 \leqslant j \leqslant n$, *be a* (± 1)*-valued* n*-dimensional matrix of size* $a_1 \times a_2 \times \cdots \times a_n$. A *is called a perfect binary array, abbrivated as a* $\mathrm{PBA}(a_1, a_2, \cdots, a_n)$ *if its* n*-dimensional cyclic autocorrelation is a* δ*-function, i.e., for all* $(u(1), \cdots, u(n)) \neq (0, \cdots, 0)$, *with* $0 \leqslant u(i) \leqslant a_i - 1$, *we have*

$$\begin{aligned} R_A(u(1), \cdots, u(n)) &= \sum_{j_1=0}^{a_1-1} \cdots \sum_{j_n=0}^{a_n-1} A(j_1, \cdots, j_n) \cdot A(j_1 + u(1), \cdots, j_n + u(n)) \\ &= 0, \end{aligned}$$

where $j_i + u(i) \equiv (j_i + u(i)) \mathrm{mod} a_i$, $1 \leqslant i \leqslant n$.

This definition is clearly a natural generalization of the PBAs studied in the chapter of three-dimensional Hadamard matrices. In this book we consider only the non-trival cases, i.e., at least one of a_i is larger than 1. Note that if τ is a permutation of $\{1, 2, \cdots, n\}$, then $A = [A(a(1), \cdots, a(n))]$ is a $\mathrm{PBA}(a_1, a_2, \cdots, a_n)$ if and only if $A(\tau) =: [A(a(\tau(1)), \cdots, a(\tau(r)))]$ is a $\mathrm{PBA}(a_{\tau(1)}, \cdots, a_{\tau(n)})]$.

Theorem 6.2.1 *Let* $A = [A(a(1), \cdots, a(n))]$, $0 \leqslant a(i) \leqslant a_i - 1$, $1 \leqslant i \leqslant n$, *be a* $\mathrm{PBA}(a_1, a_2, \cdots, a_n)$. *And let* a_{n+1} *be an integer satisfying* $a_{n+1} \leqslant \mathrm{LCM}(a_1, a_2, \cdots, a_n)$, *the least common multiple of* $a_1, a_2, \cdots, a_n$. *Then the following* $(n+1)$*-dimensional matrix* $H = [H(h(1), \cdots, h(n+1))]$, $0 \leqslant h(i) \leqslant a_i - 1$, $1 \leqslant i \leqslant n+1$, *defined by*

$$\begin{aligned} H(h(1), \cdots, h(n+1)) = A(&(h(1) + h(n+1)) \mathrm{mod} a_1, \cdots, \\ &(h(n) + h(n+1)) \mathrm{mod} a_n) \end{aligned}$$

is an $(n+1)$*-dimensional generalized Hadamard matrix of size* $a_1 \times \cdots \times a_{n+1}$.

Particularly, if $a_1 = a_2 = \cdots = a_n = m$, *then* a_{n+1} *can be chosen to be the integer* m. *Hence the corresponding* H *is an* $(n+1)$*-dimensional Hadamard matrix of order* m.

Proof On one hand, if $0 \leqslant a \neq b \leqslant a_{n+1} - 1$, then there exists an integer i, $1 \leqslant i \leqslant n$, such that $a \not\equiv b \mathrm{mod}(a_i)$, which is due to $a_{n+1} \leqslant \mathrm{LCM}(a_1, a_2, \cdots, a_n)$. Thus

$$\begin{aligned} &\sum_{h(1)=0}^{a_1-1} \cdots \sum_{h(n)=0}^{a_n-1} H(h(1), \cdots, h(n), a) H(h(1), \cdots, h(n), b) \\ &= \sum_{h(1)=0}^{a_1-1} \cdots \sum_{h(n)=0}^{a_n-1} A((h(1) + a) \mathrm{mod} a_1, \cdots, (h(i) + a) \mathrm{mod} a_i, \cdots, \\ &\quad (h(n) + a) \mathrm{mod} a_n) \end{aligned}$$

$$\times A((h(1)+b)\text{mod}a_1,\cdots,(h(i)+b)\text{mod}a_i,\cdots,(h(n)+b)\text{mod}a_n)$$

$$=0.$$

The last equation is due to the facts that (1) A is a PBA$(a_1,a_2,\cdots,a_n)$, and (2) $a\not\equiv b\text{mod}a_i$.

On the other hand, if $1\leqslant r\leqslant n$, and $0\leqslant a\neq b\leqslant a_r-1$, then

$$\sum_{h(1)=0}^{a_1-1}\cdots\sum_{h(r-1)=0}^{a_{r-1}-1}\sum_{h(r+1)=0}^{a_{r+1}-1}\cdots\sum_{h(n+1)=0}^{a_{n+1}-1}H(h(1),\cdots,h(r-1),a,$$

$$h(r+1),\cdots,h(n+1))H(h(1),\cdots,h(r-1),b,h(r+1),\cdots,h(n+1))$$

$$=\sum_{h(1)=0}^{a_1-1}\cdots\sum_{h(r-1)=0}^{a_{r-1}-1}\sum_{h(r+1)=0}^{a_{r+1}-1}\cdots\sum_{h(n+1)=0}^{a_{n+1}-1}A((h(1)+h(n+1))\text{mod}a_1,$$

$$\cdots,(h(r-1)+h(n+1))\text{mod}a_{r-1},(h(r)+a)\text{mod}a_r,$$

$$(h(r+1)+h(n+1))\text{mod}a_{r+1},\cdots,(h(n)+h(n+1))\text{mod}a_n)$$

$$\times A((h(1)+h(n+1))\text{mod}a_1,\cdots,(h(r-1)+h(n+1))\text{mod}a_{r-1},$$

$$(h(r)+b)\text{mod}a_r,(h(r+1)+h(n+1))\text{mod}a_{r+1},$$

$$\cdots,(h(n)+h(n+1))\text{mod}a_n)$$

$$=\sum_{h(n+1)=0}^{a_{n+1}-1}\left[\sum_{h(1)=0}^{a_1-1}\cdots\sum_{h(r-1)=0}^{a_{r-1}-1}\sum_{h(r+1)=0}^{a_{r+1}-1}\cdots\sum_{h(n)=0}^{a_n-1}A(h(1),\cdots,h(r-1),\right.$$

$$h(r)+a,h(r+1),\cdots,h(n))$$

$$\left.\times A(h(1),\cdots,h(r-1),h(r)+b,h(r+1),\cdots,h(n))\right]$$

$$=0.$$

Thus H is indeed a generalized Hadamard matrix. □

Up to now, Theorem 6.1.22, Lemma 6.1.4, and Theorem 6.1.24, together with Theorem 6.2.1 imply the following important assertation, which motivates us to construct as many higher-dimensional perfect binary arrays as possible.

Corollary 6.2.1 *Each non-trivial higher-dimensional perfect binary array of any size produces infinite families of higher-dimensional Hadamard matrices.*

The following theorem is also true which can be proved in the same way as Theorem 6.1.22.

Theorem 6.2.2 *Let* $A=[A(a(1),\cdots,a(n))]$, $0\leqslant a(i)\leqslant m-1$, $1\leqslant i\leqslant n$, *be a* PBA$(m,m,\cdots,m)$. *Then the following matrix* $H=[H(h(1),\cdots,h(n+1))]$, $0\leqslant h(i)\leqslant m-1$, $1\leqslant i\leqslant n+1$, *defined by*

$$H(h(1),\cdots,h(n+1))=A(h(1)+h(2),h(2)+h(3),\cdots,h(n)+h(n+1)),$$

is an $(n+1)$-dimensional Hadamard matrix of order m.

6.2.2 Construction and Existence of Higher-Dimensional PBAs

Let $A = [A(a(1), \cdots, a(n))]$, $0 \leqslant a(i) \leqslant a_i - 1$, $1 \leqslant i \leqslant n$, be a PBA$(a_1, \cdots, a_n)$. Its n-dimensional discrete Fourier transform (DFT) is defined by[4]

$$F(f(1), \cdots, f(n)) = \sum_{a(1),\cdots,a(n)} A(a(1), \cdots, a(n)) \exp\left[-j2\pi \left(\sum_{k=1}^{n} \frac{f(k)a(k)}{a_k}\right)\right], \tag{6.15}$$

where $0 \leqslant f(i) \leqslant a_i - 1$, $1 \leqslant i \leqslant n$.

Thus the mean value, M, of the array A is calculated by

$$M = \sum_{a(1),\cdots,a(n)} A(a(1), \cdots, a(n)) = F(0, \cdots, 0). \tag{6.16}$$

Furthermore, the volume, E, of the array A results from the DFT of the periodic autocorrelation function $R_A(u(1), \cdots, u(n))$, which is a $\delta(u(1), \cdots, u(n))$, since

$$DFT(R_A(u(1), \cdots, u(n))) = |F(f(1), \cdots, f(n))|^2 = E. \tag{6.17}$$

Using Equations (6.16) and (6.17) as well as the definition of voulme $E = \prod_{k=1}^{n} a_k$, the mean value M is specified by

$$M = \sqrt{E} = \sqrt{\prod_{k=1}^{n} a_k}. \tag{6.18}$$

Equation (6.18) makes the following results to be true[4,5]:

(1) PBAs are not (± 1)-balanced, i.e, their mean value M is a non-zero integer;

(2) The voulme of each PBA is a square number, i.e., $E = m^2$, for some integer m;

(3) The number of 1 and -1 in a PBA is given by $(E \pm \sqrt{E})/2$;

(4) In order to vanish the out-of-phase correlations of a PBA, the number of elements contained in each PBA must be even. Thus $E = 4m^2 \in \{4, 16, 36, 64, 100, 144, \cdots\}$. The following Theorem 6.2.3 will show a more comprehensive proof for this result.

Theorem 6.2.3[3,4] *If the (± 1)-valued matrix $A=[A(i_1, i_2, \cdots, i_r)]$, $0 \leqslant i_j \leqslant s_j - 1$, $1 \leqslant j \leqslant r$, is a PBA$(s_1, s_2, \cdots, s_r)$, then its volume is $\prod_{i=1}^{r} s_i = 4N^2$ for an integer N.*

Proof At first we prove that the volume must be a square of an integer. In fact, on one hand, because the out-of-phase autocorrelation of A is zero,

$$\sum_{u_1=0}^{s_1-1} \cdots \sum_{u_r=0}^{s_r-1} R_A(u_1, \cdots, u_r) = R_A(0, \cdots, 0) = \prod_{i=1}^{r} s_i.$$

On the other hand,

$$\begin{aligned}\sum_{u_1=0}^{s_1-1} \cdots \sum_{u_r=0}^{s_r-1} R_A(u_1, \cdots, u_r) &= \sum_{u_1=0}^{s_1-1} \cdots \sum_{u_r=0}^{s_r-1} \sum_{j_1=0}^{s_1-1} \cdots \sum_{j_r=0}^{s_r-1} A(j_1, \cdots, j_r) \\ &\quad \times A(j_1 + u_1, \cdots, j_r + u_r) \\ &= \left[\sum_{u_1=0}^{s_1-1} \cdots \sum_{u_r=0}^{s_r-1} A(u_1, \cdots, u_r)\right]^2.\end{aligned}$$

Therefore the volume is a perfect square. Now it is sufficient to prove that the volume is even. In fact, by the definition of a PBA, we have $R_A(1,0,\cdots,0)=0$. However, by the definition of autocorrelation, we know that $R_A(1,0,\cdots,0)$ is the sum of $\prod_{i=1}^{r} s_i$ terms, each of which is ± 1. Thus this number of terms must be even. □

Examples of PBA(2, 2) and PBA(6, 6) have been presented in the chapter on three-dimensional Hadamard matrices. In addition, $(+\ +\ +\ -)$ is a PBA(4), and

$$\begin{bmatrix} - & + & + & - & + & + & + & + & + & - & + & - \\ + & + & + & + & - & + & + & - & + & - & + & - \\ + & + & - & - & + & + & - & - & - & - & - & + \end{bmatrix}$$

is a PBA(3, 12). Starting from these four smaller PBAs, we can construct infinite families of PBAs by the following constructions.

Theorem 6.2.4[4,5] *Let* $A=[A(a(1),\cdots,a(n+k))]$ *be a* PBA$(a_1,\cdots,a_n,d_1,\cdots,d_k)$, *and* $B=[B(b(1),\cdots,b(k+m))]$ *a* PBA$(e_1,\cdots,e_k,b_1,\cdots,b_m)$. *If* $\gcd(e_i,d_i)=1$, $1\leqslant i\leqslant k$, *then the matrix* $C=[C(c(1),\cdots,c(n+m+k))]$, $0\leqslant c(i)\leqslant a_i-1$, *if* $1\leqslant i\leqslant n$; $0\leqslant c(i)\leqslant d_{i-n}e_{i-n}-1$, *if* $1+n\leqslant i\leqslant n+k$; $0\leqslant c(i)\leqslant b_{i-n-k}-1$, *if* $1+n+k\leqslant i\leqslant n+m+k$, *defined by*

$$\begin{aligned} &C(c(1),\cdots,c(n+m+k)) \\ =&A(c(1),\cdots,c(n),(c(n+1))\bmod d_1,\cdots,(c(n+k))\bmod d_k) \\ &\times B((c(n+1))\bmod e_1,\cdots,(c(n+k))\bmod e_k, \\ &c(n+k+1),\cdots,c(n+k+m)) \end{aligned}$$

is a PBA$(a_1,\cdots,a_n,(d_1e_1),\cdots,(d_ke_k),b_1,\cdots,b_m)$.

Proof The periodic correlation of C is

$$\begin{aligned} R(u)=&\sum_{c(\cdot)} C(c(1),\cdots,c(n+m+k)) \\ &\times C(c(1)+u_1,\cdots,c(n+m+k)+u_{n+m+k}). \end{aligned} \tag{6.19}$$

Case 1. If $(u_1,\cdots,u_n)\neq(0,\cdots,0)$, then

$$\begin{aligned} R(u)=&\sum_{c(n+1),\cdots,c(n+k+m)} B((c(n+1))\bmod e_1,\cdots,(c(n+k))\bmod e_k, c(n+k+1), \\ &\cdots,c(n+k+m))B((c(n+1)+u_{n+1})\bmod e_1, \\ &\cdots,(c(n+k)+u_{n+k})\bmod e_k, \\ &c(n+k+1)+u_{n+k+1},\cdots,c(n+k+m)+u_{n+k+m}) \\ &\times\left[\sum_{c(1),\cdots,c(n)} A(c(1),\cdots,c(n),(c(n+1))\bmod d_1,\cdots,(c(n+k))\bmod d_k)\right. \\ &A(c(1)+u_1,\cdots,c(n)+u_n,(c(n+1)+u_{n+1})\bmod d_1, \\ &\left.\cdots,(c(n+k)+u_{n+k})\bmod d_k)\right] \\ =&0\ \ (\text{for } A \text{ is a PBA}). \end{aligned}$$

Case 2. If $(u_{n+k+1}, \cdots, u_{n+k+m}) \neq (0, \cdots, 0)$, then in the same way as in Case 1, it can be proved that $R(u) = 0$.

Case 3. Otherwise $(u_{n+1}, \cdots, u_{n+k}) \neq (0, \cdots, 0)$. Because $0 \leqslant u_{n+i} \leqslant e_i d_i - 1$, $1 \leqslant i \leqslant k$, and $\gcd(e_i, d_i) = 1$, we know that either $(u_{n+1} \bmod e_1, \cdots, u_{n+k} \bmod e_k) \neq (0, \cdots, 0)$ or $(u_{n+1} \bmod d_1, \cdots, u_{n+k} \bmod d_k) \neq (0, \cdots, 0)$. Without loss of the generality, we assume that $(u_{n+1} \bmod d_1, \cdots, u_{n+k} \bmod d_k) \neq (0, \cdots, 0)$. Then

$$\begin{aligned} R(u) = \rho & \left[\sum_{b(\cdot)} B(b(1), \cdots, b(k), b(k+1), \cdots, b(k+m)) \right. \\ & \left. \times B(b(1) + u_{n+1}, \cdots, b(k) + u_{n+k}, b(k+1), \cdots, b(k+m)) \right] \\ & \times \left[\sum_{a(\cdot)} A(a(1), \cdots, a(n), a(n+1), \cdots, a(n+k)) \right. \\ & \left. \times A(a(1), \cdots, a(n), a(n+1) + u_{n+1}, \cdots, a(n+k) + u_{n+k}) \right] \\ = 0 & \text{ (the summation in the second bracket is zero)} . \end{aligned}$$

Therefore the identity $R(u) = 0$ has been proved in each case. □

The construction stated in Theorem 6.2.4 is called "the periodic production." Particularly, if $k = 0$, then this theorem results in the following corollary.

Corollary 6.2.2 [4,5] *Let* $A = [A(a(1), \cdots, a(n))]$ *be a* $\mathrm{PBA}(a_1, \cdots, a_n)$, *and* $B = [B(b(1), \cdots, b(m))]$ *a* $\mathrm{PBA}(b_1, \cdots, b_m)$. *Then the following matrix* $C = [C(c(1), \cdots, c(n+m))]$, $0 \leqslant c(i) \leqslant a_i - 1$, *if* $1 \leqslant i \leqslant n$; $0 \leqslant c(i) \leqslant b_{i-n} - 1$, *if* $1 + n \leqslant i \leqslant n + m$, *defined by*

$$C(c(1), \cdots, c(n+m)) = A(c(1), \cdots, c(n)) B(c(n+1), \cdots, c(n+m))$$

is a $\mathrm{PBA}(a_1, \cdots, a_n, b_1, \cdots, b_m)$.

Applying this corollary to the known PBA(4), we find $\mathrm{PBA}(4 \times \cdots \times 4)$; applying it to PBA(6, 6) we find $\mathrm{PBA}(6 \times \cdots \times 6)$ of even dimension; applying it to PBA(12, 3) and PBA(4) we find PBA(12, 12); applying it to PBA(2, 2) we find $\mathrm{PBA}(2, \cdots, 2)$ of even dimension, and so on.

Theorem 6.2.5 *Let* $A = [A(a(1), \cdots, a(n))]$, $0 \leqslant a(i) \leqslant a_i - 1$, $1 \leqslant i \leqslant n$, *be a* $\mathrm{PBA}(a_1, \cdots, a_n)$. *And let* $b_1, \cdots, b_n$ *and* $c_1, \cdots, c_n$ *be integers such that* $\gcd(b_i, a_i) = 1$, $1 \leqslant i \leqslant n$. *Then the following matrix* $H = [H(h(1), \cdots, h(n))]$, $0 \leqslant h(i) \leqslant a_i - 1$, $1 \leqslant i \leqslant n$, *defined by*

$$H(h(1), \cdots, h(n)) = A(b_1 h(1) + c_1, \cdots, b_n h(n) + c_n)$$

is also a $\mathrm{PBA}(a_1, \cdots, a_n)$.

The construction described in this theorem is called "sampling" or "decimation." Its proof is trivial.

Besides the sampling, there are many other invariance operations, which produce further perfect binary arrays of the same volume as the mother array. The example invarance operations include cyclic shifts, inverting all signs, reflection, rotation, permutation of the coordinates, and so on.

Let $A = [A(a(1), \cdots, a(n))]$, $0 \leqslant a(i) \leqslant a_i - 1$, $1 \leqslant i \leqslant n$, be an n-dimensional (± 1)-valued array of size $a_1 \times \cdots \times a_n$. If $a_1 = b_1 \times b_2$, where $\gcd(b_1, b_2) = 1$, $b_1 > 1$ and $b_2 > 1$, then, by the Chinese remainder theorem, for each $a(1) \in Z_{a_1}$, there exists one and only one pair of integers $(b(1), b(2)) \in Z_{b_1} \times Z_{b_2}$ such that $a(1) \equiv b(1) \bmod b_1$ and simultaneously $a(1) \equiv b(2) \bmod b_2$, and vice versa. Thus the n-dimensional matrix A corresponds to an $(n+1)$-dimensional matrix $B = [B(b(1), \cdots, b(n+1))]$ of size $b_1 \times b_2 \times a_2 \times \cdots \times a_n$ since

$$B(b(1), \cdots, b(n+1)) = A(a(1), b(3), \cdots, b(n+1)),$$

where $a(1) \equiv b(1) \bmod b_1$ and $a(1) \equiv b(2) \bmod b_2$. This matrix B is called "the folding of A", or equivalently the matrix A is called "the refolding of B".

Theorem 6.2.6 *An n-dimensional matrix A is a perfect binary array if and only if its folding (resp. refolding) is a perfect binary array. Hence, if $\gcd(s,t) = 1$, then there exists a PBA$(s, t, s_1, \cdots, s_n)$ if and only if there exists a PBA$((st), s_1, \cdots, s_n)$.*

Proof We prove the forward assertation only, i.e., the folding of a PBA is also a PBA. In fact, if $(t(1), \cdots, t(n+1)) \neq (0, \cdots, 0)$, $0 \leqslant t(1) \leqslant b_1 - 1$, $0 \leqslant t(2) \leqslant b_2 - 1$, $0 \leqslant t(i) \leqslant a_i - 1$, $3 \leqslant i \leqslant n+1$, then the periodic correlation is

$$\begin{aligned}
&\sum_{b(1), \cdots, b(n+1)} B(b(1), \cdots, b(n+1)).B(b(1)+t(1), \cdots, b(n+1)+t(n+1)) \\
&= \sum_{a(1)=0}^{a_1-1} \sum_{b(3), \cdots, b(n+1)} A(a(1), b(3), \cdots, b(n+1)) \\
&\quad \times A(a(1)+t'(1), b(3)+t(3), \cdots, b(n+1)+t(3)) \\
&= 0,
\end{aligned}$$

where in the last two equations, $t'(1)$ is determined from $t(1)$ and $t(2)$ by $t'(1) \equiv t(1) \bmod b_1$ and $t'(2) \equiv t(2) \bmod b_2$. Thus $t'(1) \bmod a_1 = 0$ if and only if $t(1) \bmod b_1 = t(2) \bmod b_2 = 0$. The last equation results from A begin a PBA. □

With the fact that the array's volume remains unchanged, Theorem 6.2.6 transforms a PBA of larger dimension and smaller size to a PBA of smaller dimension and larger size, and vice versa. For example, applying the forward part of Theorem 6.2.6, a PBA$(12, 3)$ leads to a PBA$(3, 4, 3)$, and a PBA$(6, 6)$ leads to a PBA$(2, 3, 2, 3)$, etc.

Recall that a (v, k, λ)-difference set, D, is a subset of an additive group, G, of order v such that D contains k elements and the number of solutions to the equation

$$d' - d = g, \quad \text{for } d, d' \in D, d \neq d',$$

is λ for each no-zero $g \in G$. In particular, a $(4N^2, 2N^2 - N, N^2 - N)$-difference set is called "a Menon difference set"[6]. An integer t is called "a (numerical) multiplier of D" if

$$\{td : d \in D\} = \{d + h : d \in D\} \text{ for some } h \in G,$$

and if $h = 0$, then D is fixed by the multiplier t. A difference set fixed by the multiplier -1 is called "symmetric." A well-known identity for a (v, k, λ)-difference set states that

$$k(k-1) = \lambda(v-1). \tag{6.20}$$

Definition 6.2.2 [3,6] *Let $A = [A(j_1, \cdots, j_n)]$ be a binary array of size $s_1 \times \cdots \times s_n$. The set equivalent of A is the subset $\nu(A)$ of the Abelian group $Z_{s_1} \times \cdots \times Z_{s_n}$ given by*

$$\nu(A) = \{(j_1, \cdots, j_n) : A(j_1, \cdots, j_n) = -1\}.$$

If $(u_1, \cdots, u_n) \in Z_{s_1} \times \cdots \times Z_{s_n}$, denote by $\lambda_A(u_1, \cdots, u_n)$ the number of solutions to the equation

$$(j_1', \cdots, j_n') - (j_1, \cdots, j_n) = (u_1, \cdots, u_n), \ \ for \ (j_1', \cdots, j_n'), (j_1, \cdots, j_n) \in \nu(A).$$

The mapping from A to $\nu(A)$ is invertible. The following lemma shows that a binary array with two-valued autocorrelation is equivalent to a difference set in $Z_{s_1} \times \cdots \times Z_{s_n}$.

Lemma 6.2.1 [3,6] *Let $A = [A(j_1, \cdots, j_n)]$ be a binary array of size $s_1 \times \cdots \times s_n$. And let $| \nu(A) | = k$. Then, for all $0 \leqslant u_i \leqslant s_i - 1$, the periodic autocorrelation of A is*

$$R_A(u_1, \cdots, u_n) = \prod_{i=1}^{n} s_i - 4(k - \lambda_A(u_1, \cdots, u_n)).$$

Proof By Definition 6.2.2, $\lambda_A(u_1, \cdots, u_n)$ is equal to the number of occurrences of the product of -1 with -1 on the right hand side of

$$R_A(u_1, \cdots, u_n) = \sum_{j_1=0}^{s_1-1} \cdots \sum_{j_n=0}^{s_n-1} A(j_1, \cdots, j_n) A(j_1 + u_1, \cdots, j_n + u_n) = 0.$$

The number of occurrences of the product of -1 with 1, of 1 with -1, and of 1 with 1 is then respectively $k - \lambda_A(u_1, \cdots, u_n)$, $k - \lambda_A(u_1, \cdots, u_n)$, and $\prod_{i=1}^{n} s_i - 2k + \lambda_A(u_1, \cdots, u_n)$.
□

The following theorem shows that a PBA is equivalent to a Menon difference set in an Abelian group.

Theorem 6.2.7 [3,6] *A is a PBA$(s_1, \cdots, s_n)$ if and only if $\nu(A)$ is a $(4N^2, 2N^2 - N, N^2 - N)$-difference set in the Abelian group $Z_{s_1} \times \cdots \times Z_{s_n}$, where $4N^2 = \prod_{i=1}^{n} s_i$.*

Proof Suppose A is a PBA$(s_1, \cdots, s_n)$, and let $| \nu(A) | = k$. Then by Lemma 6.2.1 and Definition 6.2.1, for all $(u_1, \cdots, u_n) \neq (0, \cdots, 0)$, $0 \leqslant u_i \leqslant s_i - 1$, we have

$$\lambda_A(u_1, \cdots, u_n) = k - \prod_{i=1}^{n} s_i/4.$$

Therefore by the definitions of difference set and $\nu(A)$, we know that $\nu(A)$ is a $(\prod_{i=1}^{n} s_i, k, k - \prod_{i=1}^{n} s_i/4)$-difference set in $Z_{s_1} \times \cdots \times Z_{s_n}$. By Theorem 6.2.3, $\prod_{i=1}^{n} s_i = 4N^2$ for an integer N. Then by Equation (6.20), $k = 2N^2 - N$. Hence $\nu(A)$ is a Menon difference set.

Conversely, suppose that $\nu(A)$ is a $(4N^2, 2N^2 - N, N^2 - N)$-difference set in $Z_{s_1} \times \cdots \times Z_{s_n}$, where $4N^2 = \prod_{i=1}^{n} s_i$. Then by the definitions of difference set and $\nu(A)$, we know that for all $(u_1, \cdots, u_n) \neq (0, \cdots, 0)$, $0 \leqslant u_i \leqslant s_i - 1$, we have

$$\lambda_A(u_1, \cdots, u_n) = N^2 - N.$$

The result follows from Lemma 6.2.1 and Definition 6.2.1. □

The equivalence between PBAs and Menon difference sets is useful for the construction of PBAs and especially for the proof of the existence and non-existence results. Now we turn to introducing some Menon difference sets in the Abelian group $Z_{s_1} \times \cdots \times Z_{s_n}$.

Definition 6.2.3 [6,7,15] *Let A, B, C, D be (± 1)-valued arrays of size $s_1 \times \cdots \times s_n$. $\{A,B,C,D\}$ is called "an $s_1 \times \cdots \times s_n$ binary supplementary quadruple* (BSQ)*" if the following is satisfied for all $0 \leqslant u_i \leqslant s_i - 1$, $1 \leqslant i \leqslant n$,*

(1) $(R_A + R_B + R_C + R_D)(u_1, \cdots, u_n) = 0$ *if* $(u_1, \cdots, u_n) \neq (0, \cdots, 0)$;

(2) $(R_{WX} + R_{YZ})(u_1, \cdots, u_n) = 0$ *for all* $\{W,X,Y,Z\} = \{A,B,C,D\}$,

where $R_X(\cdot)$ and $R_{XY}(\cdot)$ refer to the periodic auto- and cross-correlations, respectively.

In the next subsection, the following useful theorem will be proved (see Corollary 6.2.5).

Theorem 6.2.8 [3,6,8] *If there exists an $s_1 \times \cdots \times s_n$* BSQ, *then there exists a Menon difference in the group*

$$Z_{2^{a_1}} \times Z_{2^{a_2}} \times \cdots \times Z_{2^{a_k}} \times Z_{s_1} \times Z_{s_2} \times \cdots \times Z_{s_n},$$

where $\sum_i a_i = 2a + 2 \geqslant 2$ and $a_i \leqslant a + 2$ for all i. Thus we have a PBA$(2^{a_1}, \cdots, 2^{a_k}, s_1, \cdots, s_n)$, *from Theorem* 6.2.7.

Let G be the group of the form

$$Z_{3^b}^2 \cong \langle y, z \rangle, \quad y^{3^b} = z^{3^b} = 1.$$

And let

$$D_{1,i} = \langle yz^i \rangle, \quad i = 0, 1, \cdots, \frac{3^{b+1} - 1}{3 - 1} - 2,$$

and

$$D_{3j,1} = \langle y^{3j} z \rangle, \quad j = 0, 1, \cdots, \frac{3^b - 1}{3 - 1}.$$

Thus

$$D_{1,m} = D_{1,m+3^b} \text{ and } D_{3r,1} = D_{3(r+3^{b-1}),1}.$$

In the following, we will use D_0, D_1, D_2, and D_3 to respectively represent the

$$D_k = \bigcup_{i=0}^{(3^b-1)/2-1} z^i D_{1,3i+k}, \quad \text{if } k = 0, 1, 2$$

and

$$D_3 = \bigcup_{j=0}^{(3^b-1)/2} y^j D_{3j,1}.$$

Lemma 6.2.2 [6,9,10] *The above D_k has no repeated elements.*

Proof We will prove the case of $k = 0$ only, because the other cases can be proved in the same way. Suppose there was a repeated element. Then there exist i, i', m, m' such that $z^i (yz^{3i})^m = z^{i'} (yz^{3i'})^{m'}$. In order for this to occur, $m = m'$, since the same power of y must be present. Considering the powers of z, we have

$$z^{i+3mi} = z^{i'+3mi'} \text{ or } i(1+3m) \equiv i'(1+3m) \pmod{3^b}.$$

Since $1 + 3m$ is invertible mod 3^b, we can conclude that $i \equiv i' \pmod{3^b}$; the restrictions on the i and i' imply that they are the same. Thus, these two elements are not really distinct, so no elements are repeated. □

Lemma 6.2.3[6,8] *If χ is a character of order 3^b on $\langle y,z\rangle$, then $|\chi(D_k)| = 3^b$ for one value of k, and 0 for the others.*

Proof Note that $G/\mathrm{Ker}(\chi)$ is a cyclic group. Let $x\mathrm{Ker}(\chi)$ be a generator of $G/\mathrm{Ker}(\chi)$; the order of $x\mathrm{Ker}(\chi)$ must be 3^b since it is the size of the factor group. Thus the order of x is also 3^b since it is the maximum order in G. The subgroups $\langle x\rangle$ and $\mathrm{Ker}(\chi)$ intersect only in the identity (no power of x smaller than 3^b can be in $\mathrm{Ker}(\chi)$), and their product is all of G; thus G must be a direct product of $\langle x\rangle$ and $\mathrm{Ker}(\chi)$. The fact that G has rank 2 implies that $\mathrm{Ker}(\chi)$ must be a cyclic group of order 3^b.

Since we have established the fact that $\mathrm{Ker}(\chi)$ is cyclic, we need to observe that all of the cyclic subgroups of G of order 3^b are of the form $D_{i,j}$ for some i,j. Thus χ is principal on one $D_{i',j'}$ and non-principal on all of the others. If that $D_{i',j'}$ appears only once, then the character sum (in modulus) is the size of the set which is equal to 3^b. If the $D_{i',j'}$ appears twice, then we suppose that χ has order 3^b on the element z (it must have order 3^b on either y or z, and the y argument is the same) . The character sum is 3^b times $\chi(z^i)+\chi(z^{i+3^{b-1}})$. Since χ is a homomorphism, this can be rewritten as $\chi(z^i)(1+\chi(z^{3^{b-1}}))$; the $\chi(z^{3^{b-1}})$ is a primitive third root of unity, so $\left|\chi(z^i)(1+\chi(z^{3^{b-1}}))\right| = 1$. Thus the character sum is (in modulus) 3^b. □

Lemma 6.2.4[6,8,10] *If χ is a character of $< y,z >$ that is nonprincipal but of order less than 3^b, then $|\chi(D_k)| = 3^b$ for one k, and 0 for the others.*

Proof Let ξ be a primitive 3^bth root of unity, and suppose that χ is a character of order less than 3^b. Then

$$\chi(y) = \xi^{e3^v} \text{ and } \chi(z) = \xi^{f3^t}, \quad 1 \leqslant t,v \leqslant b, \quad (t,v) \neq (b,b),$$

where e and f are nonzero integers which are not divisible by 3.

Consider the case of $v < t$ (the other cases of $v > t$ and $v = t$ are similar) . Nothing of the form yz^x is in the kernel of the character because there is no way to satisfy the equation $e3^v + xf3^t \equiv 0 \pmod{3^b}$ when $v < t$. Thus χ is nonprincipal on every subgroup $D_{i,j}$ contained in the sets D_0, D_1, and D_2, so the character sum is 0 over these parts. The kernel does contain elements of the form $y^{3x}z$ whenever $3xe3^v + f3^t \equiv 0 \pmod{3^b}$, or whenever $x \equiv -fe^{-1}3^{t-v-1} \pmod{3^{b-v-1}}$. The character χ is principal on the subgroups $D_{3j,1}$ associated with those solutions, and non-principal on all the other subgroups $D_{3j,1}$ in D_3. There are q solutions x to this equation with $0 \leqslant x \leqslant (3^b-1)/(3-1)$, where q is either $3^v + 3^{v-1} + \cdots + 3 + 1$ or $3^v + 3^{v-1} + \cdots + 3 + 1 + 1$ (in general, the number of solutions to the congruence $x \equiv a \pmod{b}$ with $0 \leqslant x \leqslant c$ is $\lfloor c/b\rfloor$ or $\lfloor c/b\rfloor+1$). Hence the character sum over D_3 is

$$\begin{aligned}&3^b(\chi(y^x) + \chi(y^{x+3^{b-v-1}}) + \chi(y^{x+2.3^{b-v-1}}) + \cdots + \chi(y^{x+(q-1)\cdot 3^{b-v-1}}))\\ =&3^b\chi(y^x)(1+\chi(y^{3^{b-v-1}}) + \cdots + \chi(y^{(q-1)3^{b-v-1}})).\end{aligned}$$

Since $\chi(y) = \xi^{e3^v}$, we get

$$\chi(D_3) = \begin{cases} 3^b\chi(y^x)(1+\xi^{e3^{b-1}} + \xi^{2e3^{b-1}} + 1 + \cdots + \xi^{2e3^{b-1}} + 1), \\ \hfill q \equiv 1 \pmod{3}; \\ 3^b\chi(y^x)(1+\xi^{e3^{b-1}} + \xi^{2e3^{b-1}} + 1 + \cdots + \xi^{2e3^{b-1}} + 1 + \xi^{e3^{b-1}}), \\ \hfill q \equiv 2 \pmod{3} \end{cases}$$

$$= \begin{cases} 3^b\chi(y^x), & q \equiv 1(\text{mod}3); \\ 3^b\chi(y^x)(1+\xi^{e3^{b-1}}), & q \equiv 2(\text{mod}3). \end{cases}$$

In either case, the modulus of this sum is 3^b since $(1+\xi^{e3^{b-1}}) = -\xi^{2e3^{b-1}}$ has modulus 1. Thus $|\chi(D_3)| = 3^b$ and all the others are 0. □

Lemma 6.2.5 [6,8] *Let $A = v^{-1}(D_0)$, $B = v^{-1}(D_1)$, $C = v^{-1}(D_2)$, and $D = v^{-1}(D_3)$ be four subsets of G Then the sets $\{A, B, C, D\}$ form a $3^b \times 3^b$ BSQ.*

Proof We need to translate the definition of BSQ into character theoretic terms. The first condition $(R_A + R_B + R_C + R_D)(u_1, u_2) = 0$, can be shown by considering the group ring expression as

$$D_0D_0^{(-1)} + D_1D_1^{(-1)} + D_2D_2^{(-1)} + D_3D_3^{(-1)} - 3^{2b}.$$

In using Lemmas 6.2.3 and 6.2.4, any nonprincipal character on G has a sum of 3^b (in modulus) on one of the D_k, and it will be 0 on the others. Thus, the character sum on the expression is 0, which implies that the group ring expression is cG for some c. The fact that

$$|D_0| = |D_1| = |D_2| = 3^b(3^b - 1)/2 \ \text{ and } \ |D_3| = 3^b(3^b + 1)/2$$

implies that

$$\begin{aligned} c &= \left\{\sum |D_i|^2 - 3^{2b}\right\}/|G| \\ &= \{3.3^{2b}(3^b - 1)^2/4 + 3^{2b}(3^b + 1)^2/4 - 3^{2b}\}/3^{2b} \\ &= 3^{2b} - 3^b. \end{aligned}$$

The number of times that the nonidentity element (u_1, u_2) appears in the expression is $3^{2b} - 3^b$, and this corresponds to the number of times that $X[j_1, j_2] = X[j_1 + u + 1, j_2 + u_2] = -1$ in the autocorrelation equation for $X = A, B, C$, or D (we call these $(-1, -1)$ pairs) . There are $2.3^{2b} - 3^b$ times when $X[j_1, j_2] = -1$, so there are $2.3^{2b} - 3^b - (3^{2b} - 3^b) = 3^{2b}$ pairs of the form $(-1, 1)$ and $(1, -1)$. Finally, there are a total of 4.3^{2b} pairs and so there are $4.3^{2b} - 3^{2b} - 3^{2b} - (3^{2b} - 3^b) = 3^{2b} + 3^b$ $(1, 1)$ pairs. Thus, the autocorrelation equation becomes

$$\begin{aligned} (R_A + R_B + R_C + R_D)(u_1, u_2) = & (3^{2b} - 3^b)(-1)(-1) + 3^{2b}(-1)(1) \\ & + 3^{2b}(1)(-1) + (3^{2b} + 3^b)(1)(1) = 0 \end{aligned}$$

for all $(u_1, u_2) \neq (0, 0)$.

The proof of the second condition involves studying the group ring expression $D_0D_1^{(-1)} + D_2D_3^{(-1)}$. Using the same arguments as above, we find that the character sum over this equation is 0 for any nonprincipal character, which implies that $D_0D_1^{(-1)} + D_2D_3^{(-1)} = cG$ for some c. A counting argument yields that $c = (3^{2b} - 3^b)/2$, and this number is also the number of $(-1, -1)$ pairs (by this we mean the number of times that $X[j_1, j_2] = Y[j_1 + u_1, j_2 + u_2] = -1$ for $(X, Y) = (A, B)$ or (C, D)) in the sum $R_{AB} + R_{CD}$. Since $|D_0| = (3^{2b} - 3^b)/2$, and $|D_2| = (3^{2b} - 3^b)/2$, there are

$$((3^{2b} - 3^b)/2 + (3^{2b} - 3^b)/2) - (3^{2b} - 3^b)/2 = (3^{2b} - 3^b)/2$$

$(-1,1)$ pairs. Similar counts yield $(3^{2b}+3^b)/2$ pairs of both $(1,-1)$ and $(1,1)$. Thus

$$\begin{aligned}(R_{AB}+R_{CD})(u_1,u_2) &= (3^{2b}-3^b)/2(-1)(-1)+(3^{2b}-3^b)/2(-1)(1)\\ &\quad +(3^{2b}-3^b)/2(1)(-1)+(3^{2b}-3^b)/2(1)(1)\\ &= 0 \quad (\text{for every } (u_1,u_2)) .\end{aligned}$$

We can shuffle the four sets any way we want, and we will get the same result for the autocorrelation equation. Thus, these four sets satisfy the definition of BSQ.

□

Now we are ready to state one of the main results about the existence of PBAs.

Theorem 6.2.9 [3,5] *There exists*

$$\text{PBA}(2^{a_1},\cdots,2^{a_k},3^{b_1},3^{b_1},\cdots,3^{b_n},3^{b_n}),$$

where $\sum_i a_i = 2a+2$, $a \geqslant 0$, $a_i \leqslant a+2$. *Particularly, there exists* PBA$(2^{1+a}3^b, 2^{1+a}3^b)$ *for all* $a,b \geqslant 0$ *(this result has been used to construct the* 3*-dimensional Hadamard matrices of order* 2.3^b *in the chapter on* 3*-dimensional Hadamard matrices).*

Proof Lemma 6.2.5 together with Theorems 6.2.8 and 6.2.7 implies the existence of PBA$(2^{a_1},\cdots,2^{a_k},3^{b_1},3^{b_1})$. Then the proof is finished by Theorem 6.2.4. Particularly, the PBA$(2^{1+a}3^b,2^{1+a}3^b)$ is constructed by using Theorem 6.2.6 to form a PBA$(2^a,2^a,3^b,3^b)$.
□

Another important existence result about PBAs is based on the following lemma, which has been proved in [11] or reviewed in [6].

Lemma 6.2.6[11] *There exists* BSQ *of size* $p\times p\times p\times p$ *for each positive integer* p *satisfying* $p\equiv 3 \bmod 4$.

This lemma together with Lemma 6.2.5, Theorems 6.2.8, 6.2.7, and 6.2.4 results in the following larger class of PBAs.

Theorem 6.2.10 *There exists*

$$\text{PBA}(2^{a_1},\cdots,2^{a_k},3^{b_1},3^{b_1},\cdots,3^{b_n},3^{b_n},p_1^{c_1},p_1^{c_1},p_1^{c_1},p_1^{c_1},\cdots,p_m^{c_m},p_m^{c_m},p_m^{c_m},p_m^{c_m}),$$

where $\sum_i a_i = 2a+2$, $a\geqslant 0$, $a_i \leqslant a+2$, *and* $p_i \equiv 3 \bmod 4$.

Besides the above existence results, there are also many results about the nonexistence of PBAs. For example, it has been proved[3,4,11–16]

(1) There exists no PBAs of volume $4p^2$ for any prime $p>3$;
(2) Suppose there exists a Menon difference in $Z_{p^a}\times Z_{p^{2y-a}}\times H$, where $0\leqslant a\leqslant 2y$, H an Abelian group of even order h, $p>2$, a prime self-conjugate mod $\exp(H)$, and $p\nmid h$. Then $a=y$, and if $p>3$, then
   (a) if $a=1$, $p^2<h$ and $(p+1)|h$;
   (b) if $a>1$, $p<h$.
(3) If there exists a PBA$(4N^2)$ with $N>1$, then N is odd, not a prime power and $N\geqslant 55$;
(4) Suppose there exists a $(4N^2,2N^2-N,N^2-N)$-difference set in an Abelian group G. If N is even, then the Sylow 2-subgroup of G is not cyclic;

(5) There does not exist PBA(s,t), if $(s,t)=$ (2, 32), (4, 64), (2, 18), (6, 54), (12, 27), (36, 81), (2, 72), (4, 36), (8, 18), (4, 100), (8, 50), (8, 98), (24, 54), (32, 98), (6, 96), (18, 32), (32, 50), (36, 64), (54, 96), (64, 100), (10, 90), (12, 75), (18, 50), (20, 45), (18, 98), (28, 63), (50, 98), (36, 100), (50, 72), (72, 98), (28, 28), (44, 44), (76, 76), (92, 92), (30, 30), (42, 42), (70, 70), (90, 90), (21, 84), (84, 84), (10, 10), (14, 14), (7, 28), (14, 56), (66, 66), (50, 50), (98, 98), and so on;

(6) There does not exist PBA(2, 2, 5, 5), PBA(2, 2, 3, 3, 9), PBA(4, 3, 3, 9), and so on.

6.2.3 Generalized Perfect Arrays

For details of the generalized perfect arrays the readers are recommended to see the papers [3] and [10].

Generalized perfect array (GPA) is a generalization of the perfect, quasiperfect and doubly quasiperfect arrays. This subsection generalizes the elementary recursive constructions for the 2-dimensional cases. Another main aim of this subsection is to show Theorem 6.2.8, one of the most important results on the construction of PBAs. The main contents of this subsection are from [3] and [10]. In order to save space, many tedious proofs have to be omitted.

An n-dimensional m-ary array of size $a_1 \times \cdots \times a_n$ is a matrix $A = [A(a(1), \cdots, a(n))]$, $0 \leqslant a(i) \leqslant a_i - 1$, $1 \leqslant i \leqslant n$, such that

$$A(a(1), \cdots, a(n)) \in \begin{cases} \{\pm 1, \cdots, \pm m/2\}, & \text{if } m \text{ is even,} \\ \{0, \pm 1, \cdots, \pm (m-1)/2\}, & \text{if } m \text{ is odd} \end{cases}$$

(when $m = 2$ the matrix A becomes the binary array studied in last subsections). The energy and sum of the m-ary array A are defined by[3,10]

$$E_A = \sum_{a(1)=0}^{a_1-1} \cdots \sum_{a(n)=0}^{a_n-1} (A(a(1), \cdots, a(n)))^2$$

and

$$S_A = \sum_{a(1)=0}^{a_1-1} \cdots \sum_{a(n)=0}^{a_n-1} A(a(1), \cdots, a(n))$$

respectively. Thus in binary cases, $E_A = \prod_{i=1}^{n} a_i$, which is also the volume of this matrix.

Definition 6.2.4[3,10] *Let $A = [A(a(1), \cdots, a(n))]$ and $B = [B(b(1), \cdots, b(n))]$, $0 \leqslant a(i), b(i) \leqslant s_i - 1$, $1 \leqslant i \leqslant n$, be two m-ary arrays of the same size $s_1 \times \cdots \times s_n$.*

(1) *The periodic cross-correlation function of A and B is defined by*

$$R_{AB}(u_1, \cdots, u_n) = \sum_{j_1=0}^{s_1-1} \cdots \sum_{j_n=0}^{s_n-1} A(j_1, \cdots, j_n) B(j_1 + u_1, \cdots, j_n + u_n),$$

where $0 \leqslant u_i \leqslant s_i - 1$, $j_i + u_i = (j_i + u_i) \bmod s_i$, $1 \leqslant i \leqslant n$. If $A = B$, then $R_{AB}(\cdot) =: R_A(\cdot)$ is the periodic autocorrelation function of A.

(2) *The adjoining of A with B in dimension r, $1 \leqslant r \leqslant n$, is an m-ary array, $\alpha^{(r)}(A, B) = [C(j_1, \cdots, j_n)]$, of size $s_1 \times \cdots \times s_{r-1} \times (2s_r) \times s_{r+1} \times \cdots \times s_n$ defined by*

$$C(j_1, \cdots, j_{r-1}, j_r + ys_r, j_{r+1}, \cdots, j_n) = \begin{cases} A(j_1, \cdots, j_n), & \text{if } y = 0 \\ B(j_1, \cdots, j_n), & \text{if } y = 1. \end{cases}$$

Clearly, the energy of $\alpha^{(r)}(A,B)$ is E_A+E_B, the summation of its parent's energy.

Definition 6.2.5 [3,10] *Let $A=[A(a(1),\cdots,a(n))]$, $0\leqslant a(i)\leqslant a_i-1$, $1\leqslant i\leqslant n$, be an m-ary array of size $a_1\times\cdots\times a_n$, and let $z=(z_1,\cdots,z_n)$ be a $(0,1)$-valued vector, which is called "the type vector."*

(1) *The expansion of A with respect to the type vector z is the following m-ary array, $\varepsilon(A,z)=: E=[E(e(1),\cdots,e(n))]$, of size $(z_1+1)a_1\times\cdots\times(z_n+1)a_n$, defined by*

$$E(j_1+y_1a_1,\cdots,j_n+y_na_n)=(-1)^{\sum_i y_i}A(j_1,\cdots,j_n),$$

where $0\leqslant j_i\leqslant a_i-1$, $0\leqslant y_i\leqslant z_i$, $1\leqslant i\leqslant n$.

(2) *The m-ary array A is called "a generalized perfect array (GPA) of type z," abbreviated as a GPA$(m;a_1,\cdots,a_n)$ of type z, if the periodic autocorrelation of $\varepsilon(A,z)=E$ satisfies*

$$R_E(u_1,\cdots,u_n)\neq 0,\ \textit{only if}\ u_i\equiv 0(\mathrm{mod}\,a_i)\ \textit{for all}\ i,$$

where $0\leqslant u_i\leqslant(z_i+1)a_i-1$. Particularly, if $m=2$, the GPA is called GPBA.

For example, if A is a 2-dimensional m-ary array of size $a_1\times a_2$, then

$$B_0=: A=\varepsilon(A;0,0);\quad B_1=:\alpha^{(1)}(A,-A)=\begin{bmatrix}A\\-A\end{bmatrix}=\varepsilon(A;1,0);$$

$$B_2=:\alpha^{(2)}(A,-A)=[A\ \ -A]=\varepsilon(A;0,1)$$

and

$$B_3=:\alpha^{(1)}(\alpha^{(2)}(A,-A),\alpha^{(2)}(-A,A))=\begin{bmatrix}A&-A\\-A&A\end{bmatrix}=\varepsilon(A;1,1).$$

Thus a PBA is a GPA of type $(0,\cdots,0)$; a rowwise quasiperfect binary array (RQPBA) is a 2-dimensional GPBA of type $(1,0)$; a columnwise quasiperfect binary array (CQPBA) is a 2-dimensional GPBA of type $(0,1)$; a doubly quasiperfect binary array (DQPBA) is a 2-dimensional GPBA of type $(1,1)$.

Using the above definitions, it is not difficult to prove that[3,10]

(1) If $A=[A(a(1),\cdots,a(n))]$ is a PBA$(a_1,\cdots,a_n)$, then the matrix

$$B=[B(b(1),\cdots,b(n-1))]=\left[\sum_{a(n)=0}^{a_n-1}A(a(1),\cdots,a(n))\right]$$

is a GPA;

(2) If π is a permutation of $\{1,\cdots,n\}$, then $A=[A(a(1),\cdots,a(n))]$ is a GPA$(m;a_1,\cdots,a_n)$ of type $(z_1,\cdots,z_n)$ if and only if $[A(a(\pi(1)),\cdots,a(\pi(n)))]$ is a GPA$(m;a_{\pi(1)},\cdots,a_{\pi(n)})$ of type $(z_{\pi(1)},\cdots,z_{\pi(n)})$, i.e., the dimensions may be reordered.

Lemma 6.2.7 [3,10] *Let A be an m-ary array of size $a_1\times\cdots\times a_n$, and $z=(z_1,\cdots,z_n)$ be a type vector. Then*

$$\varepsilon(A;1,z)=\alpha^{(1)}(\varepsilon(A;0,z),-\varepsilon(A;0,z)).$$

From this lemma, it is known that the array $\varepsilon(A;z_1,\cdots,z_n)$ may be formed by recursively adjoining A with its negative copy $-A$ for each dimension i satisfying $z_i=1$.

Lemma 6.2.8 [3,10] *Let A be an m-ary array of size $a_1 \times \cdots \times a_n$, $z = (z_1, \cdots, z_n)$ be a type vector, and $E = \varepsilon(A; z)$. Then*

$$R_E(y_1 a_1, \cdots, y_n a_n) = E_E(-1)^{\sum_i y_i} = E_A(-1)^{\sum_i y_i} \prod_i (z_i + 1),$$

where $0 \leqslant y_i \leqslant z_i$, $1 \leqslant i \leqslant n$, E_X the energy function of X, and R_E the periodic autocorrelation of E.

Lemma 6.2.9 [3,10] *Let $A = [A(j, j_1, \cdots, j_n)]$ be an m-ary array of size $(st) \times a_1 \times \cdots \times a_n$, $z = (z_1, \cdots, z_n)$ be a type vector. And let $B = [B(j, j_1, \cdots, j_n)]$ be the $(1 + 2s\lfloor m/2 \rfloor)$-ary array defined by*

$$B(j, j_1, \cdots, j_n) = \sum_{r=0}^{s-1} A(j + rt, j_1, \cdots, j_n),$$

where $0 \leqslant j \leqslant t - 1$, $0 \leqslant j_i \leqslant a_i - 1$, $1 \leqslant i \leqslant n$. If A is a GPA$(m; (st), a_1, \cdots, a_n)$ *of type $(0, z)$, then B is a* GPA$(1 + 2s\lfloor m/2 \rfloor; t, a_1, \cdots, a_n)$ *of type $(0, z)$ and $E_B = E_A$, $S_A = S_B$, i.e., the energy and sum functions remain unchanged.*

Definition 6.2.6 [3,10] *Let G be an additive group of order mn, and H a subgroup of G of order n. And let D be a subset of G containing k elements. D is called an $(n, m, k, \lambda', \lambda)$-relative difference set in G relative to H if the number of solutions of the equation*

$$d' - d = g, \quad for \quad d, d' \in D, \ d \neq d',$$

(1) *is λ' for each nonzero $g \in H$, and*

(2) *is λ for each $g \in G - H$.*

When H is a normal subgroup and $\lambda' = 0$, denote D by $D(m, n, k, \lambda)$.

Theorem 6.2.11 [3,10] *Let A be a binary array of size $a_1 \times \cdots \times a_n$, $z = (z_1, \cdots, z_n) \neq (0, \cdots, 0)$ a type vector, and $E = \varepsilon(A; z)$. Define the following groups G, H, and K, where H is a subgroup of G and K is a subgroup of H:*

$$\begin{aligned} G &= Z_{(z_1+1)a_1} \times \cdots \times Z_{(z_n+1)a_n}, \\ H &= \{(h_1, \cdots, h_n) : h_i = y_i a_i \ and \ 0 \leqslant y_i \leqslant z_i\}, \\ K &= \{(k_1, \cdots, k_n) : k_i = y_i a_i, \ 0 \leqslant y_i \leqslant z_i, \ and \ \sum_i y_i \ is \ even\}. \end{aligned}$$

Let D be the subset of the factor group G/K given by

$$D = \{K + (j_1, \cdots, j_n) : (j_1, \cdots, j_n) \in \nu(E)\},$$

where $\nu(E)$ is the set equivalence of E defined in the last subsection.

Then A is a non-trivial GPBA$(a_1, \cdots, a_n)$ *of type z if and only if D is a $D(E_A, 2, E_A, E_A/2)$ in G/K relative to H/K, where E_A is the energy of A.*

Proof It is easy to check that H is a subgroup of G, and K is a subgroup of H. Since G is Abelian, the subgroups H and K are normal and the factor groups G/K and H/K are well-defined. The arrays A and E are binary and so by using Lemma 6.2.8 and the definition of energy,

$$E_E = E_A \prod_i (z_i + 1) = \prod_i (z_i + 1) a_i. \tag{6.21}$$

Suppose that A is a non-trivial GPBA$(a_1, \cdots, a_n)$ of type z. Then Lemma 6.2.8 together with the definition of GPA results in

$$R_E(u_1, \cdots, u_n) = \begin{cases} E_E, & \text{if } (u_1, \cdots, u_n) \in K, \\ -E_E, & \text{if } (u_1, \cdots, u_n) \in H - K, \\ 0, & \text{if } (u_1, \cdots, u_n) \in G - H. \end{cases}$$

From Definitions 6.2.5 and 6.2.2, $\nu(E)$ is a subset of G. Since $z \neq (0, \cdots, 0)$, from Lemma 6.2.7 we may write $E = \alpha^{(r)}(B, -B)$ for some binary array B and $1 \leqslant r \leqslant n$. Then from Definition 6.2.4, exactly half of the elements of E are -1 and so by Equation (6.21) , $|\nu(E)| = E_E/2$. Therefore from Lemma 6.2.1,

$$\lambda_E(u_1, \cdots, u_n) = \begin{cases} E_E/2, & \text{if } (u_1, \cdots, u_n) \in K, \\ 0, & \text{if } (u_1, \cdots, u_n) \in H - K, \\ E_E/4, & \text{if } (u_1, \cdots, u_n) \in G - H. \end{cases}$$

Recall from Definition 6.2.2 that for $(u_1, \cdots, u_n) \in G$, $\lambda_E(u_1, \cdots, u_n)$ is the number of solutions of the equation

$$(i_1, \cdots, i_n) - (j_1, \cdots, j_n) = (u_1, \cdots, u_n)$$

for $(j_1, \cdots, j_n), (i_1, \cdots, i_n) \in \nu(E)$.

Therefore for $K + (u_1, \cdots, u_n) \in G/K$, the number of solutions to the equation

$$(K + (i_1, \cdots, i_n)) - (K + (j_1, \cdots, j_n)) = K + (u_1, \cdots, u_n)),$$

for $K + (j_1, \cdots, j_n),\ K + (i_1, \cdots, i_n) \in D$, is equal to

$$\begin{cases} 0, & \text{if } K + (u_1, \cdots, u_n)) \in H/K,\ (u_1, \cdots, u_n) \neq (0, \cdots, 0), \\ E_E/(4\,|K|), & \text{if } K + (u_1, \cdots, u_n)) \in (G/K) - (H/K). \end{cases}$$

Therefore from Definition 6.2.6, D is a $D(|G/K|/|H/K|,\ |H/K|,\ |D|,\ E_E/(4|K|))$ in G/K relative to H/K.

Now $|H| = 2|K| = \prod_i (z_i + 1)$ and so $|H/K| = |H|/|K| = 2$. From Equation (6.21), $E_E/(4|K|) = E_A/2$ and $|G/K|/|H/K| = |G|/|H| = E_E/|H| = E_A$. Also, if $(j_1, \cdots, j_n) \in \nu(E)$, then from Definition 6.2.5,

$$(k_1, \cdots, k_n) + (j_1, \cdots, j_n) \in \nu(E) \quad \text{for all } (k_1, \cdots, k_n) \in K,$$

so $|D| = |\nu(E)|/|K| = E_A$.

Conversely, suppose that D is a $D(E_A, 2, E_A, E_A/2)$ in G/K relative to H/K. Then by using Definitions 6.2.2 and 6.2.6, and Equation (6.21),

$$\lambda_E(u_1, \cdots, u_n) = |K|E_A/2 = E_E/4, \quad \text{if } (u_1, \cdots, u_n) \in G - H.$$

The result follows from Lemma 6.2.1 and Definition 6.2.5. □

Definition 6.2.7 [3,10] *Let $A = [A(a(1), \cdots, a(n))]$ and $B = [B(b(1), \cdots, b(n))]$ be two m-ary arrays of size $a_1 \times \cdots \times a_n$, and let $1 \leqslant r \leqslant n$. The interleaving of A with B in dimension r is the m-ary array, $\iota^{(r)}(A, B) =: C = [C(c(1), \cdots, c(n))]$, of size $a_1 \times \cdots \times a_{r-1} \times (2a_r) \times a_{r+1} \times \cdots \times a_n$ defined by*

$$C(j_1, \cdots, j_{r-1}, 2j_r + y_r, j_{r+1}, \cdots, j_n) = \begin{cases} A(j_1, \cdots, j_r), & \textit{if} \ \ y_r = 0, \\ B(j_1, \cdots, j_r), & \textit{if} \ \ y_r = 1, \end{cases}$$

where $0 \leqslant j_i \leqslant a_i - 1$, $1 \leqslant i \leqslant n$. Clearly the energy of $\iota^{(r)}(A, B)$ is equal to $E_A + E_B$.

Definition 6.2.8 [3,10] *Let $A = [A(a(1), \cdots, a(n))]$ be an m-ary array of size $a_1 \times \cdots \times a_n$, and let $1 \leqslant r \leqslant n$. The alternate sign-change of A in dimension r is the m-ary array, $\phi^{(r)}(A) =: B = [B(b(1), \cdots, b(n))]$, of size $a_1 \times \cdots \times a_n$ defined by*

$$B(j_1, \cdots, j_n) = (-1)^{j_r} A(j_1, \cdots, j_n),$$

where $0 \leqslant j_i \leqslant a_i - 1$, $1 \leqslant i \leqslant n$. Clearly the energy of $\phi^{(r)}(A)$ is also E_A.

Definition 6.2.9 [3,10] *Let $A = [A(i_1, i_2, a(1), \cdots, a(n))]$ be an m-ary array of size $s \times t \times a_1 \times \cdots \times a_n$, where $\gcd(s,t) = 1$, so $xt + sy = 1$ for some (x, y). The folding of A is the m-ary array, $\mu(A) =: B = [B(j, b(1), \cdots, b(n))]$, of size $(st) \times a_1 \times \cdots \times a_n$ defined by*

$$B(j, j_1, \cdots, j_n) = A(jx, jy, j_1, \cdots, j_n),$$

where $0 \leqslant j \leqslant st - 1$, $0 \leqslant j_i \leqslant s_i - 1$, $1 \leqslant i \leqslant n$, jx and jy are regarded as reduced modulo s and t, respectively.

Note that if $x't + y's = 1$, then $x \equiv x'(\text{mod} s)$ and $y \equiv y'(\text{mod} t)$, so $\mu(A)$ is independent of the particular pair (x, y) used. Note also that the mapping from A to $\mu(A)$ is invertible. The energy of $\mu(A)$ is E_A.

Definition 6.2.10 [3,10] *Let $A = [A(a(1), \cdots, a(n))]$ be an m-ary array of size $a_1 \times \cdots \times a_n$. The enlargement of A is the m-ary array, $I(A) =: B = [B(0, j_1, \cdots, j_n)]$, of size $1 \times a_1 \times \cdots \times a_n$ defined by*

$$B(0, j_1, \cdots, j_n) = A(j_1, \cdots, j_n), \quad for\ 0 \leqslant j_i \leqslant a_i - 1.$$

The energy of $I(A)$ is also E_A. It is easy to see that A is a GPA$(m; a_1, \cdots, a_n)$ of type z if and only if $I(A)$ is a GPA$(m; 1, a_1, \cdots, a_n)$ of type (z_0, z)

Definition 6.2.11 [3,10] *Let $A = [A(j_1, \cdots, j_n)]$, and $B = [B(j_{n+1}, \cdots, j_{n+m})]$ be respectively an $a_1 \times \cdots \times a_n$ m_1-ary and an $a_{n+1} \times \cdots \times a_{n+m}$ m_2-ary array. The tensor product of A with B is the $a_1 \times \cdots \times a_{n+m}$ $f(m_1, m_2)$-ary array $\prod(A, B) =: C = [C(j_1, \cdots, j_{n+m})]$ defined by*

$$C(j_1, \cdots, j_{n+m}) = A(j_1, \cdots, j_n) B(j_{n+1}, \cdots, j_{n+m})$$

for all $0 \leqslant j_i \leqslant a_i - 1$, $1 \leqslant i \leqslant m + n$. From now on, we use $f(x, y)$ to represent

$$f(x, y) = \begin{cases} xy/2, & \text{if } x \text{ and } y \text{ are even,} \\ 1 + 2\lfloor x/2 \rfloor \cdot \lfloor y/2 \rfloor, & \text{otherwise.} \end{cases}$$

Note that $I(A) = \prod([+], A)$. The energy of $\prod(A, B)$ is $E_A E_B$.

Definition 6.2.12 [3,10] *Let $A = [A(r, j_1, \cdots, j_n)]$, and $B = [B(j, j_{n+1}, \cdots, j_{n+m})]$ be respectively an $s \times a_1 \times \cdots \times a_n$ m_1-ary and an $t \times a_{n+1} \times \cdots \times a_{n+m}$ m_2-ary array. The K-product of A with B is the $f(m_1, m_2)$-ary array, $\kappa(A, B) =: C = [C(j, c(1), \cdots, c(n+m))]$, of size $(st) \times a_1 \times \cdots \times a_{n+m}$ defined by*

$$C(rt + j, j_1, \cdots, j_{n+m}) = A(r, j_1, \cdots, j_n) B(j, j_{n+1}, \cdots, j_{n+m}),$$

where $0 \leqslant r \leqslant s - 1$, $0 \leqslant j \leqslant t - 1$, $0 \leqslant j_i \leqslant a_i - 1$, $1 \leqslant i \leqslant n + m$.

Note that if $t = 1$, then $\kappa(A, B) = \prod(A, B')$, where $B = I(B')$. The energy of $\kappa(A, B)$ is clearly $E_A E_B$.

Definition 6.2.13 [3,10] *Let $A = [A(r, j, j_1, \cdots, j_n)]$ be an m-ary array of size $s \times t \times a_1 \times \cdots \times a_n$. And let c be an integer such that $ct \equiv 0(\text{mod} s)$. The shear of A and the transform of A with respect to c are respectively the m-ary arrays, $\sigma(A; c) =: [B(r, j, j_1, \cdots, j_n)]$ and $\tau(A; c) = D = [D(r, j, j_1, \cdots, j_n)]$ of size $s \times t \times a_1 \times \cdots \times a_n$ defined by*

$$B(r, j, j_1, \cdots, j_n) = A(r - cj, j, j_1, \cdots, j_n)$$

and

$$D(r, j, j_1, \cdots, j_n) = (-1)^y A(r - cj, j, j_1, \cdots, j_n),$$

where

$$y = \begin{cases} 0, & \text{if } (r - cj)\text{mod}(2s) < s, \\ 1, & \text{if } (r - cj)\text{mod}(2s) \geqslant s, \end{cases}$$

$0 \leqslant r \leqslant s - 1$, $0 \leqslant j \leqslant t - 1$, $0 \leqslant j_i \leqslant a_i - 1$, $1 \leqslant i \leqslant n$, and $r - cj$ is regarded as $(r - cj)\text{mod} s$.

Note that the energy of both $\sigma(A; c)$ and $\tau(A; c)$ is the same as that of A. The condition $ct \equiv 0(\text{mod} s)$ is necessary for the mappings τ and σ to be well-defined.

In the next two lemmas, we examine the effect of composing certain pairs of mappings, including the expansion mapping ε and the adjoining mapping $\alpha^{(r)}$.

Lemma 6.2.10 [3,10] *Let A, B be m-ary arrays of size $a_1 \times \cdots \times a_n$, and $z = (z_1, \cdots, z_n)$ a type vector. Let $A' = \varepsilon(A; z)$ and $B' = \varepsilon(B; z)$. Then*

(1) $\varepsilon(\alpha^{(1)}(A, B); z) = \alpha^{(1)}(A', B')$, *if* $z_1 = 0$;

(2) $\varepsilon(\iota^{(1)}(A, B); z) = \iota^{(1)}(A', B')$;

(3) $\varepsilon(\phi^{(1)}(A); z) = \phi^{(1)}(A')$, *if* $z_1 = 0$;

(4) $\phi^{(1)}(\alpha^{(1)}(A, B)) = \alpha^{(1)}(\phi^{(1)}(A), (-1)^{a_1}\phi^{(1)}(B))$;

(5) $I(\iota^{(1)}(A, B))\iota^{(2)}(I(A), I(B))$.

Lemma 6.2.11 [3,10] *Let $z = (z_1, \cdots, z_n)$ and $z' = (z_{n+1}, \cdots, z_{n+m})$ be two type vectors.*

(1) *Let A be an m-ary array of size $s \times a_1 \times \cdots \times a_n$. Then*

$$\varepsilon(\phi^{(1)}(A); 1, z) = \phi^{(1)}(\alpha^{(1)}(\varepsilon(A; 0, z), (-1)^{s+1}\varepsilon(A; 0, z))).$$

(2) *Let A be an m-ary array of size $s \times t \times a_1 \times \cdots \times a_n$ and suppose $\gcd(s, t) = 1$. Then*

$$\varepsilon(\mu(A); 0, z) = \mu(\varepsilon(A; 0, 0, z)).$$

(3) *Let A and B be respectively an $s \times a_1 \times \cdots \times a_n$ m_1-ary and an $a_{n+1} \times \cdots \times a_{n+m}$ m_2-ary array. Then*

$$\varepsilon\left(\prod(A, B); z, z'\right) = \prod(\varepsilon(A; z), \varepsilon(B, z')).$$

(4) *Let A and B be respectively an $s \times a_1 \times \cdots \times a_n$ m_1-ary and a $t \times a_{n+1} \times \cdots \times a_{n+m}$ m_2-ary array. Let $z_0 = 0$ or 1. Then*

$$\varepsilon(\kappa(A, B); z_0, z, z') = \kappa(\varepsilon(A; z_0, z), \varepsilon(B; 0, z')).$$

(5) *Let A be an m-ary array of size $s \times t \times a_1 \times \cdots \times a_n$ and c an integer such that $ct \equiv s(\text{mod}2s)$. Then*

$$\varepsilon(\tau(A;c);1,1,z) = \sigma(\alpha^{(2)}(\varepsilon(A;1,0,z),\varepsilon(A;1,0,z));c).$$

Now we begin to establish expressions for the correlation functions of various array mappings. These expressions will be used soon to prove the construction theorems for the generalized perfect arrays.

Lemma 6.2.12[3,10] *Let A, B, C, and D be four m-ary arrays of size $a_1 \times \cdots \times a_n$. Then for all $0 \leqslant u_i \leqslant a_i - 1$, $y = 0$ or 1,*

(1) *The periodic cross-correlation between $E = \alpha^{(1)}(A,(-1)^w A)$ and $F = \alpha^{(1)}(B,B)$ satisfies*

$$R_{EF}(u_1 + ya_1, u_2, \cdots, u_n) = \begin{cases} 2R_{AB}(u_1, \cdots, u_n), & \text{if } w = 0, \\ 0, & \text{if } w = 1; \end{cases}$$

(2) *The periodic cross-correlation between $E = \iota^{(1)}(A,B)$ and $F = \iota^{(1)}(C,D)$ satisfies*

$$\begin{aligned} &R_{EF}(2u_1 + y, u_2, \cdots, u_n) \\ &= \begin{cases} R_{AC}(u_1, \cdots, u_n) + R_{BD}(u_1, \cdots, u_n), & \text{if } y = 0 \\ R_{AD}(u_1, \cdots, u_n) + R_{BC}(u_1 + 1, u_2, \cdots, u_n), & \text{if } y = 1; \end{cases} \end{aligned}$$

(3) *The periodic cross-correlation between $E = \phi^{(1)}(A)$ and $F = \phi^{(1)}(B)$ satisfies*

$$R_{EF}(u_1, \cdots, u_n) = (-1)^{u_1} R_{AB}(u_1, \cdots, u_n);$$

(4) *The periodic cross-correlation between $E = I(A)$ and $F = I(B)$ satisfies*

$$R_{EF}(0, u_1, \cdots, u_n) = R_{AB}(u_1, \cdots, u_n);$$

(5) *The periodic autocorrelation of $E = \alpha^{(1)}(A,A)$ satisfies*

$$R_E(u_1 + ya_1, u_2, \cdots, u_n) = 2R_A(u_1, \cdots, u_n);$$

(6) *The periodic autocorrelation of $E = \iota^{(1)}(A,B)$ satisfies*

$$\begin{aligned} &R_E(2u_1 + y, u_2, \cdots, u_n) \\ &= \begin{cases} R_A(u_1, \cdots, u_n) + R_B(u_1, \cdots, u_n), & \text{if } y = 0, \\ R_{AB}(u_1, \cdots, u_n) + R_{BA}(u_1 + 1, u_2, \cdots, u_n), & \text{if } y = 1; \end{cases} \end{aligned}$$

(7) *The periodic autocorrelation of $E = \phi^{(1)}(A)$ satisfies*

$$R_E(u_1, \cdots, u_n) = (-1)^{u_1} R_A(u_1, \cdots, u_n);$$

(8) *The periodic autocorrelation of $E = I(A)$ satisfies*

$$R_E(0, u_1, \cdots, u_n) = R_A(u_1, \cdots, u_n);$$

(9) *The periodic autocorrelation of* $E = \alpha^{(1)}(A, -A)$ *satisfies*

$$R_E(u_1 + ya_1, u_2, \cdots, u_n) = \begin{cases} 2(-1)^y[P_A^{(1)}(u_1, \cdots, u_n) \\ \qquad -P_A^{(1)}(a_1 - u_1, \cdots, a_n - u_n)], & \text{if } u_1 \neq 0, \\ 2(-1)^y P_A^{(1)}(u_1, \cdots, u_n), & \text{if } u_1 = 0, \end{cases}$$

where $P_A^{(1)}(u_1, \cdots, u_n)$ *is the semi-periodic autocorrelation of* A *in dimension* 1 *which is defined by*

$$P_A^{(1)}(u_1, \cdots, u_n) = \sum_{j_1=0}^{a_1-u_1-1} \sum_{j_2=0}^{a_2-1} \cdots \sum_{j_n=0}^{a_n-1} A(u_1, \cdots, u_n) \times A(u_1 + j_1, \cdots, u_n + j_n);$$

(10) *The periodic cross-correlation between* $E = \alpha^{(1)}(I(A), I(B))$ *and* $F = \alpha^{(1)}(I(C), I(D))$ *satisfies*

$$R_{EF}(u, u_1, \cdots, u_n) = \begin{cases} (R_{AC} + R_{BD})(u_1, \cdots, u_n), & \text{if } u = 0 \\ (R_{AD} + R_{BC})(u_1, \cdots, u_n), & \text{if } u = 1; \end{cases}$$

(11) *The periodic autocorrelation of* $E = \alpha^{(1)}(I(A), I(B))$ *satisfies*

$$R_E(u, u_1, \cdots, u_n) = \begin{cases} (R_A + R_B)(u_1, \cdots, u_n), & \text{if } u = 0 \\ (R_{AB} + R_{BA})(u_1, \cdots, u_n), & \text{if } u = 1; \end{cases}$$

(12) *The semi-periodic autocorrelation of* $E = \alpha^{(1)}(I(A), I(B))$ *in dimension* 1 *satisfies*

$$P_E^{(1)}(u, u_1, \cdots, u_n) = \begin{cases} (R_A + R_B)(u_1, \cdots, u_n), & \text{if } u = 0 \\ R_{AB}(u_1, \cdots, u_n), & \text{if } u = 1. \end{cases}$$

The next lemma gives expressions for the periodic correlation function of further array mappings.

Lemma 6.2.13 [3,10] *The following equations are true:*

(1) *Let* A *be an* m*-ary array of size* $s \times t \times a_1 \times \cdots \times a_n$ *and suppose that* $xt + ys = 1$. *Then for all* $0 \leqslant u \leqslant st - 1$, $0 \leqslant u_i \leqslant a_i - 1$,

$$R_{\mu(A)}(u, u_1, \cdots, u_n) = R_A(ux, uy, u_1, \cdots, u_n).$$

(2) *Let* A *and* C *be* m_1*-ary arrays of size* $a_1 \times \cdots \times a_n$ *and* B *and* D *be* m_2*-ary arrays of size* $a_{n+1} \times \cdots \times a_{n+m}$. *Then for all* $0 \leqslant u_i \leqslant a_i - 1$, $1 \leqslant i \leqslant n + m$,

$$R_{\prod(A,B)\prod(C,D)}(u_1, \cdots, u_{n+m}) = R_{AC}(u_1, \cdots, u_n)R_{BD}(u_{n+1}, \cdots, u_{n+m}),$$
$$R_{\prod(A,B)}(u_1, \cdots, u_{n+m}) = R_A(u_1, \cdots, u_n)R_B(u_{n+1}, \cdots, u_{n+m}).$$

(3) *Let A be an m_1-ary array of size $s \times a_1 \times \cdots \times a_n$ and B an m_2-ary array of size $t \times a_{n+1} \times \cdots \times a_{n+m}$. Then for all $0 \leqslant u \leqslant s-1$, $0 \leqslant w \leqslant t-1$, $0 \leqslant u_i \leqslant a_i - 1$, $1 \leqslant i \leqslant n+m$,*

$$R_{\kappa(A,B)}(ut+w, u_1, \cdots, u_{n+m}) = \begin{cases} R_A(u, u_1, \cdots, u_n) P_B^{(1)}(w, u_{n+1}, \cdots, u_{n+m}) + R_A(u+1, u_1, \cdots, u_n) \\ \quad \times P_B^{(1)}(t-w, a_{n+1}-u_{n+1}, \cdots, a_{n+m}-u_{n+m}), & \text{if } w \neq 0 \\ \\ R_A(u, u_1, \cdots, u_n) P_B^{(1)}(w, u_{n+1}, \cdots, u_{n+m}), & \text{if } w = 0. \end{cases}$$

(4) *Let A be an m-ary array of size $s \times t \times a_1 \times \cdots \times a_n$ and c an integer such that $ct \equiv 0 (\text{mod} s)$. Then for all $0 \leqslant u \leqslant s-1$, $0 \leqslant w \leqslant t-1$, $0 \leqslant u_i \leqslant a_i - 1$, $1 \leqslant i \leqslant n$,*

$$R_{\sigma(A;c)}(u, w, u_1, \cdots, u_n) = R_A(u - cw, w, u_1, \cdots, u_n).$$

Now we are ready to present construction theorems for the generalized perfect arrays.

Theorem 6.2.12 [3,10] *Let $z = (z_0, \cdots, z_n)$ be a type vector. And let s and t be integers coprimed with each other, i.e., $\gcd(s,t) = 1$. Then A is a $\mathrm{GPA}(m; s, t, a_1, \cdots, a_n)$ of type $(0, 0, z)$ if and only if $\mu(A)$ is a $\mathrm{GPA}(m; (st), a_1, \cdots, a_n)$ of type $(0, z)$.*

Proof Use the definition of GPAs, the second statement of Lemma 6.2.11 and the first statement of Lemma 6.2.13. Note that $u \equiv 0(\text{mod} st)$ if and only if $ux \equiv 0(\text{mod} s)$ and $uy \equiv 0(\text{mod} t)$, where x and y are determined by $tx + ys = 1$. □

In the binary cases, Theorem 6.2.12 states that there exists a $\mathrm{PBA}(s, t, a_1, \cdots, a_n)$ if and only if there exists a $\mathrm{PBA}((st), a_1, \cdots, a_n)$, whenever $\gcd(s,t) = 1$. Thus Theorem 6.2.12 generalizes the known folding construction of PBAs.

The following theorem is a generalized version of the periodic production construction of PBAs.

Theorem 6.2.13 [3,10] *Let $z = (z_1, \cdots, z_n)$ and $z' = (z_{n+1}, \cdots, z_{n+m})$ be type vectors. Then the following first two statements hold if and only if the third one holds:*

(1) *A is a $\mathrm{GPA}(m_1; a_1, \cdots, a_n)$ of type z;*

(2) *B is a $\mathrm{GPA}(m_2; a_{n+1}, \cdots, a_{n+m})$ of type z';*

(3) *$\prod(A, B)$ is a $\mathrm{GPA}(f(m_1, m_2); a_1, \cdots, a_{n+m})$ of type (z, z').*

Proof Let $A' = \varepsilon(A; z)$, $B' = \varepsilon(B, z')$, $C = \prod(A, B)$, and $C' = \varepsilon(C; z, z')$. Then from the third statement of Lemma 6.2.11, $C' = \prod(A', B')$, and so from the second statement of Lemma 6.2.13, we have, for all $0 \leqslant u_i \leqslant (z_i + 1)a_i - 1$, $1 \leqslant i \leqslant n+m$,

$$R_{C'}(u_1, \cdots, u_{n+m}) = R_{A'}(u_1, \cdots, u_n) R_{B'}(u_{n+m}, \cdots, u_{n+m}). \tag{6.22}$$

Therefore from the definition of GPAs, the first two statements imply the third one.

Conversely, suppose that the third statement is true. Then Equation (6.22) implies

$$R_{C'}(u_1, \cdots, u_n, 0, \cdots, 0) = x R_{A'}(u_1, \cdots, u_n),$$

where

$$x = R_{B'}(0, \cdots, 0) = E_B \prod_{i=n+1}^{n+m} (z_i + 1) \neq 0.$$

Therefore the first statement follows. The second statement follows similarly.

We shall find it useful to express, in terms of the semi-periodic autocorrelation function, conditions equivalent to an array being a GPA of two related types simultaneously. □

Lemma 6.2.14[3,10] *Let $A = [A(j, a(1), \cdots, a(n))]$ be an m-ary array of size $s \times a_1 \times \cdots \times a_n$, and $z = (z_1, \cdots, z_n)$ a type vector. Let $A' = \varepsilon(A; 0, z)$. Then the following are equivalent:*

(1) *A is a GPA$(m; s, a_1, \cdots, a_n)$ of type $(0, z)$ and type $(1, z)$;*

(2) *For $0 \leqslant u \leqslant s - 1$, $0 \leqslant u_i \leqslant (z_i + 1)a_i - 1$, $1 \leqslant i \leqslant n$,*

$$P_{A'}^{(1)}(u, u_1, \cdots, u_n) \neq 0 \ only \ if \ u = 0 \ and \ u_i \equiv 0(\text{mod} a_i) \ for \ all \ i.$$

Theorem 6.2.14[3,10] *Let $z = (z_1, \cdots, z_n)$, $z' = (z_{n+1}, \cdots, z_{n+m})$ and (z_0) be three type vectors and let $s > 1$. Then the following (1) and (2) hold if and only if (3) holds:*

(1) *A is a GPA$(m_1; s, a_1, \cdots, a_n)$ of type (z_0, z);*

(2) *B is a GPA$(m_2; t, a_{n+1}, \cdots, a_{n+m})$ of type $(0, z')$ and type $(1, z')$;*

(3) *$\kappa(A, B)$ is a GPA$(f(m_1, m_2); (st), a_1, \cdots, a_{n+m})$ of type (z_0, z, z').*

Proof Let $A' = \varepsilon(A; z_0, z)$, $B' = \varepsilon(B; 0, z')$, $C = \kappa(A, B)$, and $C' = \varepsilon(C; z_0, z, z')$. From the fourth statement of Lemma 6.2.11, $C' = \kappa(A', B')$. From the third statement of Lemma 6.2.13, for all $0 \leqslant u \leqslant (z_0 + 1)s - 1$, $0 \leqslant w \leqslant t - 1$, $0 \leqslant u_i \leqslant (z_i + 1)a_i - 1$, $1 \leqslant i \leqslant n + m$,

$$\begin{aligned} &R_{C'}(ut + w, u_1, \cdots, u_{n+m}) \\ &= \begin{cases} P_{B'}^{(1)}(t - w, (z_{n+1} + 1)a_{n+1} - u_{n+1}, \cdots, \\ \quad (z_{n+m} + 1)a_{n+m} - u_{n+m}) \times R_{A'}(u + 1, u_1, \cdots, u_n) \\ \quad + R_{A'}(u, u_1, \cdots, u_n) P_{B'}^{(1)}(w, u_{n+1}, \cdots, u_{n+m}), & \text{if } w \neq 0, \\ \\ R_{A'}(u, u_1, \cdots, u_n) P_{B'}^{(1)}(w, u_{n+1}, \cdots, u_{n+m}), & \text{if } w = 0. \end{cases} \end{aligned} \tag{6.23}$$

From the definition of GPAs, the first statement is equivalent to

$$R_{A'}(u, u_1, \cdots, u_n) \neq 0 \text{ only if } u \equiv 0(\text{mod} s), u_i \equiv 0(\text{mod} a_i). \tag{6.24}$$

From Lemma 6.2.14, the second statement is equivalent to

$$P_{B'}^{(1)}(w, u_{n+1}, \cdots, u_{n+m}) \neq 0 \text{ only if } w \equiv 0(\text{mod} t), u_i \equiv 0(\text{mod} a_i). \tag{6.25}$$

Suppose (1) and (2) hold. Then using Equation (6.23)

$$R_{C'}(ut + w, u_1, \cdots, u_{n+m}) \neq 0$$

only if

$$R_{A'}(u, u_1, \cdots, u_n) \neq 0 \text{ and } P_{B'}^{(1)}(w, u_{n+1}, \cdots, u_{n+m}) \neq 0,$$

which hold only if

$$u \equiv 0(\text{mod} s), \ w = 0 \text{ and } u_i \equiv 0(\text{mod} a_i).$$

Therefore the third statement holds.

Conversely, suppose the third statement holds. If we let $w = u_i = 0$ for all $n+1 \leqslant i \leqslant n+m$, then Equation (6.23) becomes

$$R_{C'}(ut, u_1, \cdots, u_n, 0, \cdots, 0) = xR_{A'}(u, u_1, \cdots, u_n),$$

where $x = P_{B'}^{(1)}(0, \cdots, 0) = R_{B'}(0, \cdots, 0) \neq 0$. Thus the first statement is true. Similarly, the second statement is also true. □

Because if $t = 1$, then $\kappa(A, B) = \prod(A, B')$, where $B = I(B')$, so that Theorem 6.2.13 is a special case of Theorem 6.2.14.

Theorem 6.2.15 [3,10] *Let s be an odd integer, $z = (z_1, \cdots, z_n)$ be a type vector. Then A is a $\mathrm{GPA}(m; s, a_1, \cdots, a_n)$ of type $(0, z)$ if and only if $\phi^{(1)}(A)$ is a $\mathrm{GPA}(m; s, a_1, \cdots, a_n)$ of type $(1, z)$.*

Proof Let $A' = \varepsilon(A; 0, z)$, $B' = \varepsilon(\phi^{(1)}(A); 1, z)$, and $C = \alpha^{(1)}(A', A')$. From the first statement of Lemma 6.2.11, $B' = \phi^{(1)}(C)$. From the fifth and seventh statements of Lemma 6.2.12, for all $0 \leqslant u \leqslant s-1$, $0 \leqslant u_i \leqslant (z_i+1)a_i - 1$, $1 \leqslant i \leqslant n$, $y = 0$ or 1,

$$\begin{aligned} R_{B'}(u+ys, u_1, \cdots, u_n) &= (-1)^{u+ys} R_C(u+ys, u_1, \cdots, u_n) \\ &= 2(-1)^{u+y} R_{A'}(u, u_1, \cdots, u_n). \end{aligned}$$

□

We next show under what conditions a GPA may be transformed into another GPA which is of the same type but of different size by altering the interleaving of its component arrays.

Lemma 6.2.15 [3,10] *Let A and B be m-ary arrays of size $a_1 \times \cdots \times a_n$, $n \geqslant 2$, and $z = (z_1, \cdots, z_n)$ be a type vector. Let $A' = \varepsilon(A; z)$ and $B' = \varepsilon(B; z)$. Then any two of the following equations imply the other:*

(1) *$C = \iota^{(1)}(A, B)$ is a $\mathrm{GPA}(m; 2a_1, \cdots, a_n)$ of type z;*

(2) *$D = \iota^{(2)}(A, B)$ is a $\mathrm{GPA}(m; a_1, 2a_2, a_3, \cdots, a_n)$ of type z;*

(3) *$R_{A'B'}(u_1, \cdots, u_n) = R_{A'B'}(u_1+1, u_2-1, u_3, \cdots, u_n)$ for all $0 \leqslant u_i \leqslant (z_i+1)a_i - 1$, $1 \leqslant i \leqslant n$.*

The following two theorems form the heart of the recursive constructions of GPAs.

Theorem 6.2.16 [3,10] *Let A and B be m-ary arrays of size $a_1 \times \cdots \times a_n$, $n \geqslant 2$, and $z = (z_1, \cdots, z_n)$ be a type vector. Let $C = \alpha^{(1)}(A, A)$, $D = \alpha^{(1)}(B, -B)$. Then the following equations are equivalent:*

(1) *A is a $\mathrm{GPA}(m; a_1, \cdots, a_n)$ of type $(0, z)$, B is a $\mathrm{GPA}(m; a_1, \cdots, a_n)$ of type $(1, z)$ and $E_A = E_B$;*

(2) *$E = \iota^{(1)}(C, D)$ is a $\mathrm{GPA}(m; (4a_1), \cdots, a_n)$ of type $(0, z)$;*

(3) *$F = \iota^{(2)}(C, D)$ is a $\mathrm{GPA}(m; (2a_1), (2a_2), a_3, \cdots, a_n)$ of type $(0, z)$.*

Proof Let $A' = \varepsilon(A; 0, z)$, $B' = \varepsilon(B; 1, z)$, $C' = \varepsilon(C; 0, z)$, $D' = \varepsilon(D; 0, z)$, $\overline{B} = \varepsilon(B; 1, z)$, and $E' = \varepsilon(E; 0, z)$.

At first we show that (2) is equivalent to (3). From the first statement of Lemma 6.2.10, $C' = \alpha^{(1)}(A', A')$ and $D' = \alpha^{(1)}(\overline{B}, -\overline{B})$. Therefore from the sixth statement of Lemma 6.2.12 and the known identity $R_{XY}(u_1, \cdots, u_n) = R_{YX}(a_1 - u_1, \cdots, a_n - u_n)$, we have

$$0 = R_{D'C'}(u_1, \cdots, u_n) = R_{C'D'}(u_1, \cdots, u_n) \tag{6.26}$$

for all $0 \leqslant u_1 \leqslant 2a_1 - 1$, $0 \leqslant u_i \leqslant (z_i + 1)a_i - 1$, $i \geqslant 2$. Hence from Lemma 6.2.15, (2) is equivalent to (3) .

Then we now show that (1) is equivalent to (2). From the second statement of Lemma 6.2.10, $E' = \iota^{(1)}(C', D')$. From Lemma 6.2.7, $D' = B'$. Therefore from the sixth statement of Lemma 6.2.12, we have

$$\begin{aligned} &R_{E'}(2u_1 + 2ya_1 + 1, u_2, \cdots, u_n) \\ =& R_{C'D'}(u_1 + ya_1, u_2, \cdots, u_n) + R_{D'C'}(u_1 + ya_1 + 1, u_2, \cdots, u_n) \\ =& 0 \quad \text{(from Equation (6.26))}, \end{aligned} \tag{6.27}$$

for all $0 \leqslant u_1 \leqslant a_1 - 1$, $y = 0$ or 1, $0 \leqslant u_i \leqslant (z_i + 1)a_i - 1$, $i \geqslant 2$.

Because of the fifth statement of Lemma 6.2.12 and $D' = B'$, we have

$$\begin{aligned} &R_{E'}(2u_1 + 2ya_1, u_2, \cdots, u_n) \\ =& R_{C'}(u_1 + ya_1, u_2, \cdots, u_n) + R_{D'}(u_1 + ya_1, u_2, \cdots, u_n) \\ =& 2R_{A'}(u_1, \cdots, u_n) + R_{B'}(u_1 + ya_1, u_2, \cdots, u_n). \end{aligned} \tag{6.28}$$

Letting, in Equation (6.28) , $u_1 = 0$ $y = 1$, $u_i = y_i a_i$, $0 \leqslant y_i \leqslant z_i$, $i \geqslant 2$, we have

$$\begin{aligned} R_{E'}(2a_1, y_2a_2, \cdots, y_na_n) =& 2R_{A'}(0, y_2a_2, \cdots, y_na_n) \\ &+ R_{B'}(a_1, y_2a_2, \cdots, y_na_n). \end{aligned} \tag{6.29}$$

Suppose that (1) holds. Then from Equation (6.28) , for all $0 \leqslant u_1 \leqslant a_1 - 1$, $y = 0$ or 1, $0 \leqslant u_i \leqslant (z_i + 1)a_i - 1$, $i \geqslant 2$,

$$R_{E'}(2u_1 + 2ya_1, u_2, \cdots, u_n) \neq 0$$

only if

$$R_{A'}(u_1, \cdots, u_n) \neq 0 \quad \text{or} \quad R_{B'}(u_1 + ya_1, u_2, \cdots, u_n) \neq 0,$$

which occurs only if

$$u_i \equiv 0 (\text{mod} a_i) \text{ for all } i \text{ and } y = 0 \text{ or } 1.$$

In the case $u_i \equiv 0(\text{mod} a_i)$ for all i and $y = 1$, because of Equation (6.29) and Lemma 6.2.8, for all $0 \leqslant y_i \leqslant z_i$, $i \geqslant 2$,

$$\begin{aligned} &R_{E'}(2a_1, y_2a_2, \cdots, y_na_n) \\ =& 2E_A(-1)^{\sum_{i\geqslant 2} y_i} \prod_{i\geqslant 2}(z_i + 1) + 2E_B(-1)^{1+\sum_{i\geqslant 2} y_i} \prod_{i\geqslant 2}(z_i + 1) \\ =& 0 \text{ (since } E_A = E_B). \end{aligned}$$

Hence, using Equation (6.27), the statement (2) holds.

Conversely, suppose that (2) holds. Then from Equation (6.28), for all $0 \leqslant u_1 \leqslant a_1 - 1$, $0 \leqslant u_i \leqslant (z_i + 1)a_i - 1$, $i \geqslant 2$,

$$R_{E'}(2u_1, u_2, \cdots, u_n) = 2R_{A'}(u_1, \cdots, u_n) + R_{B'}(u_1, \cdots, u_n), \tag{6.30}$$

$$\begin{aligned}R_{E'}(2u_1+2a_1,u_2,\cdots,u_n)=&2R_{A'}(u_1,\cdots,u_n)+R_{B'}(u_1+a_1,u_2,\cdots,u_n)\\&\times 2R_{A'}(u_1,\cdots,u_n)-R_{B'}(u_1,\cdots,u_n).\end{aligned}\tag{6.31}$$

Rearrange Equations (6.30) and (6.31) as

$$\begin{aligned}4R_{A'}(u_1,\cdots,u_n)=&R_{E'}(2u_1,u_2,\cdots,u_n)+R_{E'}(2u_1+2a_1,u_2,\cdots,u_n),\\2R_{B'}(u_1,\cdots,u_n)=&R_{E'}(2u_1,u_2,\cdots,u_n)-R_{E'}(2u_1+2a_1,u_2,\cdots,u_n).\end{aligned}$$

Putting $u_i=0$ for all i, we have

$$4R_{A'}(0,\cdots,0)=2R_{B'}(0,\cdots,0)$$

and so from Lemma 6.2.8, $E_A=E_B$. Also for all $0\leqslant u_1\leqslant a_1-1$, $0\leqslant u_i\leqslant (z_i+1)a_i-1$, $i\geqslant 2$,

$$R_{A'}(u_1,\cdots,u_n)\neq 0$$

only if

$$R_{E'}(2u_1,u_2,\cdots,u_n)\neq 0 \text{ or } R_{E'}(2u_1+2a_1,u_2,\cdots,u_n)\neq 0$$

which occurs only if $u_i\equiv 0(\text{mod}a_i)$ for all i. Applying a similar argument to $R_{B'}$, we have (1). □

Theorem 6.2.17 [3,10] *Let A be an m-ary array of size $s\times t\times a_1\times\cdots\times a_n$, $z=(z_1,\cdots,z_n)$ a type vector, and c an integer such that $ct\equiv s(\text{mod}2s)$. Then A is a $\text{GPA}(m;s,t,a_1,\cdots,a_n)$ of type $(1,0,z)$ if and only if $\tau(A;c)$ is a $\text{GPA}(m;s,t,a_1,\cdots,a_n)$ of type $(1,1,z)$.*

Proof Let $A'=\varepsilon(A;1,0,z)$, $B=\tau(A;c)$, $B'=\varepsilon(B;1,1,z)$, and $D=\alpha^{(2)}(A',A')$. From the fifth statement of Lemma 6.2.11, $B'=\sigma(D;c)$. Therefore from the fourth statement of Lemma 6.2.13, by using the fifth statement of Lemma 6.2.12 and $ct\equiv s(\text{mod}2s)$, for all $0\leqslant u\leqslant 2s-1$, $0\leqslant w\leqslant t-1$, $y=0$ or 1, $0\leqslant u_i\leqslant (z_i+1)a_i$,

$$\begin{aligned}R_{B'}(u,w+yt,u_1,\cdots,u_n)=&R_D(u-c(w+yt),w+yt,u_1,\cdots,u_n)\\=&2R_{A'}(u-cw-ys,w,u_1,\cdots,u_n).\end{aligned}$$

The result follows from the definition of GPAs. □

Corollary 6.2.3 [3,10] *Let A be an m-ary array of size $s\times t\times a_1\times\cdots\times a_n$, $z=(z_1,\cdots,z_n)$ be a type vector, and $t/\gcd(s,t)$ odd. Then A is a $\text{GPA}(m;s,t,a_1,\cdots,a_n)$ of type $(1,0,z)$ if and only if $\tau(A;s/\gcd(s,t))$ is a $\text{GPA}(m;s,t,a_1,\cdots,a_n)$ of type $(1,1,z)$.*

Proof An integer c satisfies $ct\equiv s(\text{mod}2s)$ if and only if $t/\gcd(s,t)$ is odd and c is an odd multiple of $s/\gcd(s,t)$. □

Corollary 6.2.4 [3,10] *Let $z=(z_1,\cdots,z_n)\neq(0,\cdots,0)$ be a type vector. For $1\leqslant i\leqslant n$, let $b_i\geqslant 0$ and let a_i be odd. Then there exists a $\text{GPA}(m;2^{b_1}a_1,\cdots,2^{b_n}a_n)$ of type z and energy E if and only if there exists a $\text{GPA}(m;2^{b_1}a_1,\cdots,2^{b_n}a_n)$ of type $(0^{(i'-1)},1,0^{(n-i')})$ and energy E, where $z_{i'}=1$ and for all $1\leqslant i\leqslant n$, $b_{i'}\geqslant b_i$ whenever $z_i=1$.*

Proof It can be proved by using the last corollary. □

This corollary shows that the existence of a GPA of type $z\neq(0,\cdots,0)$ implies the existence of a GPA of type $(1,0,\cdots,0)$ for some permutations of the original dimensions of the array.

Theorem 6.2.18 [3,10] *Let A, B, C, and D be m-ary arrays of size $a_1 \times \cdots \times a_n$. Let $E = \alpha^{(1)}(I(A), I(B))$, and $F = \alpha^{(1)}(I(C), I(D))$. Then the following* (1) *and* (2) *hold if and only if* (a), (b), (c) *and* (d) *hold for all $0 \leqslant u_i \leqslant a_i - 1$:*

(1) *$G = \alpha^{(1)}(I(E), I(F))$ is a $\mathrm{GPA}(m; 2, 2, a_1, \cdots, a_n)$ of type $(0, \cdots, 0)$ and of also simultaneously of type $(1, 0, \cdots, 0)$;*

(2) *$H = \iota^{(1)}(E, F)$ is a $\mathrm{GPA}(m; 4, a_1, \cdots, a_n)$ of type $(0, \cdots, 0)$.*

(a) *$(R_A + R_B + R_C + R_D)(u_1, \cdots, u_n) \neq 0$ only if $u_i = 0$ for all i;*

(b) *$(R_{AB} + R_{BA} + R_{CD} + R_{DC})(u_1, \cdots, u_n) = 0$;*

(c) *$(R_{AC} + R_{BD})(u_1, \cdots, u_n) = 0$;*

(d) *$(R_{AD} + R_{BC})(u_1, \cdots, u_n) = 0$.*

Proof Note that $\varepsilon(G; 0, \cdots, 0) = G$. Then from Lemma 6.2.14, (1) is equivalent to

$$P_G^{(1)}(u, w, u_1, \cdots, u_n) \neq 0, \quad \text{where } u, w = 0 \text{ or } 1, 0 \leqslant u_i \leqslant a_i - 1,$$

only if

$$u = w = u_i = 0 \quad \text{for all } i.$$

But from the twelfth statement of Lemma 6.2.12, for all $u, w = 0$ or 1, $0 \leqslant u_i \leqslant a_i - 1$,

$$\begin{aligned}
&P_G^{(1)}(u, w, u_1, \cdots, u_n) \\
&= \begin{cases} (R_E + R_F)(w, u_1, \cdots, u_n), & \text{if } u = 0 \\ R_{EF}(w, u_1, \cdots, u_n), & \text{if } u = 1 \end{cases} \\
&= \begin{cases} (R_A + R_B + R_C + R_D)(u_1, \cdots, u_n), & \text{if } (u, w) = (0, 0), \\ (R_{AB} + R_{BA} + R_{CD} + R_{DC})(u_1, \cdots, u_n), & \text{if } (u, w) = (0, 1), \\ (R_{AC} + R_{BD})(u_1, \cdots, u_n), & \text{if } (u, w) = (1, 0), \\ (R_{AD} + R_{BC})(u_1, \cdots, u_n), & \text{if } (u, w) = (1, 1) \end{cases}
\end{aligned}$$

from the tenth and eleventh statements of Lemma 6.2.12. Therefore (1) holds if and only if (a), (b), (c) and (d) hold for all $0 \leqslant u_i \leqslant a_i - 1$.

(2) holds if and only if $I(H)$ is a $\mathrm{GPA}(m; 1, 4, a_1, \cdots, a_n)$ of type $(0, \cdots, 0)$. In fact, X is a $\mathrm{GPA}(m; a_1, \cdots, a_n)$ of type z if and only if $I[A]$ is a $\mathrm{GPA}(m; 1, a_1, \cdots, a_n)$ of type (z_0, z). Now using the fifth statement of Lemma 6.2.10, $I(H) = \iota^{(2)}(I(E), I(F))$, and since $I(E)$ and $I(F)$ each has size $1 \times 2 \times a_1 \times \cdots \times a_n$ we may write $G = \iota^{(1)}(I(E), I(F))$. Therefore if (1) is given, then from Lemma 6.2.15, (2) is equivalent to

$$R_{I(E)I(F)}(0, w, u_1, \cdots, u_n) = R_{I(E)I(F)}(0, w - 1, u_1, \cdots, u_n),$$
$$w = 0, 1, \quad 0 \leqslant u_i \leqslant a_i - 1,$$

which, from the fourth statement of Lemma 6.2.12, is equivalent to

$$R_{EF}(w, u_1, \cdots, u_n) = R_{EF}(w - 1, u_1, \cdots, u_n), \quad w = 0, 1, \quad 0 \leqslant u_i \leqslant a_i - 1,$$

and from the tenth statement of Lemma 6.2.12 it is equivalent to

$$(R_{AC} + R_{BD})(u_1, \cdots, u_n) = (R_{AD} + R_{BC})(u_1, \cdots, u_n), \quad 0 \leqslant u_i \leqslant a_i - 1.$$

Therefore (2) follows from (1), (c) and (d). □

The following definition generalizes the concept of BSQ from the case of $m = 2$ to the case of general integer m.

Definition 6.2.14[3,10] *Let A, B, C, and D be m-ary arrays of size $a_1 \times \cdots \times a_n$. $\{A,B,C,D\}$ is called an $a_1 \times \cdots \times a_n$ m-ary supplementary quadruple of energy $E_A + E_B + E_C + E_D$ if the following two conditions are satisfied for all $0 \leqslant u_i \leqslant a_i - 1$:*

(1) $(R_A + R_B + R_C + R_D)(u_1, \cdots, u_n) \neq 0$ *only if* $u_i = 0$ *for all* i;

(2) $(R_{WX} + R_{YZ})(u_1, \cdots, u_n) = 0$ *for all* $\{W, X, Y, Z\} = \{A, B, C, D\}$.

Corollary 6.2.5[3,10] *Let $\{A,B,C,D\}$ be an $a_1\times\cdots\times a_n$ m-ary supplementary quadruple. Then there exists a* GPA$(m;2,2,u_1,\cdots,u_n)$ *of type $(0,\cdots,0)$ and also of type $(1,0,\cdots,0)$, and a* GPA$(m;4,u_1,\cdots,u_n)$ *of type $(0,\cdots,0)$, whose energy is $E_A+E_B+E_C+E_D$.*

Proof It can be proved immediate by using Theorem 6.2.18 and Definition 6.2.14.

□

It is worth pointing out that Corollary 6.2.5 implies Theorem 6.2.8 if $m = 2$.

Theorem 6.2.19[3,10] *Let $\{A_1,B_1,C_1,D_1\}$ and $\{A_2,B_2,C_2,D_2\}$ be respectively an $a_1 \times \cdots \times a_n$ m_1-ary and an $a_{n+1} \times \cdots \times a_{n+m}$ m_2-ary supplementary quadruple. Let*

$$\begin{aligned}
A &= \prod(A_1 + B_1, A_2) + \prod(A_1 - B_1, B_2),\\
B &= \prod(A_1 + B_1, C_2) + \prod(A_1 - B_1, D_2),\\
C &= \prod(C_1 + D_1, A_2) + \prod(C_1 - D_1, B_2),\\
D &= \prod(C_1 + D_1, C_2) + \prod(C_1 - D_1, D_2).
\end{aligned}$$

Then $\{A,B,C,D\}$ is an $a_1\times\cdots\times a_{n+m}$ $(1+4\lfloor m_1/2\rfloor\lfloor m_2/2\rfloor)$-ary supplementary quadruple. If $m_1 = m_2 = 2$, then the elements of A,B,C,D take only the values 2 or -2.

Proof This theorem can be directly verified. □

Theorem 6.2.20[3,10] *Let $z = (z_1, \cdots, z_n) \neq (0, \cdots, 0)$ be a type vector and let A be a* GPBA$(a_1, \cdots, a_n)$ *of type z with energy $E_A > 2$. Then*

(1) $E_A \equiv 0 \pmod 4$;

(2) *If $z_i = 1$ and a_i is even for some i, then a_j is even for some $j \neq i$.*

Proof If $z_i = 0$ whenever a_i is even, then from Theorem 6.2.15 there exists a PBA$(a_1, \cdots, a_n)$ of energy $E_A = \prod_{i=1}^{n} a_i$, so $E_A \equiv 0 \pmod 4$. Assume for the rest of the proof that $z_i = 1$ and a_i is even, for some i. Then from Corollary 6.2.4, without loss of generality, we may assume that there exists a GPBA$(a_1, \cdots, a_n)$ of type $(1, 0, \cdots, 0)$, whose energy is E_A and a_1 is even. Call this array B and let $B' = \varepsilon(B; 1, 0, \cdots, 0)$. From Lemma 6.2.7, $B' = \alpha^{(1)}(B, -B)$ and so by putting $y = 0$ in Lemma 6.2.12, for all $0 \leqslant u_i \leqslant a_i - 1$,

$$\begin{aligned}
&(R_{B'}(u_1, \cdots, u_n))/2\\
&= \begin{cases} P_B^{(1)}(u_1, \cdots, u_n) - P_B^{(1)}(a_1 - u_1, \cdots, a_n - u_n), & \text{if } u_1 \neq 0,\\ P_B^{(1)}(u_1, \cdots, u_n), & \text{if } u_1 = 0 \end{cases}\\
&\equiv (a_1 - 2u_1) \prod_{i=2}^{n} a_i (\text{mod} 4),
\end{aligned}$$

By the definition of GPAs we therefore have

$$(a_1 - 2u_1)\prod_{i=2}^{n} a_i \not\equiv 0(\mathrm{mod}4), \text{ only if } u_i = 0 \text{ for all } i. \tag{6.32}$$

Firstly note that

$$if \prod_{i=2}^{n} a_i \text{ is odd, then } a_1 = 2, \tag{6.33}$$

otherwise we may take $(u_1, \cdots, u_n) = (1, 0, \cdots, 0)$ and $(2, 0, \cdots, 0)$ in Equation (6.32), which gives a contradiction. Therefore

$$\prod_{i=2}^{n} a_i > 1, \tag{6.34}$$

otherwise $E_A = a_1 \prod_{i=2}^{n} a_i = 2$, contrary to hypothesis. Then using Equation (6.34), we may take $u_1 = 0$ and $(u_2, \cdots, u_n) \neq (0, \cdots, 0)$ in Equation (6.32), implying the Statement (1). Therefore $\prod_{i=2}^{n} a_i$ is even, otherwise from Equation (6.33) $E_A/2 = \prod_{i=2}^{n} a_i$ is odd, which contradicts statement (1). This implies that s_j is even for some $j \geqslant 2$, and by this assumption a_1 is even. Hence the statement (2) holds. □

The following lemma gives conditions under which a GPA may be transformed into another GPA of the same type and size.

Lemma 6.2.16 [3,10] *Let $z = (z_1, \cdots, z_n)$ be a type vector and $A = [A(a(1), \cdots, a(n))]$ an m-ary array of size $a_1 \times \cdots \times a_n$. For each of the following $a_1 \times \cdots \times a_n$ m-ary array $B = [B(b(1), \cdots, b(n))]$ of energy E_A, A is a* GPA$(m; a_1, \cdots, a_n)$ *of type z if and only if B is a* GPA$(m; a_1, \cdots, a_n)$ *of type z :*

(1) *$B(b(1), \cdots, b(n)) = A(b(1)+c, b(2), \cdots, b(n))$ for all $0 \leqslant b(i) \leqslant a_i - 1$, $0 \leqslant c \leqslant a_i - 1$, and $z_1 = 0$;*

(2) *$B(b(1), \cdots, b(n)) = A(a_1 - b(1) - 1, b(2), \cdots, b(n))$ for all $0 \leqslant b(i) \leqslant a_i - 1$;*

(3) *$B(b(1), \cdots, b(n)) = A(yb(1), b(2), \cdots, b(n))$ for all $0 \leqslant b(i) \leqslant a_i - 1$, where* $\gcd(a_1, y) =$ *1 and $z_1 = 0$;*

(4) *$B = \phi^{(1)}(A)$, where a_1 is even;*

(5) *$B = \sigma(A; c)$, where $ca_2 \equiv 0(\mathrm{mod} a_1)$ and $z_1 = 0$;*

(6) *$B = \tau(A; c)$, where $ca_2 \equiv 0(\mathrm{mod} 2a_1)$ and $z_1 = 1$.*

Theorem 6.2.21 [3,10] *Let $u \geqslant 1$ be an integer. Then there exist binary arrays $M_1, \cdots, M_{10}$ satisfying the following properties:*

(1) *M_1 is a* GPBA(2) *of type* (1);

(2) *M_2 is a* GPBA$(2, 2)$ *of type* $(0, 0)$ *and simultaneously type* $(1, 0)$;

(3) *M_3 is a* PBA(4);

(4) *M_4 is a* GPBA$(2, 2, 3^{(2u)})$ *of type* $(0^{(2u+2)})$ *and simultaneously type* $(1, 0^{(2u+1)})$;

(5) *M_5 is a* PBA$(4, 3^{(2u)})$;

(6) *M_6 is a* GPBA$(2, 2, 2, 3^{(2u)})$ *of type* $(1, 0^{(2u+2)})$;

(7) M_7 *is a* GPBA$(2, 4, 3^{(2u)})$ *of type* $(1, 0^{(2u+1)})$;

(8) M_8 *is a* GPBA$(4, 2, 3^{(2u)})$ *of type* $(1, 0^{(2u+1)})$;

(9) M_9 *is a* GPBA$(2, 3, 2, 3^{(2u-1)})$ *of type* $(1, 0^{(2u)})$;

(10) M_{10} *is a* GPBA$(2, 3, 2, 2, 3^{(2u-1)})$ *of type* $(1, 0^{(2u+1)})$.

Proof It is easy to check the following choices for M_1, M_2, and M_3:

$$M_1 = \begin{bmatrix} + \\ + \end{bmatrix}, \quad M_2 = \begin{bmatrix} + & + \\ + & - \end{bmatrix}, \quad M_3 = \begin{bmatrix} + \\ + \\ + \\ - \end{bmatrix}.$$

Let

$$A_1 = \begin{bmatrix} - & + & + \\ - & + & + \\ - & + & + \end{bmatrix}, \quad B_1 = \begin{bmatrix} + & + & + \\ - & - & - \\ - & - & - \end{bmatrix},$$

and

$$C_1 = \begin{bmatrix} + & - & - \\ - & - & + \\ - & + & - \end{bmatrix}, \quad D_1 = \begin{bmatrix} + & - & - \\ - & + & - \\ - & - & + \end{bmatrix}.$$

Then $\{A_1, B_1, C_1, D_1\}$ is a 3×3 binary supplementary quadruple of energy 4.3^2. Now from Theorem 6.2.19, if there exist $3 \times \cdots \times 3$ binary supplementary quadruple of energy 4.3^{2u} and $4.3^{2u'}$, then there exists a $3 \times \cdots \times 3$ binary supplementary quadruple of energy $4.3^{2(u+u')}$. Therefore there exists a $3 \times \cdots \times 3$ binary supplementary quadruple of energy 4.3^{2u} for all $u \geqslant 1$. Hence arrays M_4 and M_5 exist because of Corollary 6.2.5.

From Theorem 6.2.13 we may take $M_6 = \prod(M_1, M_4)$ and $M_7 = \prod(M_1, M_5)$. From Theorem 6.2.14 we may take $M_8 = \kappa(M_1, M_4)$. To form M_9, first use Theorem 6.2.12 to form a GPBA$(2, 2, 3, 3^{(2u-1)})$ of type $(1, 0, \cdots, 0)$ from M_4. Then use Corollary 6.2.3 to change the type to $(1, 1, 0, \cdots, 0)$ and then to $(0, 1, 0, \cdots, 0)$. Reordering dimensions gives M_9. To form M_{10}, first use Theorem 6.2.13 to form a GPBA$(2, 2, 3, 2, 3^{(2u-1)})$ of type $(1, 1, 0, \cdots, 0)$ as $\prod(M_1, M_9)$. Then use Corollary 6.2.3 to change the type to $(0, 1, 0, \cdots, 0)$ and reorder dimensions. □

We shall refer to the arrays $M_1, \cdots, M_{10}$ freely from now on. In order to prove the recursive constructions, we introduce some more definitions here.

Definition 6.2.15 [3,10] *A non-negative integer sequence* $s = (s_1, \cdots, s_n)$ *is called odd or positive according to the fact that* s_i *is respectively odd or positive for all* $1 \leqslant i \leqslant n$. *This vector is called empty if* $n = 0$. *For* $x \geqslant 1$, *denote by* A_x *the set of all positive vectors* $s = (s_1, \cdots, s_n)$ *satisfying* $\sum s_i = x$.

Definition 6.2.16 [3,10] *Let* $s = (s_1, \cdots, s_n)$ *be a vector satisfying* $\sum s_i = x \geqslant 1$, *let* $a = (a_1, \cdots, a_m)$ *be a positive odd vector and let* $E \geqslant 1$. *Denote the set of all* $2^{s_1} \times \cdots \times 2^{s_n} \times a_1 \times \cdots \times a_m$ *k-ary arrays of energy* E *by* $S_x(E, k; s, a)$.

Lemma 6.2.17 [3,10] *Let* $x \geqslant 1$, *and* $s = (s_1, \cdots, s_n) \in A_x$. *Let* $a = (a_1, \cdots, a_m)$ *be a positive odd vector. Let* $z = (z_1, \cdots, z_n)$ *and* $z' = (z'_1, \cdots, z'_m)$ *be type vectors.*

(1) *Suppose that* $z_1 = 1$ *and that for all* $1 \leqslant i \leqslant n$, $s_1 \geqslant s_i$ *whenever* $z_i = 1$. *Then there exists a* GPA *of type* (z, z') *in* $S_x(E, k; s, a)$ *if and only if there exists a* GPA *of type* $(1, 0, \cdots, 0)$ *in* $S_x(E, k; s, a)$;

(2) *Let* $x \geqslant 3$ *and* $n \geqslant 2$. *Then there exists a* GPA *of type* $(1, 0, \cdots, 0)$ *in* $S_x(E, k; s, a)$ *if there exist a* GPA *of type* $(1, 0, \cdots, 0)$ *and a* GPA *of type* $(1, 1, 0, \cdots, 0)$ *in* $S_{x-2}(E/4, k; b_1, b_2, s_3, \cdots, s_n; a)$, *where* $(b_1, b_2) = (s_1 - 1, s_2 - 1)$ *or* $(s_1, s_2 - 2)$;

(3) *Let* $y \geqslant 2$ *and* $n \geqslant 2$. *Then there exists a* GPA *of type* $(0, \cdots, 0)$ *in* $S_{2y}(E, k; s, a)$ *if there exist a* GPA *of type* $(0, \cdots, 0)$ *and a* GPA *of type* $(1, 0, \cdots, 0)$ *in* $S_{2y-2}(E/4, k; b_1, b_2, s_3, \cdots, s_n; a)$, *where* $(b_1, b_2) = (s_1 - 1, s_2 - 1)$ *or* $(s_1 - 2, s_2)$;

(4) *Let* $0 \leqslant N \leqslant n - 1$ *and let* $s_i = 1$ *for all* $n - N + 1 \leqslant i \leqslant n$. *If there exists a* GPA *of type* $(1, 0, \cdots, 0)$ *in* $S_{x-N}(2^{-N}E, k, s_1, \cdots, s_{n-N}; a)$, *then there exists a* GPA *of type* $(1, 0, \cdots, 0)$ *in* $S_x(E, k; s, a)$.

Using the first statement of this lemma, the only GPAs we need to construct in $S_x(E, k; s, a)$ are those of type $(1, 0, \cdots, 0)$ or type $(0, \cdots, 0)$. We recursively construct these two types in the following propositions.

Proposition 6.2.1 [3,10] *Let* a *be a positive odd vector. Let* $w \geqslant 1$ *and* $t \geqslant 0$ *be integers and let* $x = w + 2t$. *Then the following* (1) *implies* (2):

(1) *For each* $s' = (s'_1, \cdots, s'_m) \in A_w$ *there exists a* GPA *of type* $(1, 0, \cdots, 0)$ *in* $S_w(E, k; s'; a)$ *if* $s'_1 \leqslant \lceil w/2 \rceil$ *and* $s'_i \leqslant \lceil w/2 \rceil + 1$ *for all* i.

(2) *For each* $s = (s_1, \cdots, s_m) \in A_x$ *there exists a* GPA *of type* $(1, 0, \cdots, 0)$ *in* $S_x(2^{2t}E, k; s; a)$ *if* $s_1 \leqslant \lceil x/2 \rceil$ *and* $s_i \leqslant \lceil x/2 \rceil + 1$ *for all* i.

Proposition 6.2.2 [3,10] *Let* a *be a positive odd vector. Let* $w \geqslant 1$ *and* $t \geqslant 0$ *be integers and let* $y = w + t$. *Then the following* (1) *and* (2) *imply* (3):

(1) *For each* $s' = (s'_1, \cdots, s'_m) \in A_{2w}$ *there exists a* GPA *of type* $(0, \cdots, 0)$ *in* $S_{2w}(E, k; s'; a)$ *if* $s'_1 \leqslant w + 1$ *for all* i.

(2) *For each integer* w' *satisfying* $w \leqslant w' \leqslant y - 1$ *and for each* $s'' = (s''_1, \cdots, s''_{m'}) \in A_{2w'}$ *there exists a* GPA *of type* $(1, 0, \cdots, 0)$ *in* $S_{2w'} \cdot (2^{2(w'-w)}E, k; s''; a)$ *if* $s''_1 \leqslant w'$ *and* $s''_i \leqslant w' + 1$ *for all* i.

(3) *For each* $s = (s_1, \cdots, s_n) \in A_{2y}$ *there exists a* GPA *of type* $(0, \cdots, 0)$ *in* $S_{2y}(2^{2t}E, k; s; a)$ *if* $s_i \leqslant y + 1$ *for all* i.

We wish to prove restrictions on $s = (s_1, \cdots, s_n)$ for the existence of a GPA of type $(1, 0, \cdots, 0)$ in $S_x(E, k; s; a)$. Suppose that for each y there exists a GPA of type $(0, \cdots, 0)$ in $S_{2y}(2^{2y-x}E, k; s'; a)$ if and only if $s'_i \leqslant y + 1$ for all i. We now prove that this implies $s_1 \leqslant \lceil x/2 \rceil$.

Proposition 6.2.3 [3,10] *Let* a *be a positive odd vector and let* $x \geqslant 1$ *be an integer. Then the following* (1) *and* (2) *imply* (3):

(1) *For each integer* y *satisfying* $\lceil x/2 \rceil \leqslant y \leqslant x - 1$ *and for each* $s' = (s'_1, \cdots, s'_m) \in A_{2y}$ *there exists a* GPA *of type* $(0, \cdots, 0)$ *in* $S_{2y}(2^{2y-x}E, k; s'; a)$ *if* $s'_i \leqslant y + 1$ *for all* i;

(2) *For each integer y satisfying $\lceil x/2 \rceil \leqslant y \leqslant x-1$ and for each $s'' = (s''_1, \cdots, s''_{m'}) \in A_{2y+2}$ there exists a* GPA *of type* $(0, \cdots, 0)$ *in* $S_{2y+2}(2^{2y-x+2}E, k; s''; a)$ *only if* $s''_i \leqslant y+2$ *for all* i;

(3) *For each* $s = (s_1, \cdots, s_n) \in A_x$ *there exists a* GPA *of type* $(1, 0, \cdots, 0)$ *in* $S_x(E, k; s; a)$ *only if* $a_1 \leqslant \lceil x/2 \rceil$.

Theorem 6.2.22[3,10] *Let* $n, u \geqslant 1$ *and let* $a_i \geqslant 1$ *for* $1 \leqslant i \leqslant n$. *Let* $z = (z_1, \cdots, z_n)$ *and* $z' = (z'_1, \cdots, z'_{2u})$ *be type vectors,*

(1) *There exists a* PBA$(2^{a_1}, \cdots, 2^{a_n})$, *where* $\sum a_i = 2y \geqslant 2$, *if and only if* $a_i \leqslant y+1$ *for all* i;

(2) *There exists a* GPBA$(2^{a_1}, \cdots, 2^{a_n})$ *of type* z, *where* $\sum a_i = x \geqslant 1$ *and* $z \neq (0, \cdots, 0)$,

(a) *if* $a_i \leqslant \begin{cases} \lceil x/2 \rceil, & \text{when } z_i = 1 (1 \leqslant i \leqslant n), \\ \lceil x/2 \rceil + 1, & \text{when } z_i = 0 (1 \leqslant i \leqslant n); \end{cases}$

(b) *only if* $a_i \leqslant \lceil x/2 \rceil$ *when* $z_i = 1 (1 \leqslant i \leqslant n)$;

(3) *There exists a* PBA$(2^{a_1}, \cdots, 2^{a_n}, 3^{(2u)})$, *where* $\sum a_i = 2y \geqslant 2$, *if and only if* $a_i \leqslant y+1$ *for all* i;

(4) *There exists a* GPBA$(2^{a_1}, \cdots, 2^{a_n}, 3^{(2u)})$ *of type* (z, z'), *where* $\sum a_i = x \geqslant 2$ *and* $z \neq (0, \cdots, 0)$,

(a) *if* $a_i \leqslant \begin{cases} \lceil x/2 \rceil, & \text{when } z_i = 1 (1 \leqslant i \leqslant n), \\ \lceil x/2 \rceil + 1, & \text{when } z_i = 0 (1 \leqslant i \leqslant n); \end{cases}$

(b) *only if* $a_i \leqslant \lceil x/2 \rceil$ *when* $z_i = 1 (1 \leqslant i \leqslant n)$.

Proof We first prove (1) and (2). From Definition 6.2.16, any $2^{a_1} \times \cdots \times 2^{a_n}$ binary array with $\sum a_i = x$ is a member of $S_x(2^x, 2; s; a)$, where a is empty. We shall apply Propositions 6.2.1–6.2.3 with a being empty and $k = 2$.

We may satisfy the first condition of Proposition 6.2.1 when $(E, w) = (2, 1)$ and $(4, 2)$ using the existence of M_1 and M_2, respectively. Therefore

$$\text{there exists a GPBA}(2^{a_1}, \cdots, 2^{a_n}) \text{ of type } (1, 0, \cdots, 0), \quad \text{where} \sum a_i = x \geqslant 1 \tag{6.35}$$

if

$$a_1 \leqslant \lceil x/2 \rceil \text{ and } a_i \leqslant \lceil x/2 \rceil + 1 \text{ for all } i. \tag{6.36}$$

(2.a) follows from Lemma 6.2.17(1).

We may then satisfy conditions (1) and (2) of Proposition 6.2.2 when $(E, w) = (4, 1)$ and $y \geqslant 1$ using respectively the existence of M_2, M_3 and the result that Equation (6.36) implies Equation (6.35). Therefore

$$\text{there exists a PBA}(2^{a_1}, \cdots, 2^{a_n}) \text{ where} \sum a_i = 2y \geqslant 2, \tag{6.37}$$

if

$$a_i \leqslant y + 1 \text{ for all } i. \tag{6.38}$$

Equation (6.38) is also a necessary condition for Equation (6.37), we have (1).

Finally (1) implies conditions (1) and (2) of Proposition 6.2.3 when $E = 2^x$ and $x \geqslant 1$, and therefore there exists a GPBA$(2^{a_1}, \cdots, 2^{a_n})$ of type $(1, 0, \cdots, 0)$, where $\sum a_i = x \geqslant 1$, only if $a_1 \leqslant \lceil x/2 \rceil$. (2.b) follows from Lemma 6.2.17.

We now outline the proof of (3) and (4), which closely resembles that of (1) and (2). We apply Propositions 6.2.1–6.2.3 with $a = 3^{(2u)}$ and $k = 2$. For $(E, w) = (4.3^{2u}, 2)$ and $(8.3^{2u}, 3)$ the respective existence of M_4, M_6, M_7, and M_8 satisfies the first statement of Proposition 6.2.1. For $(E, w) = (4.3^{2u}, 1)$ the existence of M_4 and M_5 satisfies the first statement of Proposition 6.2.2. We also require Proposition 6.2.3 and Lemma 6.2.17 to complete the proof. □

We may follow a similar procedure to prove existence and nonexistence results for further families of GPBAs. We start the recursive constructions with the binary arrays M_9 and M_{10}.

Theorem 6.2.23 [3,10] *Let $n, u \geqslant 1$ and let $a_i \geqslant 1$ for $1 \leqslant i \leqslant n$. Let $1 \leqslant N \leqslant \min(n, 2u)$. Let $z = (z_1, \cdots, z_n) \neq (0, \cdots, 0)$ and $z' = (z'_{N+1}, \cdots, z'_{2u})$ be type vectors. Then there exists a* GPBA$(2^{a_1} \cdot 3, \cdots, 2^{a_N} \cdot 3,\ 2^{a_{N+1}}, \cdots, 2^{a_n}, 3^{(2u-N)})$ *of type (z, z'), where $\sum a_i = x \geqslant 2$,*

(1) *if x is even and*

$$a_i \leqslant \begin{cases} x/2, & \text{when } z_i = 1 (1 \leqslant i \leqslant n), \\ x/2, +1, & \text{when } z_i = 0 (1 \leqslant i \leqslant n); \end{cases}$$

(2) *if x is odd and either $a_i \leqslant (x-1)/2$ for all $1 \leqslant i \leqslant n$ or*

$$a_i \leqslant \begin{cases} (x-3)/2, & \text{when } z_i = 1 (1 \leqslant i \leqslant n), \\ (x+1)/2, & \text{when } z_i = 0 (1 \leqslant i \leqslant n); \end{cases}$$

(3) *only if $a_i \leqslant \lceil x/2 \rceil$ when $z_i = 1 (1 \leqslant i \leqslant n)$.*

Proof We may assume that $z_i = 1$ for all $1 \leqslant i \leqslant N$, otherwise by using Theorem 6.2.12 we may transform to a GPBA of the same form but with smaller N. Reorder dimensions so that $a_1 = \max_{1 \leqslant i \leqslant N} a_i$. We may assume that for all $N+1 \leqslant i \leqslant n$, $a_1 > a_i$ whenever $z_i = 1$, otherwise by using Corollary 6.2.4 and Theorem 6.2.12 we may transform to a GPBA of the form already considered in the fourth statement of Theorem 6.2.22. Then by using Corollary 6.2.4 and Theorem 6.2.12 we may transform to a GPBA$(2^{a_1} \cdot 3, 2^{a_2}, \cdots, 2^{a_n}, 3^{(2u-1)})$ of type $(1, 0, \cdots, 0)$.

(1) Following the method of Proposition 6.2.1, we use Theorem 6.2.16 to inductively construct the desired array. We begin with M_9, and make use of the arrays constructed in the fourth part of Theorem 6.2.22 for the induction step.

(2) The method is similar, but a weaker inductive hypothesis is needed.

(3) The method is similar to that of Proposition 6.2.3 and make use the result of the third part of Theorem 6.2.22.

Up to this point, we have proved several construction theorems for generalized perfect m-ary arrays in n dimensions. By recursive application in the binary case we have the following[3,10]

(1) There exists a PBA$(2^{a_1}, \cdots, 2^{a_n}, 3^{(2u)})$, where $\sum a_i = 2y \geqslant 2$ and $u \geqslant 0$, if and only if $a_i \leqslant y + 1$ for all i;

(2) There exists a GPBA$(2^{a_1}, \cdots, 2^{a_n}, 3^{(2u)})$ of type $(z_1, \cdots, z_n, z'_1, \cdots, z'_{2u})$, where $\sum a_i = x \geqslant 1$, $u \geqslant 0$, $(z_1, \cdots, z_n) \neq (0, \cdots, 0)$, and $x \geqslant 2$ whenever $u > 0$,

(a) if $a_i \leqslant \begin{cases} \lceil x/2 \rceil, & \text{when } z_i = 1 (1 \leqslant i \leqslant n), \\ \lceil x/2 \rceil + 1, & \text{when } z_i = 0 (1 \leqslant i \leqslant n); \end{cases}$

(b) only if $a_i \leqslant \lceil x/2 \rceil$, when $z_i = 1 (1 \leqslant i \leqslant n)$.

(3) There exists a GPBA$(2^{a_1}.3, \cdots, 2^{a_N}.3, 2^{a_{N+1}}, \cdots, 2^{a_n}, 3^{(2u-N)})$ of type (z, z'), where $\sum a_i = x \geqslant 2$, $N \geqslant 1$, $z = (z_1, \cdots, z_n) \neq (0, \cdots, 0)$ and $z' = (z'_{N+1}, \cdots, z'_{2u})$,

(a) if x is even and $a_i \leqslant \begin{cases} x/2, & \text{when } z_i = 1 (1 \leqslant i \leqslant n) \\ x/2+1, & \text{when } z_i = 0 (1 \leqslant i \leqslant n); \end{cases}$

(b) if x is odd and either $a_i \leqslant (x-1)/2$ for all $1 \leqslant i \leqslant n$

or
$$a_i \leqslant \begin{cases} (x-3)/2, & \text{when } z_i = 1 (1 \leqslant i \leqslant n) \\ (x+1)/2, & \text{when } z_i = 0 (1 \leqslant i \leqslant n); \end{cases}$$

(c) only if $a_i \leqslant \lceil x/2 \rceil$, when $z_i = 1 (1 \leqslant i \leqslant n)$.

6.3 Higher-Dimensional Hadamard Matrices Based on Orthogonal Designs[17−19]

Besides the cases of $m = 1$ and $m = 2$, it has been proved that two-dimensional Hadamard matrices of order m can exist only if m is a multiple of four. The problem of finding the existence of at least one Hadamard matrix for all values of $m = 4t$ has been studied since 1892, and the most important recent results have depended entirely on the theory of orthogonal designs.

The first person to use an orthogonal design to find two-dimensional Hadamard matrices was Williamson in 1944. His work depended on using the orthogonal design of size 4×4 and type $(1, 1, 1, 1)$:
$$\begin{bmatrix} x & y & u & v \\ -y & x & v & -u \\ -u & -v & x & y \\ -v & u & -y & x \end{bmatrix} \tag{6.39}$$
and replacing the variables by four symmetric circulant (± 1)-valued matrices X, Y, U, V of order t that satisfy
$$XX^{\mathrm{T}} + YY^{\mathrm{T}} + UU^{\mathrm{T}} + VV^{\mathrm{T}} = 4tI_t. \tag{6.40}$$
Subsequently, four symmetric circulant (± 1)-valued matrices satisfying Equation (6.40), or just four (± 1)-valued matrices satisfying Equation (6.40) and satisfying $MN^{\mathrm{T}} = NM^{\mathrm{T}}$, for $M, N \in \{X, Y, U, V\}$, have come to be known as Williamson matrices of order t. Williamson matrices of order t can be used in Equation (6.39) to obtain a two-dimensional Hadamard matrix of order $4t$. These matrices will be used later to obtain higher-dimensional Hadamard matrices.

6.3.1 Definitions of Orthogonality

For the details of this subsection the readers are recommended to see the papers [1, 17].

Orthogonality for higher-dimensional matrices can be defined in several ways[1,17-19]. For example, the orthogonalities of higher-dimensional matrices can be quantified by defining an n-dimensional matrix to be orthogonal of property $(d_1, d_2, \cdots, d_n)$ with $2 \leqslant d_i \leqslant n$, where d_i indicates that in the i-th direction (i.e., the i-th coordinate) , the $(d_i - 1)$-st, d_i-th, $(d_i + 1)$-st, $\cdots$, $(n-1)$-th dimensional layers are mutually uncorrelated, but the $(d_i - 2)$-nd dimensional layer is not. $d_i = \infty$ means that not even the $(n-1)$-st dimensional layers are orthogonal. The two extreme definitions of orthogonalities can be described as follows:

Orthogonality by the first definition[17] In this case, the n-dimensional matrix has its two-dimensional layers, M, that are orthogonal in all axis-normal directions, that is, if the inner product of their rows are pairwise zero, or equivalently if $MM^{\mathrm{T}} = D_m$, a diagonal matrix of order m. In other words, an n-dimensional matrix is said to be orthogonal by the first definition if it has property $(2, 2, \cdots, 2)$.

Orthogonality by the second definition[17] In this case, the n-dimensional matrix has its $(n-1)$-dimensional sections normal to one coordinate axis that are mutually uncorrelated but are not in themselves orthogonal in any sense; moreover, the sets of $(n-1)$-dimensional layers normal to other axes are neither mutually uncorrelated nor orthogonal . In other words, an n-dimensional matrix is said to be orthogonal by the second definition if it has property $(\infty, \infty, \cdots, \infty, n, \infty, \cdots, \infty)$.

When the term "orthogonal" is used to describe an n-dimensional matrix without modifying the word orthogonal, it means that the orthogonality lies between the above extremes of the first and the second definitions of orthogonality. An absolutely proper n-dimensional Hadamard matrix is orthogonal of property $(2, 2, \cdots, 2)$; an absolutely improper n-dimensional Hadamard matrix is orthogonal of property $(n, n, \cdots, n)$.

Let a two-dimensional matrix A be denoted by

$$A = \begin{bmatrix} A(1) \\ A(2) \\ \vdots \\ A(n) \end{bmatrix},$$

where $A(i) = (A(i,1), A(i,2), \cdots, A(i,n))$, $i = 1, 2, \cdots, n$. Similarly, let

$$B = \begin{bmatrix} B(1) \\ B(2) \\ \vdots \\ B(n) \end{bmatrix}.$$

The inner product of A and B will be meant

$$A \cdot B = A(1)B(1) + A(2)B(2) + \cdots + A(n)B(n).$$

Alternatively, $A \cdot B$ can be written as the sum of the diagonal elements of AB^{T}, which is in fact $\mathrm{tr}(AB^{\mathrm{T}})$. From the rows of a given orthogonal square matrix H, we construct layers of a three-dimensional matrix that is orthogonal by the second definition in the following way:

Let H be given by

$$H = \begin{bmatrix} H(1) \\ H(2) \\ \vdots \\ H(n) \end{bmatrix} = \begin{bmatrix} H(1,1) & H(1,2) & \cdots & H(1,n) \\ H(2,1) & H(2,2) & \cdots & H(2,n) \\ \vdots & \vdots & & \vdots \\ H(n,1) & H(n,2) & \cdots & H(n,n) \end{bmatrix},$$

where $H(p)H(q) = 0$, $p, q = 1, 2, \cdots, n$, $p \neq q$. Orthogonal layers $H(i,j)$ are obtained by taking the tensor products of the vectors $H(i)$, $H(j)$:

$$H(i,j) = H(i) \otimes (H(j))^{\mathrm{T}}$$

$$= \begin{bmatrix} H(i,1)H(j,1) & H(i,1)H(j,2) & \cdots & H(i,1)H(j,n) \\ H(i,2)H(j,1) & H(i,2)H(j,2) & \cdots & H(i,2)H(j,n) \\ \vdots & \vdots & & \vdots \\ H(i,n)H(j,1) & H(i,n)H(j,2) & \cdots & H(i,n)H(j,n) \end{bmatrix}$$

$$= \begin{bmatrix} H(i,1)H(j) \\ H(i,2)H(j) \\ \vdots \\ H(i,n)H(j) \end{bmatrix}.$$

It is easy to check that $H(i,j)H(k,l) = 0$, for all $i,j,k,l = 1,2,\cdots,n$, $j \neq l$. For

$$\begin{aligned} H(i,j)H(k,l) = & (H(i,1)H(j)) \cdot (H(k,1)H(l)) + (H(i,2)H(j))(H(k,2)H(l)) \\ & + \cdots + (H(i,n)H(j)) \cdot (H(k,n)H(l)) \\ = & (H(i,1)H(k,1) + H(i,2)H(k,2) + \cdots \\ & + H(i,n)H(k,n))(H(j) \cdot H(l)) \\ = & 0 \end{aligned}$$

since $H(j) \cdot H(l) = 0$, for all $l,j = 1,2,\cdots,n$, $l \neq j$.

Fourth and higher-dimensional orthogonal matrices that are orthogonal by the second definition can be constructed in an analogous way. In fact, denote an n-dimensional matrix by $[M]_n$ and its $(n-1)$-dimensional layers by $[M_{n-1}]$. The inner product of two n-cube A_n and B_n is defined by the sum of the inner products of their respective $(n-1)$-layers parallel to the i-th coordinate plane $i = 1,2,\cdots,n$, e.g.,

$$[A]_n \cdot [B]_n = \sum_{j=1}^{n} [A_j^i]_{n-1} [B_j^i]_{n-1},$$

where A_j^i denotes the j-th layer parallel to the i-th coordinate plane.

Then $[H]_n$ is said to be orthogonal if the inner product of the $(n-1)$-dimensional layers parallel to a coordinate hyperplane is pairwise equal to zero. Again we can construct such $(n-1)$-dimensional layers by taking the tensor products of the layers of an $(n-1)$-dimensional orthogonal cube.

The second definition of orthogonality of higher-dimensional matrices is quite in agreement with the orthogonality of the two-dimensional matrices from the point of view of dimensions. In a two-dimensional matrix M, the layers are the rows (or columns) of M that can be assumed as one-dimensional matrices, i.e., one dimension lower than the dimension of M, so that in this case, too, orthogonality can be defined as the inner product of the parallel layer matrices pair-wise equal to zero.

Let $[H]_n$ be an n-cube of order h. We denote by $[H^i]_1$, $[H^i]_2$, $\cdots$, $[H^i]_{n-1}$ the 1-, 2-, $\cdots$, $(n-1)$-dimensional layer matrices of $[H]_n$, respectively, where the i represents the layers embedded in the subspaces parallel to the i-th coordinate hyperplane. If the equation $[H^i]_1 \cdot [H^i]_1' = 0$ is satisfied by each pair of distinct one-dimensional layers $[H^i]_1$ and $[H^i]_1'$ parallel to the i-th coordinate hyperplane, then $[H^i]_2 \cdot [H^i]_2' = 0$ is also true for each distinct pair of two-dimensional layers $[H^i]_2$ and $[H^i]'_2$. In general, it is easy to prove that[17]

$$[H^i]_1 \cdot [H^i]'_1 = 0 \Longrightarrow [H^i]_2 \cdot [H^i]'_2 = 0$$
$$\cdots\cdots\cdots$$
$$\Longrightarrow [H^i]_{n-1} \cdot [H^i]'_{n-1} = 0.$$

It can be observed that $[H^i]_1 \cdot [H^i]'_1 = 0$ is equivalent to the first definition of orthogonality, provided that i takes up all values from 1 to n. On the other hand, $[H^i]_{n-1} \cdot [H^i]'_{n-1} = 0$ is equivalent to the second definition of orthogonality. Thus the orthogonality by the first definition implies the orthogonality by the second definition[17].

Between the first and second definitions of orthogonalities, there are variety of orthogonalities according to the dimension of the layers, which can vary from 1 to $n-1$, and the numbers of coordinate hyperplanes to which the layers can be parallel, which can go from 1 to n. If A and B are two n-cubes orthogonal of proprieties $(a_1, a_2, \cdots, a_n)$ and $(b_1, b_2, \cdots, b_n)$ and orders a and b, respectively, then the direct multiplication (or Kronecker product) $A \otimes B$ of A and B is an n-cube of order ab with orthogonal of property $(c_1, c_2, \cdots, c_n)$, where $c_i = \max(a_i, b_i)$, $i = 1, 2, \cdots, n$. A set $\mathcal{A} = \{A_1, A_2, \cdots, A_m\}$ of n-cubes orthogonal of proprieties $(a_{i1}, a_{i2}, \cdots, a_{in})$, where $a_{ij} \leqslant \max\{a_{i1}, a_{i2}, \cdots, a_{im}\}$ for $i = 1, 2, \cdots, m$ generates a monoid under the operation $\otimes$, i.e., it satisfies the following properties[1,17]:

(1) $A_i \otimes A_j$ is orthogonal;

(2) $A_i(\otimes A_j \otimes A_k) = (A_i \otimes A_j) \otimes A_k$; and

(3) $I_{1\times 1} \in \mathcal{A}$ is the unit element.

Since $A_i \otimes A_j$ is equivalent to $A_j \otimes A_i$ by using an appropriate permutation of rows and columns, we could say that the monoid is "combinatorially commutative."

Example 6.3.1[1,17–19] Let $A = [A(i,j,k)]$, $0 \leqslant i,j,k \leqslant 1$, be the three-dimensional matrix of order 2 defined by

$$[A(i,0,k)] = \begin{bmatrix} -1 & 1 \\ -1 & -1 \end{bmatrix} \quad \text{and } [A(i,1,k)] = \begin{bmatrix} -1 & 1 \\ 1 & 1 \end{bmatrix}.$$

These two faces are, as vectors $(-1,1,-1,-1)$ and $(-1,1,1,1)$, which are orthogonal, and so these faces (or parallel two-dimensional layers in the direction of y-axes) are said to be orthogonal. It can be verified that A is orthogonal of property $(2,2,3)$.

Let $B = [B(i,j,k)]$, $0 \leqslant i,j,k \leqslant 1$, be the three-dimensional matrix of order 2 defined by

$$[B(i,0,k)] = \begin{bmatrix} 1 & 1 \\ 1 & -1 \end{bmatrix} \quad \text{and } [B(i,1,k)] = \begin{bmatrix} -1 & 1 \\ 1 & 1 \end{bmatrix}.$$

Every two-dimensional face of B is a Hadamard matrix. So B is an absolutely three-diemensional Hadamard matrix, which is orthogonal of property $(2,2,2)$[1,17].

Example 6.3.2 (The Paley Cube[1,17]) Let $q \equiv 3(\text{mod}4)$ be a prime power and z_0, z_1, $\cdots$, z_{q-1} be the elements of $GF(q)$, the Galois field. We define

$$p_{ij\cdots r} = \begin{cases} 1, & \text{if any of the subscripts is } q, \\ \mathcal{X}(z_i + z_j + \cdots + z_r), & \text{otherwise,} \end{cases}$$

where each subscript runs from zero to q, $P = [p_{ij\cdots r}]$ is a $(q+1)$-dimensional Paley cube, and $\mathcal{X}(0) = -1$. Thus

$$\mathcal{X}(z) = \begin{cases} 1, & \text{if } z \text{ is a square in GF}(q), \\ -1, & \text{otherwise.} \end{cases}$$

By the same reasoning as before, and using the two-dimensional properties of this matrix, we see that each two-dimensional face of the Paley cube except that one face containing all ones is an Hadamard matrix. So then, when $q \equiv 3(\text{mod} 4)$ is a prime power, there is an almost Hadamard $(q+1)$-dimensional cube, called the Paley cube, of order $q+1$, which has one two-dimensional layer, in each direction, consisting of all ones with every other face being a Hadamard matrix. The Paley cube is orthogonal of property $(\infty, \infty, \cdots, \infty)$, but if the two-dimensional layer consisting of all ones is removed in one direction, the remaining n-dimensional matrix (note it is no longer a cube) has all two-dimensional layers in that direction orthogonal to each other.

6.3.2 Higher-Dimensional Orthogonal Designs[1,17–19]

Two-dimensional orthogonal designs of type $(s_1, s_2, \cdots, s_t)$ are defined by [20] as a square orthogonal matrices with entries from $\{0, \pm x_1, \pm x_2, \cdots, \pm x_t\}$, where $x_1, x_2, \cdots, x_t$ are commuting variables and s_j is the number of times $\pm x_j$ occurs in each row and column, that is, in which all distinct rows and columns have scalar product zero. Hence an $m \times m (= m^2)$ matrix $[d(i,j)]$ is an orthogonal design of type $(s_1, s_2, \cdots, s_t)$ if it has entries from $\{0, \pm x_1, \pm x_2, \cdots, \pm x_t\}$ and

$$\sum_{i=0}^{m-1} d(i,a)d(i,b) = \sum_{j=0}^{m-1} d(a,j)d(b,j) = \sum_{k=0}^{m-1} s_k x_k^2 \delta_{ab}. \tag{6.41}$$

Thus, the two-dimensional Hadamard matrices are special cases in which the variables are from $\{\pm 1\}$ and $(s_1, s_2, \cdots, s_t)$ is (m). Therefore $[h(i,j)]$, $0 \leqslant i, j \leqslant m-1$, is a Hadamard matrix if

$$\sum_{i=0}^{m-1} h(i,a)h(i,b) = \sum_{j=0}^{m-1} h(a,j)h(b,j) = m\delta_{ab}. \tag{6.42}$$

In general, a proper n-cube orthogonal design $D = [D(d(1), \cdots, d(n))]$, $0 \leqslant d(i) \leqslant m-1$, of order m and type $(s_1, s_2, \cdots, s_t)^n$ is defined by [17] as such a cube in which all parallel two-dimensional layers, in any orientation parallel to a plane, are uncorrelated, which is equivalent to the requirement that $D(d(1), \cdots, d(n)) \in \{0, \pm x_1, \cdots, \pm x_t\}$, where $x_1, x_2, \cdots, x_t$ are commuting variables, and that

$$\begin{aligned} &\sum_{d(1)} \cdots \sum_{d(r-1)} \sum_{d(r+1)} \cdots \sum_{d(n)} D(d(1), \cdots, d(r-1), a, d(r+1), \cdots, d(n)) \\ &\times D(d(1), \cdots, d(r-1), b, d(r+1), \cdots, d(n)) \\ &= \left(\sum_i s_i x_i^2 \right)^{n-1} \delta_{ab}, \end{aligned} \tag{6.43}$$

where $(s_1, s_2, \cdots, s_t)$ are integers giving the occurrences of $\pm x_1, \cdots, \pm x_t$ in each row and column (called the type $(s_1, s_2, \cdots, s_t)^n$ by the first definition), i.e., it is of property $(2, 2, \cdots, 2)$.

In a fashion similar to the last subsection, it is possible to define orthogonal designs according to the second definition or according to any other property of orthogonality.

Higher-dimensional orthogonal designs may be constructed by noting that if A is an n-cube orthogonal design of order a and type $(s_1, s_2, \cdots, s_t)^n$, and H is an n-dimensional Hadamard matrix of order h, then the direct multiplication $H \otimes A$ is an n-cube orthogonal design of order ah and type $(hs_1, hs_2, \cdots, hs_t)^n$. The property depends on the property of the matrices used[17].

Example 6.3.3[17] Let $H = [H(i,j,k)]$, $0 \leqslant i,j,k \leqslant 1$, be the three-dimensional Hadamard matrix of order 2 defined by

$$[H(i,0,k)] = \begin{bmatrix} 1 & 1 \\ 1 & -1 \end{bmatrix} \quad \text{and } [H(i,1,k)] = \begin{bmatrix} -1 & 1 \\ 1 & 1 \end{bmatrix}$$

and let $A = [A(i,j,k)]$, $0 \leqslant i,j,k \leqslant 1$, be the 3-cube orthogonal design of order 2 and type $(1,1)^3$ defined by

$$[A(i,0,k)] = \begin{bmatrix} y & x \\ x & -y \end{bmatrix} \quad \text{and } [A(i,1,k)] = \begin{bmatrix} -x & y \\ y & x \end{bmatrix}.$$

Then their direct multiplication $A \otimes H =: C = [C(i,j,k)]$, $0 \leqslant i,j,k \leqslant 3$, is the 3-cube orthogonal design of order $2 \times 2 = 4$ and type $(2,2)^3$ defined by

$$[C(i,0,k)] = \begin{bmatrix} -y & -x & -y & -x \\ -x & -y & -x & -y \\ -y & -x & y & x \\ -x & y & x & -y \end{bmatrix}; \quad [C(i,1,k)] = \begin{bmatrix} x & -y & x & -y \\ -y & -x & -y & -x \\ x & -y & -x & y \\ -y & -x & y & x \end{bmatrix};$$

and

$$[C(i,2,k)] = \begin{bmatrix} y & x & -y & -x \\ x & -y & -x & y \\ -y & -x & -y & -x \\ -x & y & -x & y \end{bmatrix}; \quad [C(i,3,k)] = \begin{bmatrix} -x & y & x & -y \\ y & x & -y & -x \\ x & -y & x & -y \\ -y & -x & -y & -x \end{bmatrix}.$$

Example 6.3.4[17] The following $B = [B(i,j,k)]$, $0 \leqslant i,j,k \leqslant 3$, is a three-dimensional orthogonal design of order 4 and type $(1,1,1,1)^3$:

$$[B(i,0,k)] = \begin{bmatrix} d & c & -b & -a \\ -c & d & -a & b \\ b & a & d & c \\ -a & b & c & -d \end{bmatrix}; \quad [B(i,1,k)] = \begin{bmatrix} c & -d & a & -b \\ d & c & -b & -a \\ a & -b & -c & d \\ b & a & d & c \end{bmatrix};$$

and

$$[B(i,2,k)] = \begin{bmatrix} b & a & d & c \\ a & -b & -c & d \\ -d & -c & b & a \\ -c & d & -a & b \end{bmatrix}; \quad [B(i,3,k)] = \begin{bmatrix} -a & b & c & -d \\ b & a & d & c \\ c & -d & a & -b \\ -d & -c & b & a \end{bmatrix}.$$

Theorem 6.3.1[1,17–19] *There exists an n-dimensional orthogonal design of order 2, type $(1,1)^n$, and property $(2,2,\cdots,2)$.*

Proof Let a and b be commuting variables. Let $H = [H(h(1), \cdots, h(n))]$, $0 \leqslant h(i) \leqslant 1$, $1 \leqslant i \leqslant n$, be defined by

$$H(h(1), \cdots, h(n)) = \begin{cases} a(-1)^{w/2+1}, & w \equiv 0(\text{mod}2), \\ b(-1)^{w/2-1}, & w \equiv 1(\text{mod}2), \end{cases}$$

where $w =: \sum_{i=1}^{n} h(i)$, the weight of the subscripts.

In order to check the orthogonality of this H, we consider

$$\begin{aligned} &H(0,0,h(3),\cdots,h(n))H(0,1,h(3),\cdots,h(n)) \\ &+H(1,0,h(3),\cdots,h(n))H(1,1,h(3),\cdots,h(n)) \end{aligned} \tag{6.44}$$

and

$$\begin{aligned} &H(0,0,h(3),\cdots,h(n))H(1,0,h(3),\cdots,h(n)) \\ &+H(0,1,h(3),\cdots,h(n))H(1,1,h(3),\cdots,h(n)). \end{aligned} \tag{6.45}$$

Suppose $v = \sum_{i=3}^{n} h(i)$. Then we have the following four cases:

Case 1. If $v \equiv 0(\text{mod}4)$, then both Equations (6.44) and (6.45) become $-ab + ba = 0$;

Case 2. If $v \equiv 1(\text{mod}4)$, then both Equations (6.44) and (6.45) become $ba + a(-b) = 0$;

Case 3. If $v \equiv 2(\text{mod}4)$, then both Equations (6.44) and (6.45) become $a(-b) + (-b)(-a) = 0$;

Case 4. If $v \equiv 3(\text{mod}4)$, then both Equations (6.44) and (6.45) become $(-b)(-a) + (-a)b = 0$.

Similarly, it can be verified that all of the two-dimensional faces of H are orthogonal. So this H is the wanted n-dimensional orthogonal design of order 2, type $(1,1)^n$, and property $(2, 2, \cdots, 2)$. □

Corollary 6.3.1 [17–19] *There exist n-dimensional orthogonal designs of order 2^{t+1}, type $(2^t, 2^t)^n$, and property $(2, 2, \cdots, 2)$.*

Proof The corollary follows the direct multiplication $A \otimes H$ of an absolutely proper n-dimensional Hadamard matrix A of order 2 and the n-dimensional orthogonal design H of order 2, type $(1,1)^n$, and property $(2, 2, \cdots, 2)$ constructed in Theorem 6.3.1. □

An n-dimensional orthogonal design of type $(1,1,1,1)^n$ would be preserved under the following equivalence relations [17–19]:

(1) Each variable is replaced throughout by its negative;

(2) Rearrangement of the parallel k-dimensional hyper-planes;

(3) Multiplication of every variable of one entire k-dimensional hyper-plane by -1.

Horadam and Lin proved in *J. Combin. Math. Comp.* in 1998 that there is only one equivalent $(1,1,1,1)^2$ design of order 4 on the variables a, b, c, d, which is

$$\begin{bmatrix} a & b & c & d \\ -b & a & d & -c \\ -c & -d & a & b \\ -d & c & -b & a \end{bmatrix}.$$

Theorem 6.3.2 [17–19] *For each positive integer n, there exist n-dimensional orthogonal designs of order 4, type $(1,1,1,1)^n$, and property $(2,2,\cdots,2)$.*

Proof We proceed inductively. First, define

$$a_1 = (-d,-c,b,a); \quad a_2 = (c,-d,a,-b); \quad a_3 = (b,a,d,c); \quad a_4 = (-a,b,c,-d).$$

and note that

$$a_i \cdot a_j = 0 \text{ for all } 1 \leqslant i \neq j \leqslant 4.$$

Now we can describe the faces of the $(1,1,1,1)^3$ design as

$$b_4^{\mathrm{T}} = \begin{bmatrix} a_4 \\ a_3 \\ a_2 \\ a_1 \end{bmatrix}, \quad b_3^{\mathrm{T}} = \begin{bmatrix} a_3 \\ -a_4 \\ a_1 \\ -a_2 \end{bmatrix}, \quad b_2^{\mathrm{T}} = \begin{bmatrix} a_2 \\ -a_1 \\ -a_4 \\ a_3 \end{bmatrix}, \quad b_1^{\mathrm{T}} = \begin{bmatrix} -a_1 \\ -a_2 \\ a_3 \\ a_4 \end{bmatrix},$$

or, equivalently,

$$\begin{bmatrix} b_4 \\ b_3 \\ b_2 \\ b_1 \end{bmatrix} = \begin{bmatrix} a_4 & a_3 & a_2 & a_1 \\ a_3 & -a_4 & a_1 & -a_2 \\ a_2 & -a_1 & -a_4 & a_3 \\ -a_1 & -a_2 & a_3 & a_4 \end{bmatrix}.$$

Now there are two inequivalent two-dimensional $(1,1,1,1)^2$ orthogonal designs and both of them can be used to construct three-dimensional $(1,1,1,1)^3$ orthogonal designs. Thus, we have a $(1,1,1,1)^3$ design on the commuting variables a_1, a_2, a_3, a_4 and hence a $(1,1,1,1)^4$ design on the commuting variables a, b, c, d.

The orthogonality within the $(1,1,1,1)^3$ design is established by the construction. The orthogonality of the $(1,1,1,1)^4$ design is obtained by using the extra property $a_i . a_j = 0$ for all $1 \leqslant i \neq j \leqslant 4$.

To obtain the $(1,1,1,1)^{k+1}$ design, we assume the existence of the $(1,1,1,1)^j$ designs for all $j \leqslant k$ made by the construction. Now we have a $(1,1,1,1)^k$ design whose hyper-rows, c_1, c_2, c_3, c_4, comprise objects which are the hyper-rows of the $(1,1,1,1)^{k-1}$ design. We now write down the hyper-rows of each of the four hyper-planes containing these rows as the columns of a 4×4 matrix, D. According to the construction, D is an orthogonal design of type $(1,1,1,1)$ whose objects are the c_i. Now we complete D to form a $(1,1,1,1)^3$ design, E, with objects c_i. Now E is a $(1,1,1,1)^{k+1}$ design whose orthogonality is guaranteed by the existence assumption.

In fact, at all stages of the above construction, the property was completely preserved. Hence the theorem follows. □

6.3.3 Higher-Dimensional Hadamard Matrices from Orthogonal Designs

Let $z_1, z_2, \cdots, z_m$ be the elements of an Abelian group G of order m. A type 2 or 1 matrix $A = [A(i,j)]$ is a matrix defined by[17–19]

$$A(i,j) = \alpha(z_i \pm z_j),$$

where α is a map into a commutative ring. A circulant matrix of order m is a special case in that $G = Z_m$ (the cyclic group of order m), with $z_1 = 1$, $z_2 = 2$, $\cdots$, $z_m = m$, and so

$$A(i,j) = \alpha(z_i \pm z_j) = \begin{cases} \alpha(z_i + z_j), & \text{for type 2 matrices,} \\ \alpha(z_i - z_j), & \text{for type 1 matrices.} \end{cases} \tag{6.46}$$

A set of t matrices X_1, X_2, $\cdots$, X_t of order m with $X_k = [X_k(i,j)]$ is called t-suitable matrices[17–19] if

$$\sum_{k=1}^{t}\sum_{i=1}^{m} X_k(a,i)X_k(b,i) = f\delta(a,b) \tag{6.47}$$

and

$$\sum_{i=1}^{m} X_k(a,i)X_l(b,i) = \sum_{i=1}^{m} X_l(a,i)X_k(b,i), \tag{6.48}$$

where f is a constant or constant function. Thus the Williamson matrices of order m are 4-suitable matrices with entries ± 1 and $f = 4m$.

Suppose that $X_k = [X_k(i,j)]$, $k = 1, 2, \cdots, t$, are t-suitable matrices of type 2 and order m defined by[17–19]

$$X_k(i,j) = \psi_k(z_i + z_j).$$

Then from Equations (6.47) and (6.48), we have

$$\sum_{k=1}^{t}\sum_{i=1}^{m} \psi_k(z_a + z_i)\psi_k(z_b + z_i) = f\delta(a,b) = \sum_{k=1}^{t}\sum_{g\in G} \psi_k(z_a + g)\psi_k(z_b + g), \tag{6.49}$$

and

$$\sum_{i=1}^{m} \psi_k(z_a + z_i)\psi_l(z_b + z_i) = \sum_{i=1}^{m} \psi_l(z_a + z_i)\psi_k(z_b + z_i). \tag{6.50}$$

We define the elements of the n-dimensional cube $X_k{=}[X_k(x(1),\cdots,x(n))]$, $1 \leqslant x(i) \leqslant m$, $1 \leqslant i \leqslant n$, for each k, by[17–19]

$$X_k(x(1),\cdots,x(n)) = \psi_k(z_{x(1)} + \cdots + z_{x(n)}). \tag{6.51}$$

To consider the inner product properties of the two-dimensional faces of this cube, we let the q-th coordinate take two values a and b, and the r-th coordinate run from 1 to m, all the other coordinates being constant. Then, with $y = \sum_{i=1}^{m} z_{x(i)} - z_{x(q)} - z_{x(r)}$, we have

$$\begin{aligned}
&\sum_{x(r)=1}^{m} X_k(x(1),\cdots,x(q-1),a,x(q+1),\cdots,x(n)).\\
&\times X_k(x(1),\cdots,x(q-1),b,x(q+1),\cdots,x(n))\\
=&\sum_{x(r)=1}^{m} \psi_k(y + z_a + z_{x(r)})\psi_k(y + z_b + z_{x(r)})\\
=&\sum_{g\in G} \psi_k(y + z_a + g)\psi_k(y + z_b + g)\\
=&\sum_{h\in G} \psi_k(z_a + h)\psi_k(z_b + h).
\end{aligned} \tag{6.52}$$

To find the inner product of the rows of the corresponding two-dimensional layers in different n-dimensional matrices X_k and X_l, we let the r-th coordinate sum from one to n, the q-th coordinate, take two values (a and b) , and all other coordinates remain constant. Letting

$$y = \sum_{i=1}^{m} z_{x(i)} - z_{x(q)} - z_{x(r)},$$

we have[17–19]

$$\begin{aligned}
&\sum_{x(r)=1}^{m} X_k(x(1),\cdots,x(q-1),a,x(q+1),\cdots,x(n)).\\
&\times X_l(x(1),\cdots,x(q-1),b,x(q+1),\cdots,x(n))\\
&=\sum_{x(r)=1}^{m} \psi_k(y+z_a+z_{x(r)})\psi_l(y+z_b+z_{x(r)})\\
&=\sum_{i=1}^{m}\psi_k(z_a+z_i)\psi_l(z_b+z_i)\\
&=\sum_{i=1}^{m}\psi_l(z_a+z_i)\psi_k(z_b+z_i)\ \text{(from Equation (6.50))}\\
&=\sum_{x(r)=1}^{m} X_l(x(1),\cdots,x(q-1),a,x(q+1),\cdots,x(n))\\
&\times X_k(x(1),\cdots,x(q-1),b,x(q+1),\cdots,x(n)).
\end{aligned}\tag{6.53}$$

Higher-dimensional Hadamard matrices may be constructed by replacing the variables of an n-cube orthogonal design of type $(\overbrace{1,1,\cdots,1}^{t})^n$ by the t-suitable matrices[17–19].

Combining Equations (6.49) and (6.52), we see that if the rows of t-suitable matrices are orthogonal, the rows of the n-dimensional matrices formed from these matrices, in any direction parallel to the axis, will also be orthogonal. The contribution of the different t-suitable matrices is cancelled out in the orthogonal design by using Equation (6.53).

It is known that Williamson matrices of order m (or 4-suitable matrices in our present terminology) exist for the following orders[20]:

(1) m : where $m \in \{1,3,5,\cdots,29,37,43\}$;

(2) $(p+1)/2$: $p \equiv 1(\text{mod}4)$ a prime power;

(3) 3^c, 7.3^{c-1}: c is a natural number;

(4) $p^r(p+1)/2 : p \equiv 1(\text{mod}4)$ a prime power, r a natural number;

(5) $s(4s-1) : s$ is the order of a good matrix (see [21] for definition);

(6) $s(4s+3) : s$ is the order of a good matrix, and $4s+4$ is the order of a symmetric Hadamard matrix;

(7) $sv : s$ is the order of a good matrix, v is the order of an Abelian group G on which are defined a (v,k,λ) and a $(v,(v-1)/2,(v-s)/4)$ difference set $v-4(k-\lambda)=4s-1$;

(8) $3^{2r}(p_1^{r_1}\cdots p_n^{r_n})^4 : r, r_i$ are nonnegative integers, p_i are primes satisfying $p_i \equiv 3(\text{mod}4)$ and $p_i > 3$.

The Williamson matrices are used to form n-dimensional cubes $X_1 = [X_1(x(1),\cdots, x(n))]$, $X_2 = [X_2(x(1),\cdots, x(n))]$,$X_3 = [X_3(x(1),\cdots, x(n))]$, and $X_4 = [X_4(x(1),\cdots, x(n))]$, which are used to replace the a, b, c, and d of the orthogonal design of order 4 and type $(1,1,1,1)^3$ (see Example 2 in the last subsection). Because of the properties of these matrices, each of the faces parallel to the axes will be a Hadamard matrix of property $(2,2,2)$.

Thus when m is the order of 4 Williamson matrices, there is a three-dimensional Hadamard matrix of order $4m$ and property $(2,2,2)$, i.e., there exists an absolutely proper three-dimensional Hadamard matrix of order $4m$. In general, from Theorem 6.3.2 and the above construction, we have

Theorem 6.3.3 [17–19] *Let m be the order of 4 Williamson matrices. Then there is an n-dimensional Hadamard matrix of order $(4m)^m$ and property $(2,2,\cdots,2)$, i.e., there exists an absolutely proper n-dimensional Hadamard matrix of order $(4m)^m$.*

Definition 6.3.1 [1,17–19] *Let $A=[A(a(1),\cdots,a(n))]$ and $B=[B(b(1),\ \cdots,\ b(n))]$, $0\leqslant a(i),\ b(i)\leqslant m-1$, $1\leqslant i\leqslant n$, be two n-dimensional matrices of order m. Then A and B are called anti-amicable if the following two conditions are satisfied:*

(1) *For each p, $1\leqslant p\leqslant n$, and prefixed $0\leqslant a(1),\ \cdots,\ a(p-1),\ a(p+1),\ \cdots,\ a(n)\leqslant m-1$,*

$$\sum_{a(p)=0}^{m-1} A(a(1),\cdots,a(p-1),a(p),a(p+1),\cdots,a(n))$$
$$\times B(a(1),\cdots,a(p-1),a(p),a(p+1),\cdots,a(n))=0;$$

(2) *For each pair of $p\neq q$, $1\leqslant p,q\leqslant n$, prefixed $0\leqslant a(1),\cdots,a(p-1),a(p+1),\cdots,a(q-1),a(q+1),\cdots,a(n)\leqslant m-1$, and $0\leqslant x\neq z\leqslant m-1$, we have*

$$\sum_{a(p)=0}^{m-1}[A(a(1),\cdots,a(p-1),a(p),a(p+1),\cdots,a(q-1),x,$$
$$a(q+1),\cdots,a(n))B(a(1),\cdots,a(p-1),a(p),a(p+1),\cdots,$$
$$a(q-1),z,a(q+1),\cdots,a(n))+A(a(1),\cdots,a(p-1),a(p),a(p+1),\cdots,$$
$$a(q-1),z,a(q+1),\cdots,a(n))B(a(1),\cdots,a(p-1),a(p),a(p+1),\cdots,$$
$$a(q-1),x,a(q+1),\cdots,a(n))]=0.$$

In particular, two two-dimensional matrices $A=\begin{bmatrix} a_{11} & a_{12}\\ a_{21} & a_{22}\end{bmatrix}$ and $B=\begin{bmatrix} b_{11} & b_{12}\\ b_{21} & b_{22}\end{bmatrix}$ are anti-amicable if

$$a_{11}b_{11}+a_{12}b_{12}=0;\quad a_{21}b_{21}+a_{22}b_{22}=0$$

and

$$a_{11}b_{21}+a_{12}b_{22}+a_{21}b_{11}+a_{22}b_{12}=0.$$

The following two three-dimensional matrices $C=[C(i,j,k)]$ and $B=[B(i,j,k)]$ are three-dimensional anti-amicable Hadamard matrices of order 2 and property $(2,2,2)$[1,17–19], where

$$[C(i,j,0)]=\begin{bmatrix} 1 & 1\\ -1 & 1\end{bmatrix},\quad [C(i,j,1)]=\begin{bmatrix} 1 & -1\\ 1 & 1\end{bmatrix}.$$

and

$$[D(i,j,0)]=\begin{bmatrix} 1 & -1\\ 1 & 1\end{bmatrix},\quad [D(i,j,1)]=\begin{bmatrix} -1 & -1\\ 1 & -1\end{bmatrix}.$$

Lemma 6.3.1 [1,17–19] *There exist n-dimensional anti-amicable Hadamard matrices of order 2 and property $(2,2,\cdots,2)$.*

Proof Define two n-dimensional (± 1)-valued matrices $A = [A(a(1), \cdots, a(n))]$ and $B = [B(b(1), \cdots, b(n))]$ of order 2 by

$$A(a(1), \cdots, a(n)) = \begin{cases} (-1)^{s/2+1}, & s \text{ even} \\ (-1)^{(s+1)/2}, & s \text{ odd} \end{cases}$$

and

$$B(b(1), \cdots, b(n)) = \begin{cases} (-1)^{v/2}, & v \text{ even} \\ (-1)^{(v+1)/2}, & v \text{ odd,} \end{cases}$$

where

$$s =: \sum_{i=1}^{n} a(i), \quad v =: \sum_{i=1}^{n} b(i), \quad 0 \leqslant a(i), \quad b(i) \leqslant 1, \quad 1 \leqslant i \leqslant n.$$

It can be proved that these A and B are the wanted matrices of the lemma. In fact, each two-dimensional section of A and B is of the form $\begin{bmatrix} a & b \\ b & -a \end{bmatrix}$ $(a, b = \pm 1)$, which is clearly a Hadamard matrix. Thus both A and B are absolutely proper n-dimensional Hadamard matrices of order 2.

In order to prove the anti-amicability, without loss of the generality, we consider the following four cases according to the values of $t = \sum_{i=3}^{n} a(i)$. Let $w = (a(3), \cdots, a(n))$. Then

Case 1. $t \equiv 0(\text{mod} 4)$. Then

$$A(1,1,w)B(1,1,w) + A(1,2,w)B(1,2,w) = (-1)^2(-1) + (-1)^2(-1)^2 = 0,$$

$$A(2,1,w)B(2,1,w) + A(2,2,w)B(2,2,w) = (-1)^2(-1)^2 + (-1)(-1)^2 = 0$$

and

$$\begin{aligned} & A(1,1,w)B(2,1,w) + A(1,2,w)B(2,2,w) \\ & + A(2,1,w)B(1,1,w) + A(2,2,w)B(1,2,w) \\ = & (-1)^2(-1)^2 + (-1)^2(-1)^2 + (-1)^2(-1) + (-1)(-1)^2 \\ = & 0. \end{aligned}$$

Case 2. $t \equiv 1(\text{mod} 4)$. Then

$$A(1,1,w)B(1,1,w) + A(1,2,w)B(1,2,w) = (-1)^2(-1)^2 + (-1)(-1)^2 = 0;$$

$$A(2,1,w)B(2,1,w) + A(2,2,w)B(2,2,w) = (-1)(-1)^2 + (-1)(-1) = 0$$

and

$$\begin{aligned} & A(1,1,w)B(2,1,w) + A(1,2,w)B(2,2,w) \\ & + A(2,1,w)B(1,1,w) + A(2,2,w)B(1,2,w) \\ = & (-1)^2(-1)^2 + (-1)(-1) + (-1)(-1)^2 + (-1)(-1)^2 \\ = & 0. \end{aligned}$$

Case 3. $t \equiv 2(\text{mod} 4)$. Then

$$A(1,1,w)B(1,1,w) + A(1,2,w)B(1,2,w) = (-1)(-1)^2 + (-1)(-1) = 0,$$

$$A(2,1,w)B(2,1,w)+A(2,2,w)B(2,2,w)=(-1)(-1)+(-1)^2(-1)=0$$

and

$$\begin{aligned}&A(1,1,w)B(2,1,w)+A(1,2,w)B(2,2,w)\\&+A(2,1,w)B(1,1,w)+A(2,2,w)B(1,2,w)\\=&(-1)(-1)+(-1)(-1)+(-1)(-1)^2+(-1)^2(-1)\\=&0.\end{aligned}$$

Case 4. $t\equiv 3(\mathrm{mod}4)$. Then

$$A(1,1,w)B(1,1,w)+A(1,2,w)B(1,2,w)=(-1)(-1)+(-1)^2(-1)=0,$$
$$A(2,1,w)B(2,1,w)+A(2,2,w)B(2,2,w)=(-1)^2(-1)+(-1)^2(-1)^2=0$$

and

$$\begin{aligned}&A(1,1,w)B(2,1,w)+A(1,2,w)B(2,2,w)\\&+A(2,1,w)B(1,1,w)+A(2,2,w)B(1,2,w)\\=&(-1)(-1)+(-1)^2(-1)^2+(-1)^2(-1)+(-1)^2(-1)\\=&0.\end{aligned}$$

□

Definition 6.3.2 [1,17–19] *Two 2-dimensional Hadamard matrices (or orthogonal designs) X and Y are said to be amicable if $XY'=YX'$.*

In general, two n-dimensional orthogonal designs $H=[H(h(1),\cdots,h(n))]$ and $G=[G(g(1),\cdots,g(n))]$, $0\leqslant h(i),g(i)\leqslant m-1$, of order m are said to be amicable if for each pair $1\leqslant p\neq q\leqslant n$, $0\leqslant i,k\leqslant m-1$, and prefixed $(g(1),\cdots,g(p-1),g(p+1),\cdots,g(q-1),g(q+1),\cdots,g(n))$,

$$\begin{aligned}&\sum_{g(p)=0}^{m-1}G(g(1),\cdots,g(p-1),g(p),g(p+1),\cdots,g(q-1),i,g(q+1),\cdots,g(n))\\&\times H(g(1),\cdots,g(p-1),g(p),g(p+1),\cdots,g(q-1),k,g(q+1),\cdots,g(n))\\=&\sum_{g(p)=0}^{m-1}G(g(1),\cdots,g(p-1),g(p),g(p+1),\cdots,g(q-1),k,g(q+1),\cdots,g(n))\\&\times H(g(1),\cdots,g(p-1),g(p),g(p+1),\cdots,g(q-1),i,g(q+1),\cdots,g(n)).\end{aligned}\tag{6.54}$$

If Equation (6.54) is satisfied except for some subscripts, we will say that H and G are amicable except for those subscripts.

After the try-and-fail search, it is found that

Theorem 6.3.4 [1,17–19] *There exist no three-dimensional amicable orthogonal designs of type $(1,1)^3$ and property $(2,2,\cdots,2)$.*

Let a, and b be commuting variables. Let $G=[G(g(1),\cdots,g(n))]$, $0\leqslant g(i)\leqslant m-1$, $1\leqslant i\leqslant n$, be the matrix defined by

$$G(g(1),\cdots,g(n))=\begin{cases}(-1)^{w/2-g(n)}a, & w\text{ even},\\(-1)^{(w+1)/2-g(n)}b, & w\text{ odd},\end{cases}\tag{6.55}$$

where $w=:\sum_{i=1}^{n}g(i)$.

Lemma 6.3.2 [1,17–19] *The matrix G constructed in Equation (6.55) is an n-dimensional orthogonal design of type $(1,1)^n$ and property $(2,2,\cdots,2)$.*

Proof At first we prove the following two equations:

$$G(0,0,x)G(0,1,x)+G(1,0,x)G(1,1,x)=0 \tag{6.56}$$

and

$$G(0,0,x)G(1,0,x)+G(0,1,x)G(1,1,x)=0, \tag{6.57}$$

where $x =: (g(3),\cdots,g(n))$ is the prefixed vector of length $n-2$.

Suppose $v =: \sum_{i=3}^{n} g(i)$.

Case 1. If $v \equiv 0(\text{mod}4)$, then the left hand sides of both Equations (6.56) and (6.57) become

$$(-1)^{-g(n)}a(-1)^{1-g(n)}b+(-1)^{1-g(n)}b(-1)^{1-g(n)}a=0;$$

Case 2. If $v \equiv 1(\text{mod}4)$, then the left hand sides of both Equations (6.56) and (6.57) become

$$(-1)^{1-g(n)}b(-1)^{1-g(n)}a+(-1)^{1-g(n)}a(-1)^{2-g(n)}b=0;$$

Case 3. If $v \equiv 2(\text{mod}4)$, then the left hand sides of both Equations (6.56) and (6.57) become

$$(-1)^{1-g(n)}a(-1)^{2-g(n)}b+(-1)^{2-g(n)}b(-1)^{2-g(n)}a=0;$$

Case 4. If $v \equiv 3(\text{mod}4)$, then the left hand sides of both Equations (6.56) and (6.57) become

$$(-1)^{2-g(n)}b(-1)^{-g(n)}a+(-1)^{-g(n)}a(-1)^{1-g(n)}b=0;$$

Thus the face $A=[A(i,j)]=:[G(i,j,x)]$ is a two-dimensional orthogonal design of order 2.

Similarly, it can be proved that each two-dimensional face of this matrix G is an orthogonal design of order 2. □

Lemma 6.3.3 [1,17–19] *There exist n-dimensional orthogonal designs of order 2 and type $(1,1)^n$ that are amicable except that one distinguished coordinate (the last) is constant.*

Proof Let $G=[G(g(1),\cdots,g(n))]$ and $H=[H(h(1),\cdots,h(n))]$ be two matrices produced by Equation (6.55) starting from the commuting variables (a_1,b_1) and (a_2,b_2), respectively. Then by the same proof as that of Lemma 6.3.2, it can be proved that G and H are the wanted designs. □

Lemma 6.3.4 [1,17–19] *There exist n-dimensional amicable Hadamard matrices of order 2 and property $(2,2,\cdots,2)$.*

Proof It can be proved by replacing the variables a_i, b_i by ± 1. □

For more details about the higher-dimensional Hadamard matrices the readers are recommended to see the papers [22–31].

Bibliography

[1] Shlichta P J. Higher-dimensional Hadamard matrices. *IEEE Trans. On Inform. Theory*, 1979, IT-25(5): 566-572.

[2] Yang Y X and Lin X D. *Coding and Cryptography*. Beijing: PPT Press, 1992.

[3] Jedwab J. *Perfect Arrays, Barker Arrays and Difference Sets*. PhD Thesis. University of London, 1991.

[4] Luke H D, Bomer L and Antweiler M. *Perfect binary arrays. Signal Processing*, 1989, 17(1): 69-80.

[5] Luke H D. Sequences and arrays with perfect periodic correlation. *IEEE Trans. AES*, 1988, 24, (2): 287-294.

[6] Davis J A and Jedwab J. A summary of Menon difference sets. *Congressus Numerantium*, 1993, 93: 203-207.

[7] Launey W. A note on n-dimensional Hadamard matrices of order 2^t and Reed-Muller codes. *IEEE Trans. on Inform. Theory*, 1991, 27(3): 664-667.

[8] Arasu K T, Davis J A, Jedwab J and Sehgal S K. New constructions of Menon difference sets. *J. Combin. Theory* (A), 1993, 64(2): 329-336.

[9] Jedwab J, Mitchell C J. Constructing new perfect binary arrays. *Electron. Lett.*, 1988, 24(11): 650-652.

[10] Jedwab J. Generalized perfect arrays and Menon difference sets. *Designs, Codes and Cryptography*, 1992, 2: 19-68.

[11] Xia M Y. Some infinite classes of special Williamson matrices and difference sets. *J. Combin. Theory (A)*, 1992, 61: 230-242.

[12] Yang Y X. Quasi-perfect binary arrays. *Chinese J. of Electronics*, 1992, 20(4): 37-44.

[13] Jedwab J. Nonexistence of perfect binary arrays. *Electron. Lett.*, 1991, 27(14): 1252-1253.

[14] Kopilovich. On perfect binary arrays. *Electron. Lett.*, 1988, 24(9): 566-567.

[15] Wild P. Infinite families of perfect binary arrays. *Electron. Lett.*, 1988, 24(14): 845-847.

[16] Chan W K, Siu M K, and Tong P. Two-dimensional binary arrays with good autocorrelation. *Inform. and Control*, 1979, 42: 125-130.

[17] Hammer J and Seberry J. Higher-dimensional orthogonal designs and applications. *IEEE Trans. Inform. Theory*, 1981, 27(6): 772-779.

[18] Seberry J. Higher-dimensional orthogonal designs and Hadamard matrices. *Combinatorics VII: Proc. Seventh Australian Conf., Lecture Notes in Mathematics*, New York: Springer-Verlag, 1980.

[19] Hammer J and Seberry J. Higher-dimensional orthogonal designs and Hadamard matrices (II). *Proc. Ninth Conf. on Numerical Mathematics. Congressus Numerantium, Utilitas Mathematics.* Winnipeg, 1979: 23-29.

[20] Geramita A V and Seberry J. *Orthogonal Designs: Quadratic Forms and Hadamard Matrices.* New York: Marcel Dekker, 1979.

[21] Seberry Wallis J. On the existence of Hadamard matrices. *J. Combinatorial Theory, Ser. A*, 1976, 21: 188-195.

[22] Yang Y X. Proofs of some conjectures on higher-dimensional Hadamard matrices. *Chinese Science Bulletin*, 1986, 31(24): 1662-1667.

[23] Yang Y X. Existence, construction methods and enumeration of higher-dimensional Hadamard matrices. *IEEE ISIT*'90. U.S.A, 1990.

[24] Yang Y X. On the perfect binary arrays. *J. of Electronics* (*China*), 1990, 7(2): 175-181.

[25] Yang Y X. Dyadic methods in communication theory. *Advances in Mathematics*, 1990, 19(2): 254-255.

[26] Li S Q, and Yang Y X. On the circulant Hadamard conjecture. *Select Papers for the J. of BUPT*, 1990: 80-83.

[27] Yang Y X. New proofs for the conjectures of higher-dimensional Hadamard matrices. *J. Systems Science and Mathematical Sciences*, 1988, 8(1): 52-55.

[28] Li S Q, and Yang Y X. The analysis of array sampling and folding. *J. of Beijing Univ. of Posts and Telecomm*, 1989, 12(1): 28-34.

[29] Yang Y X. On the constructions of higher-dimensional Hadamard matrices. *J. of Beijing Univ. of Posts and Telecomm*, 1988, 11(2): 31-38.

[30] Yang Y X. Applications of higher-dimensional matrices to cryptography. *J. of Beijing Univ. of Posts and Telecomm*, 1989, 12(4): 41-46.

[31] Calabro D, Wolf J K. On the synthesis of two-dimensional arrays with desirable correlation properties. *Information and Control*, 1968, 11: 537-560.

Part IV

Applications to Signal Design and Analysis

Chapter 7

Design and Analysis of Sequences

From Chapter 1, we know that each row of a Walsh matrix W of order $2^n \times 2^n$ is a binary (± 1) sequence of length 2^n, such that its dyadic correlation function $c(k) = \sum_{m=0}^{2^n-1} f(m)f(m \oplus k)$ is zero if and only if $k \neq 0$, i.e., the $C(k)$ is an impulse function. Thus the Walsh rows can be used as the perfect signal sequences in dyadic filtering. From Chapter 2, we know that the rows of circulant Hadamard matrices are perfect signal sequences with mismatched correlation functions. Chapter 1 also tells us that the rows of Walsh matrices are in fact the truth tables of Boolean functions. From both Chapter 3 and Chapter 5, we know that the n-dimensional Hadamard matrices of order 2 can be treated as the special Boolean functions too.

In a word, the rows of Hadamard matrices play an important role in the design and analysis of signals of cryptographic significance. This chapter will present more results to show how to share the Hadamard ideas in the design and analysis of signal sequences.

7.1 Sequences of Cryptographic Significance

7.1.1 Enumerating Boolean Functions of Cryptographic Significance[1,3]

A function from $\mathrm{GF}(2)^m$ to $\mathrm{GF}(2)^n$ (where $m \geqslant n \geqslant 1$) is said to be an (m,n)-Boolean function or abbreviated as an (m,n)-LUT in the following text. We restate five of the important conditions that might be required of an (m,n)-LUT f used in a stream cipher system[2].

C1. Balance. Over the complete set of possible inputs, each possible n-bit output should occur 2^{m-n} times, i.e., if y is any n-bit vector, then $| f^{-1}(y) |= 2^{m-n}$.

C2. Nonlinearity/affinity. f must be a nonlinear and nonaffine function for all n outputs, i.e., for every $i(1 \leqslant i \leqslant n)$ there must not be a vector h in $\mathrm{GF}(2)^m$ and a fixed scalar such that $f(x) |_i = x \cdot h + a$ for every x in $\mathrm{GF}(2)^m$, where $y |_i$ denotes the bit in the ith position of the vector y.

C3. Nondegeneracy. Each of the n outputs of f must depend on all the inputs, i.e., if each of the n output variables is expressed as an equation in the m input variables, then each equation must involve all of the m input variables.

C4. Uncorrelatedness. Given any vectors $x = (x_1, x_2, \cdots, x_m)$ and $y = (y_1, y_2, \cdots, y_n)$ such that $f(x) = y$, then $Pr(x_i = y_j) = 0.5$ for any $i, j(1 \leqslant i \leqslant m, 1 \leqslant j \leqslant n)$.

C5. Symmetry. If P is any permutation matrix, then $f(x) = f(xP)$ for any x in $\mathrm{GF}(2)^m$.

Let $A_{mn}(i)(1 \leqslant i \leqslant 5)$ be the number of (m,n)-LUTs satisfying condition Ci. The exact values of $A_{mn}(1), A_{mn}(2), A_{mn}(3), A_{mn}(5)$ are easy to calculate, but, up to now, only a lower bound for $A_{mn}(4)$ is known[2]. In this subsection, we will give an improved bound for $A_{mn}(4)$.

Lemma 7.1.1[2] *Suppose x_m is the number of $(m,1)$-LUTs satisfying some combination of conditions C2, C3, C4, and C5. Then the number of (m,n)-LUTs satisfying the same set of conditions is simply $(x_m)^n$.*

Lemma 7.1.2 *The $(m,1)$-LUT f satisfies condition C4 if and only if $W(f(x_1, x_2, \cdots, x_m) + x_i) = 2^{m-1}$ for every $i(1 \leqslant i \leqslant m)$, where $W(g(.))$ denotes the Hamming weight of the $(m,1)$-LUT g, i.e., the number of ones in the set of 2^m entries of the look-up table for g.*

Proof Suppose i satisfies $1 \leqslant i \leqslant m$. Then $Pr(x_i = f(x_1, x_2, \cdots, x_m)) = 0.5$ if and only if $Pr(x_i + f(x_1, x_2, \cdots, x_m) = 0) = 0.5$ if and only if $Pr(x_i + f(x_1, x_2, \cdots, x_m) = 1) = 0.5$ if and only if $W(x_i + f(x_1, x_2, \cdots, x_m)) = 2^{m-1}$. □

Lemma 7.1.3 *Suppose h is an $(m,1)$-LUT, with the property that*

$$h(x_1, x_2, \cdots, x_m) = f(x_1, x_2, \cdots, x_{m-r}) + g(x_{m-r+1}, x_{m-r+2}, \cdots, x_m),$$

where f is an $(m-r,1)$-LUT and g is an $(r,1)$-LUT. If f and g satisfy condition C4, then so does h.

Proof This result follows from Lemma 7.1.2. □

Now define three set of $(m,1)$-LUTs as follows. Let

$$S_1^m = \{f : f(x_1, x_2, \cdots, x_m) = f(x_1 + 1, x_2 + 1, \cdots, x_m + 1)\},$$

i.e., S_1^m is the set of all $(m,1)$-LUTs with the property that inverting the input leaves the output unchanged. Let

$$S_2^m = \left\{ g : g(x_1, x_2, \cdots, x_m) = a + \sum_{i=1}^{m} b_i x_i \right\},$$

where $a \in \{0,1\}$ and $\sum_{i=1}^{m} b_i$ is odd and greater than 1. If $1 \leqslant r \leqslant m-1$, then let

$$\begin{aligned} S_3^{m,r} = \{h : h(x_1, x_2, \cdots, x_m) &= f(x_1, x_2, \cdots, x_{m-r}) \\ &+ g(x_{m-r+1}, x_{m-r+2}, \cdots, x_m)\}, \end{aligned}$$

where $f \in S_1^{m-r}$, f is nonlinear, and $g \in S_2^r$ with the property that $a = 0$.

Lemma 7.1.4 *Suppose $m > 0$ and $1 \leqslant r \leqslant m-1$,*

(i) *S_1^m, S_2^m, and S_3^m are mutually disjointed.*

(ii) *The LUTs in S_1^m, S_2^m, and S_3^m satisfy condition C4.*

(iii) $| S_1^m |= 2^{2^m - 1}$.

(iv) $| S_2^m |= 2^m - 2m$.

(v) $| S_3^{m,r} |= (2^{2^{m-r-1}} - 2^{m-r})(2^{r-1} - r)$.

Proof (i) Suppose $f \in S_2^m$, i.e., $f(x_1, x_2, \cdots, x_m) = a + \sum_{i=1}^m b_i x_i$. Then since $\sum_{i=1}^m b_i$ is odd, $f(x_1+1, x_2+1, \cdots, x_m+1) = f(x_1, x_2, \cdots, x_m) + 1$, i.e., $f \notin S_1^m$.

Now suppose $h \in S_3^{m,r}$, i.e.,

$$h(x_1, x_2, \cdots, x_m) = f(x_1, x_2, \cdots, x_{m-r}) + g(x_{m-r+1}, x_{m-r+2}, \cdots, x_m),$$

where f is nonlinear (and nonconstant) and g is linear. Hence h is nonlinear and nonaffine, and thus $h \notin S_2^m$. In addition

$$\begin{aligned} & h(x_1+1, x_2+1, \cdots, x_m+1) \\ = & f(x_1+1, x_2+1, \cdots, x_{m-r}+1) \\ & + g(x_{m-r+1}+1, x_{m-r+2}+1, \cdots, x_m+1) \\ = & f(x_1, x_2, \cdots, x_{m-r}) + g(x_{m-r+1}, x_{m-r+2}, \cdots, x_m) + 1 \\ = & h(x_1, x_2, \cdots, x_m) + 1. \end{aligned}$$

Hence $h \notin S_1^m$.

(ii) Suppose $f \in S_1^m$. Then

$$\begin{aligned} W(f(x_1, x_2, \cdots, x_m) + x_i) &= W(f(x_1+1, x_2+1, \cdots, x_m+1) + x_i + 1) \\ &= W(f(x_1, x_2, \cdots, x_m) + x_i + 1) \\ &= 2^m - W(f(x_1, x_2, \cdots, x_m) + x_i). \end{aligned}$$

Hence $W(f(x_1, x_2, \cdots, x_m) + x_i) = 2^{m-1}$, and so f satisfies condition C4 from Lemma 7.1.2. Next suppose $g \in S_2^m$. Then

$$W(g(x_1, x_2, \cdots, x_m) + x_i) = W\left(a + \sum_{j=i} b_j x_j + (1+b_i)x_i\right) = 2^{m-1}.$$

Since the weight of any nonconstant linear or affine $(m, 1)$-LUT equals 2^{m-1}, the desired result again follows from Lemma 7.1.2.

Now suppose $h \in S_3^{m,r}$. Then

$$h(x_1, x_2, \cdots, x_m) = f(x_1, x_2, \cdots, x_{m-r}) + g(x_{m-r+1}, x_{m-r+2}, \cdots, x_m),$$

where both f and g satisfy condition C4. Then h satisfies condition C4 from Lemma 7.1.3.

(iii) This follows directly from Theorem 3.4.2 of [2].

(iv) Since there are two choices for a, it is clear that

$$| S_2^m | = 2 \sum_{i=1}^{\lfloor (m-1)/2 \rfloor} \binom{m}{2i+1}.$$

However, it is also well known that

$$\sum_{i=1}^{\lfloor (m-1)/2 \rfloor} \binom{m}{2i+1} = 2^{m-1}$$

and (iv) follows.

(v) There are 2^{m-r} linear $(m-r,1)$-LUTs which are also members of S_1^{m-r}. Hence there are $2^{2^{m-r-1}} - 2^{m-r}$ nonlinear $(m-r,1)$-LUTs f which satisfy

$$f(x_1, x_2, \cdots, x_{m-r}) = f(x_1+1, x_2+1, \cdots, x_{m-r}+1),$$

as in the proof of (iv) above, there are $2^{r-1} - r$ $(r,1)$-LUTs g such that

$$g(x_1, x_2, \cdots, x_r) = \sum_{i=1}^{r} b_i x_i,$$

where $\sum_{i=1}^{r} b_i x_i$ is odd and greater than 1. The desired result follows. □

The following theorem is an immediate corollary of the previous lemma (since choosing $r=3$, clearly, maximizes the size of $S_3^{m,r}$ given m is not too small).

Theorem 7.1.1[2] *Suppose $m > 3$. Then*

$$A_{mn}(4) \geqslant (2^{2^{m-1}} + 2^m - 2m + 2^{2^{m-4}} - 2^{m-3})^n.$$

This lower bound improves the previously known lower bound $A_{mn}(4) \geqslant (2^{2^{n-1}})^n$.

We now establish an upper bound which is also useful in establishing certain other lower bounds. First we need the following lemma.

Lemma 7.1.5 *Suppose f is an $(m,1)$-LUT, $m \geqslant 1$. Then $W(f(x_1, x_2, \cdots, x_m) + x_i) = 2^{m-1}$ if and only if*

$$\begin{aligned} W(f) &= 2W(f(x_1, x_2, \cdots, x_{i-1}, 0, x_{i+1}, \cdots, x_m)) \\ &= 2W(f(x_1, x_2, \cdots, x_{i-1}, 1, x_{i+1}, \cdots, x_m)). \end{aligned}$$

Proof First observe that

$$\begin{aligned} & W(f(x_1, x_2, \cdots, x_m) + x_i) \\ =& W(f(x_1, x_2, \cdots, x_{i-1}, 0, x_{i+1}, \cdots, x_m)) \\ & + W(f(x_1, x_2, \cdots, x_{i-1}, 1, x_{i+1}, \cdots, x_m) + 1) \\ =& W(f(x_1, x_2, \cdots, x_{i-1}, 0, x_{i+1}, \cdots, x_m)) \\ & + 2^{m-1} - W(f(x_1, x_2, \cdots, x_{i-1}, 1, x_{i+1}, \cdots, x_m)). \end{aligned}$$

Hence

$$2^{m-1} = W(f(x_1, x_2, \cdots, x_m) + x_i),$$

if and only if

$$\begin{aligned} & W(f(x_1, x_2, \cdots, x_{i-1}, 0, x_{i+1}, \cdots, x_m)) \\ =& W(f(x_1, x_2, \cdots, x_{i-1}, 1, x_{i+1}, \cdots, x_m)) \end{aligned}$$

and the result follows. □

Theorem 7.1.2 *Suppose $m > 1$. Then*

$$A_{mn}(4) \leqslant \sum_{k=0}^{2^{m-1}} \sum_{r=0}^{k} \binom{2^{m-2}}{r}^2 \binom{2^{m-2}}{k-r}^2.$$

Proof Let T be the set of $(m,1)$-LUTs satisfying both $W(f(x_1, x_2, \cdots, x_m)+x_1) = 2^{m-1}$ and $W(f(x_1, x_2, \cdots, x_m)+x_2) = 2^{m-1}$, by Lemma 7.1.5, T is equal to the set of $(m,1)$-LUTs satisfying

$$\begin{aligned} W(f(0, x_2, \cdots, x_m)) &= W(f(1, x_2, \cdots, x_m)) = W(f(x_1, 0, \cdots, x_m)) \\ &= W(f(x_1, 1, \cdots, x_m)). \end{aligned}$$

There are

$$\sum_{r=0}^{k} \binom{2^{m-2}}{r}^2 \binom{2^{m-2}}{k-r}^2,$$

$(m,1)$-LUTs with Hamming weight $2k$ satisfying the above equation and hence

$$|T| = \sum_{k=0}^{2^{m-1}} \sum_{r=0}^{k} \binom{2^{m-2}}{r}^2 \binom{2^{m-2}}{k-r}^2.$$

Finally observe that, from Lemma 7.1.2, every $(m,1)$-LUT satisfying C4 is a member of T, and the result follows. □

The following theorem is an immediate corollary of Lemma 7.1.2.

Theorem 7.1.3 *Suppose f_1, f_2, f_3, f_4 are four $(m,1)$-LUTs satisfying C4 where $W(f_1) = W(f_3)$ and $W(f_2) = W(f_4)$. Then the following $(m+2,1)$-LUT g also satisfies condition C4:*

$$\begin{aligned} & g(y_1, y_2, x_1, x_2, \cdots, x_m) \\ &= (1+y_1)(1+y_2)f_1(x_1, x_2, \cdots, x_m) + (1+y_1)y_2 f_2(x_1, x_2, \cdots, x_m) \\ &\quad + y_1(1+y_2)f_3(x_1, x_2, \cdots, x_m) + y_1 y_2 f_4(x_1, x_2, \cdots, x_m). \end{aligned}$$

Clearly Theorem 7.1.3 gives a recursive construction for LUTs satisfying C4.

Let $A_{mn}(i_1, \cdots, i_s)$ be the number of (m,n)-LUTs satisfying the conditions $Ci_1, \cdots, Ci_s$. In the following, we consider the values of $A_{mn}(i_1, \cdots, i_s)$ for every $\{i_1, \cdots, i_s\}$, $s \geqslant 2$. The exact values of $A_{mn}(2,5)$, $A_{mn}(3,5)$, $A_{mn}(1,2)$, $A_{mn}(1,2,3)$, and $A_{mn}(2,3,5)$, and lower bounds for $A_{mn}(1,4)$, $A_{mn}(1,5)$, $A_{mn}(1,3,5)$, and $A_{mn}(1,2,3,5)$ are known in [1]. Now we will consider the values of $A_{mn}(i_1, \cdots, i_s)$ for all other subsets $\{i_1, \cdots, i_s\}$ of $\{1,2,3,4,5\}$.

Theorem 7.1.4 *If $m > 1$, then*

$$A_{m1}(1,3) = \binom{2^m}{2^{m-1}} + \sum_{r=1}^{m} (-1)^r \binom{m}{r} \binom{2^{m-r}}{2^{m-r-1}}.$$

Proof Let E_i be the following set of $(m,1)$-LUTs:

$$E_i = \{f(x_1, x_2, \cdots, x_m) : f \text{ is independent of } x_i \text{ and } W(f) = 2^{m-1}\}.$$

There are $\binom{2^m}{2^{m-1}}$ $(m,1)$-LUTs satisfying $W(f) = 2^{m-1}$. Therefore we have

$$A_{m1}(1,3) = \binom{2^m}{2^{m-1}} - |E_1 \cup E_2 \cup \cdots \cup E_m|. \tag{7.1}$$

For any $1 \leqslant i_1 < i_2 < \cdots < i_r \leqslant m (1 \leqslant r \leqslant m)$, the following equations holds:

$$\begin{aligned}|E_{i_1} \cap E_{i_2} \cap \cdots \cap E_{i_r}| &= |\{g(y_1, y_2, \cdots, y_{m-r}) : W(g) = 2^{m-r-1}\}| \\ &= \binom{2^{m-r}}{2^{m-r-1}}.\end{aligned}$$

Then, by the well known Polyas enumeration formula, we have

$$\begin{aligned}|E_1 \cup E_2 \cup \cdots \cup E_m| &= \sum_{r=1}^{m} (-1)^{r-1} \sum_{1 \leqslant i_1 < i_2 < \cdots < i_r \leqslant m} |E_{i_1} \cap E_{i_2} \cap \cdots \cap E_{i_r}| \\ &= \sum_{r=1}^{m} (-1)^{r-1} \binom{m}{r} \binom{2^{m-r}}{2^{m-r-1}}.\end{aligned}$$

The proof is finished by substituting this equation into Equation (7.1). □

Theorem 7.1.5 *If $m, n > 1$, then*

$$A_{mn}(2,3) = (A_{m1}(3) - 2)^n.$$

Proof There are only two linear or affine nondegenerate $(m,1)$-LUTs, namely, $x_1 + x_2 + \cdots + x_m$ and $x_1 + x_2 + \cdots + x_m + 1$, i.e.,

$$A_{m1}(2,3) = (A_{m1}(3) - 2).$$

The result follows immediately from Lemma 7.1.1. □

A recurrence relation for $A_{mn}(3)$ is given in [2].

Theorem 7.1.6 *If $m, n > 1$, then*

$$A_{mn}(2,4) = (A_{m1}(4) - 2(2^m - m))^n.$$

Proof The linear/affine $(m,1)$-LUT $a + \sum_{i=1}^{m} b_i x_i$ satisfies C4 if and only if $\sum_{i=1}^{m} b_i \neq 1$. The number of such $(m,1)$-LUTs is equal to

$$2\left(\binom{m}{0} + \binom{m}{2} + \binom{m}{3} + \cdots + \binom{m}{m}\right) = 2(2^m - m).$$

Hence $A_{m1}(2,4) = A_{m1}(4) - 2(2^m - m)$, and the result follows from Lemma 7.1.1. □

Theorem 7.1.7 *If $m, n > 1$, then*

$$A_{mn}(3,4) = \left(\sum_{r=0}^{m} (-1)^r \binom{m}{r} A_{m-r,1}(4)\right)^n.$$

Proof Let F_i be the following set of $(m,1)$-LUTs:

$$F_i = \{f(x_1, x_2, \cdots, x_m) : f \text{ is independent of } x_i \text{ and } f \text{ satisfies C4}\}.$$

Then

$$A_{m1}(3,4) = A_{m1}(4) - \mid F_1 \cup F_2 \cup \cdots \cup F_m \mid, \tag{7.2}$$

for any $1 \leqslant i_1 < i_2 < \cdots < i_r \leqslant m$ $(1 \leqslant r \leqslant m)$, the following equation holds:

$$\begin{aligned} \mid F_{i_1} \cap F_{i_2} \cap \cdots F_{i_r} \mid &= \mid \{g(y_1, y_2, \cdots, y_{m-r}) : g \text{ satisfies C4}\} \mid \\ &= A_{m-r,1}(4). \end{aligned}$$

Then, by Polyas enumeration formula, we have

$$\begin{aligned} \mid F_1 \cup F_2 \cup \cdots \cup F_m \mid &= \sum_{r=1}^{m} (-1)^{r-1} \sum_{1 \leqslant i_1 < i_2 \cdots < i_r \leqslant m} \mid F_{i_1} \cap F_{i_2} \cap \cdots \cap F_{i_r} \mid \\ &= \sum_{r=1}^{m} (-1)^{r-1} \binom{m}{r} A_{m-r,1}(4). \end{aligned}$$

The proof is finished by substituting this equation into (7.2) and using Lemma 7.1.1. □

Theorem 7.1.8 *If $m, n > 1$, then*

$$A_{mn}(4,5) = [2^{\lfloor m/2 \rfloor + 1}]^n,$$

where $\lfloor x \rfloor$ denotes the integer part of x.

Proof First observe that an $(m,1)$-LUT f satisfies condition C5 if and only if f is uniquely determined by the values of $a_0 = f(0,0,\cdots,0)$, $a_1 = f(1,0,\cdots,0)$, $a_2 = f(1,1,\cdots,0)$, $\cdots$, $a_m = f(1,1,\cdots,1)$, where $f(x_1,x_2,\cdots,x_m) = a_i$ if and only if the vector $(x_1,x_2,\cdots,x_m)$ has Hamming weight i.

Now consider those functions f satisfying C5 and where $a_i = a_{m-i}$ for every $i (0 \leqslant i \leqslant \lfloor m/2 \rfloor)$. There are $2^{\lfloor m/2 \rfloor + 1}$ such functions. Moreover, if f is a function of this type, then $f(x_1,x_2,\cdots,x_m) = f(x_1+1, x_2+1, \cdots, x_m+1)$ and hence f satisfies C4 from Theorem 3.4.2 of [2]. The result follows immediately from Lemma 7.1.1. □

Theorem 7.1.9 *If $m, n > 1$, then*

(i) $A_{m1}(1,2,4) = A_{m1}(1,4) - 2(2^m - m - 1)$.
(ii) $A_{m1}(1,2,5) = A_{m1}(1,5) - 2$.
(iii) $A_{m1}(1,3,4) = \sum_{r=0}^{m} (-1)^r \binom{m}{r} A_{m-r,1}(4)$.
(iv) $A_{1,4,5} \geqslant 2$.
(v) $A_{mn}(2,3,4) = (A_{m1}(3,4) - 2)^n$.
(vi) $A_{mn}(2,4,5) = (A_{m1}(4,5) - 2)^n$.
(vii) $A_{mn}(3,4,5) = (\sum_{r=0}^{m} (-1)^r \binom{m}{r} A_{m-r,1}(4,5))^n$.
(viii) $A_{m1}(1,2,3,4) = A_{m1}(1,3,4) - 2$.
(ix) $A_{m1}(1,2,4,5) = A_{m1}(1,4,5) - 2$.
(x) $A_{m1}(1,3,4,5) = \sum_{r=0}^{m} (-1)^r \binom{m}{r} A_{m-r,1}(1,4,5)$.

(xi) $A_{mn}(2,3,4,5) = (A_{m1}(3,4,5) - 2)^n$.

(xii) $A_{m1}(1,2,3,4,5) = A_{m1}(1,3,4,5) - 2$.

Proof All these results use proofs analogous to those given above. The proofs of (ii), (v), (vi), (viii), (ix), and (xii) follow Theorem 7.1.5; (i) follows Theorem 7.1.6; (iii), (vii), and (x) follow Theorem 7.1.7; and (iv) follows by noting that the two constant functions satisfy C1, C4, and C5. □

Among the conditions C1–C5, the condition C4 is more useful than the others. Among the (m, n)-LUT the special case of $(m, 1)$-LUT is most important. So in the following text of this section, we concentrate on improving the enumeration lower bound of $A_{m,1}(4)$, which will be abbreviated as $N(m)$.

Definition 7.1.1 *Let $f(x_1, x_2, \cdots, x_n) =: f(x)$ be a Boolean function. For $b = 0$ or 1, let $E_b =: \{s \in \mathrm{GF}(2)^n : f(x+s)+f(x) = b,\ for\ each\ x \in \mathrm{GF}(2)^n\}$. The union set $E =: E_0 \cup E_1$ is said to be the linear kernel of $f(\cdot)$.*

It is easy to see that both E_0 and E are subspaces of $\mathrm{GF}(2)^n$, moreover E_0 is also a subspace of E. While the set E_1 is perhaps empty. If $E_1 \neq \varnothing$, then E_1 is the coset of the subspace E_0 in E.

Definition 7.1.2 *Let $f(x_1, x_2, \cdots, x_n)$ be a Boolean function with its linear kernel $E =: E_0 \cup E_1$. Then $f(x)$ is called a function with linear, denoted by $f(x) \in \mathrm{LS}(n)$, if E is a subspace with positive dimension; otherwise $f(x)$ is called a function with non-linear, denoted by $f(x) \in \mathrm{NLS}(n)$. If $f(x) \in \mathrm{LS}(n)$ and $E_1 = \varnothing$, then $f(x)$ is called a function with the first type linear, denoted by $f(x) \in L_1(n)$. If $f(x) \in \mathrm{LS}(n)$ and $E_1 \neq \varnothing$, then $f(x)$ is called a function with the second type linear, denoted by $f(x) \in L_2(n)$.*

Definition 7.1.3 *A Boolean function $f(x_1, x_2, \cdots, x_n)$ is said to be of correlational immune of m-th order, if $f(\cdot)$ is independent of each set of m variables $x_{i_1}, x_{i_2}, \cdots, x_{i_m}$, $1 \leqslant m \leqslant n$. Particularly, if $m = 1$, then $f(x)$ is said to be of correlational immune of 1-st order.*

Lemma 7.1.6 *$f(x)$ is of correlational immune of 1-st order, if and only if the condition C4 is satisfied for $(n, 1)$-LUT.*

Theorem 7.1.10 *The following statements are true:*

(1) Let $f(x_1, x_2, \cdots, x_n) \in \mathrm{LS}(n)$ be a Boolean function with its linear kernel $E =: E_0 \cup E_1$. If $(1, 1, \cdots, 1) \in E_0$, then $f(x)$ is of correlational immune of 1-st order.

(2) Let $f(x_1, x_2, \cdots, x_n) \in L_2(n)$ be a Boolean function with its linear kernel $E =: E_0 \cup E_1$. If there exists a pair $c_1 = (c_{1,1}, c_{1,2}, \cdots, c_{1,n}), c_2 = (c_{2,1}, c_{2,2}, \cdots, c_{2,n}) \in E_1$ such that $c_{1,i} = 0$, if $c_{2,i} = 1$ and $c_{2,i} = 0$, if $c_{1,i} = 1$, $1 \leqslant i \leqslant n$, then $f(x)$ is of correlational immune of 1-st order.

Proof The first statement was proved by [2]. It is sufficient to prove the second statement.

If $f(x) \in L_2(n)$, then $E_1 \neq \varnothing$, thus $W_H(f(x)) = 2^{n-1}$. Let $1 \leqslant i \leqslant n$ be a fixed integer. Without loss of generality, we assume that $c_{1,i} = 0$. Thus the function $g_i(x_1, x_2, \cdots, x_n) =:$

$f(x_1, x_2, \cdots, x_n) + x_i$ satisfies $g_i(x) + g_i(x + c_1) = f(x) + f(x + c_1) = 1$, i.e., $g_i(x) \in L_2(n)$. Hence $W_H(g_i(x)) = 2^{n-1}$, or equivalently, $W_H(f(x_1, x_2, \cdots, x_n) + x_i) = 2^{n-1}$. The proof is finished by Lemma 7.1.6. □

This theorem provides us with a large family of Boolean functions of correlational immune of 1-st order. Now we concentrate on the estimation of the enumeration of such functions in Theorem 7.1.10.

Let $\sigma : z_2^n \longrightarrow z_2^n$ be a nonsingular linear transformation, $f(x_1, x_2, \cdots, x_n)$ a Boolean function. Define σ^* by

$$\sigma^*(f(x_1, x_2, \cdots, x_n) = f(\sigma(x_1, x_2, \cdots, x_n)).$$

Lemma 7.1.7 *If the linear kernel, $E =: E_0 \cup E_1$, of $f(x_1, x_2, \cdots, x_n) \in L_2(n)$ is a subspace of dimension k, then there exists a nonsingular linear transformation $\sigma : z_2^n \longrightarrow z_2^n$ and a $g(x_1, x_2, \cdots, x_{n-k}) \in \mathrm{NLS}(n-k)$ such that*

$$\sigma^*(f(x_1, x_2, \cdots, x_n) = g(x_1, x_2, \cdots, x_{n-k}) + x_{n-k+1}.$$

Proof Because of that $f(\cdot) \in L_2(n)$ and that $E =: E_0 \cup E_1$ is of dimension k, then we know that E_0 is a subspace of dimension $k-1$. Let E' be the subspace satisfying $E' \oplus E = z_2^n$, where " $\oplus$ " stands for the direct sum of two subspaces. Let $\{\alpha_1, \alpha_2, \cdots, \alpha_{n-k}\}$ and $\{\alpha_{n-k+2}, \alpha_{n-k+3}, \cdots, \alpha_n\}$ be the bases of the subspaces E' and E_0, respectively. Thus for each $\alpha_{n-k+1} \in E_1$, the set $\{\alpha_{n-k+1}, \alpha_{n-k+2}, \cdots, \alpha_n\}$ forms a base of E. Define a $\sigma : z_2^n \longrightarrow z_2^n$ by $\sigma(x_1, x_2, \cdots, x_n) =: \sum_{i=1}^n x_i\alpha_i$, which is clearly a nonsingular linear transformation. Thus $\sigma^*(f(x_1, x_2, \cdots, x_n) = f(\sigma(x_1, x_2, \cdots, x_n)) = f(\sum_{i=1}^n x_i\alpha_i) = f(\sum_{i=1}^{n-k+1} x_i\alpha_i) = x_{n-k+1} + f(\sum_{i=1}^{n-k} x_i\alpha_i) =: g(x_1, x_2, \cdots, x_{n-k}) + x_{n-k+1}$.

Now it is sufficient to prove that $g(x_1, x_2, \cdots, x_{n-k}) \in \mathrm{NLS}(n-k)$. In fact, if otherwise there should exist a $s = (s_1, s_2, \cdots, s_{n-k}) \neq 0$ such that

$$g(x_1 + s_1, x_2 + s_2, \cdots, x_{n-k} + s_{n-k}) + g(x_1, x_2, \cdots, x_{n-k}) = b$$

for every $(x_1, x_2, \cdots, x_{n-k}) \in z_2^{n-k}$ and $b \in \{0, 1\}$. Thus $f(\sum_{i=1}^{n-k}(x_i+s_i)\alpha_i)+f(\sum_{i=1}^{n-k} x_i\alpha_i) = b$. Let $\alpha =: \sum_{i=1}^{n-k} s_i\alpha_i$, for every $x = \sum_{i=1}^n x_i\alpha_i \in z_2^n$, we have $f(x + \alpha) + f(x) = f(\sum_{i=1}^{n-k}(x_i + s_i)\alpha_i + \sum_{i=n-k+1}^n x_i\alpha_i) + f(\sum_{i=1}^n x_i\alpha_i) = b$, i.e., $\alpha \in E$, which is impossible, because $0 \neq \alpha = \sum_{i=1}^{n-k} s_i\alpha_i \in E'$. Thus $g(x) \in \mathrm{NLS}(n-k)$. □

Lemma 7.1.8 *Let $f(x_1, x_2, \cdots, x_n)$ be a Boolean function with linear kernel E, and let $\sigma : z_2^n \longrightarrow z_2^n$ be a nonsingular linear transformation, then the linear kernel of $\sigma^*(f)$ is equal to $\sigma^{-1}(E)$.*

Proof This lemma follows the definition of $\sigma^*(f)$ and its linear kernel. □

Lemma 7.1.9 (see page 225 of [4]) *The number of Boolean functions contained in* LS(n) *is equal to*

$$|\mathrm{LS}(n)| = \sum_{k=1}^{n} 2^{2^{n-k}+k(k+1)/2} \cdot \frac{\prod_{i=0}^{k-1}(2^n - 2^i)}{\prod_{i=0}^{k-1}(2^k - 2^i)} < (2^n - 1) \cdot 2^{2^{n-1}+1}.$$

Let $E =: E_0 \cup E_1$ be the linear kernel of $f(x_1, x_2, \cdots, x_n)$. Let I_n be the set of all correlation-immune Boolean functions, $f(x_1, x_2, \cdots, x_n)$, of order one. Define

$$S_1 = \{f(x_1, x_2, \cdots, x_n) : (1, 1, \cdots, 1) \in E_0\}$$

and

$$\begin{aligned} P_n = \{f : f \in L_2(n), \dim(E) = 2, (1, 1, \cdots, 1) \notin E, \text{ and there exist} \\ c_1, c_2 \in E_1 \text{ such that } c_{1,i} = 0, \ if\ c_{2,i} = 1, \text{ and } c_{2,i} = 0, \ if\ c_{1,i} = 1\}. \end{aligned}$$

It is easy to see that $E = \{0, c_1, c_2, c_1 + c_2\}$, $c_1 + c_2 \neq (1, 1, \cdots, 1)$, $E_1 = \{c_1, c_2\}$, and $P_n \subseteq I_n$.

A lower bound of $|I_n|$ will be found by the estimation of $|P_n|$.

From Lemma 7.1.7, we know that, for each $f \in P_n$, there exists a nonsingular linear transformation $\sigma : z_2^n \longrightarrow z_2^n$ and a $g(x_1, x_2, \cdots, x_{n-2}) \in \text{NLS}(n-2)$ such that $\sigma^*(f(x_1, x_2, \cdots, x_n)) = g(x_1, x_2, \cdots, x_{n-2}) + x_{n-1}$, which is called the standard form of the linear kernel space of dimension 2. Clearly, (1) there are exactly $|\text{NLS}(n-2)|$ such standard forms, and (2) the linear kernel of the standard form is $E = \{0, e_{n-1}, e_{n-1} + e_n, e_n\}$, $E_1 = \{e_{n-1}, e_{n-1} + e_n\}$, here and after e_i stands for the unit vector with its i-th compoment 1 and the others 0. From Lemma 7.1.8, we know that a standard form can generate different functions in $L_2(n)$, if different nonsingular linear transformations are used. The necessary and sufficient conditions of $\sigma(f) \in P_n$ are that $\sigma^{-1}(e_{n-1}) =: a$ and $\sigma^{-1}(e_{n-1} + e_n) =: b$ are two binary vectors satisfying $a_i = 0$, if $b_i = 1$, and $b_i = 0$, if $a_i = 1$. Hence there are $\sum_{j=1}^{n-2}(C_n^j \sum_{i=1}^{n-j-1} C_{n-j}^i)$ approaches to choose the $\sigma^{-1}(e_{n-1})$ and $\sigma^{-1}(e_{n-1} + e_n)$. The nonsingularity of σ implies that $\sigma^{-1}(e_1)$, $\sigma^{-1}(e_2)$, $\cdots$, $\sigma^{-1}(e_{n-1})$, and $\sigma^{-1}(e_{n-1} + e_n)$ is a linear independent base, while it is known that there are $(2^n - 2^2)(2^n - 2^3) \cdots (2^n - 2^{n-1})$ approaches to choose such base, thus there are

$$\sum_{j=1}^{n-2} \left(C_n^j \sum_{i=1}^{n-j-1} C_{n-j}^i \right) (2^n - 2^2)(2^n - 2^3) \cdots (2^n - 2^{n-1})$$

nonsingular linear transformations that can transform the standard forms into the functions in P_n.

Definition 7.1.4 *Let g_1 and g_2 be two standard forms. If there exists a nonsingular linear transformation σ such that $g_2 = \sigma^* g_1$, then we say that g_1 and g_2 are equivalent to each other.*

Let $B_i, i = 1, 2$, be the set of functions produced by the standard form g_i under different nonsingular linear transformations.

Lemma 7.1.10 *If g_1 and g_2 are two standard forms equivalent to each other, then $B_1 = B_2$.*

Proof For every $f \in B_1$, there exists a nonsingular linear transformation σ_1 such that $\sigma_1^* g_1 = f$. While we have known that $g_2 = \sigma^* g_1$, thus $\sigma_1^*(\sigma^*)^{-1} g_2 = f$, i.e., $f \in B_2$, so $B_1 \subseteq B_2$. By the same way, it can be proved that $B_2 \subseteq B_1$. Thus $B_1 = B_2$. □

Lemma 7.1.11 *If $B_1 \cap B_2 \neq \varnothing$, then g_1 and g_2 are equivalent to each other.*

Proof If $f \in B_1$ (resp. $f \in B_2$), then there exists a nonsingular linear transformation σ_1 (resp. σ_2) such that $\sigma_1^* g_1 = f$(resp. $\sigma_2^* g_2 = f$). Thus $\sigma_1^* g_1 = \sigma_2^* g_2$, i.e., $g_2 = ((\sigma_2^*)^{-1}\sigma_1^*)g_1$. Let $\sigma^* = (\sigma_2^*)^{-1}\sigma_1^*$, then $g_2 = \sigma^* g_1$. □

From Lemma 7.1.11, it is known that $B_1 \cap B_2 = \varnothing$ if g_1 is not equivalent to g_2.

Theorem 7.1.11 $|P_n| \geqslant 2^{n-2}(3^n - 3.2^n + 5) \cdot |\text{NLS}(n-2)|$.

Proof At first, we enumerate the nonsingular linear transformations from one standard form to the other standard form. Equivalently, we try to enumerate those σ satisfying

$$\sigma^*(g(x_1, x_2, \cdots, x_{n-2}) + x_{n-1}) = g'(x_1, x_2, \cdots, x_{n-2}) + x_{n-1}, \ g, g' \in \text{NLS}(n-2).$$

Two necessary conditions of this equation are (1) $\sigma(\sum_{i=1}^{n-2} x_i e_i) \subset L(e_1, e_2, \cdots, e_{n-2})$, the linear space spanned by $e_1, e_2, \cdots, e_{n-2}$; and (2) $\sigma(e_{n-1}) \in L(e_1, e_2, \cdots, e_{n-2}, e_n) - L(e_1, e_2, \cdots, e_{n-1})$. It is not difficult to see that the enumeration of σ satisfying the Condition 7.2 is $(2^{n-2}-2^0)(2^{n-2}-2^1)\cdots(2^{n-2}-2^{n-3})(2^{n-1}-2^{n-2})$, which is also the enumeration of standard forms equivalent to each other. Divide all $|\text{NLS}(n-2)|$ standard forms into the following

$$\frac{|\text{NLS}(n-2)|}{(2^{n-2}-2^0)(2^{n-2}-2^1)\cdots(2^{n-2}-2^{n-3})(2^{n-1}-2^{n-2})}$$

equivalent classes. From Lemma 7.1.11, we know that at least

$$\frac{|\text{NLS}(n-2)|(2^n-2^2)(2^n-2^3)\cdots(2^n-2^{n-1})}{(2^{n-2}-2^0)(2^{n-2}-2^1)\cdots(2^{n-2}-2^{n-3})(2^{n-1}-2^{n-2})} \sum_{j=1}^{n-2}\left(C_n^j \sum_{i=1}^{n-j-1} C_{n-j}^i\right)$$

different elements of P_n can be generated by the

$$\sum_{j=1}^{n-2}\left(C_n^j \sum_{i=1}^{n-j-1} C_{n-j}^i\right)(2^n-2^2)(2^n-2^3)\cdots(2^n-2^{n-1})$$

nonsingular linear transformations. Thus

$$\begin{aligned}|P_n| &\geqslant \frac{|\text{NLS}(n-2)|(2^n-2^2)(2^n-2^3)\cdots(2^n-2^{n-1})}{(2^{n-2}-2^0)(2^{n-2}-2^1)\cdots(2^{n-2}-2^{n-3})(2^{n-1}-2^{n-2})} \sum_{j=1}^{n-2}\left(C_n^j \sum_{i=1}^{n-j-1} C_{n-j}^i\right) \\ &= 2^{n-2}(3^n - 3 \cdot 2^n + 5) \cdot |\text{NLS}(n-2)|\end{aligned}$$

because $\sum_{j=1}^{n-2}(C_n^j \sum_{i=1}^{n-j-1} C_{n-j}^i) = 3^n - 3.2^n + 5$. □

Lemma 7.1.12 *The Hamming weight of each Boolean function of the form $f(x_1, x_2, \cdots, x_n) = g(x_1, x_2, \cdots, x_{i-1}, x_{i+1}, \cdots, x_n) + x_i$ is 2^{n-1}.*

Lemma 7.1.13 *Every Boolean function of the form $f(x) = a + \sum_{i=1}^n x_i + g(x_{i_1}, \cdots, x_{i_{n-3}})$ is of correlation-immune of order one.*

Proof For each integer $1 \leqslant i \leqslant n$, the function $x_i + f(x)$ satisfies Lemma 7.1.12, thus $W_H(x_i + f(x)) = 2^{n-1}$, i.e., $f(x)$ is of correlation-immune of order one. □

Lemma 7.1.14 *Let E be the linear kernel of $f(x_1, x_2, \cdots, x_n)$, then $\dim(E) \geqslant n-1$, if and only if $f(x)$ is projective.*

Proof The " if " part is easy. We will prove only the "only if " part.

Because of $\dim(E) \geqslant n-1$, there exist linear independent vectors $c_1, c_2, \cdots, c_{n-1} \in E$ such that they can be expended to be a base, $c_1, c_2, \cdots, c_{n-1}$ and c_n of $\mathrm{GF}(2)^n$. For each $1 \leqslant i \leqslant n-1$, define $b_i =: f(x) + f(x + c_i)$, $x \in \mathrm{GF}(2)^n$. For each pair $x, y \in \mathrm{GF}(2)^n$, there exist $d_1, d_2, \cdots, d_n, a_1, a_2, \cdots, a_n$ such that $y = \sum_{i=1}^n a_i c_i$ and $x = \sum_{i=1}^n d_i c_i$, thus

$$
\begin{aligned}
f(x+y) &= f\left(\sum_{i=1}^{n}(d_i + a_i)c_i\right) \\
&= f((e_n + a_n)c_n) + \sum_{i=1}^{n-1}(d_i + a_i)b_i \\
&= f(0) + f(d_n c_n) + f(a_n c_n) + \sum_{i=1}^{n-1}(d_i + a_i)b_i \\
&= f(0) + f(x) + f(y),
\end{aligned}
$$

i.e., $f(x)$ is a projective function. □

In order to estimate the enumeration of correlation-immune Boolean function of order one, we introduce two more sets of functions as follows:

$$S_2 =: \left\{ f : f(x) = a + \sum_{i=1}^{n} b_i x_i, \text{ with} \sum_{i=1}^{n} b_i \text{ an odd larger than } 1 \right\}$$

and

$$S_3 =: \left\{ f : f(x) = a + \sum_{i=1}^{n} x_i + g(x_{i_1}, \cdots, x_{i_{n-3}}) \right\},$$

where $i_1, \cdots, i_{n-3}$ are different integers in $[1, n]$.

Theorem 7.1.12 *If $n > 3$, then $|I_n| \geqslant |P_n| + |S_1| + |S_2| + |S_3 - (S_1 \cup S_2)|$.*

Proof From the previous lemmas, we know that the functions in P_n, S_1, S_2, S_3 are of correlation-immune of order one, thus $I_n \supseteq P_n \cup S_1 \cup S_2 \cup S_3$. From the Polya's enumeration, we have

$$
\begin{aligned}
|I_n| = &|P_n| + |S_1| + |S_2| + |S_3| - |P_n \cap S_1| - |P_n \cap S_2| - |P_n \cap S_3| \\
&- |S_1 \cap S_2| - |S_1 \cap S_3| - |S_2 \cap S_3| + |S_1 \cap P_n \cap S_2| \\
&+ |S_1 \cap P_n \cap S_3| + |S_1 \cap S_2 \cap S_3| + |S_2 \cap P_n \cap S_3| \\
&- |S_1 \cap P_n \cap S_2 \cap S_3|.
\end{aligned}
\tag{7.3}
$$

While it is known from Eq. (7.2) that $S_1 \cap S_2 = \varnothing$, thus

$$|S_1 \cap S_2| = |S_1 \cap P_n \cap S_2| = |S_1 \cap S_2 \cap S_3| = |S_1 \cap P_n \cap S_2 \cap S_3| = 0. \tag{7.4}$$

Moreover, it is easy to see that $S_1 \cap P_n = \varnothing$, thus

$$|S_1 \cap P_n| = |S_1 \cap P_n \cap S_3| = 0. \tag{7.5}$$

All functions in S_2 are projective, thus from Lemma 7.1.14, we know that the linear kernel of each $f(x)$ in S_2 has its dimension larger than or equal to $n-1>2$, while the dimension of the linear kernel space of each function in P_2 is exactly 2. Thus $S_2 \cap P_n = \varnothing$, i.e.,

$$|S_2 \cap P_n| = |S_2 \cap P_n \cap S_3| = |S_1 \cap P_n \cap S_2| = 0. \tag{7.6}$$

From the fact that the dimension of the linear kernel space of each function in S_3 is larger than 3, thus $S_3 \cap P_n = \varnothing$, i.e.,

$$|S_3 \cap P_n| = 0. \tag{7.7}$$

Setting Eqs. (7.4)–(7.6) and (7.7) into Eq. (7.3), we have

$$\begin{aligned}|I_n| &\geqslant |P_n| + |S_1| + |S_2| + |S_3| - |S_1 \cap S_3| - |S_2 \cap S_3| \\ &= |P_n| + |S_1| + |S_2| + |S_3 - (S_1 \cup S_2)|, \quad n>3.\end{aligned}$$

□

Theorem 7.1.13 *The number, $|I_n|$, of correlation-immune functions of order one satisfies:*

$$\begin{aligned}|I_n| \geqslant\, & 2^{2^{n-1}} + 2^{n-3}(3^n - 3\cdot 2^n + 5)\cdot 2^{2^{n-2}} + 2C_n^3\cdot(2^{2^{n-3}} - 2^{2^{n-4}} - 2^{n-3}) \\ & +2^n - 2n + (3^n - 3\cdot 2^n + 5)\cdot(2^{2^{n-3}+n-3} - 2^{2n-3} + 2^{n-1})\cdot 2^{2^{n-3}}.\end{aligned}$$

Proof Because of that $|\text{NLS}(n-2)| = 2^{2^{n-2}} - |\text{LS}(n-2)|$, thus from Lemma 7.1.9, we have $|\text{LS}(n-2)| < (2^{2^{n-2}} - 1)\cdot 2^{2^{n-2}+1}$, thus

$$|\text{NLS}(n-2)| > 2^{2^{n-2}} - (2^{2^{n-2}} - 1)\cdot 2^{2^{n-3}+1}. \tag{7.8}$$

Moreover, from Theorem 7.1.11, we know that

$$\begin{aligned}|P_n| &\geqslant 2^{n-2}(3^n - 3\cdot 2^n + 5)\cdot|\text{NLS}(n-2)| \\ &> 2^{n-2}(3^n - 3\cdot 2^n + 5)\cdot(2^{2^{n-2}} - (2^{n-2} - 1)\cdot 2^{2^{n-3}+1}) \\ &= (3^n - 3\cdot 2^n + 5)\cdot 2^{2^{n-2}+n-2} - (3^n - 3\cdot 2^n + 5)\cdot(2^{n-2} - 1) \\ &\quad \cdot 2^{2^{n-3}+n-1}\end{aligned}$$

It has been proved: (1) $|S_1| = 2^{2^{n-1}}$ (see Ref. [2]); (2) $|S_2| = 2(2^{n-1} - n)$ and (3) $|S_3 - (S_1 \cup S_2)| \geqslant 2C_n^3.(2^{2^{n-2}} - 2^{2^{n-4}} - 2^{n-3})$. Setting these results into Theorem 7.1.12, the proof is finished. □

7.1.2 Constructing Boolean Functions of Cryptographic Significance[5,6]

A Boolean function is said to be of good cryptographic significance if is: nonlinear, non-degenerated, symmetric, balanced, correlation-immune, and/or propagatable, etc. When the enumeration of those Boolean functions simultaneously satisfying some of the previous properties is insufficient, then those Boolean function can't be applied to the cryptosystems. Thus it is an important topic to enumerate and construct the Boolean functions satisfying the wanted properties.

A linear block error-correcting code (n, k, d) is, in fact, a k-dimensional subspace of $\text{GF}(2)^n$ such that the Hamming distance between each pair of its codewords is at least d, or equivalently, the Hamming weight of each of its nonzero codewords is at least d.

Up to now, only unbalanced Boolean functions satisfying both correlational immunity and propagation criterion of degrdd 1 were found (see [7]); this section will initially find the balanced ones. Our construction is based on the theory of error-correcting codes.

Let $f(x), x \in \mathrm{GF}(2)^n$, be a Boolean function of n-variables. Then

(1) $f(x)$ is said to satisfy the propagation criterion at α, $\alpha \in \mathrm{GF}(2)^n$, iff the function $f(x) \oplus f(x \oplus \alpha)$ is balanced.

(2) $f(x)$ is said to satisfy the propagation criterion of degree k, denoted by $f(x) \in \mathrm{PC}(k)$, iff $f(x)$ satisfies the propagation criterion at each point α, $1 \leqslant w(\alpha) \leqslant k$, where $w(\cdot)$ stands for the Hamming weight function.

(3) $f(x)$ is said to be Bent iff the function $f(x) \oplus f(x \oplus \alpha)$ is balanced at each $\alpha \in \mathrm{GF}(2)^n$.

The definition of Bent function is invariant under any linear isomorphism, and we may define the Bent functions on any even-dimensional subspace E of $\mathrm{GF}(2)^n$ as the functions satisfying

$$\sum_{x \in E} (-1)^{f(x) \oplus f(x \oplus \alpha)} = 0, \quad \text{for each } \alpha \in E - \{0\}.$$

Definition 7.1.5 *A Boolean function $f(x), x \in \mathrm{GF}(2)^n$, is called a partially-Bent iff there exist two subspaces $E, F \subset \mathrm{GF}(2)^n$ and a $t \in \mathrm{GF}(2)^n$ such that:*

(1) *$\mathrm{GF}(2)^n$ is the direct sum of E and F;*

(2) *For each $u \in E$ and $v \in F$,*

$$f(x) = f^*(v) + t \cdot u, \quad x = u + v, \tag{7.9}$$

where $f^*(\cdot)$ is the restriction of $f(x)$ to F; and

(3) *The function $f^*(v)$ is Bent on F.*

It is easy to prove that

Lemma 7.1.15 *If $t \notin E^{\perp}$, then the function defined by Eq. (7.9) is balanced.*

Lemma 7.1.16[7] *A partially-Bent function $f(x)$ satisfies $\mathrm{PC}(k)$ iff the Hamming weight of each nonzero $\alpha \in E$ is greater than k, where the subspace E is defined by Definition* 7.1.5.

Definition 7.1.6[9] *A Boolean function $f(x), x \in \mathrm{GF}(2)^n$, is said to satisfy the correlation-immune of order m, or $f(x) \in \mathrm{CI}(m)$, iff $f(x)$ is statistically independent of each set of m random inputs $x_{i_1}, x_{i_2}, \cdots, x_{i_m}$.*

Lemma 7.1.17[7] *A balanced partially-Bent function $f(x)$ satisfies $\mathrm{CI}(m)$ iff the Hamming weight of each nonzero $\alpha \in t \oplus E^{\perp}$ is greater than m, where the subspace E and the vector t are defined by Definition* 7.1.5.

Theorem 7.1.14 *A balanced Boolean function $f(x)$ satisfying $\mathrm{PC}(d-1)$ can be constructed by a linear block error-correcting code (n, k, d), if $n-k$ is even.*

Proof In order to make use of Definition 7.1.5, we choose E as the linear code (n, k, d) and F as such a subspace that $\mathrm{GF}(2)^n$ is the direct summation of E and F. Thus the dimension $\dim(F) = n-k$, an even integer. Let $f^*(v), v \in F$, be a Bent function on F (the existence of $f^*(\cdot)$ is confirmed by the existence of Bent functions of even-variables). Choose a nonzero $t \in \mathrm{GF}(2)^n - E^{\perp}$, and define

$$f(x) =: f^*(v) \oplus t \cdot u, \text{ for each } x = u \oplus v, \ u \in E, \ v \in F \tag{7.10}$$

From Lemma 7.1.15 and Lemma 7.1.16, it has been proved that the function defined by Eq. (7.10) is a balanced function satisfying PC$(d-1)$. □

Theorem 7.1.15 *Let C be an (n, k, d) linear block error-correcting code with $n-k$ being even. If there exist an integer m such that*

$$2^k > \binom{n}{0} + \binom{n}{1} + \cdots + \binom{n}{m},$$

then there exists a balanced function $f(x)$ satisfying both PC$(d-1)$ *and* CI(m).

Proof Let the subspace E and F be those defined in the proof of Theorem 7.1.14. Because of $\dim E^{\perp} = \dim F = n-k$, we know that the subspace $E^{\perp}$ produces the cosets.

The enumeration of n-dimensional vectors of Hamming weight up to m is equivalent to $\binom{n}{0} + \binom{n}{1} + \cdots + \binom{n}{m}$, thus the inequality Eq. (7.11) implies the existence of at least one coset, say $t' \oplus E^{\perp}$, such that each of the vector has a Hamming weight greater than m. Hence, by the same approach as that of the proof of Theorem 7.1.14, it can be proved that the function

$$f(x) =: f^*(v) \oplus t' \cdot u, \text{ for each } x = u \oplus v, \ u \in E, \ v \in F$$

is balanced (from Lemma 7.1.15), satisfying PC$(d-1)$ (from Lemma 7.1.16), and satisfying CI(m) (from Lemma 7.1.17). □

A straightforward corollary of Theorem 7.1.15 is:

Corollary 7.1.1 *Let C be an (n, k, d) linear block error-correcting code with $n-k$ being even. If $2^k > n+1$, then there exist balanced functions satisfying both* PC$(d-1)$ *and* CI(1).

The proofs of Theorem 7.1.14 and 7.1.15 construct a large group of balanced Boolean functions satisfying higher propagation criterion and correlation-immune. The following example shows an illustration of our constructions.

Example 7.1.1 Let $n = 5, k = 3$, and E the linear $(5, 3, 2)$ code produced by the generation matrix

$$\begin{bmatrix} 1 & 0 & 0 & 1 & 1 \\ 0 & 1 & 0 & 0 & 1 \\ 0 & 0 & 1 & 1 & 0 \end{bmatrix},$$

$E^{\perp}$ is the linear block code produced by the generation matrix

$$\begin{bmatrix} 1 & 0 & 1 & 1 & 0 \\ 1 & 1 & 0 & 0 & 1 \end{bmatrix}.$$

Let $t = (1, 1, 1, 0, 0)$ and F be the linear block code generated by the matrix

$$\begin{bmatrix} 1 & 0 & 1 & 1 & 0 \\ 0 & 0 & 0 & 0 & 1 \end{bmatrix}.$$

Then the function

$$F(x) = x_1x_2 \oplus x_1x_5 \oplus x_1x_3 \oplus x_1x_4 \oplus x_2x_3 \oplus x_2x_4 \oplus x_3x_5 \oplus x_4x_5 \oplus x_1 \oplus x_2 \oplus x_4$$

is a balanced function satisfying both PC(1) and CI(1).

The concept of resilient functions was first introduced and studied by Chor et al. and Bennett et al. Some possible applications of resilient functions involve fault-tolerant distributed computing, quantum cryptographic key distribution and random sequence generation for stream ciphers. Correlation functions which were first introduced and studied by Siegenthaler possess important applications in cryptology. It turned out that the resilient functions are special cases of unbiased multi-value correlation immune functions. Therefore resilient functions were intensively investigated in recent years.

Bounds on resiliency of resilient functions were discussed in Ref.[10]. Some conjectures on resilient functions were disproved by construction of some infinite classes of counter examples of resilient functions. Constructions were studied in Refs.[11–13]. One construction for resilient functions uses linear error-correcting codes, so resilient functions constructed in this way are linear functions. Another construction for resilient functions is constructing new functions from old ones, so resilient functions constructed in this way have a lower algebraic degree. Resilient functions of high algebraic degree can be constructed from lower ones by applying highly nonlinear permutations. In the following text of this subsection, resilient functions are constructed recursively by applying orthogonal array. The enumerating of resilient functions is also discussed.

At first, we introduce some basic conceptions and results. Let

$$f(x) = f(x_1, x_2, \cdots, x_n), \quad x_i \in \mathrm{GF}(2)^n, \quad 1 \leqslant t \leqslant n,$$
$$\text{if } \forall \{i_1, i_2, \cdots, i_t\} \subseteq \{1, 2, \cdots, n\},$$
$$a = (a_1, a_2, \cdots, a_t) \in \mathrm{GF}(2)^t, \quad c \in \mathrm{GF}(2),$$
$$P(f(x) = c) = P(f(x) = c | x_{i_1} = a_1, x_{i_2} = a_2, \cdots, x_{i_t} = a_t),$$

then $f(x)$ is called t-th-order correlation immune function.

Definition 7.1.7 *Let*

$$F(x) = (f_1(x), f_2(x), \cdots, f_m(x)), \quad m \leqslant n, \quad x \in \mathrm{GF}(2)^n, \quad f_i(x) \in \mathrm{GF}(2)$$

be a multi-value Boolean function of $\mathrm{GF}(2)^n \longrightarrow \mathrm{GF}(2)^m$, *for* $1 \leqslant t \leqslant n$, *if* $\forall \{i_1, i_2, \cdots, i_t\} \subseteq \{1, 2, \cdots, n\}$,

$$a = (a_1, a_2, \cdots, a_t) \in \mathrm{GF}(2)^t, \quad c \in \mathrm{GF}(2)^m,$$
$$P(F(x) = c) = P(F(x) = c | x_{i_1} = a_1, x_{i_2} = a_2, \cdots, x_{i_t} = a_t),$$

then $f(x)$ *is called* t*-th-order multi-value correlation immune function.*

Definition 7.1.8 *Let* A *be a* $w \times n$ *matrix over* $\mathrm{GF}(2)$, $1 \leqslant t \leqslant n$, *if the rows of* A *are all distinct, and in the rows of* $w \times t$ *submatrix of* A *which consist of arbitrary* t *columns of* A, *every possible* t*-tuple of* $\mathrm{GF}(2)^t$ *is equally likely to occur* ($w/2^t$ *times*), *then* A *is called a* t*-orthogonal array, which is denoted by* $OA_\lambda(t, n, 2)$, *where,* $w = \lambda 2^t$.

Let A be a $w \times n$ matrix over $\mathrm{GF}(2)$, or the rows of A are vectors of $\mathrm{GF}(2)^n$. If the rows of A are different from each other, we think matrix A is the same as the set which consists of rows of A, e.g., $\mathrm{GF}(2)^n$ being either a $2^n \times n$ matrix over $\mathrm{GF}(2)$ or a set of all n-tuple consisting of $0, 1$.

Definition 7.1.9 *Let all $w \times n$ matrices $A_1, A_2, \cdots, A_k$ be t-orthogonal array. $A_1, A_2, \cdots, A_k$ is said to be a t-orthogonal split of $\mathrm{GF}(2)^n$, if every possible vector of $\mathrm{GF}(2)^n$ occurs in exactly one of $A_1, A_2, \cdots, A_k$ and $\mathrm{GF}(2)^n = \bigcup_{i=1}^{k} A_i$, where $k = 2^m, w = 2^n/k = 2^{n-m}, 2^t|w$.*

Definition 7.1.10 *Let*

$$F(x) : \mathrm{GF}(2)^n \longrightarrow \mathrm{GF}(2)^m, \quad \text{if } \forall \beta \in \mathrm{GF}(2)^m, \quad \#\{x|F(x) = \beta\} = 2^{n-m},$$

then $F(x)$ is said to be unbiased, where # denotes a cardinal number.

Definition 7.1.11 *Let*

$$F(x) : \mathrm{GF}(2)^n \longrightarrow \mathrm{GF}(2)^m,$$

$$t \geqslant 0, \quad \text{if } \forall a = (a_1, a_2, \cdots, a_t) \in \mathrm{GF}(2)^t,$$

$$\{i_1, i_2, \cdots, i_t\} \subseteq \{1, 2, \cdots, n\},$$

$$F(x|x_{i_1} = a_1, x_{i_2} = a_2, \cdots, x_{i_t} = a_t)$$

are unbiased, then $F(x)$ is said to be t-resilient functions, which are denoted as (n, m, t)-RF.

We will use the following conclusions:

Conclusion 7.1.1 *If $f(x) = f(x_1, x_2, \cdots, x_n), x_i \in \mathrm{GF}(2)$, then $f(x)$ is t-th-order correlation immune function if and only if A_f is a t-orthogonal array, where A_f is the matrix consisting of all the rows satisfying $\{x|f(x) = 1\}$.*

Conclusion 7.1.2 *Let $F(x) : \mathrm{GF}(2)^n \longrightarrow \mathrm{GF}(2)^m$ be a multi-value Boolean function, then $F(x)$ is a t-th-order correlation immune function, if and only if for any $r, 1 \leqslant r \leqslant t, F(x)$ is a r-th-order correlation immune function.*

Conclusion 7.1.3 *Let $F(x) : \mathrm{GF}(2)^n \longrightarrow \mathrm{GF}(2)^m$ be a multi-value Boolean function, then $F(x)$ is unbiased tth-order correlation immune function, if and only if $F(x)$ is t-resilient function.*

Unbiased function is 0-resilient function in fact, so t may be 0 in Conclusion 7.1.3.

Conclusion 7.1.4 *Let $F(x) : \mathrm{GF}(2)^n \longrightarrow \mathrm{GF}(2)^m$ be a multi-value Boolean function, then $F(x)$ is unbiased t-th-order correlation immune function, if and only if $A_{\beta_1}, A_{\beta_2}, \cdots, A_{\beta_{2^m}}$ is a t-orthogonal split of $\mathrm{GF}(2)^n$, where $A_{\beta_i} = \{x|F(x) = \beta_i\}$, $\beta_i \in \mathrm{GF}(2)^m, 1 \leqslant i \leqslant 2^m$.*

The proofs of these conclusions can be concluded directly from the definitions above. According to Conclusion 7.1.3 and Conclusion 7.1.4, constructing t-resilient function equals to constructing a t-orthogonal split of $\mathrm{GF}(2)^n$. From now on, we will give a recursive method to construct a t-orthogonal split of $\mathrm{GF}(2)^n$.

We will discuss the recursive construction of resilient functions. From the Conclusion 7.1.4, if we can construct a t-orthogonal split of $\mathrm{GF}(2)^n$: $A_1, A_2, \cdots, A_k$, for any $x \in A_i \subseteq \mathrm{GF}(2)^n$, let $F(x) = \beta_i, \beta_i \in \mathrm{GF}(2)^m$, ($F(x) \neq F(y)$ if $x \in A_i, y \in A_j, i \neq j$), then

$F(x) : \mathrm{GF}(2)^n \longrightarrow \mathrm{GF}(2)^m$ is unbiased t-th order correlation immune function. And from Conclusion 7.1.3, we know $F(x)$ is t-resilient function. Obviously, we can get $2^m!$ t-resilient functions in this way. Then the construction of t-resilient functions is coming down to the construction of t-orthogonal split.

We will discuss the construction of t-orthogonal split in the following text.

Lemma 7.1.18 *Let A, B be two $w \times n$ matrices. If A, B are both $OA_\lambda(t, n, 2)$ orthogonal array, then*

$$C = \begin{bmatrix} 1 & A \\ 0 & B \end{bmatrix}$$

is an $OA_{2\lambda}(t, n+1, 2)$ orthogonal array, where $\lambda = w/2^t$.

Lemma 7.1.19 *Let A be a $2^{n-1} \times n$ matrix. If A is an $OA_\lambda(t, n, 2)$ orthogonal array, then*

$$C = \begin{bmatrix} 1 & A \\ 0 & \overline{A} \end{bmatrix}$$

is an $OA_\lambda(t+1, n+1, 2)$ orthogonal array, where $\lambda = 2^{n-1-t}$, and $\overline{A}$ consists of all vectors of $\mathrm{GF}(2)^n$ which don't occur in the rows of A.

Lemma 7.1.20 *Let A be a $w \times n$ matrix, then the following statements are equivalent:*

(1) *A is an $OA_\lambda(t, n, 2)$ orthogonal array ($\lambda = w/2^t$).*

(2) *For any r satisfying $1 \leqslant r \leqslant t$, the number of zeros equals to the number of ones in the sum of r columns of A.*

(3) *For any r satisfying $1 \leqslant r \leqslant t$, the number of ones in the product of r columns of A is $w/2^t$.*

Where the product (sum) of two vectors refers to the vector whose components are the product (sum) of corresponding components of the two vectors.

Theorem 7.1.16 *Let $t \geqslant 1, m \geqslant 1, j \geqslant t + m$, if $\mathrm{GF}(2)^j$ has t-orthogonal split, then $\mathrm{GF}(2)^{j+m}$ has $t+1$ orthogonal split.*

Proof Let $C_1, C_2, \cdots, C_{2^m}$ be t-orthogonal split of $\mathrm{GF}(2)^j$, then $\mathrm{GF}(2)^j$ can be rewritten as

$$\mathrm{GF}(2)^j = \begin{bmatrix} C_1 \\ C_2 \\ \vdots \\ C_{2^m} \end{bmatrix},$$

where $2^{j-m} \times j$ orthogonal arrays $C_i (1 \leqslant i \leqslant 2^m)$ are $OA_{2^{j-m-t}}(t, j, 2)$ arrays. From Lemma 7.1.18–7.1.20, we can conclude $2^j \times (j+m)$ matrix

$$C = \begin{pmatrix} 1 & \cdots & 1 & C_1 \\ 1 & \cdots & 0 & C_2 \\ \vdots & & \vdots & \vdots \\ 0 & \cdots & 1 & C_{2^m-1} \\ 0 & \cdots & 0 & C_{2^m} \end{pmatrix}$$

is an $OA_{2^{j-t-1}}(t+1, j+m, 2)$ orthogonal array, where the former m columns of C are formally like all the vectors in $\mathrm{GF}(2)^m$ which are listed in alphabetic order, and 0 denotes the column vector composed by all zeros and 1 denotes the column vector composed by all ones. We can get different orthogonal arrays by reordering C_i, so there are $2^m!$ different orthogonal arrays. In all these $2^m!$ orthogonal arrays, we can find 2^m arrays which don't have same rows, and denote them as $C_1^*, C_2^*, \cdots, C_{2^m}^*$, if we choose

$$C_1^* = C, \quad C_2^* = \begin{bmatrix} 1 & \cdots & 1 & C_2 \\ 1 & \cdots & 0 & C_3 \\ \vdots & & \vdots & \vdots \\ 0 & \cdots & 0 & C_1 \end{bmatrix},$$

$$C_{2^m}^* = \begin{bmatrix} 1 & \cdots & 1 & C_{2^m} \\ 1 & \cdots & 0 & C_1 \\ \vdots & & \vdots & \vdots \\ 0 & \cdots & 0 & C_{2^m-1} \end{bmatrix},$$

then

$$\mathrm{GF}(2)^{j+m} = \begin{bmatrix} C_1^* \\ C_2^* \\ \vdots \\ C_{2^m}^* \end{bmatrix},$$

where $2^j \times (j+m)$ arrays $C_i^* (1 \leqslant i \leqslant 2^m)$ are $(t+1)$-orthogonal arrays that is, $C_1^*, C_2^*, \cdots$, $C_{2^m}^*$, are $(t+1)$-orthogonal split of $\mathrm{GF}(2)^{j+m}$. Then the theorem holds. □

Theorem 7.1.17 *For any $t \geqslant 1, m \geqslant 1, \mathrm{GF}(2)^{mt+1}$ has t-orthogonal split.*

Proof By induction

(1) If $t = 1$, the rows of $\mathrm{GF}(2)^{1+m}$ can compose 2^m conjugate pairs. Denote the 2^m different conjugate pairs as $C_1, C_2, \cdots, C_{2^m}$. Obviously, they are all 1-orthogonal arrays, so we can see the conclusion holds for $t = 1$.

(2) Let $j = mt + 1$. From Theorem 7.1.16, if the conclusion is true for t, we can infer the conclusion is true for $t + 1$. The proof is completed. □

Lemma 7.1.21 *Let A, B be two $w \times n$ matrices, if both A and B are $OA_\lambda(t, n, 2), \lambda = w/2^t$, orthogonal arrays, and they have no identical row, then*

$$\begin{bmatrix} 1 & A \\ 0 & A \end{bmatrix}, \quad \begin{bmatrix} 1 & B \\ 0 & B \end{bmatrix}$$

have no identical row too, and both of them are $OA_{2\lambda}(t, n+1, 2)$, orthogonal arrays.

The conclusion is true from Lemma 7.1.18.

Theorem 7.1.18 *For any $n \geqslant m \geqslant 1, (n-1)/m \geqslant t \geqslant 1, \mathrm{GF}(2)^n$ has t-orthogonal split.*

Proof Since $mt + 1 \leqslant n$, if $mt + 1 = n$, the conclusion is true from Theorem 7.1.17. If $mt + 1 < n$, from Theorem 7.1.17, $\mathrm{GF}(2)^{mt+1}$ has t-orthogonal split, then it can be written as:

$$\mathrm{GF}(2)^{mt+1} = \begin{bmatrix} C_1 \\ C_2 \\ \vdots \\ C_{2^m} \end{bmatrix},$$

where $C_i (1 \leqslant i \leqslant 2^m)$ are $OA_{2^{(m-1)t-m-1}}(t, mt+1, 2)$ orthogonal arrays. Denote $C_1 = C_1^0$, $C_2 = C_2^0, \cdots, C_{2^m} = C_{2^m}^0$, and let

$$C_1^1 = \begin{bmatrix} 1 & C_1^0 \\ 0 & C_1^0 \end{bmatrix}, C_2^1 = \begin{bmatrix} 1 & C_2^0 \\ 0 & C_2^0 \end{bmatrix}, \cdots, C_{2^m}^1 = \begin{bmatrix} 1 & C_{2^m}^0 \\ 0 & C_{2^m}^0 \end{bmatrix}.$$

From Lemma 7.1.21, C_1^1, $C_2^1, \cdots$, $C_{2^m}^1$ are all $OA_{2^{(m-1)t-m+2}}(t, mt+2, 2)$ orthogonal arrays, and C_1^1, $C_2^1, \cdots$, $C_{2^m}^1$ are t-orthogonal split of $\mathrm{GF}(2)^{mt+2}$. For any $1 \leqslant k \leqslant n - (mt+1)$, let

$$C_1^k = \begin{bmatrix} 1 & C_1^{k-1} \\ 0 & C_1^{k-1} \end{bmatrix}, C_2^k = \begin{bmatrix} 1 & C_2^{k-1} \\ 0 & C_2^{k-1} \end{bmatrix}, \cdots, C_{2^m}^k = \begin{bmatrix} 1 & C_{2^m}^{k-1} \\ 0 & C_{2^m}^{k-1} \end{bmatrix}$$

then C_1^k, $C_2^k, \cdots$, $C_{2^m}^k$ are t-orthogonal split of $\mathrm{GF}(2)^{mt+1+k}$. Repeat the process above, when $k = n - (mt+1)$, we get an orthognal split of $\mathrm{GF}(2)^n$: C_1^*, $C_2^*, \cdots$, $C_{2^m}^*$. □

Theorem 7.1.18 is constructive. It gives a recursive method of construting orthognal split of $\mathrm{GF}(2)^n$. It shows, for any $n \geqslant m \geqslant 1$, $\mathrm{GF}(2)^n$ have t-orthogonal split, where $t \leqslant (n - 1/m)$. Therefore, we can construct $2^m!$ t-resilient functions from $\mathrm{GF}(2)^n$ to $\mathrm{GF}(2)^m$. Let $t = 1$, then we can get:

Corollary 7.1.2 *For any $n \geqslant m \geqslant 1$, there are at least $2^m!$ $(n, m, 1)$-resilient function.*

The motive for introducing resilient function is to resist information package. Resiliency is an important parameter to measure the resistibility to information leakage. But if the number of such functions is less, the application of the function will be confined. Therefore, studying the enumeration of resilient functions (also known as (n, m, t)-functions) is a very important task. We know that the order t of correlation immune function indicates the function's resistibility to correlation attack, but when t comes up to maximum, the number of such functions is less and they are all linear functions, so, the largest t is not the best one. When it comes to resiliency t, how are things to proceed? We will answer this question in the following discussion.

Assume that $A_1, \cdots, A_{2^m}$ are t-orthogonal split of $\mathrm{GF}(2)^n, n \geqslant m \geqslant 1$, then $2^{n-m} \times n$ matrices $A_1, \cdots, A_{2^m}$ are all t-orthogonal arrays. Hence $2^t | 2^{n-m}$, therefore $t \leqslant n - m$. Since t indicates the function's resistibility to information leakage, we expect that t can be as large as possible, but when t comes up to maximum, namely $t = n - m$, the number of (n, m, t)-functions will be quite confined, so we have the following discussion.

For convenience, we introduce some notations: $N^t(\omega, n)$ denotes the number of t-orthogonal $\omega \times n$ arrays, while $N(n, m, t)$ denotes the number of (n, m, t)-functions.

Lemma 7.1.22 *Let $t \geqslant 2$, then*

$$N^t(2^t, n) = \begin{cases} 1, & n = t, \\ 2, & n = t+1, \\ 0, & n \neq t, t+1, \end{cases}$$

namely, when $n = t$, there is only one t-orthogonal array ($2^t \times n$ matrix), and when $n = t+1$, there are two, and in other circumstance, the number is zero.

Proof (1) When $n < t$, $2^t \times n$ matrix must have same rows, hence $N^t(2^t, n) = 0$.

(2) When $n = t, 2^t \times t$ matrix having different rows, that is, the matrix is constructed by all the vectors in $\mathrm{GF}(2)^t$, and this kind of matrix is unique, and it is t-orthogonal, namely, $N^t(2^t, n) = 1$.

(3) When $n = t + 1$, assume $2^t \times t$ matrix A is $(t-1)$-orthogonal, then from Lemma 7.1.19, $2^t \times (t+1)$ matrix

$$C = \begin{bmatrix} 1 & A \\ 0 & \bar{A} \end{bmatrix}$$

is t-orthogonal, and the constructed C is different for different A, thus

$$N^t(2^t, t+1) \geqslant N^{t-1}(2^{t-1}, t). \tag{7.11}$$

Conversely, suppose $2^t \times (t+1)$ matrix C is t-orthogonal, then $C = \begin{bmatrix} 1 & A_1 \\ 0 & A_2 \end{bmatrix}$, where A_1, A_2 is $(t-1)$-orthogonal array ($2^{t-1} \times t$ matrix). Because $\begin{bmatrix} A_1 \\ A_2 \end{bmatrix}$ is t columns of t-orthogonal array, that is $\begin{bmatrix} A_1 \\ A_2 \end{bmatrix} = \mathrm{GF}(2)^t$, so $A_2 = \bar{A}_1$, C must have form like $C = \begin{bmatrix} 1 & A_1 \\ 0 & \bar{A}_1 \end{bmatrix}$, that is, $(t-1)$-orthogonal array ($2^{t-1} \times t$ matrix) A_1 is constructed from t-orthogonal array $2^t \times (t+1)$ matrix C, and the constructed A_1 is different for different C, thus

$$N^t(2^t, t+1) \leqslant N^{t-1}(2^{t-1}, t). \tag{7.12}$$

Combining Eqs. (7.11) and (7.12), we have

$$N^t(2^t, t+1) = N^{t-1}(2^{t-1}, t) \tag{7.13}$$

and by the recurrence relation from Eq. (7.13):

$$N^t(2^t, t+1) = N^2(2^2, 3)$$

and $N^2(2^2, 3) = 2$, so, when $n = t + 1$,

$$N^t(2^t, n) = N^t(2^t, t+1) = 2. \tag{7.14}$$

(4) When $n > t + 1$, assume A is a t-orthogonal array ($2^t \times n$ matrix). By Definition 7.1.8, A can be written as

$$A = \begin{bmatrix} 1 & \cdots & 1 & x_{1t+1} & \cdots & x_{1n} \\ 1 & \cdots & 0 & x_{2t+1} & \cdots & x_{2n} \\ \vdots & & \vdots & \vdots & & \vdots \\ 0 & \cdots & 0 & x_{2^t t+1} & \cdots & x_{2^t n} \end{bmatrix}$$
$$= \begin{bmatrix} A_1 & X_{t+1} & \cdots & X_n \end{bmatrix},$$

where, A_1 is the former t columns of A, while $X_i(t+1 \leqslant i \leqslant n)$ is the i-th column of A. Of course, $[A_1 X_i]$ is a t-orthogonal array ($2^t \times (t+1)$ matrix), by Eq.(7.13), there is only two such arrays. Assume the two arrays are $C = [A_1 X_{t+1}]$ and $D = [A_1 X_{t+2}]$. Apparently, $\bar{C}$, which is constructed by all the vectors of $\mathrm{GF}(2)^{t+1}$ except the rows in $C = [A_1 X_{t+1}]$, is also t-orthogonal array ($2^t \times (t+1)$ matrix), so, there must be $\bar{C} = D = [A_1 X_{t+2}]$, therefore,

$$\mathrm{GF}(2)^{t+1} = \begin{bmatrix} C \\ \bar{C} \end{bmatrix} = \begin{bmatrix} A_1 & X_{t+1} \\ A_1 & X_{t+2} \end{bmatrix},$$

that is X_{t+1}, X_{t+2} are conjugated, thus $X_{t+1} + X_{t+2} = [1 \cdots 1]^{\mathrm{T}}$. This is in conflict with the assumption that X_{t+1}, X_{t+2} are two columns of t-orthogonal array ($t \geqslant 2$). Therefore, the assumption is not true. So we can conclude that when $n > t+1$, there is no $2^t \times n$ matrix which is t-orthogonal array. Hence, when $n > t+1$,

$$N^t(2^t, n) = 0 \tag{7.15}$$

□

From Conclusion 7.1.3 and Conclusion 7.1.4, we can conclude that if there exist (n, m, t)-functions, then there exist t-orthogonal splits of $\mathrm{GF}(2)^n$: $A_1, \cdots, A_{2^m}$, and all the $A_i (1 \leqslant i \leqslant 2^m)$ are t-orthogonal array ($2^{n-m} \times n$ matrices). When $t = n - m$, they become $2^t \times n$ matrices (t-orthogonal arrays). From Lemma 7.1.22, we know that only when $n = t + 1$, there exist t-orthogonal $2^t \times n$ matrices, and they are A_1, A_2 (thus $m = 1$). So when $t = m - n$, $\mathrm{GF}(2)^n$ have only one t-orthogonal split: A_1, A_2. Thus there are only $2^m! = 2^1! = 2$ (n, m, t)-functions. Then we have:

Theorem 7.1.19 *For any $n \geqslant m \geqslant 1, t \geqslant 2, t = n - m$, there are only two (m, n, t)-functions. And here $n = t + 1$, $m = 1$.*

Theorem 7.1.19 shows, when $t \geqslant 2$, there are only two (n, m, t)-functions which can arrive at maximum resiliency $n - m$. And they are $(t + 1, 1, t)$-functions, namely Boolean functions. So we have:

Corollary 7.1.3 *When*

$$t \geqslant 2, F(x) : \mathrm{GF}(2)^n \to \mathrm{GF}(2)^m (m \geqslant 2)$$

can't arrive at maximum resiliency $n - m$.

From Theorem 7.1.19, when $t \geqslant 2$, the number of resilient function which arrive at best resiliency is less. But things are much different when $t \leqslant 1$. And we have:

Theorem 7.1.20 *Assume $n \geqslant m \geqslant 1$, the number of (n, m, t)-functions is denoted by $N(n, m, t)$, if $t = n - m$, then*

(1) When $t = 0$, $N(n, m, t) = 2^n!$.

(2) When $t = 1$, $N(n, m, t) = 2^{n-1}!$.

(3) When $t \geqslant 2$, $N(n, m, t) = 2$.

Proof (1) Since $t = n - m$, then $t = 0$ means $m = n$, and (n, m, t)-functions are the permutations over $\mathrm{GF}(2)^n$, so there are $2^n!$ such functions.

(2) When $t = n-m = 1, m = n-1, A_1, \cdots, A_{2^m}$ are the 2^{n-1} conjugate pairs in $\mathrm{GF}(2)^n$, so, there are $2^m! = 2^{n-1}!$ $(n, n-1, 1)$-functions.

(3) This is just Theorem 7.1.19. □

Theorem 7.1.20 shows that the number of functions which arrive at best resiliency is much less except for $m = n$ and $m = n-1$. Therefore, from the view of cryptography, high resiliency is not the only consideration. And generically, we should have $t < n - m$. For different t, the distribution of number of (n, m, t)-functions still needs more study.

Theorem 7.1.21 *Assume $t \geqslant 2$, for any $m \geqslant 1$, there exists N, and when $n \geqslant N$, there is no (n, m, t)-functions.*

Proof There exists (n, m, t)-function if and only if there exists t-orthogonal split of $\mathrm{GF}(2)^n$: $A_1, \cdots, A_{2^m}$, and $A_1, \cdots, A_{2^m}$ are all t-orthogonal arrays ($2^{n-m} \times n$ matrices). When $t \geqslant 2$, all the columns in a t-orthogonal array are different from each other, and there is no conjugated pairs, so, there are $2^{2^{n-m}-1}$ columns at most. Let $N = 2^{2^{n-m}-1}$, then when $n \geqslant N$, there is no $2^{n-m} \times n$ matrix which is t-orthogonal array, therefore no (n, m, t)-function. □

Theorem 7.1.21 gives an upper bound of N, but how to get an exact bound still needs more study. When m is not big, practice shows N can be 2^{n-m}, namely, when $n \geqslant 2^{n-m}$, there is no $2^{n-m} \times n$ matrix which is t-orthogonal array. Therefore there is no (n, m, t)-function.

Conjecture 7.1.1 *Assume $t \geqslant 2$, then when $n \geqslant 2^{n-m}$, there is no $2^{n-m} \times n$ matrix which is t-orthogonal array. therefore there is no (n, m, t)-function.*

The above discussion shows that when $t \geqslant 2$, there are restrictive relations between n and m, especially between n and $n-m$. For example, when $t = 2$, there are $2^3 \times n$ matrices which is t-orthogonal arrays only when $n \leqslant 7$. Namely, when $t = 2$ and $n - m = 3$, there are (n, m, t)-functions only if $n \leqslant 7 < 8 = 2^3$.

Now we present a method of constructing t-resilient functions by applying orthogonal matrices. This method is simple and easy to implement. The enumerating of t-resilient functions is a very important problem, if they are used in cryptography. We give some results here. But there still needs to be further study. For any n, m, t, the existence and enumerating of (n, m, t)-functions are worthy of further study. Theorem 7.1.21 and the Conjecture 7.1.1 have significant value for the enumerating problem.

7.1.3 Correlation Immunity of Boolean Functions[14,15]

The stream-key generator is one of the most important parts in conventional cryptosystem. The combiner shown in Fig. 7.1.1 is often used as the stream-key generator.

Where $S_1, S_2, \cdots, S_n$ are shift registers, $f(\cdot)$ is some combiner (i.e., Boolean function), K is the key stream. In 1984, Dr. Siegenthaler proved that if the $f(\cdot)$ was carelessly chosen then the corresponding cryptosystem could be broken by the correlational attack method[9]. As one of the parameters for the security of the above mentioned stream-key generator, the concept of correlation-immunity was proposed and studied. The higher the correlation

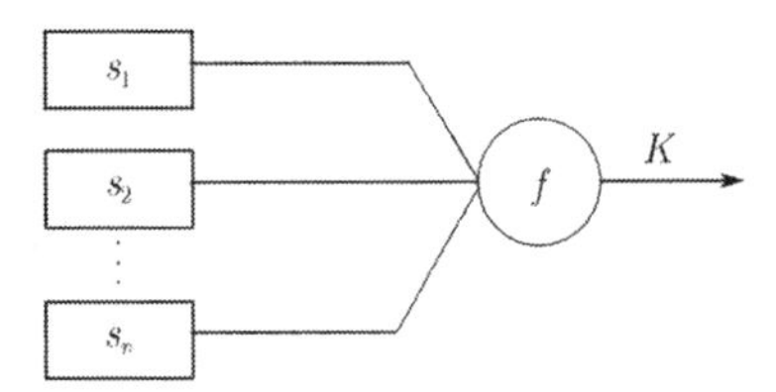

Fig.7.1.1 stream key generator

immunity of $f(\cdot)$, the stronger the corresponding cryptosystem under the correlational attack. By now, different research methods (e.g., the spectral technique from Walsh transform) have been used in the literature, but in this subsection the Hamming weight method will be originally introduced into this area. A few interesting results are presented.

Definition 7.1.12 *The Boolean function $f(x_1, \cdots, x_n)$ is called of correlation-immune of order $m(1 \leqslant m \leqslant n-1)$ iff for any $1 \leqslant i_1 < \cdots < i_m \leqslant n$ and $a_1, a_2, \cdots, a_m$ hold*

$$P\{f(x_1, \cdots, x_n) = 1 | x_{i_1} = a_1, \cdots, x_{i_m} = a_m\} = P\{f(x_1, \cdots, x_n) = 1\},$$

where x_i's are independent random variables with the same distribution and $P\{x_i = 0\} = P\{x_i = 1\} = 1/2$; $a_i = 0$ or 1; $P\{\cdot\}$, $P\{\cdot|\cdot\}$ are the probability and conditional probability, respectively.

If $f(x_1, \cdots, x_n)$ is not of correlation-immune of order m for any $1 \leqslant m \leqslant n-1$, then we say that $f(\cdot)$ has no correlation-immunity.

Lemma 7.1.23 *If $f(x_1, \cdots, x_n)$ is of correlation-immune of order $m(m \geqslant 2)$, then $f(\cdot)$ is also that of order $m-1$ but may or may not be that of order $m+1$.*

Proof For any $1 \leqslant i_1 < \cdots < i_m \leqslant n$ hold

$$\begin{aligned}
&P\left\{f(x_1, \cdots, x_n) = 1 | x_{i_1} = a_1, \cdots, x_{i_{m-1}} = a_{m-1}\right\} \\
= &P\{x_{i_m} = 0\} P\left\{f(x_1, \cdots, x_n) = 1 | x_{i_1} = a_1, \cdots, x_{i_{m-1}} = a_{m-1}, x_{i_m} = 0\right\} \\
&+ P\{x_{i_m} = 1\} P\left\{f(x_1, \cdots, x_n) = 1 | x_{i_1} = a_1, \cdots, x_{i_{m-1}} = a_{m-1}, x_{i_m} = 1\right\} \\
= &P\{x_{i_m} = 0\} P\{f(x_1, \cdots, x_n) = 1\} + P\{x_{i_m} = 1\} P\{f(x_1, \cdots, x_n) = 1\} \\
= &P\{f(x_1, \cdots, x_n) = 1\}.
\end{aligned}$$

From Definition 7.1.12, it is known that $f(\cdot)$ is also of correlation-immune of order $m-1$.

The function $f(x_1, \cdots, x_n) = x_1 + x_2 + x_3$ is one of the examples of the last part of the lemma. In fact, $f(\cdot)$ is of correlation-immune of order 2 but 3. □

Corollary 7.1.4 *If $f(\cdot)$ is not of correlation-immune of order one then it has no correlation-immunity.*

Proof Very easy. □

Because the x_i's in of Definition 7.1.12 are independent each other and have the same distribution $P\{x_i = 0\} = P\{x_i = 1\} = 1/2$, the points $(x_1, \cdots, x_n)$ are uniformly distributed in the n-dimensional space, i.e., $P\{(x_1, \cdots, x_n) =-(b_1, \cdots, b_n)\} = \dfrac{1}{2^n}$. This means that Definition 7.1.12 can be equivalently stated as:

Definition 7.1.13 *$f(x_1, \cdots, x_n)$ is said to be of correlation-immune of order $m(1 \leqslant m \leqslant n-1)$ iff for any $1 \leqslant i_1 < \cdots < i_m \leqslant n$ and $a_1, \cdots, a_m$ hold:*

$$2^m \omega \{f(\cdot)|x_{i_1} = a_1, \cdots, x_{i_m} = a_m\} = \omega \{f(x_1, \cdots, x_n)\},$$

where $\omega\{f(\cdot)\}$ be the Hamming weight of the Boolean function $f(\cdot)$ of n-variable; $\omega(f(\cdot)|x_{i_1} = a_1, \cdots, x_{i_m} = a_m)$ be the conditional Hamming weight of the $(n-m)$-variable Boolean function formed by fixing $x_{i_1} = a_1, \cdots, x_{i_m} = a_m$.

Lemma 7.1.24 *If $f(x_1, \cdots, x_n)$ is of correlation-immune of order $m(1 \leqslant m \leqslant n-1)$, then so are the $f(x_1, \cdots, x_n) + c$ and $f(t(x_1, \cdots, x_n))$, where $c = 0$ or 1; $t(\cdot)$ be any permutation of the set $\{x_1, \cdots, x_n\}$.*

From Lemma 7.1.24, we consider, in the following texts, only the correlation-immunity of Boolean functions that contain no constant term (i.e., $f(0, \cdots, 0) = 0$).

Theorem 7.1.22 *Let $f(x_1, \cdots, x_{n_1}, y_1, \cdots, y_{n_2}) = f_1(x_1, \cdots, x_{n_1}) + f_2(y_1, \cdots, y_{n_2})$, the $f_i(\cdot)(i = 1, 2)$ is of correlation-immune of order m_i and $1 \leqslant m_1 < m_2 \leqslant n-1$, then*

(1) *If $\omega\{f_2(\cdot)\} \neq 2^{n_2-1}$ then $f(\cdot)$ is of correlation-immune of order m_1 but of order $m_1 + 1$.*

(2) *If $\omega\{f_2(\cdot)\} = 2^{n_2-1}$, $\omega\{f_1(\cdot)\} \neq 2^{n_1-1}$ then $f(\cdot)$ is of correlation-immune of order m_2 but of order $m_2 + 1$.*

(3) *If $\omega\{f_2(\cdot)\} = 2^{n_2-1}$, $\omega\{f_1(\cdot)\} = 2^{n_1-1}$ then $f(\cdot)$ is of correlation-immune of order $m_1 + m_2 + 1$ but of order $m_1 + m_2 + 2$.*

Proof $\omega\{f(\cdot)\} = 2^{n_2}\omega\{f_1(\cdot)\} + 2^{n_1}\omega\{f_2(\cdot)\} - 2\omega\{f_1(\cdot)\}\omega\{f_2(\cdot)\}$.

Because $f_1(\cdot)$ is not of correlation-immune of order $m_1 + 1$, there exists some $x_{i_1} = a_1, \cdots, x_{i_{m_1+1}} = a_{m_1+1}$ such that

$$\omega\{f_1(\cdot)\} \neq 2^{m_1+1}\omega\left\{f_1(\cdot)|x_{i_1} = a_1, \cdots, x_{i_{m_1+1}} = a_{m_1+1}\right\}.$$

So

$$\begin{aligned}
&2^{m_1+1} \cdot \omega\left\{f(\cdot)|x_{i_1} = a_1, \cdots, x_{i_{m_1+1}} = a_{m_1+1}\right\} - \omega\{f(\cdot)\} \\
&= 2^{m_1+1}\left[2^{n_2}\omega\left\{f_1(\cdot)|x_{i_1} = a_1, \cdots, x_{i_{m_1+1}} = a_{m_1+1}\right\} + 2^{n_1-(m_1+1)}\right. \\
&\quad \left.\cdot\omega\{f_2(\cdot)\} - 2\omega\left\{f_1(\cdot)|x_{i_1} = a_1, \cdots, x_{i_{m_1+1}} = a_{m_1+1}\right\}\omega\{f_2(\cdot)\}\right] \\
&\quad -2^{n_2}\omega\{f_1(\cdot)\} - 2^{n_1}\omega\{f_2(\cdot)\} + 2\omega\{f_1(\cdot)\}\omega\{f_2(\cdot)\} \\
&= 2\left[2^{n_2-1} - \omega\{f_2(\cdot)\}\right]\left[2^{m_1+1}\omega\left\{f_1(\cdot)|x_{i_1} = a_1, \cdots, x_{i_{m_1+1}} = a_{m_1+1}\right\}\right. \\
&\quad \left.-\omega\{f_1(\cdot)\}\right] \\
&\neq 0.
\end{aligned}$$

The last unequality is due to $\omega\{f_2(\cdot)\} \neq 2^{n_2-1}$ and $2^{m_1+1}\omega\{f_1(\cdot)|x_{i_1} = a_1, \cdots, x_{i_{m_1+1}} = a_{m_1+1}\} \neq \omega\{f_1(\cdot)\}$. This means that $f(\cdot)$ is not of correlation-immune of order $m_1 + 1$.

Now we turn to prove that $f(\cdot)$ is of correlation-immune of order m_1.

For any $1 \leqslant j_1 < j_2 < \cdots < j_r \leqslant n_1$, $1 \leqslant i_1 < i_2 < \cdots < i_{m_1-r} \leqslant n_2$, arbitrarily fix $x_{j_1} = a_1, \cdots, x_{j_r} = a_r, y_{i_1} = b_1, \cdots, x_{i_{m_1-r}} = b_{m_1-r}$.

For $1 \leqslant m_1 < m_2$, from Lemma 7.1.23 and Definition 7.1.13, we know that

$$2^r \omega \{f_1(\cdot)|x_{j_1} = a_1, \cdots, x_{j_r} = a_r\} = \omega \{f_1(\cdot)\}$$

and

$$2^{m_1-r} \omega \left\{f_2(\cdot)|y_{i_1} = b_1, \cdots, y_{i_{m_1-r}} = b_{m_1-r}\right\} = \omega \{f_2(\cdot)\}.$$

So

$$\begin{aligned}
&2^{m_1} \omega \left\{f(\cdot)|x_{j_1} = a_1, \cdots, x_{j_r} = a_r, y_{i_1} = b_1, \cdots, y_{i_{m_1-r}} = b_{m_1-r}\right\} \\
&= 2^{m_1} \left[2^{n_2-(m_1-r)} \omega \{f_1(\cdot)|x_{j_1} = a_1, \cdots, x_{j_r} = a_r\}\right. \\
&\quad +2^{n_1-r} \omega \left\{f_2(\cdot)|y_{i_1} = b_1, \cdots, x_{i_{m_1-r}} = b_{m_1-r}\right\} \\
&\quad \left. -2\omega \{f_1(\cdot)|x_{j_1} = a_1, \cdots, x_{j_r} = a_r\} \omega \left\{f_2(\cdot)|y_{i_1} = b_1, \cdots, x_{i_{m_1-r}} = b_{m_1-r}\right\}\right] \\
&= 2^{m_1} \left[2^{n_2-m_1} \omega \{f_1(\cdot)\} + 2^{n_1-m_1} \omega \{f_2(\cdot)\} - 2^{-(m_1-r)-r} \omega \{f_1(\cdot)\} \omega \{f_2(\cdot)\}\right] \\
&= 2^{n_1} \omega \{f_1(\cdot)\} + 2^{n_1} \omega \{f_2(\cdot)\} - 2\omega \{f_1(\cdot)\} \omega \{f_2(\cdot)\} \\
&= \omega \{f(\cdot)\}.
\end{aligned}$$

So $f(\cdot)$ is indeed of correlation-immune of order m.

The other parts of the theorem can be proved in the same way. □

Theorem 7.1.23 *Let $f(x_1, \cdots, x_n)$ is of correlation-immune of order m and $\omega \{f(\cdot)\} = 2^{n-1}$, then $g(x_1, \cdots, x_n, x_{n+1}) = x_{n+1} + f(x_1, \cdots, x_n)$ is of correlation-immune of order $m + 1$.*

Proof For any $1 \leqslant i_1 < i_2 < \cdots < i_{m_1+1} \leqslant n + 1$ and $x_{i_1} = a_1, \cdots, x_{i_{m+1}} = a_{m+1}$, because of $\omega \{g(x_1, \cdots, x_n, x_{n+1})\} = 2^n$ we have:

Case 1. $i_{m+1} = n + 1$. From equality

$$2^m \omega \{f(\cdot)|x_{i_1} = a_1, \cdots, x_{i_m} = a_m\} = \omega \{f(\cdot)\} = 2^{n-1}.$$

So

$$\begin{aligned}
&2^{m+1} \omega \left\{g(\cdot)|x_{i_1} = a_1, \cdots, x_{i_{m+1}} = a_{m+1}\right\} \\
&= 2^{m+1} \omega \{a_{m+1} + f(\cdot)|x_{i_1} = a_1, \cdots, x_{i_m} = a_m\} = 2\omega \{a_{m+1} + f(\cdot)\} \\
&= 2 \cdot 2^{n-1} = 2^n = \omega \{g(\cdot)\}.
\end{aligned}$$

Case 2. $i_{m+1} < n + 1$. So

$$\begin{aligned}
&2^{m+1} \omega \left\{g(\cdot)|x_{i_1} = a_1, \cdots, x_{i_{m+1}} = a_{m+1}\right\} \\
&= 2^{m+1} \omega \left\{x_{m+1} + h(y_1, \cdots, y_{n-(m+1)})\right\} \\
&= 2^{m+1} \cdot 2^{n-(m+1)} = 2^n = \omega \{g(\cdot)\},
\end{aligned}$$

where $h(\cdot)$ is some $(n - m - 1)$-variable Boolean function which is independent of x_{n+1}.

From Cases 1 and 2 we have proved that $g(\cdot)$ is of correlation immune of order $m + 1$. □

Theorem 7.1.23 can be generalized as

Theorem 7.1.24 *Let $f(x_1, \cdots, x_{n_1}, y_1, \cdots, y_{n_2}) = f_1(x_1, \cdots, x_{n_1}) + f_2(y_1, \cdots, y_{n_2})$, where $f_1(\cdot)$ has no correlation-immunity and $f_2(\cdot)$ is of correlation-immune of order m, then*

(1) *If* $\omega\{f_2(\cdot)\} = 2^{n_2-1}$ *then* $f(\cdot)$ *is of correlation-immune of order at least* m.

(2) *If* $\omega\{f_1(\cdot)\} = 2^{n_1-1}$, $\omega\{f_2(\cdot)\} = 2^{n_2-1}$ *then* $f(\cdot)$ *is of correlation-immune of order at least* $m+1$.

(3) *If* $\omega\{f_2(\cdot)\} \neq 2^{n_2-1}$ *then* $f(\cdot)$ *has no correlation-immunity.*

Proof (1) and (2) can be proved in the same way as in Theorem 7.1.23. We prove only (3) in the following.

Because $f_1(\cdot)$ has no correlation-immunity, there exists some $x_i = a$ such that $2\omega\{f_1(\cdot)|x_i = a\} \neq \omega\{f_1(\cdot)\}$. So

$$
\begin{aligned}
&2\omega\{f(\cdot)|x_i = a\} - \omega\{f(\cdot)\} = 2\left[2^{n_2}\omega\{f_1(\cdot)|x_i = a\} + 2^{n_1-1}\omega\{f_2(\cdot)\}\right.\\
&\left.-2\omega\{f_1(\cdot)|x_i = a\}\omega\{f_2(\cdot)\}\right] - \left[2^{n_2}\omega\{f_1(\cdot)\} + 2^{n_1}\omega\{f_2(\cdot)\}\right.\\
&\left.-2\omega\{f_1(\cdot)\}\omega\{f_2(\cdot)\}\right]\\
&= 2\left[2^{n_2-1} - \omega\{f_2(\cdot)\}\right]\left[2\omega\{f_1(\cdot)|x_i = a\} - \omega\{f_1(\cdot)\}\right] \neq 0.
\end{aligned}
$$

This means that $f(\cdot)$ has no correlation-immunity. □

Theorem 7.1.25 *Let* $f(x_1,\cdots,x_{n_1},y_1,\cdots,y_{n_2}) = f_1(x_1,\cdots,x_{n_1}) + f_2(y_1,\cdots,y_{n_2})$, *where* $f_1(\cdot)$, $f_2(\cdot)$ *has no correlation-immunity, then*

(1) *If* $\omega\{f_1(\cdot)\} = 2^{n_1-1}$, $\omega\{f_2(\cdot)\} = 2^{n_2-1}$, *then* $f(\cdot)$ *is of correlation-immune of order at least one.*

(2) *If* $\omega\{f_1(\cdot)\} = 2^{n_1-1}$, $\omega\{f_2(\cdot)\} \neq 2^{n_2-1}$ *or* $\omega\{f_1(\cdot)\} \neq 2^{n_1-1}$, $\omega\{f_2(\cdot)\} = 2^{n_2-1}$, *then* $f(\cdot)$ *has no correlation-immunity.*

Theorem 7.1.26 *Let* $f(x_1,\cdots,x_{n_1},y_1,\cdots,y_{n_2}) = f_1(x_1,\cdots,x_{n_1})\cdot f_2(y_1,\cdots,y_{n_2})$, *then*

(1) *If* $f_1(\cdot)$ *is of correlation-immune of order* m_1 *but* m_1+1, $f_2(\cdot)$ *is of correlation-immune of order* m_2 *but* m_2+1 *and* $1 \leqslant m_1 \leqslant m_2$, *then* $f(\cdot)$ *is also of correlation-immune of order* m_1 *but* m_1+1.

(2) *If* $f_1(\cdot)$ *has no correlation-immune then* $f(\cdot)$ *also has none.*

Proof $\omega\{f(\cdot)\} = \omega\{f_1(\cdot)\}\omega\{f_2(\cdot)\}$.

(1) From

$$
\begin{aligned}
&2^{r+s}\omega\{f(\cdot)|x_{i_1} = a_1,\cdots,x_{i_r} = a_r, y_{j_1} = b_1,\cdots,y_{j_s} = b_s\}\\
&= 2^r\omega\{f_1(\cdot)|x_{i_1} = a_1,\cdots,x_{i_r} = a_r\}\cdot 2^s\omega\{f_2(\cdot)|y_{j_1} = b_1,\cdots,y_{j_s} = b_s\},
\end{aligned}
$$

the proof is clear.

(2) Because $f_1(\cdot)$ has no correlation-immunity, there exists some $x_i = a$ such that $\omega\{f_1(\cdot)\} \neq 2\omega\{f_1(\cdot)|x_i = a\}$. So $2\omega\{f(\cdot)|x_i = a\} = 2\omega\{f_1(\cdot)|x_i = a\}\omega\{f_2(\cdot)\} \neq \omega\{f_1(\cdot)\}\omega\{f_2(\cdot)\} = \omega\{f(\cdot)\}$. This is equivalent to say that $f(\cdot)$ has no correlation-immunity. □

Corollary 7.1.5 *Boolean functions with the form* $x_n f(x_1,\cdots,x_{n-1})$ *has no correlation-immunity.*

Theorem 7.1.27 *Let* $f(x_1,\cdots,x_n) = \sum\limits_{1\leqslant i<j\leqslant n} a_{ij}x_ix_j + \sum\limits_{i=1}^{n} b_ix_i$ *is a Boolean function with order two,* $B = [B_{ij}]$ *is an* $n\times n$ *matrix with* $B_{ii} = 0$ *and* $B_{ij} = \begin{cases} a_{ij}, & i<j \\ a_{ji}, & i>j \end{cases}$. *If*

$\omega\{f(\cdot)\} \neq 2^{n-1}$, the $f(\cdot)$ is of correlation-immune of order m iff for any $1 \leqslant i_1 < i_2 < \cdots < i_m \leqslant n$ hold rank(B)=rank(B'), *where the matrix B' is formed from B by setting all the elements in i_1'th, $\cdots$, i_m'th columns and rows to zero.*

Corollary 7.1.6 *Let $n = 2s$, $f(x_1, \cdots, x_n) = \sum\limits_{1\leqslant i<j\leqslant n} a_{ij}x_ix_j + \sum\limits_{i=1}^{n} b_ix_i$ and the matrix B is the same as that in Theorem 7.1.27, if B is nonsingular then $f(\cdot)$ has no correlation-immunity.*

Theorem 7.1.28 *Let $f(x_1, \cdots, x_n) = x_nf_1(x_1, \cdots, x_{n-1}) + f_2(x_1, \cdots, x_{n-1})$. If $\omega\{f_2(\cdot)\} \neq \omega\{f_1(\cdot) + f_2(\cdot)\}$, then $f(\cdot)$ has no correlation-immunity.*

Theorem 7.1.29 *Let $f(x_1, \cdots, x_n) = f_1(x_1, \cdots, x_n) + f_2(x_1, \cdots, x_n)$ and $f_1(x_1, \cdots, x_n)$ $f_2(x_1, \cdots, x_n) = 0$, then*

(1) *If $f_1(\cdot)$ is of correlation-immune of order m_1 but $m_1 + 1$; $f_2(\cdot)$ is of correlation-immune of order m_2 and $1 \leqslant m_1 < m_2$, then $f(\cdot)$ is of also correlation-immune of order m_1 but $m_1 + 1$.*

(2) *If $f_1(\cdot)$ has no correlation-immunity, $f_2(\cdot)$ is of correlation-immune of order at least one, then $f(\cdot)$ has no correlation-immunity.*

Theorem 7.1.30 (1) *Boolean functions consisting of one term have no correlation-immunity.*

(2) *Boolean functions consisting of two terms are of correlation-immune of order at least one iff $f(x_1, \cdots, x_n) = x_i + x_j (i < j)$.*

(3) *Let $f(x_1, \cdots, x_n)$ consist of k terms, i.e., $f(x_1, \cdots, x_n) = f_1 + f_2 + \cdots + f_k$, and there are no common variables contained by any f_i and f_j. Let $f_1f_2 \cdots f_k = x_1x_2 \cdots x_n$, and f_i is the production of n_i variables, $1 \leqslant n_1 \leqslant n_2 \leqslant \cdots \leqslant n_k$, $f(x_1, \cdots, x_n) = x_i + x_j + f_3 + \cdots + f_k$.*

Proof (1) Clear.

(2) Sufficiency. Clear.

(3) Necessity. From Lemma 7.1.24 and Corollary 7.1.5, we know that $f(\cdot)$ can be written as $f(x_1, \cdots, x_n) = x_1x_2 \cdots x_s + x_rx_{r+1} \cdots x_{r+k-1}$ (where $s < r$ and $r + k - 1 \leqslant n$). So $\omega\{f(\cdot)\} = 2^{n-s} + 2^{n-k} - 2 \cdot 2^{n-s-k}$, in another aspect $2\omega\{f(\cdot)|x_1 = 0\} = 2 \cdot 2^{n-1-k} = 2^{n-1}$ and $2\omega\{f(\cdot)|x_r = 0\} = 2 \cdot 2^{n-1-s} = 2^{n-s}$. For $\omega\{f(\cdot)\} = 2\omega\{f(\cdot)|x_1 = 0\} = 2\omega\{f(\cdot)|x_r = 0\}$, we have

$$2^{n-s} + 2^{n-k} - 2^{n+1-s-k} = 2^{n-k} = 2^{n-s},$$

so $k = s = 1$, i.e., $f(x_1, \cdots, x_n) = x_1 + x_r$.

(4) Only show the proof for necessity.

First we present a lemma as follows: Let $f(x_1, \cdots, x_{n_1}, y_1, \cdots, y_{n_2}) = f_1(x_1, \cdots, x_{n_1}) + f_2(y_1, \cdots, y_{n_2})$. If $\omega\{f(\cdot)\} = 2^{n_1+n_2-1}$, then $\omega\{f_1(\cdot)\} = 2^{n_1-1}$ or $\omega\{f_2(\cdot)\} = 2^{n_2-1}$.

Now we turn to the proof. Let x_1 is contained in f_1. Then $2\omega\{f(\cdot)|x_1 = 0\} = 2^{n_1}\omega\{f_2 + \cdots + f_k\}$, in another aspect

$$\omega\{f(\cdot)\} = 2^{n-n_1} + 2^{n_1}\omega\{f_2 + \cdots + f_k\} - 2\omega\{f_2 + \cdots + f_k\}.$$

By $2\omega\{f(\cdot)|x_1 = 0\} = \omega\{f(\cdot)\}$ know $2^{n-n_1-1} = \omega\{f_2 + \cdots + f_k\}$. So from the above lemma, $\omega\{f_2\} = 2^{n_2-1}$ or $\omega\{f_3 + \cdots + f_k\} = 2^{n-n_1-n_2-1}$.

If $\omega\{f_2\} = 2^{n_2-1}$, because of $\omega\{f_2\} = 1$, we have $n_2 = 1$, i.e., $n_1 = n_2 = 1$.

If $\omega\{f_3 + \cdots + f_k\} = 2^{n-n_1-n_2-1}$, by repeatedly using the above lemma, it can be proved that there exist some f_i such that $n_i = 1$, so $n_1 = n_2 = \cdots = n_i = 1$. □

Theorem 7.1.31 *Let $f(x_1, \cdots, x_n) = x_{n+1}f_1(x_1, \cdots, x_n) + (1 + x_{n+1})f_2(x_1, \cdots, x_n)$. If $f(\cdot)$ is of correlation-immune of order m $(m \geqslant 2)$ then $f_1(\cdot)$ and $f_2(\cdot)$ are of correlation-immune of order at least $m - 1$.*

7.1.4 Entropy Immunity of Feedforward Networks[16]

The feedforward network (FN) is an important class of stream key generators. But just as with other stream key generators, FNs also have certain disadvantages. For example, Dr. Siegenthaler found that[9], if the FN function $f(x_1, \cdots, x_n)$ has a small order of correlation immunity, then the corresponding cipher system can be broken by the so-called correlational attacks. For this reason, a "correlation immunity" concept was introduced into the literature. But in recent years it was found that during the design of FNs, there exists a trade-off between the order of correlation immunity and the linear complexity of key streams. This trade-off is an undesire phenomenon, because the linear complexity is also one of the most important parameters for the evaluation of stream ciphers. To overcome this contradiction, the FNs with memory are used to replace the memoryless ones[17]. In the following text, a brand-new approach will be presented to overcome this trade-off.

Another disadvantage of FN is the so-called "Entropy Dropping" as suggested by Prof. K. Zheng and coworkcrs. For example, if the network function $f(x_1, \cdots, x_n)$ can be approximated by some linear Boolean function, the corresponding cipher system can be broken by a new attack method named QBR attack. With this new attack method, Prof. X Z. Zheng has just successfully broken the famous Geffe's sequences. How can we design a network function $f(x_1, \cdots, x_n)$ that has enough immunity from the QBR linear approximation attack? The answer for this question is partly presented here.

Correlation immunity is a parameter for measuring the ability of cipher systems to be immune from "correlational attacks." But correlational attacks are not universally effective; for example, they fail to break a cryptosystem when little output information ε is received from inputs. Maybe for simplicity in analysis, Dr. Siegenthaler took the above-mentioned information ε as zero, i.e., $\varepsilon = 0$. But it is just this assumption $\varepsilon = 0$ that leads to the trade-off between the correlation immunity and the linear complexity. In the following, it will be found that the trade-off can be easily overcome by only taking ε as a small enough number but not zero.

Definition 7.1.14 *Feedforward network function (or Boolean function) $f(x_1, \cdots, x_n)$ is said to be of ε-correlation immune of order m iff for arbitrary $x_{i_1}, \cdots, x_{i_m}$ and $a_1, \cdots, a_m$ $(a_i = 0$ or $1)$*

$$|P(f(x_1, \cdots, x_n) = 1) - P(f(x_1, \cdots, x_n) = 1 | x_{i_1} = a_1, \cdots, x_{i_m} = a_m)| \leqslant \varepsilon,$$

where $P(\cdot)$ and $P(\cdot|\cdot)$ denote the probability and conditional probability, respectively.

It is clear that the case for $\varepsilon = 0$ in Definition 7.1.14 is the same thing as that of Siegenthaler's concept of correlation immunity.

Lemma 7.1.25 *Boolean function $f(x_1, \cdots, x_n)$ is of ε-correlation immune of order m iff*

$$\frac{1}{2^n}\left|W(f) - 2^m W(f|x_{i_1} = a_1, \cdots, x_{i_m} = a_m)\right| \leqslant \varepsilon,$$

where $W(f)$ means the Hamming weight of the Boolean function $f(\cdot)$ with n-variable. x_{i_r} and a_j means the same thing as in Definition 7.1.14. $W(f|x_{i_1} = a_1, \cdots, x_{i_m} = a_m)$ denotes the conditional Hamming weight, i.e., the weight of Boolean function with $(n-m)$-variable formed by fixing $x_{i_1} = a_1, \cdots, x_{i_m} = a_m$ in $f(x_1, \cdots, x_n)$.

Corollary 7.1.7 *If $f(x_1, \cdots, x_n)$ is of ε-correlation immune of order m, then it is also of that of order $(m-1)$.*

The reverse result of this corollary may possibly be false.

Theorem 7.1.32 *Let $f(x_1, \cdots, x_n)$ have correlation immunity of order m $(1 \leqslant m \leqslant n-1)$, $g(x_1, \cdots, x_n) = f(x_1, \cdots, x_n) + x_1 x_2 \cdots x_n$. Then $g(x_1, \cdots, x_n)$ is of $(2^m + 1)/2^n$-correlation immune of order m.*

Proof From the original Definition 7.1.12 of correlation immunity, we know that for any $x_{i_1}, \cdots, x_{i_m}$ and $a_1, \cdots, a_m$,

$$\frac{1}{2^n}\left|W(f) - 2^m W(f|x_{i_1} = a_1, \cdots, x_{i_m} = a_m)\right| = 0,$$

while $W(g) = 1 + W(f) - 2W(x_1 \cdots x_n f(\cdot)) = 1 + W(f) - 2\delta_1$ (where $\delta_1 = 0$ if $f(\cdot)$ contains even numbers of terms, $\delta_1 = 1$ otherwise).

For $a_1 = a_2 = \cdots = a_m = 1$,

$$W(g|x_{i_1} = a_1, \cdots, x_{i_m} = a_m) = 1 + W(f|x_{i_1} = 1, \cdots, x_{i_m} = 1) - 2\delta_2$$

(where $\delta_2 = 0$ if the Boolean function $f(x_1, \cdots, x_n | x_{i_1} = a_1, \cdots, x_{i_m} = a_m)$ with $(n-m)$-variable contains even numbers of terms, $\delta_2 = 1$ otherwise).

For $a_1 \cdots a_m = 0$, i.e., at least one of the a_i is zero

$$W(g|x_{i_1} = a_1, \cdots, x_{i_m} = a_m) = W(f|x_{i_1} = a_1, \cdots, x_{i_m} = a_m).$$

Let $\delta = a_1 \cdots a_m$, then

$$\begin{aligned}
&\frac{1}{2^n}\left|W(g) - 2^m W(g|x_{i_1} = a_1, \cdots, x_{i_m} = a_m)\right| \\
=& \frac{1}{2^n}\left|W(f) + 1 - 2\delta_1 - 2^m[W(f|x_{i_1} = a_1, \cdots, x_{i_m} = a_m) + \delta(1 - 2\delta_2)]\right| \\
=& \frac{1}{2^n}\left|1 - 2\delta_1 + 2^m\delta(1 - 2\delta_2)\right| \leqslant \frac{2^m + 1}{2^n}.
\end{aligned}$$

□

Note In practice the number 2^{-32} can be thought of as zero. From Theorem 7.1.32, if $n - m \geqslant 32$, $g(x_1, \cdots, x_n)$ is of ε-correlation immune of order m with very small ε and at the same time it has also the largest possible linear complexity. Thus under the concept of generalized correlation immunity in Definition 7.1.14 there is no more trade-off between linear complexity and correlation immunity within a certain toleration. Of course this $g(x_1, \cdots, x_n)$ is not also the ideal feedforward network function.

Lemma 7.1.26 *Let*

$$f_1(x_1,\cdots,x_n)=\sum\nolimits_{i=1}^{n} a_i x_1\cdots x_{i-1}x_{i+1}\cdots x_n + b x_1 x_2\cdots x_n$$

and $S=\sum_{i=1}^{n} a_i \geqslant 1$, *then*

$$W(f_1)=2\left[\frac{S+1-b}{2}\right]+b\leqslant 2\left[\frac{n+1-b}{2}\right]+b.$$

Theorem 7.1.33 *Assume that* $f(x_1,\cdots,x_n)$ *is of correlation immune of order* m, $f(x_1,\cdots,x_n)$ *has small Hamming weight. Let* $g(x_1,\cdots,x_n)=f(x_1,\cdots,x_n)+f_1(x_1,\cdots,x_n)$, *then* $g(x_1,\cdots,x_n)$ *is of* $[(1/2^n)(3+2^{m+1})W(f_1)]$*-correlation immune of order* m.

Theorem 7.1.34 *If* $W(f(x_1,\cdots,x_n)+f(1+x_1,\cdots,1+x_n))\leqslant\varepsilon$, *then* $f(x_1,\cdots,x_n)$ *is of* $\varepsilon/2^n$*-correlation immune of order* 1.

Proof

$$\begin{aligned}
&\frac{1}{2^n}\left|W(f)-2W(f|x_i=a_i)\right|\\
&=\frac{1}{2^n}\left|W(f|x_i=a_i)+W(f|x_i=1+a_i)-2W(f|x_i=a_i)\right|\\
&=\frac{1}{2^n}\left|W(f|x_i=1+a_i)-W(f|x_i=a_i)\right|\\
&=\frac{1}{2^n}\sum_x [f(x_1,\cdots,x_{i-1},0,x_{i+1},\cdots,x_n)+f(1+x_1,\cdots,1\\
&\quad+x_{i-1},1,1+x_{i+1},\cdots,1+x_n)]\\
&\leqslant\frac{1}{2^n}W(f(x_1,\cdots,x_n)+f(1+x_1,\cdots,1+x_n))\leqslant\frac{\varepsilon}{2^n}.
\end{aligned}$$

□

Theorem 7.1.35 *Assume that* $f_1(x_1,\cdots,x_n)$ *is of* ε_1*-correlation immune of order* m, $f_2(x_1,\cdots,x_n)$ *is of* ε_2*-correlation immune of order* m *and* $W(f_1)=W(f_2)$. *Let* $f(x_1,\cdots,x_n, x_{n+1})=x_{n+1}f_1(x_1,\cdots,x_n)+(1+x_{n+1})f_2(x_1,\cdots,x_n)$, *then* $f(x_1,\cdots,x_n,x_{n+1})$ *is of* ε*-correlation immune of order* m, *where* $\varepsilon=\max(\varepsilon_1,\varepsilon_2)$.

Proof When $i_m\neq n+1(i_1<i_2<\cdots<i_m)$,

$$\begin{aligned}
&W(f|x_{i_1}=a_1,\cdots,x_{i_m}=a_m)\\
&=W(f_1|x_{i_1}=a_1,\cdots,x_{i_m}=a_m)+W(f_2|x_{i_1}=a_1,\cdots,x_{i_m}=a_m),
\end{aligned}$$

then

$$\begin{aligned}
&\frac{1}{2^{n+1}}\left|W(f)-2^m W(f|x_{i_1}=a_1,\cdots,x_{i_m}=a_m)\right|\\
&=\frac{1}{2^{n+1}}|W(f_1)+W(f_2)-W(f_1|x_{i_1}=a_1,\cdots,x_{i_m}=a_m)\\
&\quad-W(f_2|x_{i_1}=a_1,\cdots,x_{i_m}=a_m)|\\
&\leqslant\frac{1}{2}(\varepsilon_1+\varepsilon_2)\leqslant\varepsilon\overset{\Delta}{=}\max(\varepsilon_1,\varepsilon_2).
\end{aligned}$$

When $i_m=n+1$,

$$W(f|x_{i_1}=a_1,\cdots,x_{i_{m-1}}=a_{m-1},x_{i_m}=0)=W(f_2|x_{i_1}=a_1,\cdots,x_{i_{m-1}}=a_{m-1}),$$

$$W(f|x_{i_1}=a_1,\cdots,x_{i_{m-1}}=a_{m-1},x_{i_m}=1)=W(f_1|x_{i_1}=a_1,\cdots,x_{i_{m-1}}=a_{m-1}).$$

From $W(f_1) = W(f_2)$ and Lemma 7.1.25, we have

$$\begin{aligned}
&\frac{1}{2^{n+1}}\left|W(f) - 2^m W(f|x_{i_1} = a_1, \cdots, x_{i_{m-1}} = a_{m-1}, x_{i_m} = 0)\right| \\
&= \frac{1}{2^{n+1}}\left|2W(f_2) - 2^m W(f_2|x_{i_1} = a_1, \cdots, x_{i_{m-1}} = a_{m-1})\right| \\
&= \frac{1}{2^n}\left|W(f_2) - 2^{m-1} W(f_2|x_{i_1} = a_1, \cdots, x_{i_{m-1}} = a_{m-1})\right| \\
&\leqslant \varepsilon_2 \leqslant \max(\varepsilon_1, \varepsilon_2).
\end{aligned}$$

In the same way, it can be proved that

$$\frac{1}{2^{n+1}}\left|W(f) - 2^m W(f|x_{i_1} = a_1, \cdots, x_{i_{m-1}} = a_{m-1}, x_{i_m} = a_m)\right| \leqslant \max(\varepsilon_1, \varepsilon_2). \qquad \square$$

Theorem 7.1.36 *Let $f_1(x_1, \cdots, x_n), \cdots, f_{2^k}(x_1, \cdots, x_n)$ are of ε_i-correlation immune $(1 \leqslant i \leqslant 2^k)$ of order m respectively, and $W(f_1) = W(f_2) = \cdots = W(f_{2^k})$. Again assume that $\varepsilon = \max\limits_{1 \leqslant i \leqslant 2^k}\{\varepsilon_i\}$ and $g(x_1, \cdots, x_n, y_1, \cdots, y_k) = \underline{a}y_1^{(a_1)} \cdots y_k^{(a_k)} f\underline{a}(x_1, \cdots, x_n)$, then $g(x_1, \cdots, x_n, y_1, \cdots, y_k)$ is of ε-correlation immune of order m, where $a_i = 0$ or 1, $\underline{a} = (a_1, \cdots, a_k)$, $y^{(0)} = 1 + y, y^{(1)} = y$.*

When $k = 1$ Theorem 7.1.36 becomes Theorem 7.1.35.

Theorem 7.1.37 *Assume that $f(x_1, \cdots, x_n)$ is of ε-correlation immune of order m. By arbitrarily fixing $x_{r+1} = b_{1+r}, \cdots, x_n = b_n$ and letting $g(x_1, \cdots, x_r) = f(x_1, \cdots, x_r, b_{r+1}, \cdots, b_n)$, we know that $g(x_1, \cdots, x_r)$ is of ε-correlation immune of order $(m - n + r)$.*

Theorem 7.1.38 *Let $f_1(x_1, \cdots, x_{n_1})$ is of ε_1-correlation immune of order m_1, $f_2(y_1, \cdots, y_{n_2})$ is of ε_2-correlation immune of order m_2, $m_1 < m_2$. Let $f(x_1, \cdots, x_{n_1}, y_1, \cdots, y_{n_2}) = f_1(x_1, \cdots, x_{n_1}) + f_2(y_1, \cdots, y_{n_2})$, then*

(1) *$f(x_1, \cdots, x_{n_1}, y_1, \cdots, y_{n_2})$ is of (2ε)-correlation immune of order $m_1 + m_2 + 1$, if $W(f_1) = 2^{n_1 - 1}, W(f_2) = 2^{n_2 - 1}$, where $\varepsilon = \max(\varepsilon_1, \varepsilon_2)$.*

(2) *$f(\cdot)$ is of $(2\varepsilon_2)$-correlation immune of order m_2, if $W(f_1) \neq 2^{n_1 - 1}$ and $W(f_2) = 2^{n_2 - 1}$.*

(3) *$f(\cdot)$ is of $(2\varepsilon_1)$-correlation immune of order m_1, if $W(f_2) \neq 2^{n_2 - 1}$.*

Proof First, it should be noted that $W(f) = 2^{n_2}W(f_1) + 2^{n_1}W(f_2) - 2W(f_1)W(f_2)$. Because of $W(f_1) = 2^{n_1 - 1}$ and $W(f_2) = 2^{n_2 - 1}$, we know that $W(f) = 2^{n_1 + n_2 - 1}$. Two cases should be considered after arbitrarily fixing $x_{i_1} = a_1, \cdots, x_{i_r} = a_r, y_{j_1} = b_1, \cdots, y_{j_{m_1 + m_2 + 1 - r}} = b_{m_1 + m_2 + 1 - r}$.

Case 1. $r \leqslant m_1$. Under this case we have

$$\frac{1}{2^{n_1}}\left|W(f_1) - 2^{m_1} W(f_1|x_{i_1} = a_1, \cdots, x_{i_r} = a_r)\right| \leqslant \varepsilon_1.$$

Therefore

$$\begin{aligned}
&\frac{1}{2^{n_1 + n_2}}\left|W(f) - 2^{m_1 + m_2 + 1} W(f_1|x_{i_1} = a_1, \cdots, x_{i_r} = a_r, y_{j_1}\right. \\
&\left.= b_1, \cdots, y_{j_{m_1 + m_2 + 1 - r}} = b_{m_1 + m_2 + 1 - r})\right|
\end{aligned}$$

$$= \frac{1}{2^{n_1+n_2}} \left| 2^{n_1+n_2-1} - 2^{m_1+m_2+1} [2^{n_2-(m_1+m_2+1-r)} W(f_1|x_{i_1} = a_1, \cdots, x_{i_r} = a_r) \right.$$
$$+ 2^{n_1-r} W(f_2|y_{j_1} = b_1, \cdots, y_{j_{m_1+m_2+1-r}} = b_{m_1+m_2+1-r})$$
$$\left. - 2W(f_1|x_{i_1} = a_1, \cdots, x_{i_r} = a_r) W(f_2|y_{j_1} = b_1, \cdots, y_{j_{m_1+m_2+1-r}} = b_{m_1+m_2+1-r})] \right|$$
$$\leqslant \frac{1}{2^{n_1}} \left| 2^{n_1-1} - 2^r W(f_1|x_{i_1} = a_1, \cdots, x_{i_r} = a_r) \right|$$
$$+ \frac{1}{2^{n_1}} \left| 2^{n_1-1-r} - W(f_1|x_{i_1} = a_1, \cdots, x_{i_r} = a_r) \right|$$
$$\leqslant \varepsilon_1 + \frac{1}{2^r} \varepsilon_1 \leqslant 2\varepsilon_1 \leqslant 2\max(\varepsilon_1, \varepsilon_2).$$

Case 2. $r > m_1 + 1$, thus $m_1 + m_2 + 1 - r \leqslant m_2$. In the same way as that of case one, we know that

$$\frac{1}{2^{n_1+n_2}} \left| W(f) - 2^{m_1+m_2+1} \right.$$
$$\left. \times W(f|x_{i_1} = a_1, \cdots, x_{i_r} = a_r, y_{j_1} = b_1, \cdots, y_{j_{m_1+m_2+1-r}} = b_{m_1+m_2+1-r}) \right|$$
$$\leqslant 2\varepsilon_2 \leqslant 2\max(\varepsilon_1, \varepsilon_2).$$

The proof for (1) is finished by cases one and two. Results (2) and (3) can be proved by the same method. □

Theorem 7.1.39 *If $f(x_1, \cdots, x_n)$ is of ε-correlation immune of order m, then for any linear Boolean function $g(x_1, \cdots, x_n) = \sum_{i=1}^n c_i x_i$ with $\sum_{i=1}^n c_i \leqslant m$, the following inequality is true:*

$$\frac{1}{2^n} \left| W(f(x_1, \cdots, x_n) + g(x_1, \cdots, x_n)) - 2^{n-1} \right| \leqslant \varepsilon.$$

Proof In general we assume that $c_{i_1} = c_{i_2} = \cdots = c_{i_r} = 1$, $r \leqslant m$, and all the other c_i are zero. Let $A = \{(x_1, \cdots, x_n) : \sum_{i=1}^n c_i x_i = 1\}$, then $|A| = 2^{r-1}$ (where $|A|$ denotes the number of elements contained in the set A). Thus

$$\frac{1}{2^n} \left| W(f(x_1, \cdots, x_n) + g(x_1, \cdots, x_n)) - 2^{n-1} \right|$$
$$= \frac{1}{2^n} \left| W(f) + 2^{n-1} - 2W\left[\left(\sum_{i=1}^n c_i x_i \right) f(x_1, \cdots, x_n) \right] - 2^{n-1} \right|$$
$$= \frac{1}{2^n} \left| W(f) - 2 \sum_{\underline{a} \in A} W(f|x_{i_1} = a_1, \cdots, x_{i_r} = a_r) \right|$$
$$= \frac{1}{2^n} \left| 2^{-r+1} \sum_{\underline{a} \in A} [W(f) - 2^r W(f|x_{i_1} = a_1, \cdots, x_{i_r} = a_r)] \right|$$
$$\leqslant 2^{-r+1} \sum_{\underline{a} \in A} \frac{1}{2^n} \left| W(f) - 2^r W(f|x_{i_1} = a_1, \cdots, x_{i_r} = a_r) \right|$$
$$\leqslant 2^{-r+1} \sum_{\underline{a} \in A} \varepsilon = \varepsilon$$

□

The entropy dropping of correlation immunity has been studied in the above text. Now we turn to the study of entropy dropping of linear approximation. It has been pointed out that the QBR attack method is a serious threat to stream cryptosystem, if the corresponding network functions $f(x_1, \cdots, x_n)$ can be approximated by some linear Boolean function. But a cryptosystem has a powerful ability to be immune from the QBR attack provided that $W(f(x_1, \cdots, x_n)+b+\sum_{i=1}^n c_i x_i) \approx 2^{n-1}$ is true for any linear Boolean function $b+\sum_{i=1}^n c_i x_i$.

From Theorem 7.1.39 we know that if $f(x_1, \cdots, x_n)$ is of ε-correlation immune of order m, $f(x_1, \cdots, x_n)$ cannot be exactly approximated by any linear Boolean function with less than or equal to $(m+1)$ terms. Thus the entropy drops of correlation immunity and linear approximation are close related to and cannot be replaced by each other.

Definition 7.1.15 *$f(x_1, \cdots, x_n)$ is a Boolean function of n-variable. If for any linear Boolean function $b + \sum_{i=1}^n c_i x_i$ hold*

$$\frac{1}{2^n} W\left(f(x_1, \cdots, x_n) + b + \sum\nolimits_{i=1}^{n} c_i x_i\right) - 2^{n-1} \leqslant \delta,$$

then we say that $f(x_1, \cdots, x_n)$ possesses δ-linear entropy dropping.

In Definition 7.1.15 when δ becomes smaller and smaller the approximation degree of $f(x_1, \cdots, x_n)$ by some linear Boolean function becomes worse and worse, thus the corresponding cryptosystem becomes more and more powerfully immune from the QBR attack method.

Theorem 7.1.40 *Bent functions have δ-linear entropy dropping with very small δ. Exact Bent functions of n-variable possess $2^{-(\frac{n}{2}+1)}$-linear entropy dropping.*

Note $2^{-(\frac{n}{2}+1)} \to 0$ when $n \to \infty$. Therefore Theorem 7.1.40 provides us with a large group of network functions with good property in immunity from the QBR attack. By combining Theorem 7.1.40 and the following Theorem 7.1.41 further, more feedforward network functions can be received that are good at immunity from QBR attack.

Theorem 7.1.41 *Let $f_1(x_1, \cdots, x_n)$ has δ_1-linear entropy dropping and $f_2(x_1, \cdots, x_n)$ is a Boolean function with small Hamming weight, say $\frac{1}{2^n} W(f_2) = \delta_2$ (δ_2 is a very small number). Assume $g(x_1, \delta, x_n) = f_1(x_1, \delta, x_n) + f_2(x_1, \delta, x_n)$ then $g(x_1, \delta, x_n)$ has $(\delta_1 + \delta_2)$-linear entropy dropping.*

Proof

$$\begin{aligned}
&\frac{1}{2^n}\left|W(g(x_1, \cdots, x_n) + b + \sum_{i=1}^{n} c_i x_i) - 2^{n-1}\right| \\
=\ &\frac{1}{2^n}\left|W(f_2(x_1, \cdots, x_n) + f_1(x_1, \cdots, x_n) + b + \sum_{i=1}^{n} c_i x_i) - 2^{n-1}\right| \\
=\ &\frac{1}{2^n}\left|B + W(f_1(x_1, \cdots, x_n) + b + \sum_{i=1}^{n} c_i x_i) - 2^{n-1}\right| \\
\leqslant\ &\frac{1}{2^n}|B| + \frac{1}{2^n}\left|W(f_1(x_1, \cdots, x_n) + b + \sum_{i=1}^{n} c_i x_i) - 2^{n-1}\right|,
\end{aligned}$$

where $|B|$ is an integer (positive or negative) depending on $f_1(\cdot)$ and $b+\sum_{i=1}^{n} c_i x_i$. Because of $|B| \leqslant W(f_2)$ we have

$$\frac{1}{2^n}\left|W(g(x_1,\cdots,x_n)+b+\sum_{i=1}^{n} c_i x_i)-2^{n-1}\right| \leqslant \delta_1+\delta_2. \qquad \square$$

From Theorem 7.1.41 we know that network functions can possess good linear entropy dropping and have large linear complexity at the same time, thus there is no trade-off between them.

Before showing the next theorem we introduce in brief some concepts about the Walsh transform of Boolean function $f(x_1,\cdots,x_n)$.

$$\widehat{\mathrm{F}}(\underline{u}) \triangleq \widehat{\mathrm{F}}(u_1,\cdots,u_n)=\frac{1}{2^n}\sum_{\underline{v}}(-1)^{\underline{u}\cdot\underline{v}+f(\underline{v})}$$

is called the Walsh transform spectrum of $f(x_1,\cdots,x_n)$ which has the following properties:

(a) Inverse transform $(-1)^{f(\underline{v})}=\sum_{\underline{u}}(-1)^{\underline{u}\cdot\underline{v}\widehat{\mathrm{F}}(\underline{u})}$;

(b) Parseval's identity $\sum_{\underline{u}}[\widehat{\mathrm{F}}(\underline{u})]^2=1$.

Theorem 7.1.42 *$f(x_1,\cdots,x_n)$ has δ-linear entropy dropping iff $\left|\widehat{\mathrm{F}}(\underline{u})\right| \leqslant 2\delta$.*

Proof From

$$\widehat{\mathrm{F}}(\underline{u})=\frac{1}{2^n}\sum_{\underline{v}}(-1)^{\underline{u}\cdot\underline{v}+f(\underline{v})}=\frac{1}{2^n}\left[2^n-2W\left(f(x_1,\cdots,x_n)+\sum_{i=1}^{n} u_i x_i\right)\right],$$

we have that

$$W(f(x_1,\cdots,x_n)+\sum_{i=1}^{n} u_i x_i)=2^{n-1}(1-\widehat{\mathrm{F}}(\underline{u})),$$

$$W(f(x_1,\cdots,x_n)+\sum_{i=1}^{n} u_i x_i+1)=2^{n-1}(1+\widehat{\mathrm{F}}(\underline{u})),$$

$$\begin{aligned}\frac{1}{2^n}\left|W(f(x_1,\cdots,x_n)+b+\sum_{i=1}^{n} c_i x_i)\right| &= \frac{1}{2^n}\left|2^{n-1}(1\pm\widehat{\mathrm{F}}(\underline{u}))-2^{n-1}\right| \\ &= \frac{1}{2^n}\left|\widehat{\mathrm{F}}(\underline{u})\right|.\end{aligned}$$

The proof is completed by using Definition 7.1.15. $\square$

Corollary 7.1.8 *If $f(x_1,\cdots,x_n)$ can make the next equation identical*

$$\frac{1}{2^n}\left|W(f(x_1,\cdots,x_n)+b+\sum_{i=1}^{n} c_i x_i)-2^{n-1}\right|=\delta,$$

then $f(x_1,\cdots,x_n)$ is certainly a Bent function and $\delta=2^{-(\frac{n}{2}+1)}$.

This corollary shows that Bent functions play very important roles in the network functions with good property of linear entropy dropping.

Proof From the proof process in Theorem 7.1.42, we have $\left|\widehat{F}(\underline{u})\right| = 2\delta$. Because of the Parseval's identity $\sum_{\underline{u}}[\widehat{F}(\underline{u})]^2 = 1$, $4\delta^2 \cdot 2^n = 1$, $\delta = 2^{-(\frac{n}{2}+1)}$. Thus $\widehat{F}(\underline{u}) = \pm 2^{-\frac{n}{2}}$, $2^n\widehat{F}(\underline{u}) = \pm 2^{\frac{n}{2}}$. This means that the Hadamard coefficients of $f(x_1, \cdots, x_n)$ are identical to be $2^{\frac{n}{2}}$. Therefore $f(x_1, \cdots, x_n)$ is certainly a Bent function. □

Theorem 7.1.43 *If $f(x_1, \cdots, x_n)$ has δ-linear entropy dropping, then for any $\underline{y} = (y_1, \cdots, y_n) \neq (0, \cdots, 0)$ hold*

$$\frac{1}{4^n}\left|W[f(x_1, \cdots, x_n) + f(x_1 + y_1, \cdots, x_n + y_n)] - 2^{n-1}\right| \leqslant \delta^2.$$

Proof By the reverse property of Walsh transform we know that

$$\begin{aligned}
\sum_{\underline{x}}(-1)^{f(\underline{x})+f(\underline{x}+\underline{y})} &= \sum_{\underline{x}}(-1)^{f(\underline{x})}(-1)^{f(\underline{x}+\underline{y})} \\
&= \sum_{\underline{x}}\left[\sum_{\underline{u}}(-1)^{\underline{u}\cdot\underline{x}}\widehat{F}(\underline{u})\right]\left[\sum_{\underline{\omega}}(-1)^{(\underline{x}+\underline{y})\cdot\underline{\omega}}\widehat{F}(\underline{\omega})\right] \\
&= \sum_{\underline{u}}\sum_{\underline{\omega}}(-1)^{\underline{y}\cdot\underline{\omega}}\widehat{F}(\underline{u})\widehat{F}(\underline{\omega})\left[\sum_{\underline{x}}(-1)^{\underline{x}\cdot(\underline{u}+\underline{\omega})}\right] \\
&= 2^n\sum_{\underline{u}}(-1)^{\underline{y}\cdot\underline{u}}\left|\widehat{F}(\underline{u})\right|^2.
\end{aligned}$$

In another aspect

$$\sum_{\underline{x}}(-1)^{f(\underline{x})+f(\underline{x}+\underline{y})} = 2^n - 2W[f(x_1, \cdots, x_n) + f(x_1 + y_1, \cdots, x_n + y_n)],$$

therefore

$$\begin{aligned}
&\left|2^{n-1} - W[f(x_1, \cdots, x_n) + f(x_1 + y_1, \cdots, x_n + y_n)]\right| \\
=&\left|2^{n-1}\sum_{\underline{u}}(-1)^{\underline{y}\cdot\underline{u}}(\widehat{F}(\underline{u}))^2\right| \\
=&\left|2^{n-1}\sum_{\underline{u}\cdot\underline{y}=0}(\widehat{F}(\underline{u}))^2 - \sum_{\underline{u}\cdot\underline{y}=1}(\widehat{F}(\underline{u}))^2\right|.
\end{aligned}$$

Directly from $\left|\widehat{F}(\underline{u})\right| \leqslant 2\delta$ in Theorem 7.1.42, it is clear that

$$\begin{aligned}
&\left|2^{n-1} - W[f(x_1, \cdots, x_n) + f(x_1 + y_1, \cdots, x_n + y_n)]\right| \\
\leqslant\ & 2^{n-1} \cdot 2^{n-1} \cdot 4\delta^2 = 4^n\delta^2.
\end{aligned}$$

□

Theorem 7.1.44 *Let $f(x_1, \cdots, x_n)$ and $g(y_1, \cdots, y_n)$ have δ_1- and δ_2-linear entropy dropping, respectively. Let $h(x_1, \cdots, x_n, y_1, \cdots, y_n) = f(x_1, \cdots, x_n) + g(y_1, \cdots, y_n)$ then $h(x_1, \cdots, x_n, y_1, \cdots, y_n)$ has $2\delta_1(1 + \delta_2)$-linear entropy dropping.*

Proof Because of $g(y_1, \cdots, y_m)$ has δ_2-linear entropy dropping,

$$\frac{1}{2^m}\left|W\left[g(y_1,\cdots,y_m)+b+\sum_{i=1}^{m}a_iy_i\right]-2^{m-1}\right|\leqslant\delta_2,$$

i.e.,

$$2^{m-1}-2^m\delta_2\leqslant W\left[g(y_1,\cdots,y_m)+b+\sum_{i=1}^{m}a_iy_i\right]\leqslant 2^{m-1}+2^m\delta_2,$$

thus

$$\frac{1}{2^{m-1}}W\left[g(y_1,\cdots,y_m)+b+\sum_{i=1}^{m}a_iy_i\right]\leqslant 1+2\delta_2,$$

in another aspect

$$\begin{aligned}
&\frac{1}{2^{m+n}}\left|W\left[f(x_1,\cdots,x_n)+g(y_1,\cdots,y_m)+\sum_{i=1}^{n}c_ix_i+\sum_{i=1}^{m}a_iy_i+b\right]-2^{m+n-1}\right|\\
&=\frac{1}{2^{m+n}}\left|2^mW\left[f(x_1,\cdots,x_n)+\sum_{i=1}^{n}c_ix_i\right]+2^nW\left[g(y_1,\cdots,y_m)+b+\sum_{i=1}^{m}a_iy_i\right]\right.\\
&\left.-2W\left[f(x_1,\cdots,x_n)+\sum_{i=1}^{n}c_ix_i\right]+W\left[g(y_1,\cdots,y_m)+b+\sum_{i=1}^{m}a_iy_i\right]-2^{m+n-1}\right|\\
&\leqslant\frac{1}{2^n}\left|W(f(x_1,\cdots,x_n)+\sum_{i=1}^{n}c_ix_i)-2^{n-1}\right|\\
&+\frac{1}{2^{m+n-1}}W\left[g(y_1,\cdots,y_m)+b+\sum_{i=1}^{m}a_iy_i\right]\left|2^{n-1}-W\left[f(x_1,\cdots,x_n)\right.\right.\\
&\left.\left.+\sum_{i=1}^{n}c_ix_i\right]\right|\leqslant\delta_1+\delta_1(1+2\delta_2)=2\delta_1(1+\delta_2).
\end{aligned}$$

□

Theorem 7.1.45 *Assume that $f_1(x_1, \cdots, x_n)$ and $f_2(x_1, \cdots, x_n)$ have δ_1- and δ_2-linear entropy dropping, then $f(x_1, \cdots, x_n, x_{n+1}) = x_{n+1}f_1(x_1, \cdots, x_n) + (1 + x_{n+1})f_2(x_1, \cdots, x_n)$ has $\frac{1}{2}(\delta_1 + \delta_2)$-linear entropy dropping.*

Proof

$$\begin{aligned}
&\frac{1}{2^{n+1}}\left|W\left(f(x_1,\cdots,x_n)+\sum_{i=1}^{n}a_ix_i+bx_{n-1}+c\right)-2^n\right|\\
&=\frac{1}{2^{n+1}}\left|W\left[f_2(x_1,\cdots,x_n)+\sum_{i=1}^{n}a_ix_i+c\right]+W\left[f_1(x_1,\cdots,x_n)+b\right.\right.\\
&\left.\left.+c+\sum_{i=1}^{n}a_ix_i\right]-2^n\right|
\end{aligned}$$

$$\leqslant \frac{1}{2}\left\{\frac{1}{2^n}\left|W\left[f_2(x_1,\cdots,x_n)+c+\sum_{i=1}^{n}a_ix_i\right]-2^{n-1}\right|\right.$$
$$\left.+\frac{1}{2^n}\left|W\left[f_1(x_1,\cdots,x_n)+b+c+\sum_{i=1}^{n}a_ix_i\right]-2^{n-1}\right|\right\}$$
$$\leqslant \frac{1}{2}(\delta_1+\delta_2). \qquad \square$$

As with Theorem 7.1.36, Theorem 7.1.45 can be generalized to the following corollary.

Corollary 7.1.9 *Assume for $1 \leqslant i \leqslant 2^k$, $f_i(x_1,\cdots,x_n)$ has δ_i-linear entropy dropping. Let*

$$f(x_1,\cdots,x_n,y_1,\cdots,y_k)=\sum_{\underline{a}} y_1^{(a_1)}\cdots y_k^{(a_k)} f_{\underline{a}}(x_1,\cdots,x_n),$$

then $f(x_1,\cdots,x_n,y_1,\cdots,y_k)$ has $\left(2^{-k}\sum_{i=1}^{2^k}\delta_i\right)$-linear entropy dropping.

When $k=1$ this Corollary 7.1.9 is just Theorem 7.1.45.

7.2 Correlation Functions of Geometric Sequences

7.2.1 On the Correlation Functions of a Family of Gold-Geometric Sequences[18,19]

Many well-known binary and nonbinary sequences such as m-sequences, GMW-sequences, Bent sequences, No sequences, Kasami sequences (small set and large set), Kumar sequences, and even Legendre sequences, can be represented by combination trace functions defined by $\mathrm{Tr}_q^{q^n}=\sum_{i=0}^{n-1}x^{q^i}$ which is a map from $\mathrm{GF}(q^n)$ to $\mathrm{GF}(q)$. Trace function plays an important role in designing sequences with good correlations and high linear complexities. It's well known that the infinite periodic sequence S defined by $s_i=\mathrm{Tr}_q^{q^n}(\alpha^i)$ is a q-ary m-sequence if α is a primitive element of the finite field $\mathrm{GF}(q^n)$.

Geometric sequences, derived from m-sequences, are becoming more and more important in crytography and spread spectrum communications. Many papers on their correlations, partial correlations, and linear complexities have been published. Klapper et al. calculated the cross-correlation functions of geometric sequences derived from linearly or quadratically related q-ary m-sequences. In the same paper, they defined the generalized geometric sequences and proposed an open problem of calculating the correlations of these generalized geometric sequences. This open problem will be solved in this subsection by deriving explicit formulas for cross(auto)-correlation functions of linearly related generalized geometric sequences.

Throughout this subsection, we use q to stand for a fixed power of a prime p, $\mathrm{GF}(q)$ the Galois field with q elements, α and β primitive elements of $\mathrm{GF}(q^n)$.

Definition 7.2.1 *Let q be a power of prime p, n a positive integer, α and β be primitive elements of $\mathrm{GF}(q^n)$ with $\beta=\alpha^k$ and $\gcd(k,q^n-1)=1$. For a nonlinear function f from $\mathrm{GF}(q)$ to $\mathrm{GF}(2)$, a sequence $S_f^{(A,B)}=\{S_f^{(A,B)}(i):0\leqslant i\leqslant q^n-2\}$ is called a Gold-geometric sequence, if*

$$S_f^{(A,B)}(i)=f(\mathrm{Tr}_q^{q^n}(A\alpha^i+B\beta^i)),$$

where $A, B \in \mathrm{GF}(q^n)^*$, $\mathrm{Tr}(^{q^n}_q)(\cdot)$ *the trace function from* $\mathrm{GF}(q^n)$ *to* $\mathrm{GF}(q)$. *If* $\beta = \alpha^{p^e}$, *i.e.,* $k = p^e$, *th sequence* $S_f^{(A,B)}$ *is called a linearly related Gold-geometric sequence.*

Remark 7.2.1 *We use the term Gold-geometric sequence instead of generalized geometric sequence to emphasize that the sequence in Definition 7.2.1 is derived from a Gold-sequence.*

Definition 7.2.2 *For given two binary* $(0,1)$ *sequence* $S = \{s(0), s(1), \cdots\}$ *and* $T = \{t(0), t(1), \cdots\}$ *with period* L, *the periodic cross-correlation function* $\Theta_{S,T}$ *of* S *and* T *is defined by*

$$\Theta_{S,T}(\tau) = \sum_{i=0}^{L-1}(-1)^{s(i+\tau)+t(i)}, \quad \tau = 0, \cdots, L-1.$$

In particular, $\Theta_{S,S}(\tau)$ *is the autocorrelation function of* S.

Let $A, B, C, D \in \mathrm{GF}(q^n)^*$, $a, b, c, d \in \mathrm{GF}(q)$. The equation $\mathrm{Tr}_q^{q^n}(Ax) = a$ describes a hyperplane in the finite geometry $\mathrm{GF}(q^n)$. Let N denote the number of intersection points of the following four hyperplanes in $\mathrm{GF}(q^n)$,

$$\begin{cases} \mathrm{Tr}_q^{q^n}(Ax) = a, \\ \mathrm{Tr}_q^{q^n}(Bx) = b, \\ \mathrm{Tr}_q^{q^n}(Cx) = c, \\ \mathrm{Tr}_q^{q^n}(Dx) = d. \end{cases}$$

Let χ_q be the canonical additive character of $\mathrm{GF}(q)$, then

$$\sum_{x \in \mathrm{GF}(q)} \chi_q(ax) = \begin{cases} q, & a = 0, \\ 0, & a \neq 0. \end{cases}$$

Thus

$$\begin{aligned} q^4 N = & \sum_{\zeta_i \in \mathrm{GF}(q)} \sum_{x \in \mathrm{GF}(q^n)} \chi_q(\zeta_1 \mathrm{Tr}_q^{q^n}(Ax) - \zeta_1 a)\chi_q(\zeta_2 \mathrm{Tr}_q^{q^n}(Bx) - \zeta_2 b) \\ & \cdot \chi_q(\zeta_3 \mathrm{Tr}_q^{q^n}(Cx) - \zeta_3 c)\chi_q(\zeta_4 \mathrm{Tr}_q^{q^n}(Dx) - \zeta_4 d) \\ = & \sum_{\zeta_i \in \mathrm{GF}(q)} \chi_q(-\zeta_1 a - \zeta_2 b - \zeta_3 c - \zeta_4 d) \\ & \cdot \sum_{x \in \mathrm{GF}(q^n)} \chi_{q^n}((A\zeta_1 + B\zeta_2 + C\zeta_3 + D\zeta_4)x) \\ = & q^n \sum_{\zeta_i \in \mathrm{GF}(q), A\zeta_1 + B\zeta_2 + C\zeta_3 + D\zeta_4 = 0} \chi_q(-\zeta_1 a - \zeta_2 b - \zeta_3 c - \zeta_4 d), \end{aligned}$$

that is,

$$N = q^{n-4} \sum_{\zeta_i \in \mathrm{GF}(q), A\zeta_1 + B\zeta_2 + C\zeta_3 + D\zeta_4 = 0} \chi_q(-\zeta_1 a - \zeta_2 b - \zeta_3 c - \zeta_4 d).$$

Lemma 7.2.1 *When* $\frac{B}{A}, \frac{C}{A}, \frac{D}{A} \in \mathrm{GF}(q)$, *then*

$$N = \begin{cases} q^{n-1}, & if\ \frac{aB}{A} - b = \frac{aC}{A} - c = \frac{aD}{A} - d = 0, \\ 0, & otherwise. \end{cases}$$

Proof We have $-\zeta_1 = \frac{B}{A}\zeta_2 + \frac{C}{A}\zeta_3 + \frac{D}{A}\zeta_4$ since $A\zeta_1 + B\zeta_2 + C\zeta_3 + D\zeta_4 = 0$ and $\frac{B}{A}, \frac{C}{A}, \frac{D}{A} \in \mathrm{GF}(q)$, so

$$\begin{aligned} N = & q^{n-4} \sum_{\zeta_2,\zeta_3,\zeta_4 \in \mathrm{GF}(q)} \chi_q\left(\left(\frac{aB}{A} - b\right)\zeta_2 + \left(\frac{aC}{A} - c\right)\zeta_3 + \left(\frac{aD}{A} - d\right)\zeta_4\right) \\ = & q^{n-4} \sum_{\zeta_2 \in \mathrm{GF}(q)} \chi_q\left(\left(\frac{aB}{A} - b\right)\zeta_2\right) \sum_{\zeta_3 \in \mathrm{GF}(q)} \chi_q\left(\left(\frac{aC}{A} - c\right)\zeta_3\right) \\ & \cdot \sum_{\zeta_4 \in \mathrm{GF}(q)} \chi_q\left(\left(\frac{aD}{A} - d\right)\zeta_4\right) \\ = & \begin{cases} q^{n-1}, & \text{if } \frac{aB}{A} - b = \frac{aC}{A} - c = \frac{aD}{A} - d = 0, \\ 0, & \text{otherwise.} \end{cases} \end{aligned}$$

□

Lemma 7.2.2 *When only one of $\frac{B}{A}, \frac{C}{A}$ and $\frac{D}{A}$ is in $\mathrm{GF}(q^n) - \mathrm{GF}(q)$.*

Case (L2.1). *If $\frac{D}{A} \notin \mathrm{GF}(q)$, then*

$$N = \begin{cases} q^{n-2}, & \text{if } \frac{aB}{A} - b = \frac{aC}{A} - c = 0, \\ 0, & \text{otherwise.} \end{cases}$$

Case (L2.2). *If $\frac{C}{A} \notin \mathrm{GF}(q)$, then*

$$N = \begin{cases} q^{n-2}, & \text{if } \frac{aB}{A} - b = \frac{aD}{A} - d = 0, \\ 0, & \text{otherwise.} \end{cases}$$

Case (L2.3). *If $\frac{B}{A} \notin \mathrm{GF}(q)$, then*

$$N = \begin{cases} q^{n-2}, & \text{if } \frac{aC}{A} - c = \frac{aD}{A} - d = 0, \\ 0, & \text{otherwise.} \end{cases}$$

Proof Case (L2.1). Suppose $(\zeta_1, \zeta_2, \zeta_3, \zeta_4) \in \mathrm{GF}(q)^4$ such that $A\zeta_1 + B\zeta_2 + C\zeta_3 + D\zeta_4 = 0$, then

$$-\frac{D}{A}\zeta_4 = \zeta_1 + \frac{B}{A}\zeta_2 + \frac{C}{A}\zeta_3 \in \mathrm{GF}(q).$$

We have that $\zeta_4 = 0$, $\zeta_1 = -\frac{B}{A}\zeta_2 - \frac{C}{A}\zeta_3$ since $\frac{D}{A} \notin \mathrm{GF}(q)$, thus

$$\begin{aligned} N = & q^{n-4} \sum_{\zeta_2,\zeta_3 \in \mathrm{GF}(q)} \chi_q\left(\left(\frac{aB}{A} - b\right)\zeta_2 + \left(\frac{aC}{A} - c\right)\zeta_3\right) \\ = & q^{n-4} \sum_{\zeta_2 \in \mathrm{GF}(q)} \chi_q\left(\left(\frac{aB}{A} - b\right)\zeta_2\right) \sum_{\zeta_3 \in \mathrm{GF}(q)} \chi_q\left(\left(\frac{aC}{A} - c\right)\zeta_3\right) \end{aligned}$$

$$= \begin{cases} q^{n-2}, & \text{if } \dfrac{aB}{A} - b = \dfrac{aC}{A} - c = 0, \\ 0, & \text{otherwise.} \end{cases}$$

Case (L2.2) and Case (L2.3) can be proved similarly. □

Let $x, y, z \in \mathrm{GF}(q^n)^*$, if $z/x \notin \mathrm{GF}(q)$, that is $(z/x)^q \neq z/x$, we denote $(x, y, z)_q = \dfrac{(y/x)^q - y/x}{(z/x)^q - z/x}$.

Lemma 7.2.3 *When only one of* $\dfrac{B}{A}, \dfrac{C}{A}, \dfrac{D}{A}$ *is in* $\mathrm{GF}(q)$.

Case (L3.1.1). *If* $\dfrac{C}{A}, \dfrac{D}{A} \notin \mathrm{GF}(q)$, $(A, D, C)_q \notin \mathrm{GF}(q)$, *then*

$$N = \begin{cases} q^{n-3}, & \textit{if } \dfrac{aB}{A} - b = 0, \\ 0, & \textit{otherwise.} \end{cases}$$

Case (L3.1.2). *If* $\dfrac{C}{A}, \dfrac{D}{A} \notin \mathrm{GF}(q)$, $(A, D, C)_q \in \mathrm{GF}(q)$, *then*

$$N = \begin{cases} q^{n-2}, & \textit{if } \dfrac{aB}{A} - b = \left(\dfrac{aD}{A} - d\right) - \left(\dfrac{aC}{A} - c\right)(A, D, C)_q = 0, \\ 0, & \textit{otherwise.} \end{cases}$$

Case (L3.2.1). *If* $\dfrac{B}{A}, \dfrac{D}{A} \notin \mathrm{GF}(q)$, $(A, D, B)_q \notin \mathrm{GF}(q)$, *then*

$$N = \begin{cases} q^{n-3}, & \textit{if } \dfrac{aC}{A} - c = 0, \\ 0, & \textit{otherwise.} \end{cases}$$

Case (L3.2.2). *If* $\dfrac{B}{A}, \dfrac{D}{A} \notin \mathrm{GF}(q)$, $(A, D, B)_q \in \mathrm{GF}(q)$, *then*

$$N = \begin{cases} q^{n-2}, & \textit{if } \dfrac{aC}{A} - c = \left(\dfrac{aD}{A} - d\right) - \left(\dfrac{aB}{A} - b\right)(A, D, B)_q = 0, \\ 0, & \textit{otherwise.} \end{cases}$$

Case (L3.3.1). *If* $\dfrac{B}{A}, \dfrac{C}{A} \notin \mathrm{GF}(q)$, $(A, C, B)_q \notin \mathrm{GF}(q)$, *then*

$$N = \begin{cases} q^{n-3}, & \textit{if } \dfrac{aD}{A} - d = 0, \\ 0, & \textit{otherwise.} \end{cases}$$

Case (L3.3.2). *If* $\dfrac{B}{A}, \dfrac{C}{A} \notin \mathrm{GF}(q)$, $(A, C, B)_q \in \mathrm{GF}(q)$, *then*

$$N = \begin{cases} q^{n-2}, & \textit{if } \dfrac{aD}{A} - d = \left(\dfrac{aC}{A} - c\right) - \left(\dfrac{aB}{A} - b\right)(A, C, B)_q = 0, \\ 0, & \textit{otherwise.} \end{cases}$$

Proof Case (L3.1.1). Suppose $(\zeta_1,\zeta_2,\zeta_3,\zeta_4)\in \mathrm{GF}(q)^4$ such that $A\zeta_1+B\zeta_2+C\zeta_3+D\zeta_4=0$, then

$$-\zeta_1-\frac{B}{A}\zeta_2=\frac{C}{A}\zeta_3+\frac{D}{A}\zeta_4\in \mathrm{GF}(q),$$

therefore $\left(\frac{C}{A}\zeta_3+\frac{D}{A}\zeta_4\right)^q=\frac{C}{A}\zeta_3+\frac{D}{A}\zeta_4$, that is $\zeta_3=-(A,D,C)_q\zeta_4$. Thus $\zeta_3=\zeta_4=0$, $\zeta_1=-\frac{B}{A}\zeta_2$ since $\zeta_3,\zeta_4\in\mathrm{GF}(q)$ and $(A,D,C)_q\notin \mathrm{GF}(q)$. So

$$\begin{aligned}N&=q^{n-4}\sum_{\zeta_2\in\mathrm{GF}(q)}\chi_q\left(\left(\frac{aB}{A}-b\right)\zeta_2\right)\\&=\begin{cases}q^{n-3}, & \text{if } \frac{aB}{A}-b=0,\\ 0, & \text{otherwise.}\end{cases}\end{aligned}$$

Case (L3.1.2). Similar to the proof of Case (L3.1.1), $(\zeta_1,\zeta_2,\zeta_3,\zeta_4)\in \mathrm{GF}(q)^4$ such that $A\zeta_1+B\zeta_2+C\zeta_3+D\zeta_4=0$ if and only $\zeta_1=-\frac{B}{A}\zeta_2+\left[\frac{C}{A}(A,D,C)_q-\frac{D}{A}\right]\zeta_4, \zeta_3=-(A,D,C)_q\zeta_4$, and $\zeta_2,\zeta_4\in\mathrm{GF}(q)$. It's easy to prove that $\left(\frac{aD}{A}-d\right)-\left(\frac{aC}{A}-c\right)(A,D,C)_q\in\mathrm{GF}(q)$, thus

$$\begin{aligned}N=&q^{n-4}\sum_{\zeta_2,\zeta_4\in\mathrm{GF}(q)}\chi_q\left(\left(\frac{aB}{A}-b\right)\zeta_2\right.\\&\left.+\left[\left(\frac{aD}{A}-d\right)-\left(\frac{aC}{A}-c\right)(A,D,C)_q-\frac{D}{A}\right]\zeta_4\right)\\=&\begin{cases}q^{n-2}, & \text{if } \frac{aB}{A}-b=\left(\frac{aD}{A}-d\right)-\left(\frac{aC}{A}-c\right)(A,D,C)_q=0,\\ 0, & \text{otherwise.}\end{cases}\end{aligned}$$

Other cases can be proved in the same way. □

Lemma 7.2.4 *When* $\frac{B}{A},\frac{C}{A}$ *and* $\frac{D}{A}$ *are in* $\mathrm{GF}(q^n)-\mathrm{GF}(q)$.

Case (L4.1). *If* $(A,C,B)_q,(A,D,B)_q\in\mathrm{GF}(q)$, *then*

$$N=\begin{cases}q^{n-2}, & if \left(\frac{aC}{A}-c\right)-\left(\frac{aB}{A}-b\right)(A,C,B)_q\\ & \quad=\left(\frac{aD}{A}-d\right)-\left(\frac{aB}{A}-b\right)(A,D,B)_q=0,\\ 0, & otherwise.\end{cases}$$

Case (L4.2). *If* $(A,C,B)_q\notin\mathrm{GF}(q)$ *and* $(A,D,B)_q\in\mathrm{GF}(q)$, *then*

$$N=\begin{cases}q^{n-3}, & if \left(\frac{aD}{A}-d\right)-\left(\frac{aB}{A}-b\right)(A,D,B)_q=0,\\ 0, & otherwise.\end{cases}$$

If $(A,D,B)_q \notin \mathrm{GF}(q)$ *and* $(A,C,B)_q \in \mathrm{GF}(q)$, *then*

$$N = \begin{cases} q^{n-3}, & if \left(\frac{aC}{A} - c\right) - \left(\frac{aB}{A} - b\right)(A,C,B)_q = 0, \\ 0, & otherwise. \end{cases}$$

Case (L4.3). *If* $(A,C,B)_q \notin \mathrm{GF}(q)$ *and* $(A,D,B)_q \in \mathrm{GF}(q^n) - \mathrm{GF}(q)$, *and*

$$if\ \delta(A,D,B,C) = \frac{(A,D,B)_q^q - (A,C,B)_q}{(A,C,D)_q^q - (A,C,B)_q} \notin \mathrm{GF}(q),\ then\ N = q^{n-4};$$

$$if\ \delta(A,D,B,C) = \frac{(A,D,B)_q^q - (A,C,B)_q}{(A,C,D)_q^q - (A,C,B)_q} \in \mathrm{GF}(q),\ then$$

$$N = \begin{cases} q^{n-3}, & if\ \Gamma = 0, \\ 0, & otherwise, \end{cases}$$

where $\Gamma = \left(\frac{aD}{A} - d\right) - \left(\frac{aC}{A} - c\right)\delta(A,D,B,C)_q + \left(\frac{aB}{A} - b\right)(\delta(A,D,B,C)(A,C,B)_q - (A,D,B)_q)$.

Proof Suppose $(\zeta_1, \zeta_2, \zeta_3, \zeta_4) \in \mathrm{GF}(q)^4$ such that $A\zeta_1 + B\zeta_2 + C\zeta_3 + D\zeta_4 = 0$, then

$$-\zeta_1 = \frac{B}{A}\zeta_2 + \frac{C}{A}\zeta_3 + \frac{D}{A}\zeta_4 \in \mathrm{GF}(q),$$

therefore $\left[\left(\frac{B}{A}\right)^q - \frac{B}{A}\right]\zeta_2 + \left[\left(\frac{C}{A}\right)^q - \frac{C}{A}\right]\zeta_3 + \left[\left(\frac{D}{A}\right)^q - \frac{D}{A}\right]\zeta_4$, that is

$$-\zeta_2 = (A,C,B)_q\zeta_3 + (A,D,B)_q\zeta_4 \in \mathrm{GF}(q).$$

Case (L4.1). Since $(A,C,B)_q, (A,D,B)_q \in \mathrm{GF}(q)$, then $\zeta_1 = \left[\frac{B}{A}(A,C,B)_q - \frac{C}{A}\right]\zeta_3 - \left[\frac{B}{A}(A,D,B)_q - \frac{D}{A}\right]\zeta_4$, $\zeta_2 = -(A,C,D)_q\zeta_3 - (A,D,B)_q\zeta_4$, and $\zeta_3, \zeta_4 \in \mathrm{GF}(q)$.

Thus

$$\begin{aligned} N &= q^{n-4} \sum_{\zeta_3,\zeta_4 \in \mathrm{GF}(q)} \chi_q\Bigg(\left(\frac{aC}{A} - \frac{aB}{A}(A,C,B)_q + b(A,C,B)_q - c\right)\zeta_3 \\ &\quad + \left(\frac{aD}{A} - \frac{aB}{A}(A,D,B)_q + b(A,D,B)_q - d\right)\zeta_4\Bigg) \\ &= \begin{cases} q^{n-2}, & \text{if } \left(\frac{aC}{A} - c\right) - \left(\frac{aB}{A} - b\right)(A,C,B)_q \\ & \quad = \left(\frac{aD}{A} - d\right) - \left(\frac{aB}{A} - b\right)(A,D,B)_q = 0, \\ 0, & \text{otherwise.} \end{cases} \end{aligned}$$

Case (L4.2). Since $(A,C,B)_q \notin \mathrm{GF}(q)$, then $(A,C,B)_q\zeta_3 = -\zeta_2 - (A,D,B)_q\zeta_4 \in \mathrm{GF}(q)$, we have $\zeta_3 = 0$. Thus $\zeta_1 = \left[\frac{B}{A}(A,D,B)_q - \frac{D}{A}\right]\zeta_4, \zeta_2 = -(A,D,B)_q\zeta_4, \zeta_3 = 0$, and $\zeta_4 \in$

GF(q). Therefore

$$N = q^{n-4} \sum_{\zeta_4 \in \mathrm{GF}(q)} \chi_q\left(\left(\frac{aD}{A} - \frac{aB}{A}(A,D,B)_q + b(A,D,B)_q - d\right)\zeta_4\right)$$
$$= \begin{cases} q^{n-3}, & \text{if } \left(\dfrac{aD}{A} - d\right) - \left(\dfrac{aB}{A} - b\right)(A,D,B)_q = 0, \\ 0, & \text{otherwise.} \end{cases}$$

If $(A,D,B)_q \notin \mathrm{GF}(q)$, the conclusion can be proved similarly.

Case (L4.3). Because $-\zeta_2 = (A,C,B)_q\zeta_3 + (A,D,B)_q\zeta_4 \in \mathrm{GF}(q)$, we have $[(A,C,B)_q^q - (A,C,B)_q]\zeta_3 + [(A,D,B)_q^q - (A,D,B)_q]\zeta_4 = 0$, that is $-\zeta_3 = \delta(A,D,B,C)\zeta_4$.

If $\delta(A,D,B,C) \notin \mathrm{GF}(q)$, then $\zeta_i = 0,\ for\ i = 1,2,3,4$. Thus $N = q^{n-4}$.

If $\delta(A,D,B,C) \in \mathrm{GF}(q)$, then $\zeta_1 = \left[\dfrac{B}{A}(A,D,B)_q - \dfrac{B}{A}(A,C,B)_q\delta(A,D,B,C) + \dfrac{C}{A}\delta(A,D,B,C) - \dfrac{D}{A}\right]\zeta_4$, $\zeta_2 = [(A,C,B)_q\delta(A,D,B,C) - (A,D,B)_q]\zeta_4$, $\zeta_3 = -\delta(A,D,B,C)\zeta_4$, $\zeta_4 \in \mathrm{GF}(q)$.

Therefore

$$N = q^{n-4} \sum_{\zeta_4 \in \mathrm{GF}(q)} \chi_q(\Gamma\zeta_4) = \begin{cases} q^{n-3}, & \text{if } \Gamma = 0, \\ 0, & \text{otherwise,} \end{cases}$$

where $\Gamma = \left(\dfrac{aD}{A} - d\right) - \left(\dfrac{aC}{A} - c\right)\delta(A,D,B,C)_q + \left(\dfrac{aB}{A} - b\right)(\delta(A,D,B,C)(A,C,B)_q - (A,D,B)_q)$. □

In the following text of this subsection, we will suppose that $k = p^e$. Let α and β be primitive elements of GF(q^n) with $\beta = \alpha^k$, $A, B, A_1, B_1 \in \mathrm{GF}(q^n)^*$, and for any two nonlinear functions f and g form GF(q) to GF(2), let $S_f^{(A,B)}(i) = f(\mathrm{Tr}_q^{q^n}(A\alpha^i + B\beta^i))$ and $S_g^{(A_1,B_1)}(i) = g(\mathrm{Tr}_q^{q^n}(A_1\alpha^i + B_1\beta^i))$ be two linearly related Gold-geometric sequences, which will be abbreivated by S and T in the sequel, respectively.

By the definition of periodic cross-correlation function of sequences, we have

$$\Theta_{S,T}(\tau) = \sum_{i=0}^{q^n-2} F(\mathrm{Tr}(A\alpha^i) + \mathrm{Tr}(B\alpha^{ki}))G(\mathrm{Tr}(A_1\alpha^{i+\tau}) + \mathrm{Tr}(B_1\alpha^{ki+k\tau}))$$
$$= \sum_{u,v \in \mathrm{GF}(q)} N_\tau(u,v)F(u)G(v) - F(0)G(0), \tag{7.16}$$

where $N_\tau(u,v)$ denotes the number of intersection points of the following two hypersurfaces in GF(q^n),

$$\begin{cases} \mathrm{Tr}_q^{q^n}(Ax + Bx^k) = u, \\ \mathrm{Tr}_q^{q^n}(A_1\alpha^\tau x + B_1\alpha^{k\tau}x^k) = v, \end{cases}$$

which is equivalent to

$$\begin{cases} \mathrm{Tr}_q^{q^n}(Ax) + \mathrm{Tr}_q^{q^n}(B^{1/k}x)^k = u, \\ \mathrm{Tr}_q^{q^n}(A_1\alpha^\tau x) + \mathrm{Tr}_q^{q^n}(B_1^{1/k}\alpha^\tau x)^k = v. \end{cases}$$

It's easy to prove that

Lemma 7.2.5

$$N_\tau(u,v)=\sum_{u_1,v_1\in \mathrm{GF}(q)} M_\tau(u_1,v_1),$$

where $M_\tau(u_1,v_1)$ denotes the number of intersection points of the following four hyperplanes in $\mathrm{GF}(q^n)$,

$$\begin{cases} \mathrm{Tr}_q^{q^n}(Ax)=u-u_1,\\ \mathrm{Tr}_q^{q^n}(B^{1/k}x)^k=u_1^{1/k},\\ \mathrm{Tr}_q^{q^n}(A_1\alpha^\tau x)=v-v_1,\\ \mathrm{Tr}_q^{q^n}(B_1^{1/k}\alpha^\tau x)=v_1^{1/k}. \end{cases}$$

For simplicity, let $A_0=A$, $B_0=B^{1/k}$, $C_0=A_1\alpha^\tau$, $D_0=B_1^{1/k}\alpha^\tau, a_0=u-u_1$, $b_0=u_1^{1/k}$, $c_0=v-v_1$, $d_0=v_1^{1/k}$, $W(a,b)$ the number of solutions of $ax^k+x=b$ in $\mathrm{GF}(q)$, $0\neq a,b\in \mathrm{GF}(q)$, and for a nonlinear function f from $\mathrm{GF}(q)$ to $\mathrm{GF}(2)$, $F(x)=(-1)^{f(x)}, I(f)=\sum_{x\in\mathrm{GF}(q)}F(x)$.

Theorem 7.2.1 *When* $\dfrac{B_0}{A_0},\dfrac{C_0}{A_0},\dfrac{D_0}{A_0}\in \mathrm{GF}(q)$.

If $\dfrac{C_0}{A_0}-\left(\dfrac{D_0}{B_0}\right)^k=0$, *then*

$$N_\tau(u,v)=\begin{cases} W\left(\left(\dfrac{B_0}{A_0}\right)^k,\left(\dfrac{uB_0}{A_0}\right)^k\right)q^{n-1}, & \textit{if } \dfrac{C_0}{A_0}u-v=0,\\ 0, & \textit{if } \dfrac{C_0}{A_0}u-v\neq 0, \end{cases}$$

$$\Theta_{S,T}(\tau)=q^{n-1}\sum_{u\in\mathrm{GF}(q)} W\left(\left(\frac{B_0}{A_0}\right)^k,\left(\frac{uB_0}{A_0}\right)^k\right)F(u)G(C_0u/A_0)-F(0)G(0).$$

If $\dfrac{C_0}{A_0}-\left(\dfrac{D_0}{B_0}\right)^k\neq 0$, *then*

$$N_\tau(u,v)=\begin{cases} q^{n-1}, & \textit{if } \left[u-\dfrac{\frac{C_0}{A_0}u-v}{\frac{C_0}{A_0}-\left(\frac{D_0}{B_0}\right)^k}\right]\dfrac{B_0}{A_0}=\left[\dfrac{\frac{C_0}{A_0}u-v}{\frac{C_0}{A_0}-\left(\frac{D_0}{B_0}\right)^k}\right]^{1/k},\\ 0, & \textit{otherwise}, \end{cases}$$

$$\Theta_{S,T}(\tau)=q^{n-1}\sum_{\left[u-\frac{\frac{C_0}{A_0}u-v}{\frac{C_0}{A_0}-(\frac{D_0}{B_0})^k}\right]\frac{B_0}{A_0}=\left[\frac{\frac{C_0}{A_0}u-v}{\frac{C_0}{A_0}-(\frac{D_0}{B_0})^k}\right]^{1/k}} F(u)G(v)-F(0)G(0).$$

Proof It's trivial that

$$M_\tau(u_1, v_1) = \begin{cases} q^{n-1}, & \text{if } \dfrac{a_0 B_0}{A_0} - b_0 = \dfrac{a_0 C_0}{A_0} - c_0 = \dfrac{a_0 D_0}{A_0} - d_0 = 0, \\ 0, & \text{otherwise} \end{cases}$$

and $N_\tau(u, v) = q^{n-1}V$ from Lemma 7.2.1 and Lemma 7.2.5, where V denotes the number of solution pairs $(x, y) \in \mathrm{GF}(q)^2$ of the following system

$$\begin{cases} \dfrac{B_0}{A_0}(u - x) = x^{1/k}, \\ \dfrac{C_0}{A_0}(u - x) = v - y, \\ \dfrac{D_0}{A_0}(u - x) = y^{1/k}. \end{cases}$$

We know that x and y are uniquely determined by each other since $\dfrac{C_0}{A_0}(u - x) = v - y$, so V is equal to the number of solutions of

$$\begin{cases} \dfrac{B_0}{A_0}(u - x) = x^{1/k}, \\ \left(\dfrac{C_0}{A_0} - \left(\dfrac{D_0}{B_0}\right)^k\right) x = \dfrac{C_0}{A_0} u - v. \end{cases}$$

If $\dfrac{C_0}{A_0} - \left(\dfrac{D_0}{B_0}\right)^k = 0$, then

$$V = \begin{cases} W\left(\left(\dfrac{B_0}{A_0}\right)^k, \left(\dfrac{uB_0}{A_0}\right)^k\right) q^{n-1}, & \text{if } \dfrac{C_0}{A_0} u - v = 0, \\ 0, & \text{if } \dfrac{C_0}{A_0} u - v \neq 0. \end{cases}$$

If $\dfrac{C_0}{A_0} - \left(\dfrac{D_0}{B_0}\right)^k \neq 0$, then

$$V = \begin{cases} 1, & \text{if } \left[u - \dfrac{\dfrac{C_0}{A_0} u - v}{\dfrac{C_0}{A_0} - \left(\dfrac{D_0}{B_0}\right)^k}\right] \dfrac{B_0}{A_0} = \left[\dfrac{\dfrac{C_0}{A_0} u - v}{\dfrac{C_0}{A_0} - \left(\dfrac{D_0}{B_0}\right)^k}\right]^{1/k}, \\ 0, & \text{otherwise.} \end{cases}$$

The cross-correlation functions can be easily obtained from (7.16) and the above equation. □

Similarly we have

Theorem 7.2.2 *When only one of* $\dfrac{B_0}{A_0}, \dfrac{C_0}{A_0}, \dfrac{D_0}{A_0}$ *is in* $\mathrm{GF}(q^n) - \mathrm{GF}(q)$.

Case (T2.1). *If* $\frac{D_0}{A_0} \notin \mathrm{GF}(q)$, *then*

$$N_\tau(u,v) = W\left(\left(\frac{B_0}{A_0}\right)^k, \left(\frac{uB_0}{A_0}\right)^k\right) q^{n-2},$$

$$\Theta_{S,T}(\tau) = q^{n-2} I(g) \sum_{u \in \mathrm{GF}(q)} W\left(\left(\frac{B_0}{A_0}\right)^k, \left(\frac{uB_0}{A_0}\right)^k\right) F(u) - F(0)G(0).$$

Case (T2.2). *If* $\frac{C_0}{A_0} \notin \mathrm{GF}(q)$, *then*

$$N_\tau(u,v) = W\left(\left(\frac{B_0}{A_0}\right)^k, \left(\frac{uB_0}{A_0}\right)^k\right) q^{n-2},$$

$$\Theta_{S,T}(\tau) = q^{n-2} I(g) \sum_{u \in \mathrm{GF}(q)} W\left(\left(\frac{B_0}{A_0}\right)^k, \left(\frac{uB_0}{A_0}\right)^k\right) F(u) - F(0)G(0).$$

Case (T2.3). *If* $\frac{B_0}{A_0} \notin \mathrm{GF}(q)$, *then*

$$N_\tau(u,v) = W\left(\left(\frac{D_0}{C_0}\right)^k, \left(\frac{vD_0}{C_0}\right)^k\right) q^{n-2},$$

$$\Theta_{S,T}(\tau) = q^{n-2} I(f) \sum_{u \in \mathrm{GF}(q)} W\left(\left(\frac{D_0}{C_0}\right)^k, \left(\frac{vD_0}{C_0}\right)^k\right) G(u) - F(0)G(0).$$

Remark 7.2.2 *If f and g are balanced, i.e., $I(f) = I(g) = 0$, and $f(0) = g(0)$, then $\Theta_{S,T}(\tau) = -1$ when only one of $\frac{B_0}{A_0}, \frac{C_0}{A_0}, \frac{D_0}{A_0}$ is in $\mathrm{GF}(q^n) - \mathrm{GF}(q)$ from Theorem 7.2.2.*

Theorem 7.2.3 *When only one of $\frac{B_0}{A_0}, \frac{C_0}{A_0}, \frac{D_0}{A_0}$ is in $\mathrm{GF}(q)$.*

Case (T3.1.1). *If* $\frac{C_0}{A_0}, \frac{D_0}{A_0} \notin \mathrm{GF}(q)$, $(A_0, D_0, C_0)_q \notin \mathrm{GF}(q)$, *then*

$$N_\tau(u,v) = W\left(\left(\frac{B_0}{A_0}\right)^k, \left(\frac{uB_0}{A_0}\right)^k\right) q^{n-3},$$

$$\Theta_{S,T}(\tau) = q^{n-3} I(f) \sum_{u \in \mathrm{GF}(q)} W\left(\left(\frac{B_0}{A_0}\right)^k, \left(\frac{uB_0}{A_0}\right)^k\right) F(u) - F(0)G(0).$$

Case (T3.1.2). *If* $\frac{C_0}{A_0}, \frac{D_0}{A_0} \notin \mathrm{GF}(q)$, $(A_0, D_0, C_0)_q \in \mathrm{GF}(q)$, *then*

$$N_\tau(u,v) = W((A_0, D_0, C_0)_q^k, 0)\overline{M}(u,v) q^{n-2},$$

$$\Theta_{S,T}(\tau) = q^{n-2} W((A_0, D_0, C_0)_q^k, 0) \sum_{u,v \in \mathrm{GF}(q)} \overline{M}(u,v) F(u) G(v) - F(0)G(0).$$

Case (T3.2.1). *If* $\frac{B_0}{A_0}, \frac{D_0}{A_0} \notin \mathrm{GF}(q)$, $(A_0, D_0, B_0)_q \notin \mathrm{GF}(q)$, *then*

$$N_\tau(u,v) = q^{n-2},$$

$$\Theta_{S,T}(\tau) = q^{n-2} I(f)I(g) - F(0)G(0).$$

Case (T3.2.2). *If* $\frac{B_0}{A_0}, \frac{D_0}{A_0} \notin \mathrm{GF}(q)$, $(A_0, D_0, B_0)_q \in \mathrm{GF}(q)$, *then*

$$N_\tau(u,v) = q^{n-2} W\left(\left[\frac{D_0}{A_0} - \frac{B_0}{A_0}(A_0, D_0, B_0)_q\right]^k \Big/ \left(\frac{C_0}{A_0} - (A_0, D_0, B_0)_q\right)^k, \right.$$
$$\left. \left(\left[\frac{uD_0}{A_0} - \frac{uB_0}{A_0}(A_0, D_0, B_0)_q\right]^k + \frac{uC_0}{A_0} - v\right) \Big/ \left(\frac{C_0}{A_0} - (A_0, D_0, B_0)_q\right)^k \right)$$

$$\Theta_{S,T}(\tau) = q^{n-2} \sum_{u,v \in \mathrm{GF}(q)} W\left(\left[\frac{D_0}{A_0} - \frac{B_0}{A_0}(A_0, D_0, B_0)_q\right]^k \Big/ \left(\frac{C_0}{A_0} - (A_0, D_0, B_0)_q\right)^k, \right.$$
$$\left(\left[\frac{uD_0}{A_0} - \frac{uB_0}{A_0}(A_0, D_0, B_0)_q\right]^k + \frac{uC_0}{A_0} - v\right)$$
$$\left. \Big/ \left(\frac{C_0}{A_0} - (A_0, D_0, B_0)_q\right)^k \right) F(u)G(v) - F(0)G(0).$$

Case (T3.3.1). *If* $\frac{B_0}{A_0}, \frac{C_0}{A_0} \notin \mathrm{GF}(q)$, $(A_0, C_0, B_0)_q \notin \mathrm{GF}(q)$, *then*

$$N_\tau(u,v) = q^{n-2},$$

$$\Theta_{S,T}(\tau) = q^{n-2} I(f)I(g) - F(0)G(0).$$

Case (T3.3.2). *If* $\frac{B_0}{A_0}, \frac{C_0}{A_0} \notin \mathrm{GF}(q)$, $(A_0, C_0, B_0)_q \in \mathrm{GF}(q)$, *then*

$$N_\tau(u,v) = q^{n-2} S(u,v),$$

$$\Theta_{S,T}(\tau) = q^{n-2} \sum_{u,v \in \mathrm{GF}(q)} S(u,v) F(u)G(v) - F(0)G(0),$$

where $\overline{M}(u,v)$ *denotes the number of those solutions of* $(u-x)\frac{B_0}{A_0} = x^{1/k}$ *in* GF(q) *making the equation* $(u-x)\frac{D_0}{A_0} - y^{1/k} = ((u-x)\frac{C_0}{A_0} - v + y)(A_0, D_0, C_0)_q$ *solvable in* y *in* GF(q), $S(u,v)$ *denotes the number of solutions of* $(u-x)\frac{C_0}{A_0} - v + (u-x)^k \left(\frac{D_0}{A_0}\right)^k = \left[(u-x)\frac{B_0}{A_0} - x^{1/k}\right](A_0, C_0, B_0)_q$ *in* GF(q).

Theorem 7.2.4 *When* $\frac{B_0}{A_0}, \frac{C_0}{A_0}$ *and* $\frac{D_0}{A_0}$ *are not in* GF(q).

Case (T4.1). *If* $(A_0, C_0, B_0)_q$ *and* $(A_0, D_0, B_0)_q \in \mathrm{GF}(q)$, *then*

$$N_\tau(u,v) = q^{n-2} \Omega(u,v),$$

$$\Theta_{S,T}(\tau) = q^{n-2} \sum_{u,v \in \mathrm{GF}(q)} \Omega(u,v)F(u)G(v) - F(0)G(0).$$

Case (T4.2). *If* $(A_0, C_0, B_0)_q \notin \mathrm{GF}(q)$ *and* $(A_0, D_0, B_0)_q \in \mathrm{GF}(q)$, *then*

$$N_\tau(u,v) = q^{n-3}\Omega'(u,v),$$

$$\Theta_{S,T}(\tau) = q^{n-3} \sum_{u,v \in \mathrm{GF}(q)} \Omega'(u,v)F(u)G(v) - F(0)G(0).$$

If $(A_0, D_0, B_0)_q \notin \mathrm{GF}(q)$ *and* $(A_0, C_0, B_0)_q \in \mathrm{GF}(q)$, *then*

$$N_\tau(u,v) = q^{n-3}\Omega''(u,v),$$

$$\Theta_{S,T}(\tau) = q^{n-3} \sum_{u,v \in \mathrm{GF}(q)} \Omega''(u,v)F(u)G(v) - F(0)G(0).$$

Case (T4.3). *If* $(A_0, C_0, B_0)_q$ *and* $(A_0, D_0, B_0)_q$ *are not in* $\mathrm{GF}(q)$, *then*

when $\delta(A_0, D_0, B_0, C_0) \notin \mathrm{GF}(q)$, *then*

$$N_\tau(u,v) = q^{n-2},$$

$$\Theta_{S,T}(\tau) = q^{n-2} I(f)I(g) - F(0)G(0);$$

when $\delta(A_0, D_0, B_0, C_0) \in \mathrm{GF}(q)$, *then*

$$N_\tau(u,v) = q^{n-3}\Omega'''(u,v),$$

$$\Theta_{S,T}(\tau) = q^{n-3} \sum_{u,v \in \mathrm{GF}(q)} \Omega'''(u,v)F(u)G(v) - F(0)G(0),$$

where $\Omega(u,v)$ *denotes the number of solution pairs* $(x,y) \in \mathrm{GF}(2)^2$ *of*

$$\begin{cases} \dfrac{(u-x)C_0}{A_0} - (v-y) = \left(\dfrac{(u-x)B_0}{A_0} - x^{1/k}\right)(A_0, C_0, B_0)_q, \\ \dfrac{(u-x)D_0}{A_0} - y^{1/k} = \left(\dfrac{(u-x)B_0}{A_0} - x^{1/k}\right)(A_0, D_0, B_0)_q, \end{cases}$$

$\Omega'(u,v)$, $\Omega''(u,v)$ *and* $\Omega'''(u,v)$ *the numbers of solution pairs* $(x,y) \in \mathrm{GF}(q)^2$ *of* $\dfrac{(u-x)D_0}{A_0} - y^{1/k} = \left(\dfrac{(u-x)B_0}{A_0} - x^{1/k}\right)(A_0, D_0, B_0)_q$, $\dfrac{(u-x)C_0}{A_0} - (v-y) = \left(\dfrac{(u-x)B_0}{A_0} - x^{1/k}\right)(A_0, C_0, B_0)_q$ *and* $\dfrac{(u-x)D_0}{A_0} - y^{1/k} = \left[\dfrac{(u-x)C_0}{A_0} - (v-y)\delta(A_0, D_0, B_0, C_0)\right] - \left(\dfrac{(u-x)B_0}{A_0} - x^{1/k}\right)$ $(\delta(A_0, D_0, B_0, C_0)(A_0, C_0, B_0)_q - (A_0, C_0, B_0)_q)$, *respectively.*

Remark 7.2.3 *If q is small, for any given* $u, v \in \mathrm{GF}(q)$, $\overline{M}(u,v)$, $S(u,v)$, $\Omega(u,v)$, $\Omega'(u,v)$, $\Omega''(u,v)$, $\Omega'''(u,v)$ *in Theorem 7.2.3 and Theorem 7.2.4 can be easily computed.*

Now we can obtain the following corollaries of Theorem 7.2.1–7.2.4 on autocorrelation functions of linearly related Gold-geometric sequences.

Corollary 7.2.1 *Let $S_f^{(A,B)}$ be a linearly related Gold-geometric sequence defined by $S_f^{(A,B)}(i) = f(\mathrm{Tr}_q^{q^n}(A\alpha^i + B\alpha^{ki}))$ with $k = p^e$, and $\dfrac{B^{1/k}}{A} \in \mathrm{GF}(q)$. Then the autocorrelation function $\Theta_{S,S}(\tau)$ of $S_f^{(A,B)}$ is as follows:*

If $\alpha^\tau \in \mathrm{GF}(p^e)$, then

$$\Theta_{S,S}(\tau) = q^{n-1} \sum_{u\in\mathrm{GF}(q)} W\left(\frac{B}{A^k}, \frac{u^k B}{A^k}\right) F(u)G(\alpha^\tau u) - F(0)G(0).$$

If $\alpha^\tau \in \mathrm{GF}(q) - \mathrm{GF}(p^e)$, then

$$\Theta_{S,S}(\tau) = q^{n-1} \sum_{\left[u-\frac{\alpha^\tau - v}{\alpha^\tau - \alpha^{k\tau}}\right]\frac{B^{1/k}}{A} = \left[\frac{\alpha^\tau - v}{\alpha^\tau - \alpha^{k\tau}}\right]^{1/k}} F(u)G(u) - F(0)G(0).$$

If $\alpha^\tau \notin \mathrm{GF}(q)$, then

$$\begin{aligned}&\Theta_{S,S}(\tau)\\ &= q^{n-2} W\left(\frac{B}{A^k}, 0\right) \sum_{u\in\mathrm{GF}(q)} W\left(\frac{B}{A^k}, \frac{u^k B}{A^k}\right) F(u) \sum_{W\left(\frac{B}{A^k}, \frac{v^k B}{A^k}\right)\neq 0} G(v) - F(0)G(0).\end{aligned}$$

Corollary 7.2.2 *Let $S_f^{(A,B)}$ be a linearly related Gold-geometric sequence defined by $S_f^{(A,B)}(i) = f(\mathrm{Tr}_q^{q^n}(A\alpha^i + B\alpha^{ki}))$ with $k = p^e$, $\dfrac{B^{1/k}}{A} \notin \mathrm{GF}(q)$. Then the autocorrelation function $\Theta_{S,S}(\tau)$ of $S_f^{(A,B)}$ is as follows:*

Case (C1). *If $\alpha^\tau \in \mathrm{GF}(q)$, then*

$$\Theta_{S,S}(\tau) = q^{n-2} I(f) I(g) - F(0)G(0).$$

Case (C2). *If $\alpha^\tau \notin \mathrm{GF}(q)$, $\dfrac{B^{1/k}}{A}\alpha^\tau \in \mathrm{GF}(q)$ and $\left(\dfrac{B^{1/k}}{A}\right)^q \alpha^{-\tau} \notin \mathrm{GF}(q)$, then*

$$\Theta_{S,S}(\tau) = q^{n-2} I(f) I(g) - F(0)G(0).$$

Case (C3). *If $\alpha^\tau \notin \mathrm{GF}(q)$, $\dfrac{B^{1/k}}{A}\alpha^\tau \in \mathrm{GF}(q)$ and $\left(\dfrac{B^{1/k}}{A}\right)^q \alpha^{-\tau} \in \mathrm{GF}(q)$, then*

$$\Theta_{S,S}(\tau) = q^{n-2} \sum_{u,v\in\mathrm{GF}(q)} S(u,v) F(u)G(v) - F(0)G(0).$$

Case (C4). *If $\alpha^\tau \notin \mathrm{GF}(q)$, $\dfrac{B^{1/k}}{A}\alpha^\tau \notin \mathrm{GF}(q)$.*

(1) *If $(A, A\alpha^\tau, B^{1/k})_q \in \mathrm{GF}(q)$ and $(A, B^{1/k}\alpha^\tau, B^{1/k})_q \in \mathrm{GF}(q)$, then*

$$\Theta_{S,T}(\tau) = q^{n-2} \sum_{u,v\in\mathrm{GF}(q)} \Omega(u,v) F(u)G(v) - F(0)G(0).$$

(2) *If* $(A, A\alpha^\tau, B^{1/k})_q \notin \mathrm{GF}(q)$ *and* $(A, B^{1/k}\alpha^\tau, B^{1/k})_q \in \mathrm{GF}(q)$, *then*

$$\Theta_{S,T}(\tau) = q^{n-3} \sum_{u,v\in \mathrm{GF}(q)} \Omega'(u,v)F(u)G(v) - F(0)G(0).$$

(3) *If* $(A, B^{1/k}\alpha^\tau, B^{1/k})_q \notin \mathrm{GF}(q)$ *and* $(A, A\alpha^\tau, B^{1/k})_q \in \mathrm{GF}(q)$, *then*

$$\Theta_{S,T}(\tau) = q^{n-3} \sum_{u,v\in \mathrm{GF}(q)} \Omega''(u,v)F(u)G(v) - F(0)G(0).$$

(4) *If* $(A, A\alpha^\tau, B^{1/k})_q \notin \mathrm{GF}(q)$ *and* $(A, B^{1/k}\alpha^\tau, B^{1/k})_q \notin \mathrm{GF}(q)$, *when* $\delta(A, B^{1/k}\alpha^\tau, B^{1/k}, \alpha^\tau) \notin \mathrm{GF}(q)$, *then*

$$\Theta_{S,T}(\tau) = q^{n-2}I(f)I(g) - F(0)G(0);$$

when $\delta(A, B^{1/k}\alpha^\tau, B^{1/k}, \alpha^\tau) \in \mathrm{GF}(q)$, *then*

$$\Theta_{S,T}(\tau) = q^{n-3} \sum_{u,v\in \mathrm{GF}(q)} \Omega'''(u,v)F(u)G(v) - F(0)G(0),$$

where $S(u,v)$ *denotes the number of solutions of* $(u-x)\alpha^\tau - v + (u-x)^k(B\alpha^{k\tau}/A^k) = [(u-x)B^{1/k}/A - x^{1/k}](A, A\alpha^\tau, B^{1/k})_q$ *in* $\mathrm{GF}(q)$, $\Omega(u,v)$ *the number of solution pairs* $(x,y) \in \mathrm{GF}(2)^2$ *of*

$$\begin{cases} (u-x)\alpha^\tau - (v-y) = ((u-x)B^{1/k}/A - x^{1/k})(A, A\alpha^\tau, B^{1/k})_q, \\ (u-x)B^{1/k}\alpha^\tau/A - y^{1/k} = ((u-x)B^{1/k}/A - x^{1/k})(A, B^{1/k}\alpha^\tau, B^{1/k})_q, \end{cases}$$

$\Omega'(u,v)$, $\Omega''(u,v)$ *and* $\Omega'''(u,v)$ *the number of solution pairs* $(x,y) \in \mathrm{GF}(q)^2$ *of* $(u-x)B^{1/k}\alpha^\tau/A - y^{1/k} = ((u-x)B^{1/k}/A - x^{1/k})(A, B^{1/k}\alpha^\tau, B^{1/k})_q$, $(u-x)\alpha^\tau - (v-y) = ((u-x)B^{1/k}/A - x^{1/k})(A, A\alpha^\tau, B^{1/k})_q$, *and* $(u-x)B^{1/k}\alpha^\tau/A - y^{1/k} = ((u-x)\alpha^\tau - (v-y)\delta(A, B^{1/k}\alpha^\tau, B^{1/k}, \alpha^\tau)) - ((u-x)B^{1/k}/A - x^{1/k})(\delta(A, B^{1/k}\alpha^\tau, B^{1/k}, \alpha^\tau)(A, A\alpha^\tau, B^{1/k})_q - (A, B^{1/k}\alpha^\tau, B^{1/k})_q)$, *respectively.*

Up to now, explicit formulas for cross (auto)-correlation functions of linearly related Gold-geometric sequences are derived, thus the open problem posed by Klapper et al. is completed solved.

7.2.2 On the Correlation Functions of a Family of Generalized Geometric Sequences[20]

For $A, B \in \mathrm{GF}(q^e)$, the sequence $S_f^{(A,B)}$ studied in this subsection is defined by

$$S_f^{(A,B)}(i) = f(\mathrm{tr}_q^{q^e}(A\alpha^i + B\alpha^{ki})) \tag{7.17}$$

for $i = 1, 2, \cdots$, which is called a generalized geometric sequence based on α and k.

At first, we recall several results on exponential sums that will be useful in what follows. Let

$$\chi_q(x) = e^{2\pi i_q^{q^e}(x)/p}, \quad x \in \mathrm{GF}(q)$$

be a canonical character of $\mathrm{GF}(q)$ (with values in the complex numbers).

Lemma 7.2.6[22] *Let $a \in \mathrm{GF}(q)$, Then*

$$\sum_{x\in \mathrm{GF}(q)} \chi_q(ax) = \begin{cases} q, & \text{if } a = 0, \\ 0, & \text{if } a \neq 0. \end{cases}$$

We use the following notation of exponential sums, where $a, b \in \mathrm{GF}(q)$,

$$S(a,b) = \sum_{x\in \mathrm{GF}(q)} \chi_q(ax^{p+1} + bx).$$

When $b = 0$, we define $S(a) = S(a, 0)$.

Lemma 7.2.7[23] *Let $q = p^n$ and $n = 2m$.*

(1) *If $a \neq 0, b = 0$, then*

$$S(a) = \begin{cases} (-1)^{m+1}p^{m+1}, & \text{if } a^{(p^n-1)/(p+1)} = (-1)^m, \\ (-1)^m p^m, & \text{otherwise}. \end{cases}$$

(2) *If $a \neq 0, b \neq 0$, then $S(a,b) = 0$ if $a^p x^{p^2} + ax + b^p = 0$ is unsolvable in* GF(q) *and*

$$S(a,b) = (-1)^m p^m \chi_q(ax_0^{p+1} + bx_0);$$

otherwise, where x_0 is an arbitrary solution in GF(q) *of the equation $a^p x^{p^2} + ax + b^p = 0$.*

It is known that if $a^{(q-1)/(p+1)} \neq (-1)^m$, then the equation $a^p x^{p^2} + ax + b^p = 0$ has a unique solution. If $a^{(q-1)/(p+1)} = (-1)^m$, then $a^p x^{p^2} + ax + b^p = 0$ is solvable in GF(q) if and only if

$$\sum_{j=0}^{m-1} (a^{-1}bc^{-p})^{p^{2j}} = 0, \quad a^{1-p} = -c^{p^2-1}.$$

Lemma 7.2.8[23] *Let $q = p^n, n = 2m + 1$.*

(1) *If $a \neq 0, b = 0$, then*

$$S(a) = S(1)\psi(a)$$

and

$$S(1) = (-1)^{m(p-1)/2} i^{(p-1)^2/4} p^{(2m+1)/2}.$$

(2) *If $a \neq 0, b \neq 0$, then*

$$S(a,b) = \chi_q(ax_0^{p+1} + bx_0)S(1)\psi(a),$$

where ψ is the quadratic character of GF(q) *and x_0 is the unique solution in* GF(q) *to the equation $a^p x^{p^2} + ax + b^p = 0$.*

If $q^e = p^{2m+1}$, that is $en = 2m + 1$. Let f and g be possibly nonlinear functions from $\mathrm{GF}(q)$ to $\mathrm{GF}(r)$, $A, B, A', B' \in \mathrm{GF}(q^e)$, and $S_f^{(A,B)}, S_f^{(A',B')}$ be generalized geometric sequences based on the same α and k. Then

$$C_{S_f^{(A,B)}, S_g^{(A',B')}}(\tau) = \sum_{i=0}^{q^e-2} \omega^{S_f^{(A,B)}(i) + S_g^{(A',B')}(i+\tau)}.$$

We also let $F(u) = \omega^{f(u)}, G(v) = \omega^{g(v)}$ by definition.

Theorem 7.2.5 *Let $S_f^{(A,B)}, S_g^{(A',B')}$ be generalized geometric sequences based on the same α and k, as defined in Equation (7.17). Assume that $A, B, A', B' \neq 0, k = p+1, q^e = p^{en} = p^{2m+1}$, and $0 \leqslant \tau \leqslant q^e - 2$. Let $A_1 = A'\alpha^\tau$ and $B_1 = B'\alpha^{(p+1)\tau}$.*

(1) *If $A/A_1 \in \mathrm{GF}(q)$ and $A_1 B = AB_1$, then*

$$|C_{S_f^{(A,B)}, S_g^{(A',B')}}(\tau) - q^{e-1} \sum_{u \in \mathrm{GF}(q)} F(u) G\left(\frac{A_1}{A} u\right) + F(0)G(0)| \leqslant q^{e/2}(q^2 - q).$$

(2) *If $A/A_1 \in \mathrm{GF}(q)$ and $A_1 B \neq AB_1$, then*

$$\begin{aligned} &|C_{S_f^{(A,B)}, S_g^{(A',B')}}(\tau) - q^{e-2} I(f) I(g) + F(0)G(0) \\ &- q^{(e-3)/2} \mu \sum_{u,v \in \mathrm{GF}(q)} \psi((AB_1 - A_1 B))(A_1 u - Av)) F(u) G(v)| \\ &\leqslant \begin{cases} q^{e/2}(q^2 - 2q + 1), & \text{if } B/B_1 \in \mathrm{GF}(q), \\ q^{e/2}(q^2 - q), & \text{if } B/B_1 \notin \mathrm{GF}(q), \end{cases} \end{aligned}$$

where $\mu \in \{1, -1, i, -i\}$.

(3) *If $A/A_1 \notin \mathrm{GF}(q)$, then*

$$\begin{aligned} &|C_{S_f^{(A,B)}, S_g^{(A',B')}}(\tau) - q^{e-2} I(f) I(g) + F(0)G(0)| \\ &\leqslant \begin{cases} q^{e/2}(q^2 - q), & \text{if } B/B_1 \in \mathrm{GF}(q), \\ q^{e/2}(q^2 - 1), & \text{if } B/B_1 \notin \mathrm{GF}(q). \end{cases} \end{aligned}$$

Proof For any τ we have

$$C_{S_f^{(A,B)}, S_g^{(A',B')}}(\tau) = \sum_{u \in \mathrm{GF}(q)} N_\tau(u,v) F(u) G(v) - F(0)G(0),$$

where $N_\tau(u,v)$ denotes the number of solutions in $\mathrm{GF}(q^e)$ to the equations

$$\mathrm{tr}_q^{q^e}(Ax + Bx^{p+1}) = u$$

and

$$\mathrm{tr}_q^{q^e}(A'\alpha^\tau x + B'\alpha^{\tau(p+1)} x^{p+1}) = v.$$

It follows from Lemma 7.2.6 that

$$q^2 N_\tau(u,v) = \sum_{x\in \mathrm{GF}(q^e)} \sum_{\lambda,\xi\in \mathrm{GF}(q)} \chi_q[\xi \mathrm{tr}_q^{q^e}(Ax + Bx^{p+1}) - \xi u]$$
$$\cdot \chi_q[\lambda \mathrm{tr}_q^{q^e}(A_1 x + B_1 x^{p+1}) - \lambda u]$$
$$= \sum_{\lambda,\xi\in \mathrm{GF}(q)} \chi_q(-\xi u - \lambda v)$$
$$\cdot \sum_{x\in \mathrm{GF}(q^e)} \chi_q[\mathrm{tr}_q^{q^e}((A\xi + A_1\lambda)x + (B\xi + B_1\lambda)x^{p+1})]$$
$$= \sum_{\lambda,\xi\in \mathrm{GF}(q)} \chi_q(-\xi u - \lambda v)$$
$$\cdot \sum_{x\in \mathrm{GF}(q^e)} \chi_{q^e}((A\xi + A_1\lambda)x + (B\xi + B_1\lambda)x^{p+1}).$$

Thus

$$N_\tau(u,v) = q^{-2} \sum_{\lambda,\xi\in \mathrm{GF}(q)} \chi_q(-\xi u - \lambda v) \sum_{x\in \mathrm{GF}(q^e)}$$
$$\cdot \chi_{q^e}((A\xi + A_1\lambda)x + (B\xi + B_1\lambda)x^{p+1})$$
$$= q^{-2} \sum_{\lambda,\xi\in \mathrm{GF}(q)} \chi_{q^e}(-\xi u - \lambda v) T_\tau(\xi,\lambda),$$

where

$$T_\tau(\xi,\lambda) = \sum_{x\in \mathrm{GF}(q^e)} \chi_{q^e}((A\xi + A_1\lambda)x + (B\xi + B_1\lambda)x^{p+1}).$$

There are two cases:

(1) When $B\xi + B_1\lambda = 0$, we have

$$T_\tau(\xi,\lambda) = \begin{cases} q^e, & \text{if } A\xi + A_1\lambda = 0, \\ 0, & \text{if } A\xi + A_1\lambda \neq 0. \end{cases}$$

(2) When When $B\xi + B_1\lambda \neq 0$, we have

$$T_\tau(\xi,\lambda) = \psi(B\xi + B_1\lambda)S(1),$$

if $A\xi + A_1\lambda = 0$ and

$$T_\tau(\xi,\lambda) = \chi_{q^e}[(A\xi + A_1\lambda)x_0 + (B\xi + B_1\lambda)x_0^{p+1}] \cdot \psi(B\xi + B_1\lambda)S(1),$$

if $A\xi + A_1\lambda \neq 0$, where x_0 is the unique solution in $\mathrm{GF}(q^e)$ to the equation

$$(B\xi + B_1\lambda)^p x^{p^2} + (B\xi + B_1\lambda)x + (A\xi + A_1\lambda)p = 0.$$

Therefore,

$$\begin{aligned}N_\tau(u,v) = q^{-2}\Big(& \sum_{\lambda,\xi\in \mathrm{GF}(q), B\xi+B_1\lambda=0, A\xi+A_1\lambda=0} q^e\chi_q(-\xi u-\lambda v)\\ & +S(1)\sum_{\lambda,\xi\in \mathrm{GF}(q), B\xi+B_1\lambda\neq 0, A\xi+A_1\lambda\neq 0} \chi_q(-\xi u-\lambda v)\psi(B\xi+B_1\lambda)\\ & +S(1)\sum_{\lambda,\xi\in \mathrm{GF}(q), B\xi+B_1\lambda\neq 0, A\xi+A_1\lambda\neq 0} \chi_q(-\xi u-\lambda v)\\ & \cdot\chi_{q^e}((A\xi+A_1\lambda)x_0+(B\xi+B_1\lambda)x_0^{p+1})\psi(B\xi+B_1\lambda)\Big)\\ = & q^{-2}(W_1+W_2+W_3).\end{aligned}$$

A. Calculation of W_1

There are two cases for W_1:

(1) If $A/A_1 \in \mathrm{GF}(q)$, then

$$\begin{aligned}W_1 &= q^e\Big(\sum_{\lambda,\xi\in \mathrm{GF}(q), B\xi+B_1\lambda=0, A\xi+A_1\lambda=0} \chi_q(-\xi u-\lambda v)\Big)\\ &= \begin{cases} q^e \displaystyle\sum_{\lambda\in \mathrm{GF}(q)} \chi_q\left[\left(\frac{B_1}{B}u-v\right)\lambda\right], & \text{if } A_1B=AB_1,\\ q^e, & \text{if } A_1B\neq AB_1\end{cases}\\ &= \begin{cases} q^{e+1}, & \text{if } A_1B=AB_1 \text{ and } B_1u=Bv,\\ 0, & \text{if } A_1B=AB_1 \text{ and } B_1u\neq Bv,\\ q^e, & \text{if } A_1B\neq AB_1.\end{cases}\end{aligned}$$

(2) If $A/A_1 \notin \mathrm{GF}(q)$, then $W_1 = q^e$.

B. Calculation of W_2

For W_2, if $A\neq 0$, $A/A_1\in \mathrm{GF}(q)$, $A_1u\neq Av$, and $B_1A\neq A_1B$, then

$$\begin{aligned}W_2 &= S(1)\sum_{\lambda,\xi\in \mathrm{GF}(q), A\xi+A_1\lambda=0} \chi_q(-\xi u-\lambda v)\psi(B\xi+B_1\lambda)\\ &= S(1)\sum_{\lambda\in \mathrm{GF}(q)}\chi_q\left[\left(\frac{A_1}{A}u-v\right)\lambda\right]\psi\left[\left(B_1-\frac{A_1}{A}B\right)\lambda\right]\\ &= S(1)\psi\left(B_1-\frac{A_1}{A}B\right)\sum_{\lambda\in \mathrm{GF}(q)}\chi_q\left[\left(\frac{A_1}{A}u-v\right)\lambda\right]\psi(\lambda)\\ &= S(1)\Gamma(\psi,\chi_q)\psi\left(\frac{AB_1-A_1B}{A_1u-Av}\right),\end{aligned}$$

where (see [22], Theorem 5.15)

$$\begin{aligned}\Gamma(\psi,\chi_q) &\stackrel{\text{def}}{=} \sum_{u\in \mathrm{GF}(q)}\psi(u)\chi_q(u)\\ &= \begin{cases}(-1)^{n-1}q^{1/2}, & \text{if } p\equiv 1 \pmod 4,\\ (-1)^{n-1}i^n q^{1/2}, & \text{if } p\equiv 3 \pmod 4.\end{cases}\end{aligned}$$

Note that $\psi(x/y) = \psi(xy)$ for any $y\neq 0$. We have $W_2 = 0$ in all other cases.

C. Estimation of W_3

Finally, we must compute W_3. Let

$$V = |\{(\lambda, \xi) \in \mathrm{GF}(q)^2 | B\xi + B_1\lambda \neq 0, A\xi + A_1\lambda \neq 0\}|.$$

There are three possibilities for V:

(1) If $A/A_1 \in \mathrm{GF}(q)$ and $B/B_1 \in \mathrm{GF}(q)$, then

$$V = \begin{cases} q^2 - 2q + 1, & \text{if } AB_1 \neq A_1B, \\ q^2 - q, & \text{if } AB_1 = A_1B. \end{cases}$$

(2) If $A/A_1 \in \mathrm{GF}(q)$ and $B/B_1 \notin \mathrm{GF}(q)$, or $A/A_1 \notin \mathrm{GF}(q)$ and $B/B_1 \in \mathrm{GF}(q)$, then $V = q^2 - q$.

(3) If $A/A_1 \notin \mathrm{GF}(q)$ and $B/B_1 \notin \mathrm{GF}(q)$, then $V = q^2 - 1$.

This leads to four cases of upper bounds for $|W_3|$, where

$$\begin{aligned} |W_3| &= S(1) \sum_{\substack{\lambda,\xi \in \mathrm{GF}(q), \\ A\xi + A_1\lambda \neq 0, \\ B\xi + B_1\lambda \neq 0}} \chi_q(-\xi u - \lambda v) \\ &\quad \cdot \chi_{q^e}[(A\xi + A_1\lambda)x_0 + (B\xi + B_1\lambda)x_0^{p+1}]\psi(B\xi + B_1\lambda) \\ &= q^{e/2} \sum_{\substack{\lambda,\xi \in \mathrm{GF}(q), \\ A\xi + A_1\lambda \neq 0, \\ B\xi + B_1\lambda \neq 0}} \chi_q(-\xi u - \lambda v) \\ &\quad \cdot \chi_{q^e}[(A\xi + A_1\lambda)x_0 + (B\xi + B_1\lambda)x_0^{p+1}]\psi(B\xi + B_1\lambda), \end{aligned} \tag{7.18}$$

since $|S(1)| = p^{(2m+1)/2} = p^{en/2} = q^{e/2}$.

(1) If $A/A_1 \in \mathrm{GF}(q)$, $B/B_1 \in \mathrm{GF}(q)$, and $AB_1 \neq A_1B$, then $|W_3| \leqslant q^{e/2}(q^2 - 2q + 1)$.

(2) If $A/A_1 \in \mathrm{GF}(q)$ and $AB_1 = A_1B$, then $|W_3| \leqslant q^{e/2}(q^2 - q)$.

(3) $A/A_1 \in \mathrm{GF}(q)$ and $B/B_1 \notin \mathrm{GF}(q)$, or $A/A_1 \notin \mathrm{GF}(q)$ and $B/B_1 \in \mathrm{GF}(q)$, then $|W_3| \leqslant q^{e/2}(q^2 - q)$.

(4) If $A/A_1 \notin \mathrm{GF}(q)$ and $B/B_1 \notin \mathrm{GF}(q)$, then $|W_3| \leqslant q^{e/2}(q^2 - 1)$.

D. Bounds for Cross-Correlations

In every case we have

$$|N(u, v) - q^{-2}(W_1 + W_2)| = q^{-2}|W_3| < \varepsilon$$

for some error term ε. It follows that

$$\begin{aligned} &|C_{S_f^{(A,B)}, S_g^{(A',B')}}(\tau) - q^{-2} \sum_{u,v \in \mathrm{GF}(q)} (W_1 + W_2)F(u)G(v) + F(0)G(0)| \\ &< q^{-2} \sum_{u,v \in \mathrm{GF}(q)} |\varepsilon F(u)G(v)| \\ &= q^{-2} \sum_{u,v \in \mathrm{GF}(q)} \varepsilon = \varepsilon. \end{aligned}$$

Of course W_1 and W_2 depend on u and v.

Evaluating these sums in various cases completes the proof of Theorem 7.2.5. □

The bound on $|W_3|$ appears to be rather weak–we have simply assumed that every term in a sum of plus ones and minus ones is a plus one, the worst possible case. It is quite possible that tighter bounds hold. The given bound arises from a sum involving χ_{q^e}, from Equation (7.18). We have $\chi_{q^e}(x) = \chi_q(\mathrm{Tr}_q^{q^e}(x))$. This can be used to rewrite the right hand side of Equation (7.18) in the form

$$|W_3| = q^{e/2}|\sum_{\lambda,\xi\in \mathrm{GF}(q), A\xi+A_1\lambda=0} \chi_q(C\xi + D\lambda)\psi(B\xi + B_1\lambda)|$$

which looks like a character sum of a type that has been analyzed. However, the terms C and D depend on x_0, which in turn depends on ξ and λ so there is no apparent way to simplify this expression and obtain a tighter bound.

If $q^e = p^{2m}$, that is $en = 2m$. Then we have

Theorem 7.2.6 *Let $S_f^{(A,B)}, S_g^{(A',B')}$ be generalized geometric sequences based on the same α and k, as defined in Equation (7.17). Assume that $A, B, A', B', \neq 0, k = p+1, q^e = p^{en} = p^{2m}$, and $0 \leqslant \tau \leqslant q^e - 2$. Let $A_1 = A'\alpha^\tau, B_1 = B'\alpha^{\tau(p+1)}$, and let Δ be as in Equation (7.21).*

(1) *If $A/A_1 \in \mathrm{GF}(q)$ and $AB_1 = A_1B$, then*

$$\left|C_{S_f^{(A,B)},S_g^{(A',B')}}(\tau) + F(0)G(0) - q\, q^{e-1} \sum_{u\in \mathrm{GF}(q)} F(u)G\left(\frac{A_1}{A}u\right)\right| \leqslant q^{e/2}(q^2 - q).$$

(2) *If e is even, $A/A_1 \in \mathrm{GF}(q)$, $AB_1 \neq A_1B$, and $(B_1A - A_1B/A)^{(q^e-1)/(p+1)} \neq (-1)^m$, then*

$$\begin{aligned}&|C_{S_f^{(A,B)},S_g^{(A',B')}}(\tau) - (q^{e-2} - (-1)^m q^{e/2-2})I(f)I(g)\\ &-(-1)^m q^{e/2-1} \sum_{u\in \mathrm{GF}(q)} F(u)G\left(\frac{A_1}{A}u\right) + F(0)G(0)|\\ &\leqslant \begin{cases} q^{e/2}(q^2 - 2q + 1), & \text{if } B/B_1 \in \mathrm{GF}(q),\\ q^{e/2}(q^2 - q), & \text{if } B/B_1 \notin \mathrm{GF}(q).\end{cases}\end{aligned}$$

(3) *If e is even, $A/A_1 \in \mathrm{GF}(q)$, $AB_1 \neq A_1B$, and*

$$(B_1 - A_1B/A)^{(q^e-1)/(p+1)} = (-1)^m,$$

then

$$\begin{aligned}&|C_{S_f^{(A,B)},S_g^{(A',B')}}(\tau) - (q^{e-2} + (-1)^m pq^{e/2-2})I(f)I(g)\\ &-(-1)^m pq^{e/2-1}(p+1) \sum_{u\in \mathrm{GF}(q)} F(u)G\left(\frac{A_1}{A}u\right) + F(0)G(0)|\\ &\leqslant \begin{cases} q^{e/2}(q^2 - 2q + 1), & \text{if } B/B_1 \in \mathrm{GF}(q),\\ q^{e/2}(q^2 - q), & \text{if } B/B_1 \notin \mathrm{GF}(q).\end{cases}\end{aligned}$$

(4) *If e is odd, $A/A_1 \in \mathrm{GF}(q)$, and $AB_1 \neq A_1B$, then*

$$\begin{aligned}
&|C_{S_f^{(A,B)},S_g^{(A',B')}}(\tau) - (-1)^m q^{(e-3)/2} \sum_{u \in \mathrm{GF}(q)} F(u)G\left(\frac{A_1}{A}u\right) \\
&-(q^{e-2} - q^{(e-3)/2} + (1+(-1)^m)q^{e/2-2})I(f)I(g) \\
&-q^{(e-3)/2}(p+1) \sum_{\substack{Av \neq Av_1, \\ \left(\frac{A_1}{A}u - v\right)\left(\frac{B_1}{B}u-v\right) a\ (p+1)\ power}} F(u)G(v) + F(0)G(0)| \\
&\leqslant \begin{cases} q^{e/2}(q^2-2q+1), & \textit{if } B/B_1 \in \mathrm{GF}(q), \\ q^{e/2}(q^2-q), & \textit{if } B/B_1 \notin \mathrm{GF}(q). \end{cases}
\end{aligned}$$

(5) *If $A/A_1 \notin \mathrm{GF}(q)$, then*

$$\begin{aligned}
&|C_{S_f^{(A,B)},S_g^{(A',B')}}(\tau) - q^{e-2}I(f)I(g) + F(0)G(0)| \\
&\leqslant \begin{cases} q^{e/2}(q^2-1), & \textit{if } AB_1 = A_1B \textit{ or } B/B_1 \notin \mathrm{GF}(q), \\ q^{e/2}(q^2-q), & \textit{if } AB_1 \neq A_1B \textit{ and } B/B_1 \notin \mathrm{GF}(q). \end{cases}
\end{aligned}$$

Proof For any τ, the correlations can be expressed in terms of $N_\tau(u,v)$ and $T_\tau(\xi,\lambda)$ whose definitions are the same as in above. We proceed by a similar analysis.

When $B\xi + B_1\lambda = 0$, we have

$$T_\tau(\xi,\lambda) = \begin{cases} q^e, & \text{if } A\xi + A_1\lambda = 0, \\ 0, & \text{if } A\xi + A_1\lambda \neq 0. \end{cases}$$

When $B\xi + B_1\lambda \neq 0$ and $A\xi + A_1\lambda = 0$, from Lemma 7.2.7 we have

$$T_\tau(\xi,\lambda) = \begin{cases} (-1)^{m+1}p^{m+1}, & \text{if } (B\xi + B_1\lambda)^{\frac{p^{2m}-1}{p+1}} = (-1)^m, \\ (-1)^m p^m, & \text{otherwise.} \end{cases}$$

When $B\xi + B_1\lambda \neq 0$ and $A\xi + A_1\lambda \neq 0$, we have $T_\tau(\xi,\lambda) \neq 0$ if and only if the equation

$$(B\xi + B_1\lambda)^p x^{p^2} + (B\xi + B_1\lambda)x + (A\xi + A_1\lambda)^p = 0 \tag{7.19}$$

is solvable in $\mathrm{GF}(q^e)$. If so, let x_0 be any solution. Then $T_\tau(\xi,\lambda) = 0$ if Equation (7.19) is unsolvable in $\mathrm{GF}(q^e)$ and

$$T_\tau(\xi,\lambda) = (-1)^m p^m \chi_{q^e}((A\xi + A_1\lambda)x_0 + (B\xi + B_1\lambda)x_0^{p+1})$$

if Equation (7.19) is solvable in $\mathrm{GF}(q^e)$.

It follows that

$$
\begin{aligned}
N_\tau(u,v) = q^{-2}\Bigg(& \sum_{\substack{\lambda,\xi\in \mathrm{GF}(q),\\ B\xi+B_1\lambda=0,\\ A\xi+A_1\lambda=0}} q^e \chi_q(-\xi u-\lambda v)\\
&+ \sum_{\substack{\lambda,\xi\in \mathrm{GF}(q),\\ B\xi+B_1\lambda\neq 0,\\ A\xi+A_1\lambda=0,\\ (B\xi+B_1\lambda)^{(q^e-1)/(p+1)}=(-1)^m}} (-1)^{m+1}p^{m+1}\chi_q(-\xi u-\lambda v)\\
&+ \sum_{\substack{\lambda,\xi\in \mathrm{GF}(q),\\ B\xi+B_1\lambda\neq 0,\\ A\xi+A_1\lambda=0,\\ (B\xi+B_1\lambda)^{(q^e-1)/(p+1)}\neq(-1)^m}} (-1)^{m}p^{m}\chi_q(-\xi u-\lambda v)\\
&+ \sum_{\substack{\lambda,\xi\in \mathrm{GF}(q),\\ B\xi+B_1\lambda\neq 0,\\ A\xi+A_1\lambda\neq 0,\\ (7.19)\text{is solvable in}\mathrm{GF}(q^e)}} (-1)^{m}p^{m}\chi_q(-\xi u-\lambda v)\\
&\cdot \chi_{q^e}((A\xi+A_1\lambda)x_0+(B\xi+B_1\lambda)x_0^{p+1})\Bigg)\\
= q^{-2}(W_1 &+ (-1)^m p^m U_1 + (-1)^{m+1}p^m(p+1)U_2 + (-1)^m p^m U_3),
\end{aligned}
$$

where

$$
W_1 = \sum_{\lambda,\xi\in \mathrm{GF}(q), B\xi+B_1\lambda=0, A\xi+A_1\lambda=0} q^e\chi_q(-\xi u-\lambda v),
$$

$$
U_1 = \sum_{\substack{\lambda,\xi\in \mathrm{GF}(q),\\ B\xi+B_1\lambda\neq 0,\\ A\xi+A_1\lambda=0}} \chi_q(-\xi u-\lambda v),
$$

$$
U_2 = \sum_{\substack{\lambda,\xi\in \mathrm{GF}(q),\\ B\xi+B_1\lambda\neq 0,\\ A\xi+A_1\lambda=0,\\ (B\xi+B_1\lambda)^{(q^e-1)/(p+1)}=(-1)^m}} \chi_q(-\xi u-\lambda v),
$$

$$
U_3 = \sum_{\substack{\lambda,\xi\in \mathrm{GF}(q),\\ B\xi+B_1\lambda\neq 0,\\ A\xi+A_1\lambda\neq 0,\\ (7.19)\text{is solvable in}\mathrm{GF}(q^e)}} \chi_{q^e}((A\xi+A_1\lambda)x+(B\xi+B_1\lambda)x_0^{p+1})\chi_q(-\xi u-\lambda v).
$$

W_1 has been calculated above.

A. Calculation of U_1

There are two cases for U_1. When $A/A_1 \notin \mathrm{GF}(q)$, we have $U_1=0$. When $A/A_1\in \mathrm{GF}(q)$, we have

$$
\begin{aligned}
U_1 &= \sum_{\lambda\in \mathrm{GF}(q),(AB_1-A_1B)\lambda\neq 0} \chi_q\left[\left(\frac{A_1}{A}u-v\right)\lambda\right]\\
&= \begin{cases} 0, & \text{if } AB_1=A_1B,\\ \displaystyle\sum_{\lambda\in \mathrm{GF}(q)^*} \chi_q\left[\left(\frac{A_1}{A}u-v\right)\lambda\right], & \text{if } AB_1\neq A_1B \end{cases}\\
&= \begin{cases} 0, & \text{if } AB_1=A_1B,\\ q-1, & \text{if } AB_1\neq A_1B \text{ and } A_1u=Av,\\ -1, & \text{if } AB_1\neq A_1B \text{ and } A_1u\neq Av. \end{cases}
\end{aligned}
$$

B. Calculation of U_2

There are two cases for U_2. When $A/A_1 \notin \mathrm{GF}(q)$, we have $U_2 = 0$. When $A/A_1 \in \mathrm{GF}(q)$, we consider the following equation in the unknown λ:

$$[(B_1 - A_1B/A)\lambda]^{(q^e-1)/(p+1)} = (-1)^m. \tag{7.20}$$

There are two subcases:

(1) If $[(B_1 - A_1B/A)\lambda]^{(q^e-1)/(p+1)} \notin \mathrm{GF}(q)$, then Equation (7.20) is unsolvable in $\mathrm{GF}(q)$, so $U_2 = 0$.

(2) If $[(B_1 - A_1B/A)\lambda]^{(q^e-1)/(p+1)} \in \mathrm{GF}(q)$, then Equation (7.20) is equivalent to

$$\lambda^{(q^e-1)/(p+1)} = (-1)^m \left(B_1 - \frac{A_1}{A}B\right)^{(1-q^e)/(p+1)}$$

and the number of solutions in $\mathrm{GF}(q)$ of Equation (7.20) is

$$\Delta = \sum_{j=0}^{d-1} \varphi^j \left((-1)^m \left(B_1 - \frac{A_1}{A}B\right)^{(1-q^e)/(p+1)}\right), \tag{7.21}$$

where φ is a multiplicative character of $\mathrm{GF}(q)$ with degree $d = \gcd((q^e-1)/(p+1), q-1)$.

Thus

$$U_2 = \sum_{\substack{\lambda \in \mathrm{GF}(q), \\ (AB_1 - A_1B)\lambda \neq 0, \\ [(B_1 - A_1B/A)\lambda]^{(q^e-1)/(p+1)} = (-1)^m}} \chi_q \left[\left(\frac{A_1}{A}u - v\right)\lambda\right],$$

we have $U_2 = 0$ if $AB_1 = A_1B$.

$$U_2 = |\{\lambda \in \mathrm{GF}(q) | [(B_1 - A_1B/A)\lambda]^{(q^e-1)/(p+1)} = (-1)^m\}|,$$

if $AB_1 \neq A_1B$ and $Av = A_1u$.

$$U_2 = \sum_{\lambda \in \mathrm{GF}(q), [(B_1 - A_1B/A)\lambda]^{(q^e-1)/(p+1)} = (-1)^m} \chi_q \left[\left(\frac{A_1}{A}u - v\right)\lambda\right],$$

if $AB_1 \neq A_1B$ and $Av \neq A_1u$.

It remains to analyze the condition

$$[(B_1 - A_1B/A)\lambda]^{(q^e-1)/(p+1)} = (-1)^m. \tag{7.22}$$

To simplify notation, let $\beta = B_1 - A_1B/A$. If m is even, then Equation (7.22) is equivalent to saying that $\beta\lambda$ is a $(p+1)$ power of an element of $\mathrm{GF}(q^e)$. If m is odd, it is equivalent to saying that $\beta\lambda$ is a $(p+1)/2$ power of an element of $\mathrm{GF}(q^e)$, but not a $(p+1)$ power. Suppose γ is a primitive element in $\mathrm{GF}(q^e)$ and $T = (q^e - 1) = /(q-1)$. We can write $\beta = \gamma^b$ and $\lambda = \gamma^{\ell T}$. Then Equation (7.20) is equivalent to

$$b + \ell T = \begin{cases} (p+1)d, & \text{for some } d \text{ if } m \text{ is even}, \\ \dfrac{(p+1)d}{2}, & \text{for some odd } d \text{ if } m \text{ is odd}. \end{cases} \tag{7.23}$$

We make use of the following lemma.

Lemma 7.2.9 *For any natural numbers p, e, and n,*

$$\gcd\left(p+1, \frac{p^{en}-1}{p^n-1}\right) = \begin{cases} (p+1), & \text{if } e \text{ is even} \\ 1, & \text{if } e \text{ is odd.} \end{cases}$$

Proof We have

$$\begin{aligned}\frac{p^{en}-1}{p^n-1} &= p^{en} + p^{(e-1)n} + \cdots + p^n + 1 \\ &= \begin{cases} (p+1)(p^{(e-2)n} + p^{(e-4)n} + \cdots + 1), & \text{if } e \text{ is even,} \\ (p+1)(p^{(e-2)n} + p^{(e-4)n} + \cdots + p) + 1, & \text{if } e \text{ is odd.} \end{cases}\end{aligned}$$

The lemma follows. □

There are four cases to consider:

(1) Suppose m and e are even. Then Equation (7.23) has a solution for a given ℓ if and only if $(p+1)$ divides b. This is equivalent to β being a $(p+1)$ power. Thus

$$U_2 = \begin{cases} 0, & \text{if } AB_1 = A_1B \text{ or } \beta \text{ is not a } p+1 \text{ power,} \\ q-1, & \text{if } AB_1 \neq A_1B, Av = A_1u \text{ and } \beta \text{ is a } p+1 \text{ power,} \\ -1, & \text{otherwise.} \end{cases}$$

(2) Suppose m is even and e is odd, so that n is even and $n/2$ is even. Then T and $(p+1)$ are relatively prime, so for each b there is at least one pair (ℓ_0, d_0) that satisfies Equation (7.23). Every other pair that satisfies this equation has the form $(\ell_0 + i(p+1), d_0 + iT)$. Thus there are $(q-1)/(p+1)$ choices of λ for which Equation (7.23) has a solution. They all have the form $\lambda_0\delta^{i(p+1)}$, where $\delta = \gamma^T$ is primitive in $\mathrm{GF}(q)$ and $\lambda_0 = \delta^{\ell_0}$. Also, for any $y, y\lambda_0$ is a $(p+1)$ power in $\mathrm{GF}(q^e)$ if and only if $y\beta$ is a $(p+1)$ power in $\mathrm{GF}(q^e)$. It follows that

$$\begin{aligned} U_2 &= \begin{cases} 0, & \text{if } AB_1 = A_1B, \\ \dfrac{q-1}{p-1}, & \text{if } AB_1 \neq A_1B \text{ and } Av = A_1u, \\ \dfrac{1}{p+1} \displaystyle\sum_{x \in \mathrm{GF}(q)^*} \chi_q\left[\left(\frac{A_1}{A}u - v\right)\lambda_0 x^{p+1}\right], & \text{otherwise} \end{cases} \\ &= \begin{cases} 0, & \text{if } AB_1 = A_1B, \\ \dfrac{q-1}{p-1}, & \text{if } AB_1 \neq A_1B \text{ and } Av = A_1u, \\ \dfrac{-1-p^{n/2+1}}{p+1}, & \text{if } AB_1 \neq A_1B,\ Av \neq A_1u \text{ and } \left(\dfrac{A_1}{A}u - v\right)\beta \\ & \text{is a } (p+1) \text{ power,} \\ \dfrac{p^{n/2}-1}{p+1}, & \text{otherwise.} \end{cases} \end{aligned}$$

(3) Suppose m is odd and e is even. Then Equation (7.23) has a solution for a given ℓ if and only if $(p+1)/2$ divides b and $(p+1)$ does not. This is equivalent to β being a

$(p+1)/2$ power but not a $(p+1)$ power. Thus

$$U_2 = \begin{cases} 0, & \text{if } AB_1 = A_1B, \ \beta \text{ is not a } (p+1)/2 \text{ power or } \beta \text{ is a } (p+1) \text{ power}, \\ q-1, & \text{if } AB_1 \neq A_1B, Av = A_1u \text{ and } \beta \text{ is a } (p+1)/2 \\ & \text{power but not a } (p+1) \text{ power}, \\ -1, & \text{otherwise.} \end{cases}$$

(4) Suppose m and e are odd. Thus n is even and $n/2$ is odd. Again, for each b there is at least one pair (ℓ_0, d_0) that satisfies the equation $b + \ell T = d(p+1)/2$. Every other pair that satisfies this equation has the form $(\ell_0 + i(p+1)/2, d_0 + iT)$. Since $(p+1)$ is even, T must be odd. In particular, there is a solution with d odd, so we may assume d_0 is odd. Every other solution to Equation (7.23) has the form $(\ell_0 + i(p+1), d_0 + 2iT)$. As in the case when m is even and e is odd,

$$U_2 = \begin{cases} 0, & \text{if } AB_1 = A_1B, \\ \dfrac{q-1}{p-1}, & \text{if } AB_1 \neq A_1B \text{ and } Av = A_1u, \\ \dfrac{p^{n/2+1}-1}{p+1}, & \text{if } AB_1 \neq A_1B, \ Av \neq A_1u \text{ and } \left(\dfrac{A_1}{A}u - v\right)\beta \text{ is} \\ & \text{a } (p+1)/2 \text{ power but not a } p+1 \text{ power}, \\ \dfrac{-p^{n/2}-1}{p+1}, & \text{otherwise.} \end{cases}$$

C. Estimation of U_3

To upper bound U_3 it is useful to first compute the following cardinalities:

$$|\{(\xi,\lambda) \in \mathrm{GF}(q)^2 | B\xi + B_1\lambda \neq 0, A\xi + A_1\lambda = 0\}|$$
$$= \begin{cases} q-1, & \text{if } A/A_1 \in \mathrm{GF}(q) \text{ and } AB_1 \neq A_1B, \\ 0, & \text{otherwise.} \end{cases}$$
$$|\{(\xi,\lambda) \in \mathrm{GF}(q)^2 | B\xi + B_1\lambda = 0, A\xi + A_1\lambda = 0\}|$$
$$= \begin{cases} q, & \text{if } A/A_1 \in \mathrm{GF}(q) \text{ and } AB_1 = A_1B, \\ 1, & \text{otherwise.} \end{cases}$$
$$|\{(\xi,\lambda) \in \mathrm{GF}(q)^2 | B\xi + B_1\lambda 0, A\xi + A_1\lambda \neq 0\}|$$
$$= \begin{cases} q-1, & \text{if } B/B_1 \in \mathrm{GF}(q) \text{ and } AB_1 \neq A_1B, \\ 0, & \text{otherwise.} \end{cases}$$

Thus,

$$\begin{aligned} |U_3| &\leqslant |\{(\xi,\lambda) \in \mathrm{GF}(q)^2 | B\xi + B_1\lambda \neq 0, A\xi + A_1\lambda \neq 0, \text{ and Equation (7.19) is solvable} \\ &\quad \text{in } \mathrm{GF}(q^e)\}| \\ &\leqslant \{(\xi,\lambda) \in \mathrm{GF}(q)^2 | B\xi + B_1\lambda \neq 0, A\xi + A_1\lambda \neq 0\}| \\ &= \begin{cases} q^2 - 2q + 1, & \text{if } A/A_1, B/B_1 \in \mathrm{GF}(q) \text{ and } AB_1 \neq A_1B, \\ q^2 - q, & \text{if } A/A_1 \in \mathrm{GF}(q), B/B_1 \notin \mathrm{GF}(q) \text{ and } AB_1 \neq A_1B, \\ q^2 - q, & \text{if } A/A_1 \in \mathrm{GF}(q) \text{ and } AB_1 = A_1B, \\ q^2 - q, & \text{if } A/A_1 \notin \mathrm{GF}(q), B/B_1 \in \mathrm{GF}(q) \text{ and } AB_1 \neq A_1B, \\ q^2 - 1, & \text{if } A/A_1 \notin \mathrm{GF}(q), B/B_1 \notin \mathrm{GF}(q) \text{ and } AB_1 \neq A_1B, \\ q^2 - 1, & \text{if } A/A_1 \notin \mathrm{GF}(q) \text{ and } AB_1 = A_1B. \end{cases} \end{aligned}$$

D. Bounds for cross-correlations

As before, we have

$$N_\tau(u,v) - q^{-2}(W_1 + (-1)^m p^m U_1 + (-1)^{m+1} p^m (p+1) U_2) = (-1)^m q^{-2} p^m U_3 \leqslant \varepsilon$$

for some error term ε. It follows that

$$\begin{aligned} &|C_{S_f^{(A,B)},S_g^{(A',B')}}(\tau) + F(0)G(0) \\ &-q^{-2} \sum_{u,v\in \mathrm{GF}(q)} (W_1 + (-1)^m p^m U_1 + (-1)^{m+1} p^m (p+1) U_2) F(u)G(v)| \\ &< q^{-2} \sum_{u,v\in \mathrm{GF}(q)} |\varepsilon F(u)G(v)| = q^{-2} \sum_{u,v\in \mathrm{GF}(q)} \varepsilon. \end{aligned}$$

Again, W_1, U_1, U_2, and ε depend on u and v.

Computing the sums for W_1 and U_1 is straightforward. For U_2, suppose first that e is even. The only nonzero case is when $A/A_1 \in \mathrm{GF}(q)$ and $\beta = B_1 - A_1 B/A$ is not a $(p+1)$ power. In this case we have

$$\sum_{u,v} U_2 F(u)G(v) = q \sum_u F(u) G\left(\frac{A_1}{A} u\right) - I(f)I(g).$$

Now suppose that e is odd. The only nonzero case is when $A/A1 \in \mathrm{GF}(q)$ and $AB_1 \neq A_1 B$. In this case we have

$$\begin{aligned} &\sum_{u,v} U_2 F(u)G(v) \\ &= \frac{q-1}{p+1} \sum_u F(u) G\left(\frac{A_1}{A} u\right) \\ &+ \frac{(-1)^{m+1} p q^{1/2} - 1}{p+1} \sum_{\substack{Av\neq A_1 u, \\ \left(\frac{A_1}{A}u - v\right)\left(B_1 - A_1 \frac{B}{A}\right) \text{a } (p+1) \text{ power}}} F(u)G(v) \\ &+ \frac{(-1)^m q^{1/2} - 1}{p+1} \sum_{\substack{Av\neq A_1 u, \\ \left(\frac{A_1}{A}u - v\right)\left(B_1 - A_1 \frac{B}{A}\right) \text{not a } (p+1) \text{ power}}} F(u)G(v) \\ &= \frac{q-1}{p+1} \sum_u F(u) G\left(\frac{A_1}{A} u\right) + \frac{(-1)^m q^{1/2} - 1}{p+1} \sum_{Av\neq A_1 u} F(u)G(v) \\ &-(-1)^m q^{1/2} \sum_{\substack{Av\neq A_1 u, \\ \left(\frac{A_1}{A}u - v\right)\left(B_1 - A_1 \frac{B}{A}\right) \text{a } (p+1) \text{ power}}} F(u)G(v) \\ &= \left(\frac{q - (-1)^m q^{1/2}}{p+1}\right) \sum_u F(u) G\left(\frac{A_1}{A} u\right) + \frac{(-1)^m q^{1/2} - 1}{p+1} I(f)I(g) \\ &-(-1)^m q^{1/2} \sum_{\substack{Av\neq A_1 u, \\ \left(\frac{A_1}{A}u - v\right)\left(B_1 - A_1 \frac{B}{A}\right) \text{a} (p+1) \text{power}}} F(u)G(v). \end{aligned}$$

This completes the proof of Theorem 7.2.6. □

A similar analysis can be given in the case when $p = 2$. The major change is that we must use different character sum results[23]. We state the results here but omit the proofs.

Theorem 7.2.7 *Suppose that $p = 2$. Let $S_f^{(A,B)}, S_g^{(A',B')}$ be generalized geometric sequences based on the same α and k, as defined in Equation (7.17). Assume that $A, B, A', B' \neq 0, k = p+1, q^e = p^{en} = p^{2m+1}$, and $0 \leqslant \tau \leqslant q^e - 2$. Let $A_1 = A'\alpha^\tau$ and $B_1 = B'\alpha^{\tau(p+1)}$.*

(1) *If $A/A_1 \in \mathrm{GF}(q)$ and $AB_1 = A_1B$, then*

$$\left| C_{S_f^{(A,B)}, S_g^{(A',B')}}(\tau) + F(0)G(0) - q^{e-1} \sum_{u \in \mathrm{GF}(q)} F(u)G(\frac{A_1}{A}u) \right| \leqslant q^{(e+3)/2}.$$

(2) *If $A/A_1 \notin \mathrm{GF}(q)$ and $AB_1 = A_1B$, then*

$$\left| C_{S_f^{(A,B)}, S_g^{(A',B')}}(\tau) + F(0)G(0) - q^{e-1} \sum_{u \in \mathrm{GF}(q)} F(u)G(\frac{A_1}{A}u) \right| \leqslant q^{e/2}(q^2 - 1).$$

(3) *If $AB_1 \neq A_1B$, then*

$$|C_{S_f^{(A,B)}, S_g^{(A',B')}}(\tau) - q^{e-2}I(f)I(g) + F(0)G(0)|$$
$$\leqslant \begin{cases} q^{(e+3)/2} + q^{(e+1)/2}, & \text{if } A/A_1 \in \mathrm{GF}(q) \text{ and } B/B_1 \in \mathrm{GF}(q), \\ q^{e/2}(q^2 - 1), & \text{if } A/A_1 \notin \mathrm{GF}(q) \text{ and } B/B_1 \notin \mathrm{GF}(q), \\ q^{(e+3)/2}, & \text{otherwise.} \end{cases}$$

Theorem 7.2.8 *Suppose that $p = 2$ and e is even. Let $S_f^{(A,B)}, S_g^{(A',B')}$ be generalized geometric sequences based on the same α and k, as defined in Equation (7.17). Assume that $A, B, A', B' \neq 0, k = p+1, q^e = p^{en} = p^{2m}$, and $0 \leqslant \tau \leqslant q^e - 2$. Let $A_1 = A'\alpha^\tau$ and $B_1 = B'\alpha^{\tau(p+1)}$.*

(1) *If $A/A_1 \in \mathrm{GF}(q)$ and $AB_1 = A_1B$, then*

$$\left| C_{S_f^{(A,B)}, S_g^{(A',B')}}(\tau) + F(0)G(0) - q^{e-1} \sum_{u \in \mathrm{GF}(q)} F(u)G\left(\frac{A_1}{A}u\right) \right| \leqslant 2q^{e/2}(q^2 - q).$$

(2) *If $A/A_1 \notin \mathrm{GF}(q), AB_1 \neq A_1B$ and $B + AB_1/A_1$ is a nonzero cube, then*

$$|C_{S_f^{(A,B)}, S_g^{(A',B')}}(\tau) + (-1)^m 2q^{e/2-1} \sum_{u \in \mathrm{GF}(q)} F(u)G\left(\frac{A_1}{A}u\right)$$
$$-q^{e-2}I(f)I(g) + F(0)G(0)|$$
$$\leqslant \begin{cases} 2q^{e/2}(q^2 - 2q + 1), & \text{if } B/B_1 \in \mathrm{GF}(q), \\ 2q^{e/2}(q^2 - q), & \text{if } B/B_1 \notin \mathrm{GF}(q). \end{cases}$$

(3) *If $A/A_1 \in \mathrm{GF}(q), AB_1 \neq A_1B$ and $B + AB_1/A_1$ is not a cube, then*

$$|C_{S_f^{(A,B)},S_g^{(A',B')}}(\tau) + (-1)^m 3q^{e/2-1} \sum_{u\in\mathrm{GF}(q)} F(u)G\left(\frac{A_1}{A}u\right)$$

$$-q^{e-2}I(f)I(g) + F(0)G(0)|$$

$$\leqslant \begin{cases} 2q^{e/2}(q^2-2q+1), & \textit{if } B/B_1 \in \mathrm{GF}(q), \\ 2q^{e/2}(q^2-q), & \textit{if } B/B_1 \notin \mathrm{GF}(q). \end{cases}$$

(4) *If* $A/A_1 \notin \mathrm{GF}(q)$, *then*

$$|C_{S_f^{(A,B)},S_g^{(A',B')}}(\tau) - q^{e-2}I(f)I(g) + F(0)G(0)|$$

$$\leqslant \begin{cases} 2q^{e/2}(q^2-q), & \textit{if } B/B_1 \in \mathrm{GF}(q) \textit{ and } AB_1 \neq A_1B, \\ 2q^{e/2}(q^2-1), & \textit{otherwise.} \end{cases}$$

Theorem 7.2.9 *Suppose that* $p=2$ *and* e *is odd. Let* $S_f^{(A,B)}, S_g^{(A',B')}$ *be generalized geometric sequences based on the same* α *and* k*, as defined in Equation* (7.17). *Assume that* $A, B, A', B' \neq 0, k = p+1, q^e = p^{en} = p^{2m}$, *and* $0 \leqslant \tau \leqslant q^e - 2$. *Let* $A_1 = A'\alpha^\tau$ *and* $B_1 = B'\alpha^{\tau(p+1)}$. *then*

(1) *If* $A/A_1 \in \mathrm{GF}(q)$ *and* $AB_1 = A_1B$, *then*

$$\left|C_{S_f^{(A,B)},S_g^{(A',B')}}(\tau) + F(0)G(0) - q^{e-1} \sum_{u\in\mathrm{GF}(q)} F(u)G\left(\frac{A_1}{A}u\right)\right| \leqslant \frac{4q^{e/2}(q^2-q)}{3}.$$

(2) *If* $A/A_1 \in \mathrm{GF}(q)$ *and* $AB_1 \neq A_1B$, *then*

$$\Big|C_{S_f^{(A,B)},S_g^{(A',B')}}(\tau) - q^{e-2}I(f)I(g) + F(0)G(0)$$

$$-(-1)^m q^{e/2-2} \sum_{u\in\mathrm{GF}(q)} F(u)G\left(\frac{A_1}{A}u\right)$$

$$-(-1)^{m+n}3q^{e/2-1} \sum_{\left(u+\frac{A}{A_1}v\right)\left(B+\frac{A}{A_1}B_1\right)\text{a nonzero cube}} F(u)G(v)\Big|$$

$$\leqslant \begin{cases} \frac{4}{3}q^{e/2}(q^2-2q+1), & \textit{if } B/B_1 \in \mathrm{GF}(q), \\ \frac{4}{3}q^{e/2}(q^2-q), & \textit{if } B/B_1 \notin \mathrm{GF}(q). \end{cases}$$

(3) *If* $A/A_1 \notin \mathrm{GF}(q)$, *then*

$$|C_{S_f^{(A,B)},S_g^{(A',B')}}(\tau) - q^{e-2}I(f)I(g) + F(0)G(0)|$$

$$\leqslant \begin{cases} \frac{4}{3}q^{e/2}(q^2-q), & \textit{if } AB_1 \neq A_1B \textit{ and } B/B_1 \in \mathrm{GF}(q), \\ \frac{4}{3}q^{e/2}(q^2-1), & \textit{otherwise.} \end{cases}$$

Now we begin to consider the linear complexity of certain generalized geometric sequences. We denote by $\lambda(S)$ the linear complexity of a sequence S. Let $p=2$ and let $n_1, n_2, \cdots, n_l$ be natural numbers with n_i at least 3. Let $n = n_1n_2\cdots n_l$, $q_0 = 2$, and

$q_i = q_{i-1}^{n_i}$, so, in particular, $q = q_l$. We also let $r_1, \cdots, r_l$ be integers, with $1 \leqslant r_i < n_i$, let $k_i = 1 + q_{i-1}^{r}$, and define $f : \mathrm{GF}(q) \to \mathrm{GF}(2)$ by

$$f(x) = \mathrm{Tr}_{q_0}^{q_1}(\mathrm{Tr}_{q_1}^{q_2}(\cdots \mathrm{Tr}_{q_{l-1}}^{q_l}(x)^{k_{l-1}} \cdots)^{k_1}).$$

If β is primitive in $\mathrm{GF}(q)$, then the sequence T whose i-th term is $T(i) = f(\beta^i)$ is a cascaded GMW sequence. This sequence has shifted autocorrelations equal to -1. Its linear complexity is

$$\lambda(T) = n_1 n_2^2 n_3^4 \cdots n^{2^{l-1}}.$$

Now we consider the generalized geometric sequence S whose i-th term is

$$S(i) = f(\mathrm{Tr}_q^{q^e}(\alpha^i + B\alpha^{3i}))$$

for some $B \neq 0$. Also, for convenience we write $n_{l+1} = e$.

Theorem 7.2.10 *If $2 \leqslant r_1 \leqslant n_1 - 2, 1 \leqslant r_j \leqslant n_j - 2$ for $j = 2, \cdots, l-1$, and $2r_j \not\equiv 0 (\mathrm{mod}\ n_j)$ for $j = 1, \cdot, l-1$, then the linear complexity of S is*

$$\lambda(S) = n_1 n_2^2 n_3^4 \cdots n^{2^{l-l}} (2e)^{2^l}.$$

Proof Key showed that the linear complexity of such a sequence equals the number of nonzero coefficients when the function

$$g(x) = f(\mathrm{Tr}_q^{q^e}(x + Bx^3)) \tag{7.24}$$

is expressed a polynomial. Let

$$K = \{\bar{t} = (t_1, t_2, \cdots, t_l) : t_i \in \{0, r_i\}\}.$$

Let

$$s : K \to \{(s_1, s_2, \cdots, s_l, s_{l+1}) : 0 \leqslant s_i < n_i\}$$

be a function with the property that whenever $\bar{t}, \bar{t'} \in K$ satisfy $t_1 = t'_1, \cdots, t_j = t'_j$ for some j, s satisfies $s(\bar{t})_1 = s(\bar{t'})_1, s(\bar{t})_2 = s(\bar{t'})_2, \cdots, s(\bar{t})_{j+1} = s(\bar{t'})_{j+1}$. For any such s and $a : K \to \{1, 3\}$, let

$$h(s, a) = \sum_{\bar{t} \in K} a(\bar{t}) \prod_{j=1}^{l} q_{j-1}^{s(\bar{t})_j + t_j} q_l^{s(\bar{t})_{l+1}} = \sum_{\bar{t} \in K} a(\bar{t}) z(\bar{t}).$$

If we expand the right hand side of equation (7.24), the resulting polynomial can be expressed as a sum of the monomials whose exponents are the $h(s, a)$'s over all possible s and a. Thus we need to see that these are all distinct.

First, suppose that for some fixed s and a and for some $\bar{t} \neq \bar{t'} \in K$, the base 2 expansions of $a(\bar{t})z(\bar{t})$ and $a(\bar{t'})z(\bar{t'})$ have some nonzero term in common. We may assume that for some j, $t_1 = t'_1, \cdots, t_{j-1} = t'_{j-1}, t_j = 0$, and $t'_j = r_j$. Then for some x (which is a power of 2), b, and c we have $z(\bar{t}) = x q_{j+1}^b$ and $z(\bar{t'}) = x q_j^{r_j} q_{j+1}^c$. There are two possibilities.

(1) If $z(\bar{t}) < z(\bar{t'})$, then we must have $a(\bar{t}) = 3$ and $3z(\bar{t}) \geqslant z(\bar{t'})$. It follows that $2z(\bar{t}) = z(\bar{t'})$. Therefore $2 = q_j^{r_j}$, which is impossible by the hypotheses.

(2) If $z(\bar{t}) > z(\bar{t'})$, then we must have $a(\bar{t'}) = 3$ and $3z(\bar{t'}) \geqslant z(\bar{t'})$. It follows that $2z(\bar{t'}) = z(\bar{t'})$. Therefore $2q_j^{r_j} = q_{j+1}$, which is impossible by the hypotheses.

Now suppose that $h(s,a) = h(s',a')$ for some s, s', a, a'. It follows first that the number of 1's among the $a(\bar{t})$'s equals the number of 1's among the $a'(\bar{t})$'s. Note also that $s_1 = s_1(\bar{t})$ and $s'_1 = s'_1(\bar{t})$ are independent of $\bar{t}$. If we take the reductions modulo n_1 of the exponents that occur on nonzero terms in the base two expansion of $h(s,a)$, we obtain $\{s_1, s_1 + r_1\}$ if all $a(\bar{t}) = 1$, and $\{s_1, 1 + s_1, s_1 + r_1, 1 + s_1 + r_1\}$ otherwise. We obtain $\{s'_1, s'_1 + r_1\}$ if all $a(\bar{t}) = 1$, and $\{s'_1, 1 + s'_1, s'_1 + r_1, 1 + s'_1 + r_1\}$ otherwise for $h(s', a')$. These sets must be equal if $h(s,a) = h(s',a')$. If $s_1 \equiv s'_1 + r_1 (\text{mod } n_1)$ and $s'_1 \equiv s_1 + r_1 (\text{mod } n_1)$ then $2r_1 \equiv 0 (\text{mod } n_1)$. This is false by hypothesis. Therefore $s_1 = s'_1$. Also, all terms with $t_1 = 1$ or $t'_1 = 1$ map to $\{s_1, 1 + s_1\}$ when we reduce exponents modulo n_1, so the sum of such terms in $h(s,a)$ equals the sum of such terms in $h(s',a')$. Similarly for the terms that map to $\{s_1 + r_1, 1 + s_1 + r_1\}$. By induction, $s = s'$ and it then follows that $a = a'$.

Consequently, the linear complexity of S is the number of pairs of functions (s,a), which is the quantity given in the statement of the theorem. □

Thus the linear complexity for these generalized geometric sequences is larger than the linear complexity of a cascaded GMW sequence based on the same tower of fields and the same exponents by a factor of 2^{2^l}.

Now we use the results of the previous sections to construct families of sequences with good pairwise correlations. We fix prime numbers p and r, natural numbers e and n, a primitive element α in $\text{GF}(p^{ne})$, and a function f from $\text{GF}(p^n)$ to $\text{GF}(r)$. As above, we let $q = p^n$ and $F(u) = \omega^{f(u)}$ where ω is a complex primitive r-th root of unity. Let $S_f^{(A,B)}$ be the sequence defined in Equation (7.17). Note that every sequence $S_f^{(A,B)}$ has a cyclic shift of the form $S_f^{(1,B)}$. Assume that f is balanced. That is, $I(f) = 0$. This is only possible if $r = p$, so we are assuming this. We also assume that f has ideal autocorrelations in the sense that

$$\sum_{u \in \text{GF}(q)} F(u)F(xu) = \begin{cases} 0, & \text{if } x \neq 1, \\ q, & \text{if } x = 1. \end{cases}$$

This is equivalent to saying that the sequence whose j-th element is $f(\beta^j)$, where β is a primitive element in $\text{GF}(q)$, has ideal autocorrelations. There are many examples of such sequences (for example, m-sequences, GMW sequences, and cascaded GMW sequences). Let $S = \{S_f^{(1,B)} : B \in \text{GF}(q^e), B \neq 0\}$. We want to show that every pair of sequences in this set is cyclically distinct. By Theorems 7.2.5 and 7.2.6, if p is odd, then for any two sequences $S_f^{(1,B)}$ and $S_f^{(1,B')}$ in S and any τ,

$$|C_{S_f^{(1,B)}, S_g^{(1,B')}}(\tau)| \leqslant q^{e/2}(q^2 - 1) + 1. \tag{7.25}$$

unless

$$\alpha^\tau \in \text{GF}(q), \quad \alpha^\tau B = \alpha^{(p+1)\tau} B'$$

and

$$\sum_{u \in \text{GF}(q)} F(u)F(\alpha^\tau u) \neq 0. \tag{7.26}$$

It follows from our assumption on the autocorrelations of f that if condition (7.26) holds, then $\tau = 0$, and therefore that $B = B'$. Thus the only correlation that fails to satisfy inequality Equation (7.25) is

$$C_{S_f^{(1,B)}, S_g^{(1,B')}}(\tau) = q^e - 1.$$

This also implies that any two sequences in S are cyclically distinct and proves the following theorem.

Theorem 7.2.11 *If p is odd, then S is a family of $q^e - 1$ $\mathrm{GF}(p)$-ary sequences of period $q^e - 1$ all of whose cross-correlations and shifted autocorrellations are bounded by*

$$|C_{S_f^{(1,B)}, S_g^{(1,B')}}(\tau)| \leqslant q^{e/2}(q^2 - 1) + 1. \tag{7.27}$$

We have a similar family when p is even.

Theorem 7.2.12 *If $p = 2$, then S is a family of $q^e - 1$ binary sequences of period $q^e - 1$ all of whose cross-correlations and shifted autocorrellations are bounded by*

$$|C_{S_f^{(1,B)}, S_g^{(1,B')}}(\tau)| \leqslant \begin{cases} 2q^{e/2}(q^2 - 1) + 1, & \text{if } e \text{ is even}, \\ \frac{4}{3}q^{e/2}(q^2 - 1) + 1, & \text{if } e \text{ is odd and } n \text{ is even}, \\ q^{e/2}(q^2 - 1) + 1, & \text{if } e \text{ and } n \text{ are odd}. \end{cases}$$

When $n = 1$ the sequences studied here reduce to one case of Gold sequences. However, in this case our estimates of the cross-correlations are too high. For example, when $p = 2$ and e is odd, they are too high by a factor of three. We conjecture, therefore, that our estimates are too high in general. In the case when p is odd and $en = 2m + 1$, some improvement would come if we could choose f so that $\psi(Cu + Dv)F(u)F(v)$ is small for every $C, D \in \mathrm{GF}(q^e)$. However, the greatest improvement would come from sharper bounds on W_3. The situation is similar when en is even and when $p = 2$.

7.2.3 On the Correlation Functions of p-Ary d-Form Sequences[19]

Binary sequences with large linear spans and good correlation functions have many important applications in spread spectrum communications and cryptography. Klapper defined a kind of so-called binary d-form sequences[24] and constructed TN sequences[24] and $(0, j)$-QF sequences which are two families of binary 2-form sequences. However, p-ary sequences with large linear span and good correlation functions are more important than binary ones in some sense. In this subsection, we generalize the binary d-form sequences to p-ary d-form sequences which have the same correlation property as the binary ones if the parameter d is well chosen and construct a family of p-ary 2-form sequences, and their cross-correlation functions are also given.

Definition 7.2.3 *Let c and e be positive integers, $q = p^e, p > 2$ an odd prime, α a primitive element of $\mathrm{GF}(q^e)$, and let $H(x)$ be a homogeneous polynomial over $\mathrm{GF}(q)$ of degree d. Let ℓ be a positive integer such that $\gcd(\ell, q-1) = 1$. Then a p-ary sequences $S = \{s(t) : 0 \leqslant t \leqslant q^e - 2\}$ is called a d-form sequence, where $s(t) = \mathrm{Tr}_p^q((H(\alpha^t))^\ell)$.*

The period of a d-form sequence is a divisor of $q^n - 1$.

Definition 7.2.4 *Let $S = \{s(i)\}$ and $T = \{t(i)\}$ be two p-ary sequences with period L, ω a p-th primitive unity root, then their periodic cross-correlation function $C_{ST}(\tau)$ of S and T is defined by $C_{ST}(\tau) = \sum_{i=0}^{L-1} \omega^{s(i+\tau)-t(i)}, \tau = 0, 1, \cdots, L-1$.*

Lemma 7.2.10 *Let χ be a nontrivial additive character of* GF$(q), n$ *a positive integer, and λ a multiplicative character of* GF(q) *of order* $d = \gcd(n, q-1)$. *Then*

$$\sum_{c \in \mathrm{GF}(q)} \chi(ac^n + b) = \chi(b) \sum_{j=1}^{d-1} \overline{\lambda}^j(a) G(\lambda^j, \chi) \tag{7.28}$$

for any $a, b \in \mathrm{GF}(q)$ with $a \neq 0$, where $\overline{\lambda}$ is the conjugate character of λ, and $G(\lambda^j, \chi) = \sum_{x \in \mathrm{GF}(q)^*} \lambda^j(x)\chi(x)$.

Lemma 7.2.11 *Let χ be a nontrivial character of* GF(q) *and* $d = \gcd(n, q-1)$. *Then*

$$\sum_{x \in \mathrm{GF}(q)} \chi(ax^n + b) = 0$$

for any $a, b \in \mathrm{GF}(q)$ with $a \neq 0$.

Theorem 7.2.13 *Let S and T be p-ary sequences determined by two homogeneous polynomial $H_1(x)$ and $H_2(x)$ of degree $d, \overline{d} = \gcd(d, q-1)$. Then their cross-correlation function is*

$$C_{ST}(\tau) = \frac{q}{q-1} Z_\tau^{(d)} - T + \sum_{j=1}^{\overline{d}-1} G(\lambda^j, \chi) \sum_{t=0}^{T-1} \overline{\lambda}^j(\delta_d(\alpha^t, \tau)),$$

where $\delta_d(x, \tau) = H_1(x)^\ell - H_2(\alpha^\tau x)^\ell, T = \dfrac{q^e - 1}{q-1}$.

Proof For any $0 \leqslant t \leqslant q^e - 2, t = t_1 + Tt_2, 0 \leqslant t_1 \leqslant T-1, 0 \leqslant t_2 \leqslant q-2$, and

$$\delta_d(\alpha^t, \tau) = \alpha^{\ell d T t_2} \delta_d(\alpha^{t_1}, \tau).$$

From Definition 7.2.4, and Lemma 7.2.10, we have

$$\begin{aligned}
C_{ST} &= \sum_{t_1=0}^{T-1} \sum_{t_2=0}^{q-2} \omega^{\mathrm{tr}_p^q(\alpha^{\ell d T t_2} \delta_d(\alpha^{t_1}, \tau))} \\
&= \sum_{t_1=0}^{T-1} \sum_{x \in \mathrm{GF}(q)} \chi(x^d \delta_d(\alpha^{t_1}, \tau)) - T \\
&= \sum_{\delta_d(\alpha^{t_1}, \tau) = 0} q + \sum_{\delta_d(\alpha^{t_1}, \tau) \neq 0} \left(\sum_{j=1}^{\overline{d}-1} \overline{\lambda}^j(\delta_d(\alpha^{t_1}, \tau)) G(\lambda^j, \chi) \right) - T \\
&= q N_\tau^{(d)} - T + \sum_{j=1}^{\overline{d}-1} G(\lambda^j, \chi) \sum_{\delta_d(\alpha^{t_1}, \tau) \neq 0} \overline{\lambda}^j(\delta_d(\alpha^{t_1}, \tau)),
\end{aligned}$$

where $N_\tau^{(d)} = |\{0 \leqslant t_1 \leqslant T-1 : \delta_d(\alpha^{t_1}, \tau) = 0\}|$.

Let $Z_\tau^{(d)} = |\{0 \leqslant t \leqslant q^e - 2 : \delta_d(\alpha^t, \tau) = 0\}|$. Then $N_\tau^{(d)} = \dfrac{Z_\tau^{(d)}}{q-1}$ since $\delta_d(\alpha^t, \tau) = \alpha^{d\ell T t_2} \delta_d(\alpha^{t_1}, \tau)$.

From that $\sum_{\delta_d(\alpha^{t_1}, \tau) \neq 0} \overline{\lambda}^j(\delta_d(\alpha^{t_1}, \tau)) = \sum_{t_1=0}^{T-1} \overline{\lambda}^j(\delta_d(\alpha^{t_1}, \tau))$, we get the desired result. □

Corollary 7.2.3 *If* $\gcd(d, q-1) = 1$, *then*

$$C_{ST} = \frac{q}{q-1} Z_T^{(d)} - T,$$

which is the same as that in [24].

In order to calculate cross-correlation functions of a family of p-ary 2-form sequences which will be constructed in the next section, several lemmas are listed.

Lemma 7.2.12[22] *Let* χ *be a nontrivial additive character of* $\mathrm{GF}(q)$ *with* q *odd, and let* $f(x) = a_2x^2 + a_1x + a_0 \in \mathrm{GF}(q)[x]$ *with* $a_2 \neq 0$. *Then*

$$\sum_{x \in \mathrm{GF}(q)} \chi(f(x)) = \chi(a_0 - a_1^2(4a_2)^{-1})\eta(a_2)G(\eta, \chi),$$

where η *is the quadratic character of* $\mathrm{GF}(q)$, *and* $G(\eta, \chi) = \sum_{c \in \mathrm{GF}(q)^*} \eta(c)\chi(c)$.

Lemma 7.2.13[22] *Let* f *be a nondegenerate quadratic form over* $\mathrm{GF}(q)$, q *odd, in an even number* n *of indeterminates. Then for* $b \in \mathrm{GF}(q)$ *the number of solutions of the equation* $f(x_1, x_2, \cdots, x_n) = b$ *in* $\mathrm{GF}(q)^n$ *is*

$$q^{n-1} + \xi(b)q^{(n-2)/2}\eta((-1)^{n/2}\det(f)),$$

where $\xi(b) = q - 1$ *when* $b = 0$ *while* $\xi(b) = -1$ *when* $b \neq 0$.

Lemma 7.2.14[22] *Let* f *be a nondegenerate quadratic form over* $\mathrm{GF}(q)$, q *odd, in an odd number* n *of indeterminates. Then for* $b \in \mathrm{GF}(q)$ *be the number of solutions of the equation* $f(x_1, \cdots, x_n) = b$ *in* $\mathrm{GF}(q)^n$ *is*

$$q^{n-1} + q^{(n-2)/2}\eta((-1)^{n/2}\det(f)b).$$

Lemma 7.2.15[21] *Let* q *be power of an odd prime* p, e *a positive integer,* $1 \leqslant r \leqslant e$ *such that* $\gcd(e, r) = 1$. *Then* $f(x) = \mathrm{Tr}_q^{q^e}(cx^{q^r+1})$ *is a quadratic form over* $\mathrm{GF}(q)$ *of rank* e *for any* $c \in \mathrm{GF}(q)^*$.

Lemma 7.2.16[21] *Let* α *be a primitive element of* $\mathrm{GF}(q^n), u \in \mathrm{GF}(q)$, $H_{\alpha^i}^u$ *denote the set of solutions of* $\mathrm{Tr}_q^{q^n}(\alpha^i x) = u$ *in* $\mathrm{GF}(q^n)$. *Then*

$$|H_{\alpha^i}^u = q^{n-1}|, \tag{7.29}$$

$$|H_{\alpha^i}^u \cap H_{\alpha^i}^v| = \begin{cases} q^{n-2}, & \alpha^{i-j} \notin \mathrm{GF}(q), \\ q^{n-1}, & \alpha^{i-j} \in \mathrm{GF}(q), \alpha^i v = \alpha^j u, \\ 0, & \textit{otherwise}. \end{cases} \tag{7.30}$$

Let n be odd, $a, b, c, a_i, b_i, c_i \in \mathrm{GF}(q), 1 \leqslant i \leqslant n$ such that $\prod_{i=1}^n a_i \neq 0, b_i \neq 0$ and $c_j \neq 0$ for some $1 \leqslant i, j \leqslant n$. Let $N_n(a, b), N_n(a, b, c)$ respectively denote the number of solutions of the following systems

$$\begin{cases} a_1x_1^2 + a_2x_2^2 + \cdots + a_nx_n^2 = a, \\ b_1x_1 + b_2x_2 + \cdots + b_nx_n = b, \end{cases} \tag{7.31}$$

$$\begin{cases} a_1x_1^2 + a_2x_2^2 + \cdots + a_nx_n^2 = a, \\ b_1x_1 + b_2x_2 + \cdots + b_nx_n = b, \\ c_1x_1 + c_2x_2 + \cdots + c_nx_n = c. \end{cases} \tag{7.32}$$

Lemma 7.2.17 *Let $n = 2k+1, E = \prod_{i=1}^n a_i \in \mathrm{GF}(q)^*$, $M = \sum_{i=1}^n b_i^2/a_i, \gamma = b^2 - Ma$. Then*

$$N_n(a,b) = \begin{cases} q^{n-2} + q^{k-1}(q-1)\eta((-1)^kEM), & M \neq 0, \gamma = 0, \\ q^{n-2} + q^{k-1}\eta((-1)^kEM), & M \neq 0, \gamma \neq 0, \\ q^{n-2}, & M = a = \gamma = 0, \\ q^{n-2} + q^k\eta((-1)^kaM), & M = \gamma = 0, a \neq 0, \\ q^{n-2}, & M = 0 \neq \gamma. \end{cases} \tag{7.33}$$

Lemma 7.2.18 *Let $n = 2k+1, E = \prod_{i=1}^n a_i \in \mathrm{GF}(q)^*$, $M = \sum_{i=1}^n b_i^2/a_i, \varphi = \sum_{i=1}^n c_i^2/a_i$, $\delta = \sum_{i=1}^n b_ic_i/a_i, \Delta = \delta^2 - M\varphi, \sigma = c\delta - b\varphi, \hat{A} = 2bc\delta - a\delta^2 - Mc^2, \hat{B} = b^2 - aM, \hat{C} = c^2 - a\varphi, \lambda = \hat{C}\Delta - \sigma^2$. Suppose that hyperplanes $\sum_{i=1}^n b_ix_i = b$ and $\sum_{i=1}^n c_ix_i = c$ are not parallel, then*

$$N_n(a,b,c) = \begin{cases} q^{n-3} + q^{k-1}\eta((-1)^{k+1}E\varphi\lambda), & \varphi \neq 0, \Delta = 0, \\ q^{n-3}, & \varphi \neq 0, \Delta = 0 \neq \sigma, \\ q^{n-3} + q^{k-1}\eta((-1)^kE\varphi)\xi(\hat{C}), & \varphi \neq 0, \Delta = 0 = \sigma, \\ q^{n-3} + q^{k-1}\eta((-1)^{k+1}E\hat{A}), & \varphi = 0, \delta \neq 0, \\ q^{n-3} + q^{k-1}\eta((-1)^kEM)\xi(\hat{B}), & \varphi = \delta = c = 0, M \neq 0, \\ q^{n-3} + q^k\eta((-1)^kaE), & \varphi = \delta = c = 0, M = b = 0, \\ q^{n-3}, & \varphi = \delta = c = 0, M = 0 \neq b, \\ q^{n-3}, & \varphi = \delta = 0, c \neq 0. \end{cases} \tag{7.34}$$

Lemma 7.2.19 *Let $f(x) = a_0 + a_1x + a_2x^2 \in \mathrm{GF}(q)[x]$ with q odd and $a_2 \neq 0$. Put $d = a_1^2 - 4a_0a_2$ and let η be the quadratic character of $\mathrm{GF}(q)$. Then*

$$\sum_{x \in \mathrm{GF}(q)} \eta((f(x))) = \xi(d)\eta(a_2).$$

Now we will construct a family of p-ary 2-form sequences, or p-ary QF(quadratic form) sequences and calculate their cross-correlation functions. Let $q = p^e$, $p > 2$ and odd prime, e an odd positive integer, α a primitive element of $\mathrm{GF}(q^e)$, and let $\beta_i \in \mathrm{GF}(q^e)$, $s_i(t) = \mathrm{Tr}_p^q\{\mathrm{tr}_q^{q^e}(\alpha^t)^2 + \mathrm{Tr}_q^{q^e}(\beta_i\alpha^{(q^r+1)t})\}$, then $S = \{s_i(t) : 0 \leqslant t \leqslant q^e - 2\}$ is a p-ary 2-form sequence since $H(x) = \mathrm{Tr}_q^{q^e}(\alpha^t)^2 + \mathrm{Tr}_q^{q^e}(\beta_i\alpha^{(q^r+1)t})$ is a quadratic form over $\mathrm{GF}(q)$.

From Theorem 7.2.13, Lemma 7.2.12 and Lemma 7.2.19, we have

$$C_{ij}(\tau) = \frac{q}{q-1}Z_\tau^{(2)} - T + \frac{1}{q-1}G(\eta,\chi)\sum_{x \in \mathrm{GF}(q^e)} \eta(\delta_2(x,\tau)), \tag{7.35}$$

where $\delta_2(x,\tau) = \mathrm{Tr}_q^{q^e}(x)^2 + \mathrm{Tr}_q^{q^e}(\beta_ix^{q^r+1}) - \mathrm{Tr}_q^{q^e}(\alpha^\tau x)^2 - \mathrm{Tr}_q^{q^e}(\beta_i\alpha^{(q^r+1)\tau}x^{q^r+1})$.

Obviously, from Equation (7.35), it is enough to calculate $Z_\tau^{(2)}$ and $\sum_{x \in \mathrm{GF}(q^e)} \eta(\delta_2(x,\tau))$.

(1) When $\tau = 0$, $i = j$, then $Z_\tau^{(2)} = q^e - 1$, and $\sum_{x \in \mathrm{GF}(q^e)} \eta(\delta_2(x,\tau)) = 0$, so $C_{ij}(\tau) = q^e - 1$.

(2) When $\tau = 0$, $i \neq j$, then $Z_\tau^{(2)}$ is the number of nonzero solutions of $\mathrm{Tr}_q^{q^e}((\beta_i - \beta_j)x^{q^\tau+1}) = 0$. $\mathrm{Tr}_q^{q^e}((\beta_i - \beta_j)x^{q^\tau+1})$ is a quadratic form of rank e over GF(q), which is equivalent to $a_1x_1^2 + a_2x_2^2 + \cdots + a_ex_e^2$, $a_i \neq 0, i = 1, 2, \cdots, e$, from Lemma 7.2.15. Thus $Z_\tau = q^{e-1} - 1$ by Lemma 7.2.14.

From Lemmas 7.2.13 and 7.2.19,

$$\begin{aligned}\sum_{x\in \mathrm{GF}(q^e)} \eta(\delta(x,\tau)) &= \sum_{x_i\in \mathrm{GF}(q), i\neq 1}\ \sum_{x_1\in \mathrm{GF}(q)} \eta\left(a_1x_1^2 + \sum_{i=2}^{e} a_ix_i^2\right)\\ &= \sum_{x_i\in \mathrm{GF}(q), i\neq 1} \eta(a_1)\xi\left(\sum_{i=2}^{e} x_i^2\right)\\ &= (q-1)q^{(e-1)/2}\eta\left((-1)^{(e-1)/2}\prod_{i=1}^{e} a_i\right).\end{aligned}$$

Therefore $C_{ij} = -1 + q^{(e-1)/2}\eta((-1)^{(e-1)/2}\prod_{i=1}^{e} a_i)G(\eta,\chi)$.

(3) When $\tau \neq 0$, $\alpha^\tau \neq -1$, $\beta_i = \beta_j\alpha^{(q^\tau+1)\tau}$, then

$$\begin{aligned}&Z_\tau^{(2)}\\ &= |\{x \in \mathrm{GF}(q^e) : \mathrm{Tr}_q^{q^e}((1+\alpha^\tau)x) = 0\}| + |\{x \in \mathrm{GF}(q^e) : \mathrm{Tr}_q^{q^e}((1-\alpha^\tau)x) = 0\}|\\ &\quad -|\{x \in \mathrm{GF}(q^e) : \mathrm{Tr}_q^{q^e}((1+\alpha^\tau)x) = 0 \text{ and } \mathrm{Tr}_q^{q^e}((1-\alpha^\tau)x) = 0\}| - 1.\end{aligned}$$

It is easy to prove that $\dfrac{1+\alpha^\tau}{1-\alpha^\tau} \in \mathrm{GF}(q)$ if and only if $\alpha^\tau \in \mathrm{GF}(q)$, thus from Lemma 7.2.16,

$$Z_\tau = \begin{cases} q^{e-1} - 1, & \alpha^\tau \in \mathrm{GF}(q),\\ (2q-1)q^{e-2} - 1, & \alpha^\tau \notin \mathrm{GF}(q).\end{cases}$$

From Lemma 7.2.13 and Lemma 7.2.14, we know that, if $\alpha^\tau \notin \mathrm{GF}(q)$, then $\delta_2(x,\tau)$ is equivalent to $b_1'x_1^2 + b_2'x_2^2$ and $\eta(b_1'b_2') = \eta(-1)$, thus $\sum_{x\in\mathrm{GF}(q^e)}\eta(\delta_2(x,\tau)) = q^{e-2}\sum_{x_2\in\mathrm{GF}(q^e)} \eta(b_1')\xi(x_2) = 0$.

If $\alpha^\tau \in \mathrm{GF}(q)$, then $\delta_2(x,\tau) = (1-\alpha^{2\tau})\mathrm{Tr}_q^{q^e}(x)^2$, $\sum_{x\in\mathrm{GF}(q^e)}\eta(\delta(x,\tau)) = \eta(1-\alpha^{2\tau})(q-1)q^{e-1}$.

Therefore $C_{ij}(\tau) = -1$ when $\alpha^\tau \notin \mathrm{GF}(q)$ while $C_{ij}(\tau) = -1 + q^{e-1}\eta(1-\alpha^{2\tau})G(\eta,\chi)$ when $\alpha^\tau \in \mathrm{GF}(q)$.

(4) When $\tau \neq 0$, $\alpha^\tau = -1$, $\beta_i = \beta_j\alpha^{(q^\tau+1)\tau}$, then $Z_\tau^{(2)} = q^e - 1$, $\sum_{x\in\mathrm{GF}(q^e)}\eta(\delta(x,\tau)) = 0$, thus $C_{ij}(\tau) = q^e - 1$.

(5) When $\tau \neq 0$, $\alpha^\tau = -1$, $\beta_i \neq \beta_j\alpha^{(q^\tau+1)\tau}$, similar to Case (2), we get $C_{ij}(\tau) = -1 + q^{(e-1)/2}\eta((-1)^{(e-1)/2}\prod_{i=1}^{e} a_i)G(\eta,\chi)$.

(6) When $\tau \neq 0$, $\alpha^\tau \neq -1$, $\beta_i \neq \beta_j\alpha^{(q^\tau+1)\tau}$, it's trivial to prove that $Z_\tau^{(2)} = \sum_{u\in\mathrm{GF}(q^e)} Z_\tau(u) - 1$, where $Z_\tau(u)$ denotes the number of solutions of the system

$$\begin{cases} \mathrm{Tr}_q^{q^e}(x)^2 - \mathrm{tr}_q^{q^e}(\alpha^\tau x)^2 = u,\\ \mathrm{Tr}_q^{q^e}((\beta_i - \beta_j\alpha^{(q^\tau+1)\tau})x^{q^\tau+1}) = -u.\end{cases} \tag{7.36}$$

If $u = 0$, let A, B, C, respectively, denote the number of solutions of the following systems

$$\begin{cases} \mathrm{Tr}_q^{q^e}((1+\alpha^\tau)x)^2 = 0,\\ \mathrm{Tr}_q^{q^e}((\beta_i - \beta_j\alpha^{(q^\tau+1)\tau})x^{q^\tau+1} = 0,\end{cases} \tag{7.37}$$

$$\begin{cases} \mathrm{Tr}_q^{q^e}((1-\alpha^\tau)x) = 0, \\ \mathrm{Tr}_q^{q^e}((\beta_i - \beta_j\alpha^{(q^r+1)\tau})x^{q^r+1} = 0, \end{cases} \tag{7.38}$$

$$\begin{cases} \mathrm{Tr}_q^{q^e}((1+\alpha^\tau)x) = 0, \\ \mathrm{Tr}_q^{q^e}((1-\alpha^\tau)x) = 0, \\ \mathrm{Tr}_q^{q^e}((\beta_i - \beta_j\alpha^{(q^r+1)\tau})x^{q^r+1} = 0, \end{cases} \tag{7.39}$$

then $Z_\tau(0) = A + B - C$.

We know that $\mathrm{Tr}_q^{q^e}((\beta_i - \beta_j\alpha^{(q^r+1)\tau})x^{q^r+1}$ can be transformed to $\sum_{i=1}^e a_i x_i^2$, $a_i \neq 0$, in some basis of $\mathrm{GF}(q^e)$ over $\mathrm{GF}(q)$, and in this basis $\mathrm{Tr}_q^{q^e}((1+\alpha^\tau)x) = \sum_{i=1}^e b_i x_i$, $\mathrm{Tr}_q^{q^e}((1-\alpha^\tau)x) = \sum_{i=1}^e c_i x_i$.

Then, from Lemma 7.2.17 and Lemma 7.2.18

$$A = \begin{cases} q^{e-2} + q^{\frac{e-3}{2}}(q-1)\eta((-1)^{\frac{e-1}{2}}EM), & M \neq 0, \\ q^{e-2}, & M = 0, \end{cases}$$

$$B = \begin{cases} q^{e-2} + q^{\frac{e-3}{2}}(q-1)\eta((-1)^{\frac{e-1}{2}}E\varphi), & \varphi \neq 0, \\ q^{e-2}, & \varphi = 0. \end{cases}$$

When $\alpha^\tau \notin \mathrm{GF}(q)$,

$$C = \begin{cases} q^{e-3}, & \varphi \neq 0, \Delta \neq 0, \\ q^{e-3} + q^{\frac{e-3}{2}}(q-1)\eta((-1)^{\frac{e-1}{2}}E\varphi), & \varphi \neq 0, \Delta = 0, \\ q^{e-3}, & \varphi = 0, \delta \neq 0, \\ q^{e-3} + q^{\frac{e-3}{2}}(q-1)\eta((-1)^{\frac{e-1}{2}}EM, & \varphi \neq 0, \delta = 0, M \neq 0, \\ q^{e-3}, & \varphi = \delta = M = 0. \end{cases}$$

When $\alpha^\tau \in \mathrm{GF}(q)$, $C = A = B$.

Therefore, when $\alpha^\tau \notin \mathrm{GF}(q)$,

$$Z_\tau(0) = \begin{cases} (2q-1)q^{e-3} + q^{\frac{e-3}{2}}(q-1)\eta((-1)^{\frac{e-1}{2}}E\varphi) & \\ \quad +\eta((-1)^{\frac{e-1}{2}}EM), & M \neq 0, \varphi \neq 0, \Delta \neq 0, \\ (2q-1)q^{e-3} + q^{\frac{e-3}{2}}(q-1)\eta((-1)^{\frac{e-1}{2}}EM), & \\ & M \neq 0, \varphi \neq 0, \Delta = 0, \\ (2q-1)q^{e-3} + q^{\frac{e-3}{2}}(q-1)\eta((-1)^{\frac{e-1}{2}}EM), & \\ & M \neq 0, \varphi = 0, \delta \neq 0, \\ (2q-1)q^{e-3}, & M \neq 0, \varphi = 0, \delta = 0, \\ (2q-1)q^{e-3} + q^{\frac{e-3}{2}}(q-1)\eta((-1)^{\frac{e-1}{2}}E\varphi), & \\ & M = \Delta = 0, \varphi \neq 0, \\ (2q-1)q^{e-3}, & \text{otherwise.} \end{cases}$$

When $\alpha^\tau \notin \mathrm{GF}(q)$,

$$Z_\tau(0) = \begin{cases} q^{e-2} + q^{\frac{e-3}{2}}(q-1)\eta((-1)^{\frac{e+1}{2}}EM), & M \neq 0, \\ q^{e-2}, & M = 0. \end{cases}$$

If $u \neq 0$, then $Z_\tau(u) = \sum_{u\in \mathrm{GF}(q^e)^*} \Gamma(v)$ denotes the number of solutions of the system

$$\begin{cases} \mathrm{Tr}_q^{q^e}((1+\alpha^\tau)x) = v, \\ \mathrm{Tr}_q^{q^e}((1-\alpha^\tau)x) = u/v, \\ \mathrm{Tr}_q^{q^e}((\beta_i - \beta_j\alpha^{(q^r+1)\tau})x^{q^r+1} = -u. \end{cases} \tag{7.40}$$

When $\alpha^\tau \notin \mathrm{GF}(q)$, from Lemma 7.2.18

$$\Gamma(v) = \begin{cases} q^{e-3} + q^{\frac{e-3}{2}}\eta((-1)^{\frac{e+1}{2}}E\varphi\lambda), & \varphi \neq 0, \Delta \neq 0, \\ q^{e-3}, & \varphi \neq 0, \Delta = 0 \neq \sigma, \\ q^{e-3} + q^{\frac{e-3}{2}}\eta((-1)^{\frac{e-1}{2}}E\varphi)\xi(\hat{C}), & \varphi \neq 0, \Delta = \sigma = 0, \\ q^{e-3} + q^{\frac{e-3}{2}}\eta((-1)^{\frac{e+1}{2}}E\hat{A}), & \varphi \neq 0, \delta \neq 0, \\ q^{e-3}, & \varphi = 0, \delta = 0, \end{cases}$$

where $\sigma = u\delta/v - v\varphi$, $\hat{A} = 2u\delta + u\delta^2 - Mu^2/v^2$, $\hat{C} = u^2/v^2 + u\varphi$, $\lambda = \hat{C}\Delta - \sigma^2$.

Thus

$$Z_\tau(u) = \begin{cases} \sum\limits_{v\in \mathrm{GF}(q^e)^*} (q^{e-3} + q^{\frac{e-3}{2}}\eta((-1)^{\frac{e+1}{2}}E\varphi\lambda)), & \varphi \neq 0, \Delta \neq 0, \\ \sum\limits_{\substack{\sigma\neq 0,\\ v\in \mathrm{GF}(q^e)^*}} q^{e-3} + \sum\limits_{\substack{\sigma=0,\\ v\in \mathrm{GF}(q^e)^*}} (q^{e-3} + q^{\frac{e-3}{2}} \cdot\eta((-1)^{\frac{e-1}{2}}E\varphi)\xi(\hat{C})), & \varphi \neq 0, \Delta = 0, \\ \sum\limits_{v\in \mathrm{GF}(q^e)^*} (q^{e-3} + q^{\frac{e-3}{2}}(q-1)\eta((-1)^{\frac{e+1}{2}}E\hat{A})), & \varphi = 0, \delta \neq 0, \\ (q-1)q^{e-3}, & \varphi = 0, \delta = 0, \end{cases}$$

$$= \begin{cases} (q-1)q^{e-3} + q^{\frac{e-3}{2}}\eta((-1)^{\frac{e+1}{2}}E\varphi) \sum\limits_{v\in \mathrm{GF}(q^e)^*} \eta(D), & \varphi\neq 0, \Delta\neq 0, \\ (q-1)q^{e-3}, & \varphi\neq 0, \Delta=0, \eta(u\delta/\varphi)=-1, \\ (q-1)q^{e-3} - 2q^{\frac{e-3}{2}}\eta((-1)^{\frac{e-1}{2}}E\varphi), & \varphi \neq 0, \Delta = 0, \eta(u\delta/\varphi) = 1, \\ (q-1)q^{e-3} + q^{\frac{e-3}{2}}\eta((-1)^{\frac{e-1}{2}}E) \cdot(\eta(u\delta^2 + 2u\delta)\xi(M\delta + 2M) - \eta(-M)), & \varphi = 0, \delta \neq 0, \\ (q-1)q^{e-3}, & \varphi = 0, \delta = 0. \end{cases}$$

where $D = -\varphi^2 v^4 + (\Delta + 2\delta)u\varphi v^2 + u^2(\Delta - \delta^2)$. After complicated calculating, we have

$$Z_\tau^{(2)} = \begin{cases} q^{e-1} + (q-1)q^{\frac{e-3}{2}}(\eta((-1)^{\frac{e-1}{2}}EM) + \eta((-1)^{\frac{e+1}{2}}E\varphi), & \\ \qquad \eta(\Delta - \delta^2)\xi(\Delta + 4\delta + 4)) - 1, & \varphi \neq 0, \Delta \neq 0, \\ q^{e-1} + (q-1)q^{\frac{e-3}{2}}(\eta((-1)^{\frac{e-1}{2}}EM), & \\ \qquad -\eta((-1)^{\frac{e-1}{2}}E\varphi) - 1 & \varphi \neq 0, \Delta = 0, \\ q^{e-1} - 1, & \varphi = 0. \end{cases}$$

When $\alpha^\tau \in \mathrm{GF}(q)$, the intersection of hyperplanes $\mathrm{tr}_q^{q^e}((1+\alpha^\tau)x) = v$ and $\mathrm{tr}_q^{q^e}((1+\alpha^\tau)x) = v$ is not empty only when $v^2 = \dfrac{1+\alpha^\tau}{1-\alpha^\tau}u$, from Lemma 7.2.17,

$$\Gamma(v) = \begin{cases} q^{e-2} + q^{\frac{e-3}{2}}(q-1)\eta((-1)^{\frac{e-1}{2}}EM), & M = -\dfrac{1+\alpha^\tau}{1-\alpha^\tau}, \\ q^{e-2} - q^{\frac{e-3}{2}}\eta((-1)^{\frac{e-1}{2}}EM), & M \neq 0, M \neq -\dfrac{1+\alpha^\tau}{1-\alpha^\tau}, \\ q^{e-2}, & M = 0. \end{cases}$$

Therefore, if $\eta\left(\dfrac{1+\alpha^\tau}{1-\alpha^\tau}u\right) = 1$, then

$$Z_\tau(v) = \begin{cases} 2q^{e-2} + 2q^{\frac{e-3}{2}}(q-1)\eta((-1)^{\frac{e-1}{2}}EM), & M = -\dfrac{1+\alpha^\tau}{1-\alpha^\tau}, \\ 2q^{e-2} - 2q^{\frac{e-3}{2}}\eta((-1)^{\frac{e-1}{2}}EM), & M \neq 0, M \neq -\dfrac{1+\alpha^\tau}{1-\alpha^\tau}, \\ 2q^{e-2}, & M = 0, \\ 0, & \text{otherwise}. \end{cases}$$

Thus when $\alpha^\tau \in \mathrm{GF}(q)$,

$$Z_\tau(v) = \begin{cases} q^{e-1} - 1, & M = 0, \\ q^{e-1} - 1 + (q-1)q^{\frac{e-1}{2}}\eta((-1)^{\frac{e-1}{2}}EM), & M = -\dfrac{1+\alpha^\tau}{1-\alpha^\tau}, \\ q^{e-1} - 1, & M \neq 0, M \neq -\dfrac{1+\alpha^\tau}{1-\alpha^\tau}. \end{cases}$$

Now we calculate $\sum_{x \in \mathrm{GF}(q^e)} \eta(\delta(x,\tau))$ in this case.

From Lemma 7.2.19, we can get

$$\begin{aligned} \sum_{x \in \mathrm{GF}(q^e)} \eta(\delta(x,\tau)) &= \sum_{x_i \in \mathrm{GF}(q^e)} \eta\left(\sum_{i=1}^{e} a_i x_i^2 + \sum_{i=1}^{e} b_i x_i \sum_{j=1}^{e} c_j x_j\right) \\ &= \sum_{\substack{x_i \in \mathrm{GF}(q^e), \\ i \neq 1}} \sum_{x_1 \in \mathrm{GF}(q^e)} \eta\Big((a_1 + b_1 c_1)x_1^2 \end{aligned}$$

$$
\begin{aligned}
&+\sum_{i=2}^{e}(b_1c_i+b_ic_1)x_ix_1\\
&+\sum_{i=2}^{e}a_ix_i^2+\sum_{i,j=2}^{e}b_ic_jx_ix_j\Big)\\
=&\,\eta(a_1+b_1c_1)\sum_{x_i\in \mathrm{GF}(q^e),i\neq 1}\xi(d(x_2,x_3,\cdots,x_e))\\
=&\,\eta(a_1+b_1c_1)(q\mathcal{R}-q^{e-1}),
\end{aligned}
$$

where

$$
\begin{aligned}
\xi(d(x_2,x_3,\cdots,x_e))=&\left[\sum_{i=2}^{e}(b_1c_i+b_ic_1)x_i\right]^2\\
&-4(a_1+b_1c_1)\left[\sum_{i=2}^{e}a_ix_i^2+\sum_{i,j=2}^{e}b_ic_jx_ix_j)\right],\\
\mathcal{R}=&\,|\{(x_2,\cdots,x_e)\in \mathrm{GF}(q)^{e-1}:d(x_2,x_3,\cdots,x_e)=0\}|.
\end{aligned}
$$

If rank $(d(x_2,x_3,\cdots,x_e))=\hat{t}$ is odd, then from Lemma 7.2.14, $d(x_2,x_3,\cdots,x_e)=0$ have q^{e-2} solutions, and $\sum_{x\in\mathrm{GF}(q^e)}\eta(\delta(x,\tau))=0$ If rank$(d(x_2,x_3,\cdots,x_e))=\hat{t}$ is even, then from Lemma 7.2.13, $d(x_2,x_3,\cdots,x_e)=0$ have $q^{e-2}+(q-1)q^{e-2-\hat{t}/2}\eta((-1)^{\hat{t}/2}\hat{d})$ solutions, and $\sum_{x\in\mathrm{GF}(q^e)}\eta(\delta(x,\tau))=(q-1)q^{e-2-\hat{t}/2}\eta((-1)^{\hat{t}/2}\hat{d}(a_1+b_1c_1))$, where $\hat{d}=\det(d(x_2,x_3,\cdots,x_e))$.

Summing up, when Case (6) holds, we have the following theorem

Theorem 7.2.14 *When $\alpha^\tau\notin\mathrm{GF}(q)$, if* rank$(d(x_2,x_3,\cdots,x_e))=\hat{t}$ *is odd, then*

$$
C_{ij}(\tau)=\begin{cases}
q^{\frac{e-1}{2}}(\eta((-1)^{\frac{e-1}{2}}EM)+\eta((-1)^{\frac{e+1}{2}}E\varphi)\eta(\Delta-\delta^2)\\
\quad\cdot\xi(\Delta+4\delta+4))-1, & \varphi\neq 0,\Delta\neq 0,\\
q^{\frac{e-1}{2}}(\eta((-1)^{\frac{e-1}{2}}EM)-\eta((-1)^{\frac{e-1}{2}}E\varphi)-1, & \\
 & \varphi\neq 0,\Delta=0,\\
-1, & \varphi=0,
\end{cases}
$$

if rank $(d(x_2,x_3,\cdots,x_e))=\hat{t}$ *is even, then*

$$
C_{ij}(\tau)=\begin{cases}
q^{\frac{e-1}{2}}(\eta((-1)^{\frac{e-1}{2}}EM)+\eta((-1)^{\frac{e+1}{2}}E\varphi)\eta(\Delta-\delta^2)\xi(\Delta+4\delta+4))\\
\quad+G(\eta,\chi)q^{e-2-\hat{t}/2}\eta((-1)^{\hat{t}/2}\hat{d}(a_1+b_1c_1))-1, & \varphi\neq 0,\Delta\neq 0,\\
q^{\frac{e-1}{2}}(\eta((-1)^{\frac{e-1}{2}}EM)-\eta((-1)^{\frac{e-1}{2}}E\varphi)-1\\
\quad+G(\eta,\chi)q^{e-2-\hat{t}/2}\eta((-1)^{\hat{t}/2}\hat{d}(a_1+b_1c_1)), & \varphi\neq 0,\Delta=0,\\
-1+G(\eta,\chi)q^{e-2-\hat{t}/2}\eta((-1)^{\hat{t}/2}\hat{d}(a_1+b_1c_1)), & \varphi=0.
\end{cases}
$$

When $\alpha^{\tau} \in \mathrm{GF}(q)$, *if* $\mathrm{rank}(d(x_2, x_3, \cdots, x_e)) = \hat{t}$ *is odd, then*

$$C_{ij}(\tau) = \begin{cases} -1, & M = 0, \\ -1 + q^{\frac{e+1}{2}}(\eta((-1)^{\frac{e-1}{2}} EM), & M = -\dfrac{1+\alpha^{\tau}}{1-\alpha^{\tau}}, \\ -1, & M \neq 0, M \neq -\dfrac{1+\alpha^{\tau}}{1-\alpha^{\tau}}, \end{cases}$$

if $\mathrm{rank}(d(x_2, x_3, \cdots, x_e)) = \hat{t}$ *is even, then*

$$C_{ij}(\tau) = \begin{cases} q^{\frac{e-1}{2}}(\eta((-1)^{\frac{e-1}{2}} EM) & \\ \quad +G(\eta, \chi) q^{e-2-\hat{t}/2} \eta((-1)^{\hat{t}/2} \hat{d}(a_1 + b_1 c_1)) - 1, & M = -\dfrac{1+\alpha^{\tau}}{1-\alpha^{\tau}}, \\ -1 + G(\eta, \chi) q^{e-2-\hat{t}/2} \eta((-1)^{\hat{t}/2} \hat{d}(a_1 + b_1 c_1)), & \text{otherwise.} \end{cases}$$

In this subsection, we generalized the binary d-form sequences to p-ary d-form sequences, constructed a family of p-ary 2-form sequences and calculated their cross-correlation functions. Moreover, if $\gcd(d, q-1) = 1$, then their cross-correlation functions are the same as those of binary d-form sequences, which have been studied extensively.

7.3 Sequence Pairs with Mismatched Filtering

7.3.1 Binary Sequences Pairs with Two-Level Autocorrelation Functions (BSPT)

Definition 7.3.1 *Let* $a = (a(0), a(1), \cdots, a(N-1))$ *and* $b = (b(0), b(1), \cdots, b(N-1))$ *be two binary sequences of length* N *with* $a(j), b(j) = \pm 1, j = 0, 1, \cdots, N-1$. *The periodic and aperiodic autocorrelation functions of the sequence pair* (a, b) *are defined, respectively, as*

$$P_{(a,b)}(u) = \sum_{j=0}^{N-1} a(j) b(j+u), \quad 0 \leqslant u \leqslant N-1, \tag{7.41}$$

$$A_{(a,b)}(u) = \sum_{j=0}^{N-1-u} a(j) b(j+u), \quad 0 \leqslant u \leqslant N-1, \tag{7.42}$$

where $j + u = (j+u) \bmod N$.

If $d(a, b)$ denotes Hamming distance of the sequences a and b, from Definition 7.3.1, we know that $P_{(a,b)}(0) = A_{(a,b)}(0) = N - 2d(a, b)$, $P_{(a,b)}(u) = A_{(a,b)}(u) + A_{(a,b)}(N-u)$, $0 \leqslant u \leqslant N-1$.

Definition 7.3.2 *The sequence pair* (a, b) *is called the binary sequence pair with two-level autocorrelation function* (BSPT) *if the autocorrelation function of sequence pair* (a, b),

$$P_{(a,b)}(u) = \begin{cases} E, & u = 0, \\ F, & u \neq 0, \end{cases} \tag{7.43}$$

where E *and* F *are two constants.* E *is the valuve of the autocorrelation function of sequence pair* (a, b) *at peak.* F *is the value of the out-phase autocorrelation function of sequence pair*

(a, b). *The sequence pair* (a, b) *is called as perfect sequence pair* (PSP) *if* $F = 0$. *Here, the autocorrelation function of sequence pair* (a, b) *is an ideal impulse.*

When $a = b$, (a, b) is reduced to the common sequence with two-level autocorrelation function, the sequence pairs with two-level autocorrelation function are extensions of the usual sequences with ideal autocorrelation functions.

Definition 7.3.3 *Let* $a = (a(0), a(1), \cdots, a(N-1))$ *be a sequence with length* N. *Then*

(1) $-a = (-a(0), -a(1), \cdots, -a(N-1))$ *denotes the negation of the sequence* a;

(2) $R(a) = (R(a)(0), R(a)(1), \cdots, R(a)(N-1))$ *denotes the reverse of the sequence* a *if*

$$R(a)(j) = \begin{cases} a(j), & j = 0 \\ a(N-j), & j \neq 0 \end{cases}, \quad j = 0, \cdots, N-1;$$

(3) $F(a) = (F(a)(0), F(a)(1), \cdots, F(a)(N-1))$ *denotes the reflected sequence* a *if* $F(a)(j) = a(N-1-j)$, $j = 0, \cdots, N-1$;

(4) $L(a) = (L(a)(0), L(a)(1), \cdots, L(a)(N-1))$ *denotes the sequence of alternately negating the elements of the sequence* a *if* $L(a)(j) = (-1)^j a(j)$, $j = 0, \cdots, N-1$;

(5) $D^q(a) = (D^q(a)(0), D^q(a)(1), \cdots, D^q(a)(N-1))$ *denotes the proper decimated sequence of the sequence* a *if* $D^q(a)(j) = a(qj \bmod N)$, *with* q *and* N *coprime;*

(6) $T^s(a) = (T^s(a)(0), \cdots, T^s(a)(N-1))$ *denotes the shifted sequence of the sequence* a *if* $T^s(a)(j) = a((s + j) \bmod N)$;

Definition 7.3.4 *Let* $a = (a(0), a(1), \cdots, a(N-1))$ *be a sequence with length* N, *then the polynomial*

$$f_a(x) = \sum_{i=0}^{N-1} a(i)x^i$$

is called the characteristic polynomial of sequence a.

Theorem 7.3.1 *If* (a, b) *is a* BSPT, *then*

(1) (b, a) *is also a* BSPT;

(2) $(-a, -b)$ *is also a* BSPT;

(3) $(R(a), R(b))$ *is also a* BSPT;

(4) $(F(a), F(b))$ *is also a* BSPT;

(5) $(L(a), L(b))$ *is also a* BSPT;

(6) $(D^{(q)}(a), D^{(q)}(b))$ *is also a* BSPT;

(7) $(T^s(a), T^s(b))$ *is also a* BSPT.

Proof From Definition 7.3.1 and Definition 7.3.3, we have

(1) $P_{(b,a)}(u) = P_{(a,b)}(N-u)$;

(2) $P_{(-a,-b)}(u) = P_{(a,b)}(u)$;

(3) $P_{(R(a),R(b))}(u) = P_{(a,b)}(u)$;

(4) $P_{(F(a),F(b))}(u) = P_{(a,b)}(u)$;

(5) $P_{(L(a),L(b))}(u) = (-1)^u P_{(a,b)}(u)$;

(6) $P_{(D^{(q)}(a),D^{(q)}(b))}(u) = P_{(a,b)}(u)$;

(7) $P_{(T^s(a),T^s(b))}(u) = P_{(a,b)}(u)$.

Therefore from Definition 7.3.2 the results of the theorem can be proved. □

From Theorem 7.3.1, it is known that the operators defined in Definition 7.3.3 are invariant operators of BSPTs. Applying an invariant operator to a BSPT can yield another BSPT.

Theorem 7.3.2 *If (a, b) is a binary sequence pair with two-level aperiodic autocorrelation function, then (a, b) is a* BSPT.

Proof This theorem can be derived straightforward from $P_{(a,b)}(u) = A_{(a,b)}(u)+A_{(a,b)}(N-u)$. □

Let $a = (a(0), a(1), \cdots, a(N-1))$ be a binary sequence with length N and entries ± 1. If

$$F_a(t) = \sum_{j=0}^{N-1} a(j)W^{tj}, \quad 0 \leqslant t \leqslant N-1, \tag{7.44}$$

then $F_a(t)$ is called Fourier transform spectra of the sequence a, where $W = \exp(2\pi i/N)$,$i = \sqrt{-1}$. Obviously $a(j) = 1/N\sum_{j=0}^{N-1} F_a(t)W^{-tj}, 0 \leqslant j \leqslant N-1$. Here $(a(0), a(1), \cdots, a(N-1))$ is called Fourier reverse transform of ($F_a(0)$, $F_a(1)$, $\cdots$, $F_a(N-1)$).

The following theorem presents the spectrum properties of PSPs.

Theorem 7.3.3 *Let (a, b) be a* PSP, *and let $F_a(t)$ and $F_b(t)$ be respectively Fourier transform spectra of a and b, then the equation:*

$$F_a(t)F_b(-t) + 2d(a, b) = N$$

holds, where $d(a, b)$ denotes the Hamming distance between a and b.

Proof Let

$$\begin{aligned} c &= (c(0), c(1), \cdots, c(N-1)) \\ &= (a(0)b(0+u), a(1)b(1+u), \cdots, a(N-1)b(N-1+u)), \end{aligned} \tag{7.45}$$

then we have $F_c(t) = 1/N\sum_{x=0}^{N-1} W^{-u(t-x)}F_a(x)F_b(t-x)$. Therefore we have

$$F_c(0) = \frac{1}{N}\sum_{x=0}^{N-1} W^{ux}F_a(x)F_b(-x). \tag{7.46}$$

But from Eq. (7.44) and Eq. (7.45), we have

$$F_c(0) = \sum_{j=0}^{N-1} c(j) = \sum_{j=0}^{N-1} a(j)b(j+u) = P_{(a,b)}(u). \tag{7.47}$$

According to Eqs. (7.46), (7.47) and the fact that(a, b) is a PSP, we have

$$\frac{1}{N}\sum_{x=0}^{N-1} W^{ux}F_a(x)F_b(-x) = R_{(a,b)}(u) = \begin{cases} N-2d(a,b), & u = 0, \\ 0, & u \neq 0. \end{cases} \tag{7.48}$$

Let $F(u) = \sum_{x=0}^{N-1} W^{ux}F_a(x)F_b(-x)$, then from Eq. (7.48),

$$F(u) = \begin{cases} N(N-2d(a,b)), & u = 0, \\ 0, & u \neq 0. \end{cases}$$

Taking Fourier reverse transform to $F(u)$, we have

$$F_a(t)F_b(-t) = \frac{1}{N}\sum_{u=0}^{N-1} F(u)W^{-ut} = N - 2d(a,b). \qquad \square$$

Using Theorem 7.3.3, we have following existent condition of PSP.

Theorem 7.3.4 *Let (a, b) be a binary* PSP, *then the equation*

$$(N - 2\mathrm{pl}(a))(N - 2\mathrm{pl}(b)) + 2d(a,b) = N$$

holds, where $\mathrm{pl}(a)$ *denotes the number of elements with value -1 in the sequence a.*

Proof Since $F_a(0) = \sum_{j=0}^{N-1} a(j) = N - 2\mathrm{pl}(a)$ and $F_b(0) = \sum_{j=0}^{N-1} b(j) = N - 2\mathrm{pl}(b)$ hold, from Theorem 7.3.3, the equation $F_a(0)F_b(0) + 2d(a,b) = (N - 2\mathrm{pl}(a))(N - 2\mathrm{pl}(b)) + 2d(a,b) = N$ holds. $\square$

Next, we give the description of the characteristic polynomial of BSPT.

Theorem 7.3.5 *Let $f_a(x)$ and $f_b(x)$ be the characteristic polynomial respectively of $a = (a(0), a(1), \cdots, a(N-1))$ and $b = (b(0), b(1), \cdots, b(N-1))$, where $a(j), b(j) = \pm 1$, $j = 0, 1, \cdots, N-1$, E and F be respectively the peak values of autocorrelation function and the values of out-phase autocorrelation function of (a, b). Then (a, b) is* BSPT *if and only if*

$$f_a(x)f_b(x^{-1}) = E - F + FT(x),$$

where $T(x) = \sum_{i=0}^{N-1} x^i$.

Proof Let Z_N indicate integral ring of mod N residue class. Since

$$\begin{aligned} f_a(x)f_b(x^{-1}) &= \sum_{i=0}^{N-1} a(i)x^i \sum_{j=0}^{N-1} b(j)x^{-j} \\ &= \sum_{i=0}^{N-1}\sum_{j=0}^{N-1} a(i)b(j)x^{i-j} \\ &= \sum_{r=0}^{N-1}\sum_{i=0}^{N-1} a(i)b(i+r)x^{-r} \\ &= \sum_{i=0}^{N-1} a(i)b(i) + \sum_{r=1}^{N-1}\sum_{i=0}^{N-1} a(i)b(i+r)x^{-r}. \end{aligned}$$

Then from Definition 7.3.2, (a, b) is BSPT if and only if $\sum_{i=0}^{N-1} a(i)b(i) + \sum_{r=1}^{N-1}\sum_{i=0}^{N-1} a(i)b(i+r)x^{-r} = E + F(T(x) - 1)$. Therefore (a, b) is BSPT if and only if $f_a(x)f_b(x^{-1}) = E - F + FT(x)$. $\square$

Next, we give another description of the polynomial of BSPT.

Theorem 7.3.6 *Let $a = (a(0), a(1), \cdots, a(N-1))$ and $b = (b(0), b(1), \cdots, b(N-1))$ be binary sequence with length N, where $a(j), b(j) = \pm 1, j = 0, 1, \cdots, N-1$. Let $g_a(x) = \sum_{i=0}^{N-1}(1 - a(i))/2x^i$ and $g_b(x) = \sum_{j=0}^{N-1}(1 - b(j))/2x^j$, then (a, b) is* BSPT *if and only if*

$$g_a(x)g_b(x^{-1}) = (E - F)/4 + (F + 2n_a + 2n_b - N)/4T(x),$$

where n_a and n_b denote respectively the number of "–1" in sequence a and b, E and F are respectively the peak values of autocorrelation function and the values of out-phase autocorrelation function of (a, b).

Proof Let Z_N indicate integral ring of mod N residue class. Since

$$\begin{aligned}
g_a(x)g_b(x^{-1}) &= \sum_{i=0}^{N-1}(1-a(i))/2x^i\sum_{j=0}^{N-1}(1-b(j))/2x^{-j} \\
&= \sum_{i=0}^{N-1}\sum_{j=0}^{N-1}((1-a(i))/2)((1-b(j))/2)x^{i-j} \\
&= \frac{1}{4}\sum_{u=0}^{N-1}\sum_{i=0}^{N-1}(1-a(i))(1-b(i+u))x^{-u} \\
&= \frac{1}{4}\Big(\sum_{u=0}^{N-1}\sum_{i=0}^{N-1}x^{-u} - \sum_{u=0}^{N-1}\sum_{i=0}^{N-1}a(i)x^{-u} \\
&\quad - \sum_{u=0}^{N-1}\sum_{i=0}^{N-1}b(i+u))x^{-u} + \sum_{u=0}^{N-1}\sum_{i=0}^{N-1}a(i)b(i+u)x^{-u}\Big) \\
&= \frac{1}{4}\Big(NT(x) - (N-2n_a)T(x) - (N-2n_b)T(x) + \sum_{u=0}^{N-1}\sum_{i=0}^{N-1}a(i)b(i+u)x^{-u}\Big).
\end{aligned}$$

Then from Definition 7.3.2, (a, b) is BSPT if and only if $\sum_{u=0}^{N-1}\sum_{i=0}^{N-1}a(i)b(i+u)x^{-u} = E - F + FT(x)$. Therefore (a, b) is BSPT if and only if

$$g_a(x)g_b(x^{-1}) = (E-F)/4 + (F + 2n_a + 2n_b - N)/4T(x). \qquad \square$$

The following theorem is presented to show the uniqueness of BSPT.

Theorems 7.3.7 *If (a, b) and (a, b') are two* BSPTs *with length N, E and F are the peak values of autocorrelation function and the values of out-phase autocorrelation function respectively of sequence pair (a, b) and sequence pair (a, b'), and $F < E$, then $b = b'$.*

Proof Let $f_a(x)$, $f_b(x)$ and $f_{b'}(x)$ be respectively the characteristic polynomials of a, b, b' in Definition 7.3.4. If (a, b) and (a, b') are BSPTs, from Theorem 7.3.5, we have

$$f_a(x)f_b(x^{-1}) = E - F + F\cdot T(x), \tag{7.49}$$

$$f_a(x)f_{b'}(x^{-1}) = E - F + F\cdot T(x), \tag{7.50}$$

by Eq. (7.49)–(7.50), we have

$$f_a(x)(f_b(x^{-1}) - f_{b'}(x^{-1})) = 0. \tag{7.51}$$

Multiplying $f_b(x^{-1})$ in both sides of Eq. (7.51) , we have

$$f_a(x)f_b(x^{-1})(f_b(x^{-1}) - f_{b'}(x^{-1})) = 0.$$

Since $f_a(x)f_b(x^{-1}) = E - F + F\cdot T(x) > 0$, then

$$f_b(x^{-1}) - f_{b'}(x^{-1}) = 0.$$

From the uniqueness of the characteristic polynomial of sequence, we have $b = b'$. $\square$

7.3.2 Difference Set Pairs

In this subsection, we introduce a new concept of combinatorial design.

Definition 7.3.5 *Let S be an addition group with v elements, D and D' be two subsets of S with k and k' elements respectively, and $e = |D \cap D'|$. If for any $\alpha \neq o \mod v$, the equation*

$$d - d' \equiv \alpha \mod v$$

has exactly λ solution pair (d, d') with $d \in D$ and $d' \in D'$, then the set pair (D, D') is called (v, k, k', e, λ)-difference set pair (DSP).

When $D = D'$, (D, D') is reduced to the usual wellknown difference set, (DS) in Ref.[25], so the concept of difference set pair is the extension of the concept of difference set.

Example 7.3.1 Let $D = \{3, 4, 5, 6, 7\}$ and $D' = \{1, 3, 4, 6, 7\}$ be two subsets in finite integer group Z_8. Observing the following difference table

$$\begin{aligned}
1 \equiv 4 - 3 \equiv 5 - 4 \equiv 7 - 6,\\
2 \equiv 3 - 1 \equiv 5 - 3 \equiv 6 - 4,\\
3 \equiv 4 - 1 \equiv 6 - 3 \equiv 7 - 4,\\
4 \equiv 5 - 1 \equiv 7 - 3 \equiv 3 - 7,\\
5 \equiv 6 - 1 \equiv 3 - 6 \equiv 4 - 7,\\
6 \equiv 7 - 1 \equiv 4 - 6 \equiv 5 - 7,\\
7 \equiv 3 - 4 \equiv 5 - 6 \equiv 6 - 7,
\end{aligned}$$

we can know that each non-zero element in Z_8 appears 3 times in the table. Therefore $(D, D') = (\{3, 4, 5, 6, 7\}, \{1, 3, 4, 6, 7\})$ are a $(8, 5, 5, 4, 3)$- difference set pair.

For the convenience of analyzing the properties of difference set pairs, the corresponding relationship is established between the set and polynomial.

Definition 7.3.6 *Let $D = \{d_1, d_2, \cdots, d_k\}$ be a subset in finite integer group Z_v. If the polynomial*

$$\theta_D(x) = \sum_{i=1}^{k} x^{d_i},$$

then the polynomial $\theta_D(x)$ is called the Hall polynomial of the set D.

Example 7.3.2 If $D = \{1, 3, 4, 6, 7\}$, then the Hall polynomial of D is $\theta_D(x) = x + x^3 + x^4 + x^6 + x^7$.

Difference set pairs have the following properties depicted by polynomials.

Theorem 7.3.8 *Let $D = \{d_1, d_2, \cdots, d_k\} \subseteq Z_v$, $D' = \{d'_1, d'_2, \cdots, d'_{k'}\} \subseteq Z_v$, $\theta_D(x) = x^{d_1} + x^{d_2} + \cdots + x^{d_k}$, and $\theta_{D'}(x) = x^{d'_1} + x^{d'_2} + \cdots + x^{d'_{k'}}$.$(D, D')$ is (v, k, k', e, λ)-difference set pair if and only if*

$$\theta_D(x)\theta_{D'}(x^{-1}) \equiv e - \lambda + \lambda T(x) \pmod{x^v - 1},$$

where $T(x) = 1 + x + x^2 + \cdots + x^{v-1}$.

Proof Since

$$\begin{aligned}
\theta_D(x)\theta_{D'}(x^{-1}) &= \left(\sum_{i=1}^{k} x^{d_i}\right)\left(\sum_{j=1}^{k'} x^{-d'_j}\right) \\
&= \sum_{1\leqslant i\leqslant k, 1\leqslant j\leqslant k'} x^{d_i-d'_j} \\
&= \sum_{g\in Z_v}\left(\sum_{\substack{g=d_i-d'_j \\ 1\leqslant i\leqslant k, 1\leqslant j\leqslant k'}} 1\right) x^g,
\end{aligned}$$

so $\theta_D(x)\theta_{D'}(x^{-1}) = e - \lambda + \lambda T(x)$ holds if and only if

$$\sum_{\substack{g=d_i-d'_j \\ 1\leqslant i\leqslant k, 1\leqslant j\leqslant k'}} 1 = \begin{cases} e, & g=0, \\ \lambda, & g\neq 0. \end{cases}$$

The above equation holds if and only if (D, D') is (v, k, k', e, λ)-difference set pair. □

Theorem 7.3.9 *If (D, D') is a (v, k, k', e, λ)- difference set pair. $\overline{D} = Z_v - D$, $\overline{D'} = Z_v - D'$, then $(\overline{D}, \overline{D'})$ is a $(v, v-k, v-k', v-k-k'+e, v-k-k'+\lambda)$- difference set pair.*

Proof From the following equations

$$\begin{aligned}
\theta_{\overline{D}}(x) &= T(x) - \theta_D(x), \\
\theta_{\overline{D'}}(x) &= T(x) - \theta_{D'}(x), \\
\theta_D(x)\theta_{D'}(x) &= e - \lambda + \lambda T(x), \\
X^i T(x) &= T(x), \\
T(x^{-1}) &= T(x),
\end{aligned}$$

we have $\theta_{\overline{D}}(x)\theta_{\overline{D'}}(x) = v-k-k'+e-(v-k-k'+\lambda)+(v-k'-k+\lambda)T(x)$. Since $v-k-k'+e = |\overline{D}\cap\overline{D'}|$, from Theorem 7.3.8, we can know that $(\overline{D}, \overline{D'})$ is a $(v, v-k, v-k', v-k-k'+e, v-k-k'+\lambda)$- difference set pair. □

Definition 7.3.7 *The difference set pair $(\overline{D}, \overline{D'})$ generated from Theorem 7.3.9 is called the complementary difference set pair of (D, D').*

Form Theorem 7.3.9, we can know that there is the difference set pair with $k, k' > v/2$ if there is the difference set pair with $k, k' < v/2$.

Theorem 7.3.10 *The parameters (v, k, k', e, λ) of difference set pairs (D, D') satisfy the following*

$$kk' = \lambda(v-1) + e. \tag{7.52}$$

Proof Applying Theorem 7.3.8 and taking $x = 1$ in $\theta_D(x)\theta_{D'}(x^{-1}) = e - \lambda + \lambda T(x)$, we can have $kk' = \lambda(v-1) + e$. □

Definition 7.3.8 *Let $a = (a(0), \cdots, a(N-1))$ be a binary sequence with length N, $a(i) = \pm 1, i = 0, \cdots, N-1$, and $D \subseteq Z_N$, If $D = \{i | a(i) = -1, i = 0, \cdots, N-1\}$, then D is called the equivalent set of the sequence a, and a is called characteristic sequence of the set D.*

Theorem 7.3.11 *Let $a = (a(0), \cdots, a(N-1))$ and $b = (b(0), \cdots, b(N-1))$ be two binary sequences of length N with $a(i), b(i) = \pm 1, i = 0, \cdots, N-1$. If D and D' are equivalent set of a and b respectively, $k = |D|, k' = |D'|$, then (a, b) are perfect binary sequence pairs if and only if (D, D') is a (N, k, k', λ) difference set pair with*

$$N - 2(k + k') + 4\lambda = 0. \tag{7.53}$$

Proof Let $a(i) = 1 - 2p(i)$ and $b(i) = 1 - 2q(i)$, $i = 0, 1, \cdots, N-1$. Since D and D' are equivalent set of a and b respectively, we have

$$p(i) = \begin{cases} 1, & i \in D \\ 0, & i \notin D \end{cases}, \quad i = 0, \cdots, N-1,$$

$$q(i) = \begin{cases} 1, & i \in D' \\ 0, & i \notin D' \end{cases}, \quad i = 0, \cdots, N-1.$$

Since

$$\begin{aligned} P_{(a,b)}(u) &= \sum_{i=0}^{N-1} a(i)b(i+u) \\ &= \sum_{i=0}^{N-1} (1 - 2p(i) - 2q(i+u) + 4p(i)q(i+u)) \\ &= N - 2k - 2k' + 4 \sum_{i \in D, i+u \in D'} 1, \end{aligned}$$

so

$$P_{(a,b)}(u) = \begin{cases} N - 2(k + k') + 4e, & u = 0 \\ 0, & u \neq 0 \end{cases}$$

If and only if

$$\sum_{i \in D, i+u \in D'} 1 = \begin{cases} e, & u = 0 \\ \lambda, & u \neq 0 \end{cases}$$

and $N - 2(k + k') + 4\lambda = 0$. Then (D, D') is a (N, k, k', λ)-difference set pairs whose parameters satisfy $N - 2(k + k') + 4\lambda = 0$. □

From Theorem 7.3.11, we know that a perfect binary sequence pair corresponds to a kind of difference set pairs, whereas (E, k, k', λ)-difference set pair with $E - 2(k + k') + 4\lambda = 0$ corresponds to a perfect binary sequence pair. This means that a perfect binary sequence pair is equivalent to a kind of difference set pair. Then we can study perfect binary sequence pairs by using the theory of difference set pairs. As a result, the theory of difference set pairs is a method to be used in the study of perfect binary sequence pairs.

By applying the properties of difference set pairs, we can get some essential conditions of perfect binary sequences.

Theorem 7.3.12 *Let $a = (a(0), \cdots, a(N-1))$ and $b = (b(0), \cdots, b(N-1))$ be two binary sequences with length N, $a(i), b(i) = \pm 1, i = 0, \cdots, N-1$, the weights of a and b, the number* *of elements* -1 *in* a *and* b*, be* k, k' *respectively, and* d *be the Hamming distance between* a *and* b*. If* (a, b) *is a perfect binary sequence pair, then*

$$N(N - 2k - 2k' - 1) + 4kk' + 2d = 0.$$

Proof Using Eq. (7.52) in Theorem 7.3.10, Eq. (7.53) in Theorem 7.3.11 and the equation $d = k + k' - 2e$, we can establish the equation group as follows

$$\begin{cases} kk' = \lambda(N-1) + e, \\ N - 2(k + k') + 4\lambda = 0, \\ d = k + k' - 2e. \end{cases}$$

Solving the above equation group and getting rid of the variable e and λ, we can have $N(N - 2k - 2k' - 1) + 4kk' + 2d = 0$. □

From the proof of Theorem 7.3.11, we can know that there is the equivalent relationship between binary sequence pairs with two-level autocorrelation functions and difference set pairs as follows.

Theorem 7.3.13 *Let* $a = (a(0), \cdots, a(N-1))$*, and* $b = (b(0), \cdots, b(N-1))$ *be two sequences with element* 1 *or* -1*. Let* D *and* D' *be two subsets of integer ring* Z_N *with* k *and* k' *elements respectively. For* $i = 0, 1, \cdots, N-1$*, let*

$$a(i) = \begin{cases} -1, & i \in D, \\ 1, & i \notin D, \end{cases} \qquad b(i) = \begin{cases} -1, & i \in D', \\ 1, & i \notin D', \end{cases}$$

(D, D') *is* (N, k, k', e, λ)*-difference set pair if and only if sequence pair* (a, b) *has a two-level autocorrelation function as*

$$R_{(a,b)}(u) = \begin{cases} N - 2(k + k') + 4e, & u = 0, \\ N - 2(k + k') + 4\lambda, & u \neq 0. \end{cases}$$

The values of the out-phase autocorrelation function of a sequence pair with two-level autocorrelation function have the forms in the following theorem.

Theorem 7.3.14 *If* (a, b) *is a sequence pair with length* N*, element* 1 *or* -1 *and two-level autocorrelation function, then the value of out-phase autocorrelation function of sequence pair* (a, b)*,*

$$F \equiv \begin{cases} 0 \bmod 2, & \textit{when } N \equiv 0 \bmod 2, \\ 1 \bmod 2, & \textit{when } N \equiv 1 \bmod 2. \end{cases}$$

Theorem 7.3.14 means that F is an even integer if N is an even integer and F is an odd integer if N is an odd integer.

Proof From Theorem 7.3.13,

$$F = N - 2(k + k') + 4\lambda \equiv \begin{cases} 0 \bmod 2, & \text{when } N \equiv 0 \bmod 2, \\ 1 \bmod 2, & \text{when } N \equiv 1 \bmod 2. \end{cases}$$

The values of the autocorrelation functions of the sequence pairs with two-level autocorrelation functions at zero-phase and out-phase have the property of the following theorem.□

Theorem 7.3.15 *E and F in Definition 7.3.2 satisfy that $E-F=4n$, where n is some integer.*

Proof From Theorem 7.3.13,

$$E-F=N-2(k+k')+4e-(N-2(k+k')+4\lambda)=4(e-\lambda).$$ □

7.3.3 Construction of BSPTs

According to the equivalent relationship between DSPs and BSPTs as mentioned above, construction of BSPTs can be realized by constructing DSPs.

First we give the infinite families of DSPs. Then we give the corresponding infinite families of BSPTs by using the equivalent relationship between DSPs and BSPTs.

Theorem 7.3.16 *Let $D=\{0,n-1,\cdots,j(n-1),\cdots,2n^2-2n\}$ and $D'=\{0,n,\cdots,in,\cdots,2n^2-2n\}$ be two subsets of integer ring Z_{2n^2-1} with $2n+1$ elements and $2n-1$ elements respectively, where n is an integer and $n>1, j=0,1,\cdots,2n, i=0,1,\cdots,2n-2$. Then (D,D') is $(2n^2-1,2n+1,2n-1,3,2)$-DSP.*

Proof Let $\theta_D(x)$ and $\theta_{D'}(x)$ be the Hall polynomial of the set D and the set D'.

$$\begin{aligned}
&\theta_D(x)\theta_{D'}(x^{-1})\\
=&\sum_{j=0}^{2n}x^{j(n-1)}\sum_{i=0}^{2n-2}x^{-in}\\
=&\sum_{j=0}^{2n}x^{j(n-1)}\sum_{i=2}^{2n}x^{in-1}\\
=&\sum_{j=1}^{2n-2}\sum_{i=1}^{2n}x^{in-j}+\sum_{i=2}^{2n}x^{in-(2n+1)}+\sum_{i=2}^{2n+1}x^{in-2n}\\
=&\left(\sum_{j=1}^{n}x^{n-j}+\sum_{j=n+1}^{2n-2}x^{n-j}\right)+\sum_{j=1}^{2n}\sum_{i=1}^{n}x^{2in-j}+\sum_{j=1}^{2n}\sum_{i=1}^{n-1}x^{(2i+1)n-j}+x^{2n^2-n}\\
=&\sum_{j=0}^{n-1}x^{n-j}+\left(\sum_{j=2}^{n-1}x^{2n^2-j}+x^{2n^2-n}\right)+\sum_{j=1}^{2n}\sum_{i=1}^{n}x^{2in-j}+\sum_{j=1}^{2n}\sum_{i=1}^{n-1}x^{(2i+1)n-j}\\
=&\sum_{j=0}^{n-1}x^{n-j}+\sum_{j=2}^{n}x^{2n^2-j}+\left(1+\sum_{i=1}^{2n^2-2}x^i\right)+\sum_{i=n}^{2n^2-2n-1}x^i\\
=&1+\sum_{i=1}^{2n^2-2}x^i+\sum_{i=0}^{n-1}x^i+\sum_{i=2n^2-2n}^{2n^2-2}x^i+\sum_{i=n}^{2n^2-2n-1}x^i\\
=&1+2\sum_{i=0}^{2n^2-2}x^i\bmod(x^{2n^2-1}),
\end{aligned}$$

i.e., $\theta_D(x)\theta_{D'}(x^{-1}) = 1 + 2T(x),\text{mod}(x^{2n^2-1} - 1)$, where $T(x) = \sum_{i=0}^{2n^2-2}$. From Theorem 7.3.8, we know that (D, D) is $(2n^2 - 1, 2n + 1, 2n - 1, 3, 2)$-DSP. □

By applying Theorem 7.3.16 and Theorem 7.3.13, we can get infinite families of BSPTs. Table 7.3.1 shows some examples of DSPs and BSPTs with length $2n^2 - 1$, where "−" denotes −1, and "+" denotes 1.

Theorem 7.3.17 *Let $D = \{0, 1, \cdots, i, \cdots, 2n\}$ and $D' = \{0, n, 2n, 3n+1, 4n+1, \cdots, (2j+1)n+j, (2j+2)n+j, \cdots, 2n^2-2n\}$ be two subsets of integer ring Z_{2n^2-1} with $2n+1$ elements and $2n - 1$ elements respectively, where n is an integer and $n > 1, i = 0, 1, \cdots, 2n, j = 0, 1, \cdots, n-2$. Then (D, D') is $(2n^2 - 1, 2n + 1, 2n - 1, 3, 2)$-DSP.*

Table 7.3.1 Examples of DSPs and corresponding BSPTs with length $N = 2n^2 - 1$

parameters	difference set pair	sequence pairs
$n = 2$ $N = 7$	$D = \{0, 1, 2, 3, 4\}$ $D' = \{0, 2, 4\}$	$a = - - - - - + +$ $b = - + - + - + +$
$n = 3$ $N = 17$	$D = \{0, 2, 4, 6, 8, 10, 12\}$ $D' = \{0, 3, 6, 9, 12\}$	$a = - + - + - + - + -$ $+ - + - + + + +$ $b = - + + - + + - + +$ $- + + - + + + +$
$n = 4$ $N = 31$	$D = \{0, 3, 6, 9, 12, 15, 18, 21, 24\}$ $D' = \{0, 4, 8, 12, 16, 20, 24\}$	$a = - + + - + + - + + - + + - + +$ $- + + - + + - + + - + + + + + +$ $b = - + + + - + + + - + + + - + +$ $+ - + + + - + + + - + + + + + +$

Proof Let $\theta_D(x)$ and $\theta_{D'}(x)$ be the Hall polynomial of the set D and the set D'.

$$
\begin{aligned}
&\theta_D(x) \cdot \theta_{D'}(x^{-1}) \\
=&\sum_{i=0}^{2n} x^i \left(1 + \sum_{j=0}^{n-2}(x^{-(2j+1)n-j} + x^{-(2j+2)n-j}\right) \\
=&\sum_{i=0}^{2n} x^i + \sum_{i=0}^{2n}\sum_{j=0}^{n-2} x^{i-(2n+1)j-n} + \sum_{i=0}^{2n}\sum_{j=0}^{n-2} x^{i-(2n+1)j-2n} \\
=&\sum_{i=0}^{2n} x^i + \sum_{i=-2n^2+2n+2}^{n} x^i + \sum_{i=-2n^2+n+2}^{0} x^i \\
=&\sum_{i=0}^{2n} x^i + \left(\sum_{i=2n+1}^{2n^2-2} x^i + \sum_{i=0}^{n} x^i\right) + \left(\sum_{i=n+1}^{2n^2-2} x^i + 1\right) \\
=&1 + 2\sum_{i=0}^{2n^2-2} x^i \bmod (x^{2n^2-1}),
\end{aligned}
$$

i.e., $\theta_D(x)\theta_{D'}(x^{-1}) = 1 + 2T(x), \text{mod}(x^{2n^2-1} - 1)$, where $T(x) = \sum_{i=0}^{2n^2-2}$. From Theorem 7.3.8, we know that (D, D) is $(2n^2 - 1, 2n + 1, 2n - 1, 3, 2)$-DSP. □

By applying Theorem 7.3.17 and Theorem 7.3.13, we can get infinite families of BSPTs. Table 7.3.2 shows some examples of DSPs and BSPTs with length $2n^2 - 1$, where "−" denotes −1, and "+" denotes 1. Note that the sequence pairs derived from Theorem 7.3.17 are not equivalent to the sequence pairs derived from Theorem 7.3.16 when $n \geqslant 3$.

Theorem 7.3.18 *Let $D = \{0, n-1, 2(n-1), 3(n-1)+1, 4(n-1)+1, \cdots, (2i+1)(n-1)+i, (2i+2)(n-1)+i, \cdots, (2n-1)(n-1)+(n-1), 2n(n-1)+(n-1)\}$ and $D' = \{0, 1, \cdots, j, \cdots, 2n-2\}$ be two subsets of integer ring Z_{2n^2-1} with $2n+1$ elements and $2n-1$ elements respectively, where n is an integer and $n > 1, i = 0, 1, \cdots, n-1, j = 0, 1, \cdots, 2n-2$. Then (D, D') is $(2n^2-1, 2n+1, 2n-1, 3, 2)$-DSP.*

Table 7.3.2 Examples of DSPs and corresponding BSPTs with length $N = 2n^2 - 1$

parameters	difference set pair	sequence pairs
$n = 3$ $N = 17$	$D = \{0, 1, 2, 3, 4, 5, 6\}$ $D' = \{0, 3, 6, 10, 113\}$	$a = - - - - - - - + +$ $+ + + + + + + +$ $b = - + + - + + - + +$ $+ - + + - + + +$
$n = 4$ $N = 31$	$D = \{0, 1, 2, 3, 4, 5, 6, 7, 8\}$ $D' = \{0, 4, 8, 13, 17, 22, 26\}$	$a = - - - - - - - - - + + + + + +$ $+ + + + + + + + + + + + + + + +$ $b = - + + + - + + + - + + + + - +$ $+ + - + + + + - + + + - + + + +$

Proof Let $\theta_D(x)$ and $\theta_{D'}(x)$ be the Hall polynomial of the set D and the set D'.

$$
\begin{aligned}
&\theta_D(x) \cdot \theta_{D'}(x^{-1}) \\
&= \left(1 + \sum_{i=0}^{n-1} (x^{(2i+1)(n-1)+i} + x^{(2i+2)(n-1)+i}\right) \sum_{j=0}^{2n-2} x^{-j} \\
&= \sum_{j=0}^{2n-2} x^{-j} + \sum_{j=0}^{2n-2} \sum_{i=0}^{n-1} x^{(2i+1)(n-1)+i-j} + \sum_{j=0}^{2n} \sum_{i=0}^{n-1} x^{(2i+2)(n-1)+i-j} \\
&= \sum_{j=0}^{2n-2} x^{2n^2-1-j} + \sum_{j=0}^{2n-2} \sum_{i=0}^{n-1} x^{(2n-1)i+(n-1)-j} + \sum_{j=0}^{2n} \sum_{i=0}^{n-1} x^{(2n-1)i+2(n-1)-j} \\
&= \left(1 + \sum_{i=2n^2-2n+1}^{2n^2-2} x^i + \sum_{i=1-n}^{2n^2-2n} x^i + \sum_{i=0}^{2n^2-2n-1} x^i\right. \\
&= 1 + \sum_{i=2n^2-2n+1}^{2n^2-2} x^i + \left(\sum_{i=2n^2-n}^{2n^2-2} x^i + \sum_{i=0}^{2n^2-2n} x^i\right) + \sum_{i=0}^{2n^2-n-1} x^i \\
&= 1 + 2 \sum_{i=0}^{2n^2-2} x^i \bmod (x^{2n^2-1}),
\end{aligned}
$$

i.e., $\theta_D(x)\theta_{D'}(x^{-1}) = 1 + 2T(x) \bmod(x^{2n^2-1} - 1)$, where $T(x) = \sum_{i=0}^{2n^2-2}$. From Theorem 7.3.8, we know that (D, D) is $(2n^2-1, 2n+1, 2n-1, 3, 2)$-DSP. □

By applying Theorem 7.3.18 and Theorem 7.3.13, we can get infinite families of BSPTs. Table 7.3.3 shows some examples of DSPs and BSPTs with length $2n^2 - 1$, where "−" denotes -1, and "+" denotes 1. Note that the sequence pairs derived from Theorem 7.3.18 are not equivalent to the sequence pairs derived from Theorem 7.3.16 and Theorem 7.3.17.

Definition 7.3.9 *Let G be the Abel Group with v elements and $A = \{a_1, a_2, \cdots, a_k\}$ be a subset of G, then*

Table 7.3.3 Examples of DSPs and corresponding BSPTs with length $N = 2n^2 - 1$

parameters	difference set pair	sequence pairs
$n = 2$ $N = 7$	$D = \{0, 1, 2, 4, 5\}$ $D' = \{0, 1, 2\}$	$a = - - - + - - +$ $b = - - - + + + +$
$n = 3$ $N = 17$	$D = \{0, 2, 4, 7, 9, 12, 14\}$ $D' = \{0, 1, 2, 3, 4\}$	$a = - + - + - + + - +$ $- + + - + - + +$ $b = - - - - - + + + +$ $+ + + + + + + +$
$n = 4$ $N = 31$	$D = \{0, 1, 2, 3, 4, 5, 6, 7, 8\}$ $D' = \{0, 4, 8, 13, 17, 22, 26\}$	$a = - - - - - - - - + + + + + + +$ $+ + + + + + + + + + + + + + + +$ $b = - + + + - + + + - + + + + - +$ $+ + - + + + + - + + + - + + + +$

$$A + \tau = \{a_1 + \tau, a_2 + \tau, \cdots, a_k + \tau\};$$

$$qA = \{qa_1, qa_2, \cdots, qa_k\}, (q, v) = 1;$$

$$A^* = \{v - a_1, v - a_2, \cdots, v - a_k\};$$

$$\overline{A} = G - A$$

are respectively the translation, the sampling, the concomitance and the complementary set of set A.

Theorem 7.3.19 *Let G be the Abel Group with v elements. if (v, k, k', e, λ)-DSP (D, D') exists, then*

(1) $(D + \tau, D' + \tau), (qD, qD'), (D^*, D'^*)$ *is a* (v, k, k', e, λ)–DSP;
(2) $(\overline{D}, D')$ *is a* $(v, v - k, k', k' - e, k' - \lambda)$-DSP; $(D, \overline{D'})$ *is a* $(v, k, v - k', k - e, k - \lambda)$-DSP;
(3) $(\overline{D}, \overline{D'})$ *is a* $(v, v - k, v - k', e + v - k - k', v - k - k' + \lambda)$-DSP.

Proof (1) From Definition 7.3.4, we know

$$f_{D+\tau}(x) = x^\tau f_D(x), \quad f_{D'+\tau}(x) = x^\tau f_{D'}(x),$$
$$f_{qD}(x) = f_D(x^q), \quad f_{qD'}(x) = f_{D'}(x^q),$$
$$f_{D^*}(x) = f_D(x^{-1}), \quad f_{D'^*}(x) = f_{D'}(x^{-1}).$$

We can prove the result of the theorem directly from Theorem 7.3.8.

(2) From Definition 7.3.4, we know

$$f_{\overline{D}}(x) = T(x) - f_D(x^{-1}), \quad f_{\overline{D'}}(x) = T(x) - f_{D'}(x),$$

where ,$T(x) = \sum_{g \in G} x^g$. Then,

$$\begin{aligned} f_D(x) f_{D'}(x) &= (T(x) - f_D(x)) f_{D'}(x^{-1}) \\ &= k'T(x) - (e - \lambda) - \lambda T(x) \\ &= (k' - e) - (k' - \lambda) + (k' - \lambda)T(x). \end{aligned}$$

Therefore, from Theorem 7.3.8 it is obvious that $(\overline{D}, D')$ is a $(v, v - k, k', k' - e, k' - \lambda)$-DSP. Similarly, $(D, \overline{D}')$ is a $(v, k, v - k', k - e, k - \lambda)$- DSP.

The proof of (3) is similar to (2), so it is omitted. □

From Theorem 7.3.19, we know that DSP can be constructed by DS. If there exists (v,k,λ)-DS , i.e., (v,k,k,k,λ)- DSP, then $(v,k,v-k,0,k-\lambda)$- DSP also exists.

Theorem 7.3.20 *Let*

$$f_D(x)=\sum_{i=0}^{sv}x^i$$

and

$$f_{D'}(x)=\sum_{i=0}^{v}x^{si}+x^s\sum_{i=0}^{v-1}x^{si}\sum_{i=1}^{t-1}x^{(sv+1)i}$$

be the Hall polynomial of set D *and* D' *respectively. Then* (D,D') *is a* $(tsv+s+t, sv+1, tv+1, v+1, v)$-DSP, *where* s,t,v *is respectively the positive integer.*

Proof Since under the condition of mod $(x^{tvs+s+t}-1)$,

$$\begin{aligned}
f_D(x)f_{D'}(x^{-1})&=\sum_{i=0}^{sv}x^i\left(\sum_{i=0}^{v}x^{-si}+x^{-s}\sum_{i=0}^{v-1}x^{-si}\sum_{i=1}^{t-1}x^{-(sv+1)i}\right)\\
&=\sum_{i=0}^{sv}x^i\left(1+x^{-s}\sum_{i=0}^{v-1}x^{-si}+x^{-s}\sum_{i=0}^{v-1}x^{-si}\sum_{i=1}^{t-1}x^{-(sv+1)i}\right)\\
&=\sum_{i=0}^{sv}x^i\left(1+x^{-s}\sum_{i=0}^{v-1}x^{-si}\sum_{i=0}^{t-1}x^{-(sv+1)i}\right)\\
&=\sum_{i=0}^{sv}x^i+x^{-s}\sum_{i=0}^{v-1}x^{-si}\sum_{i=0}^{t-1}x^{-(sv+1)i}\sum_{i=0}^{sv}x^i\\
&=\sum_{i=0}^{sv}x^i+x^{-s}\sum_{i=0}^{v-1}x^{-si}\left(\sum_{i=0}^{tsv+s+t-1}x^i-\sum_{i=0}^{s}x^{sv+i}\right)\\
&=\sum_{i=0}^{sv}x^i+v\sum_{i=0}^{tsv+s+t-1}x^i-x^{tsv+t}\sum_{i=0}^{v-1}x^{tsv+s+t-si}\sum_{i=0}^{s}x^{sv+i}\\
&=v\sum_{i=0}^{tsv+s+t-1}x^i+\sum_{i=0}^{sv}x^i-x^{tsv+t}\sum_{i=s+1}^{sv+s}x^i\\
&=v\sum_{i=0}^{tsv+s+t-1}x^i+\sum_{i=0}^{sv}x^i-\sum_{i=1}^{sv}x^i\\
&=v\sum_{i=0}^{tsv+s+t-1}x^i+1.
\end{aligned}$$

□

Example 7.3.3 According to Theorem 7.3.20, let $t=3, s=5, v=4$, there exists $(69,21,13,5,4)$- DSP.

$$D=\{0,1,2,3,4,5,6,7,8,9,10,11,12,13,14,15,16,17,18,19,20\},$$

$$D'=\{0,5,10,15,20,26,31,36,41,47,52,57,62\}.$$

Their characteristic sequences are a BSPT with length 69.

Theorem 7.3.21 *Let*

$$f_D(x)=\sum_{i=0}^{sv-2}x^i$$

and

$$f_{D'}(x)=x^{sv-2}\sum_{i=0}^{v-1}x^{si}\sum_{i=0}^{t-1}x^{(sv-1)i}-x^{sv-1}$$

be the Hall polynomial of set D and D' respectively. Then (D,D') is a $(tsv-s-t,sv-1,tv-1,v+1,v)$-DSP, where s,t,v is respectively the positive integer, and $tsv-s-t\geqslant sv-1$, $tsv-s-t\geqslant tv-1$.

Proof Under the condition of mod $(x^{tvs-s-t}-1)$,

$$\begin{aligned}
&f_D(x)f_{D'}(x^{-1})\\
&=\sum_{i=0}^{sv-2}x^i\left(x^{-(sv-2)}\sum_{i=0}^{v-1}x^{-si}\sum_{i=0}^{t-1}x^{-(sv-1)i}-x^{-(sv-1)}\right)\\
&=x^{-(sv-2)}\sum_{i=0}^{sv-2}x^i\sum_{i=0}^{v-1}x^{-si}\sum_{i=0}^{t-1}x^{-(sv-1)i}-x^{-(sv-1)}\sum_{i=0}^{sv-2}x^i\\
&=x^{-(sv-2)}\sum_{i=0}^{v-1}x^{-si}\left(\sum_{i=0}^{tsv-s-t-1}x^i+\sum_{i=-1-s}^{-2}x^{sv+i}\right)-x^{-(sv-1)}\sum_{i=0}^{sv-2}x^i\\
&=v\sum_{i=0}^{tsv-s-t-1}x^i+x^{tsv-s-t-(sv-2)}\sum_{i=0}^{v-1}x^{tsv-s-t-si}\\
&\quad\sum_{i=-1-s}^{-2}x^{sv+i}-x^{tsv-s-t-(sv-1)}\sum_{i=0}^{sv-2}x^i\\
&=v\sum_{i=0}^{tsv-s-t-1}x^i+1+\sum_{i=-(sv-1)}^{-1}x^{tsv-s-t+i}-\sum_{i=-(sv-1)}^{-1}x^{tsv-s-t+i}\\
&=v\sum_{i=0}^{tsv-s-t-1}x^i+1.
\end{aligned}$$

□

Example 7.3.4 According to Theorem 7.3.5, let $t=2,s=3,v=4$, there exists $(19,11,7,5,4)$- DSP:

$$D=\{0,1,2,3,4,5,6,7,8,9,10\},\quad D'=\{0,2,5,8,10,13,16\}.$$

Their characteristic sequences are a BSPT with length 19.

Theorem 7.3.22 *Let $q=4v+1$ be a prime, then in the finite field F_q, there exist $(4v+1,2v,2v,0,v)$-DSP and $(4v+1,2v+1,2v,2v,v)$-DSP.*

Proof Since $q=4v+1\equiv 1(\bmod 4)$ is a prime, then "-1" is the square element in F_q. Let α be a primitive element of F_q, then the set D with all the nonzero square elements is $\{\alpha^{2i},0\leqslant i\leqslant 2v-1\}$, the set D' with all nonsquare elements is $\{\alpha^{2i+1},0\leqslant i\leqslant 2v-1\}$. D

can be written as $\{-\alpha^{2i}, 0 \leqslant i \leqslant 2v-1\}$, and D' can be written as $\{-\alpha^{2i+1}, 0 \leqslant i \leqslant 2v-1\}$ according to the fact that "-1" is the square element in F_q. If $1 = x - y, x \in D, y \in D'$, then $\alpha^{2i} = \alpha^{2i}x - \alpha^{2i}y$, $-\alpha^{2i+1} = \alpha^{2i+1}y - \alpha^{2i+1}x$; on the other hand, if $\alpha^{2i} = x - y, x \in D, y \in D'$, then $1 = \alpha^{-2i}x - \alpha^{-2i}y$. If $-\alpha^{2i+1} = x - y, x \in D, y \in D'$, then $1 = \alpha^{-(2i+1)}y - \alpha^{-(2i+1)}x$. So if "1" denotes the element for λ time, which is the difference of different elements in D and D', then the nonzero element in F_q is denoted as the element for λ time, which is the difference of the different elements in D and D'. Both number of nonzero square element and number of nonsquare element is $2v$. Then (D, D') is a $(4v+1, 2v, 2v, 0, v)$- DSP. From 2) in Theorem 7.3.19, we can get $(4v+1, 2v+1, 2v, 2v, v)$-DSP. □

Example 7.3.5 According to Theorem 7.3.22, we can construct $(13, 6, 6, 0, 3)$- DSP and $(17, 8, 8, 0, 4)$- DSP.

$$(\{1, 3, 9, 4, 12, 10\}, \{2, 6, 5, 7, 8, 11\}),$$

$$(\{1, 2, 4, 8, 16, 15, 13, 9\}, \{3, 6, 12, 7, 14, 11, 5, 10\}).$$

Their characteristic sequences are BSPTs with the length 13 and 19, respectively.

In the following subsection, the construction of mismatched sequence by shifting sequence is given. The ideas of this method are as follows.

We change sequence pairs into the form of matrix. If the initial phase of one column in the original sequence agrees with the corresponding one in the mismatched sequence, the value of corresponding shift sequence e_j is 0; otherwise, if the initial phase of one column in the original sequence does not agree with the corresponding one in the mismatched sequence, the value of e_j is equal to the initial phase of this column minus the initial phase of corresponding column in the mismatched sequence. So we can see that, if we know the original sequence of sequence pairs and the corresponding value of shift sequence e_j, we can get the mismatched sequence rapidly. This is the basic idea for constructing shift sequences. Changing the perfect binary sequence pairs into the form of matrix with $N/4$ rows and 4 columns, we will find that the column vector of mismatched sequence, corresponding to its value of shift sequence, would be $0, [N/8], 0, [N/8]$ (when $N/4$ is even) or $\{0, [N/8], 0, [N/8]+1\}$ (when $N/4$ is odd) if we use the column vector of original sequence as the original sequence.

Theorem 7.3.23 *In the original sequence X of a perfect binary sequence pairs (X, Y) whose length is N, the number of 1 is $(N/2) - 1$. Let the original sequence*

$$\underbrace{1, 1, \cdots 1}_{N/2-1}, \underbrace{0, 0, \cdots, 0}_{N/2+1} \tag{7.54}$$

become the matrix with $N/4$ rows and 4 columns. Let its 4 column vectors be the original sequence, and then taking shifting sequence $e_j = (0, [N/8], 0, [N/8])$ (when $N/4$ is even) or $e_j = (0, [N/8], 0, [N/8]+1)$ (when $N/4$ is odd) as a mismatched sequence, we can have the binary sequence pairs as the perfect binary sequence pairs.

Proof Change the original sequence and the shifted mismatched sequence into the form of $2 \times (N/2)$ matrices:

$$X = \begin{bmatrix} 1 & 1 & & 1 & 0 \\ 0 & 0 & & 0 & 0 \end{bmatrix}, \tag{7.55}$$

$$Y = \begin{bmatrix} 1 & 0 & & 1 & 0 \\ 0 & 1 & & 0 & 0 \end{bmatrix}. \tag{7.56}$$

From Definition 7.3.2, we can see that calculating the autocorrelation function of sequence pairs (X, Y) can be changed into calculating the sum of the cross-correlation of the corresponding column in sequence pairs. As the value of correlation function of sequences $(1, 0)$ and $(1, 0)$ or $(0, 0)$ and $(0, 0)$ is 2, the value of correlation function of sequences $(1, 0)$ and $(0, 1)$ is -2, and the sequences $(0, 0)$ and $(1, 0)$ or$(0, 0)$ and $(0, 1)$ is 0. As N is an integer multiple of 4, $N/2$ is even. When $j = 0$, we can easily see that $R_{(X,Y)}(j) = 4 \neq 0$ by observing the corresponding column of Equations (7.55) and (7.56). When $j \neq 0$, in the numbers of $N/2$, there are 2 kinds of cases corresponding to $(0, 0)$ and $(1, 0)$ or $(0, 0)$ and $(0, 1)$. In the other numbers of $(N/2) - 2$, there are $(N/4) - 1$ kinds of cases corresponding to $(1, 0)$ and $(1, 0)$ and $(1, 0)$ and $(0, 1)$. Therefore

$$R_{(X,Y)}(j) = 0 \times 2 + 2 \times (N/4 - 1) + (-2) \times (N/4 - 1) = 0.$$

So binary sequence pairs acquired by the above ideas are perfect binary sequence pairs. □

The steps of constructing mismatched sequence by using Theorem 7.3.23 are as follows.

In original sequence X of perfect binary sequence pairs (X, Y) whose length is N, the number of 1 is $(N/2) - 1$. The constructing steps are as follows.

(1) Construct original sequence $X : \underbrace{1, 1, \cdots, 1}_{N/2-1}, \underbrace{0, 0, \cdots 0}_{N/2+1}$ change the original sequence X into the form of matrix with $N/4$ rows and 4 columns and work out 4 column vectors of X.

(2) Then work out the shifting sequence e_j , $e_j = (0, [N/8], 0, [N/8])$ (when $N/4$ is even); or $e_j = (0, [N/8], 0, [N/8] + 1)$ (when $N/4$ is odd) .

(3) Use the columns of X as the original columns, work out each column of mismatched sequence Y by the shifting sequence $\{e_j\}$, i.e., the value of initial phase of each column in Y is equal to that of the initial phase of the corresponding column of X plus the value of corresponding e_j $(j = 0, 1, 2, 3)$.

Example 7.3.6 Construct a mismatched sequence whose length is 16.

The number of 1 in the original sequence X is 7, $X = (1111111000000000)$. Change the original sequence into the matrix with 4 rows and 4 columns. we can get

$$X = \begin{bmatrix} 1 & 1 & 1 & 1 \\ 1 & 1 & 1 & 0 \\ 0 & 0 & 0 & 0 \\ 0 & 0 & 0 & 0 \end{bmatrix}.$$

Since $N/4 = 4$ is even, the shifting sequence e_j in the column vectors of Y corresponding to the column vectors of original sequence X is

$$e_j = (0, [N/8], 0, [N/8]) = (0, 2, 0, 2),$$

we can get mismatched sequence Y:

$$Y = \begin{bmatrix} 1 & 0 & 1 & 0 \\ 1 & 0 & 1 & 0 \\ 0 & 1 & 0 & 1 \\ 0 & 1 & 0 & 0 \end{bmatrix}.$$

That is $Y = (1010101001010100)$

Example 7.3.7 Construct a mismatched sequence whose length is 28.

The number of 1 in the original sequence X is 13, $X = (1111111111111 000000000000000)$. Change the original sequence into the matrix with 7 rows and 4 columns. we can get

$$X = \begin{bmatrix} 1 & 1 & 1 & 1 \\ 1 & 1 & 1 & 1 \\ 1 & 1 & 1 & 1 \\ 1 & 0 & 0 & 0 \\ 0 & 0 & 0 & 0 \\ 0 & 0 & 0 & 0 \\ 0 & 0 & 0 & 0 \end{bmatrix}.$$

Since $N/4 = 7$ is odd, the shifting sequence $\{e_j\}$ in the column vectors of Y corresponding to the column vectors of original sequence X is

$$e_j = (0, [N/8], 0, [N/8] + 1) = (0, 3, 0, 4),$$

we can get mismatched sequence Y:

$$Y = \begin{bmatrix} 1 & 0 & 1 & 0 \\ 1 & 0 & 1 & 0 \\ 1 & 0 & 1 & 0 \\ 1 & 0 & 0 & 1 \\ 0 & 1 & 0 & 1 \\ 0 & 1 & 0 & 1 \\ 0 & 1 & 0 & 0 \end{bmatrix}.$$

That is $Y = (1010101010101001010101010100)$.

In the above text, we put forward a method to rapidly construct the perfect binary sequence pairs whose length is an integer multiple of 4 by studying the properties of the perfect binary sequence pairs, based on the idea of shifting sequences. Comparing with other methods, this method is easy to be used and understood and it is very helpful for engineers to construct perfect binary sequence pairs with high speed and accuracy. Therefore, it could be used in relevant communication systems.

7.3.4 Periodic Complementary Binary Sequence Pairs

In this subsection, we develop a new set of the sequences, periodic complementary binary sequence pairs (PCSP). PCSP is composed of the sequence pairs when the sum of their autocorrelation function is a delta function. PCSPs have ideal periodic correlation propeties and include the set of periodic complementary binary sequences as a special case. Next, we give the definitions of PCSP and the difference family pair (DFP), transform properties, Fourier spectrum values, and existence conditions of PCSPs. Also, we point out the relationship between PCSP and the subclass of DFP. Finally we give the recursive constructions for PCSP and the recursive constructions for DFP as well.

Definition 7.3.10 *Let $\{(a_i, b_i)|0 \leqslant i \leqslant Q-1\}$ be a set of Q binary sequence pairs, each sequence has length N and entries ± 1. Then $\{(a_i, b_i)|0 \leqslant i \leqslant Q-1\}$ is called a set of* *periodic complementary binary sequence pairs and termed as* PCSP_Q^N *if*

$$\sum_{i=0}^{Q-1} P_{(a_i,b_i)}(u) = \begin{cases} C, & u = 0, \\ 0, & u \neq 0, \end{cases}$$

where C *is a certain constant.* $\{(a_i, b_i)|0 \leqslant i \leqslant Q-1\}$ *is called a set of aperiodic complementary binary sequence pairs* (ACSP) *and termed as* ACSP_Q^N *if*

$$\sum_{i=0}^{Q-1} A_{(a_i,b_i)}(u) = \begin{cases} C, & u = 0, \\ 0, & u \neq 0, \end{cases}$$

where C *is a certain constant.*

When $a_i = b_i$, PCSP and ACSP are respectively the periodic complementary binary sequence (PCS) and aperiodic complementary binary sequence (ACS) in Refs. [26,27]. Let $A = [A(i,j)]$ and $B = [B(i,j)]$ be two binary arrays with order $Q \times N$. (A, B) is called a perfect array pair if

$$\sum_{i=0}^{Q-1}\sum_{j=0}^{N-1} A(i,j)B(i+u, j+u) = \begin{cases} C, & (u,v) = (0,0), \\ 0, & (u,v) \neq (0,0). \end{cases}$$

If (A, B) is a perfect binary array pair with order $Q \times N$, the rows of A and B build PCSP_Q^N. Therefore PCSs and perfect binary arrays are a special case of PCSP.

Definition 7.3.11 *Let* $\{a_1, a_2, \cdots, a_t\}$ *be a set of the sequences with length* N*, then the sequence* b *is called the cascade sequence of* $\{a_1, a_2, \cdots, a_t\}$ *and termed as* $b = a_1 \oplus a_2 \oplus \cdots \oplus a_t$ *if*

$$b = (a_1(0), \cdots, a_1(N-1), a_2(0), \cdots, a_2(N-1), \cdots, a_t(0), \cdots, a_t(N-1)).$$

Definition 7.3.12 *Let* $[h(l,j)]$ *be a matrix of size* $Q \times R$*. Two columns* $h(,i)$ *and* $h(,j)$ *for* $0 \leqslant i, j \leqslant R-1$ *and* $i \neq j$ *of* $[h(l,j)]$ *are orthogonal if* $\sum_{l=0}^{Q-1} h(l,i)h(l,j) = 0$.

Now we present a new concept of block design, difference family pair (DFP), and give the relationship between PCSP and DFP. It is shown that the PCSP is a subclass of the more general DFP.

Definition 7.3.13 *Let* $S = \{D_0, D_1, \cdots, D_{Q-1}\}$ *and* $S' = \{D'_0, D'_1, \cdots, D'_{Q-1}\}$ *be two families of* Q *base blocks, where the base block* $D_l = \{d_{l,0}, d_{l,1}, \cdots, d_{l,k_l-1}\}$ *and* $D'_l = \{d'_{l,0}, d'_{l,1}, \cdots, d'_{l,k'_l-1}\}$ *are respectively the collections of* k_l *and* k'_l *residues modulo* v*. Let* $K = \{k_0, k_1, \cdots, k_{Q-1}\}$ *and* $K' = \{k'_0, k'_1, \cdots, k'_{Q-1}\}$ *contain respectively the lengths of the blocks in* S *and* S'*. Let* E *denotes a set* $\{e_0, e_1, \cdots, e_{Q-1}\}$ *with* $e_l = |D_l \cap D'_l|$, $l = 0, \cdots, Q-1$, *then the family pair* (S, S') *is called a* (v, K, K', E, λ)*-difference family pair* (DFP), *if for any residue* $\alpha \not\equiv 0$ *mod* v*, the congruence:*

$$\sum_{l=1}^{Q-1} (d_{l,i} - d'_{l,j}) \equiv \alpha \bmod v \tag{7.57}$$

has exactly λ *solution pairs* $(d_{l,i}, d'_{l,j})$ *with* $d_{l,i} \in D_l, d'_{l,i} \in D'_l, D_l \in S$*, and* $D'_l \in S', l = 0, \cdots, Q-1$.

When $S = S'$ and $Q = 1$, DFP is the difference set in Ref.[25].

The parameters of the $(v.K, K', E, \lambda)$-DFP satisfy the following condition.

Theorem 7.3.24 *If there exists a* (v, K, K', E, λ)-DFP (S, S'), *then*

$$\sum_{l=0}^{Q-1}(k_l k_l' - e_l) = \lambda(v-1). \tag{7.58}$$

Proof Since in (S, S') there are $\sum_{l=0}^{Q-1}(k_l k_l' - e_l)$ pairs $(d_{l,i}, d_{l,j}')$ appearing on the left side of Eq. (7.57) and there are $\lambda(v-1)$ values of α appearing on the right side of Eq. (7.57), so these two numbers should be the same. Therefore $\sum_{l=0}^{Q-1}(k_l k_l' - e_l) = \lambda(v-1)$ holds. □

If the set of binary sequence pairs $\{(a_l, a_l')|0 \leqslant l \leqslant Q-1\}$ with length $N = v$ is formed from the DFP (S, S') by

$$a_l(j) = \begin{cases} 1, & j \in D_l, \\ -1, & j \notin D_l, \end{cases} \quad a_l'(j) = \begin{cases} 1, & j \in D_l', \\ -1, & j \notin D_l', \end{cases} \tag{7.59}$$

then the sum of correlation function $P_{(a_l, a_l')}(u)$ satisfies

$$\sum_{l=0}^{Q-1} P_{(a_l, a_l')}(u) = \begin{cases} Qv - 2\sum_{l=0}^{Q-1}(k_l + k_l' - 2e_l), & u = 0, \\ Qv - 2\sum_{l=0}^{Q-1}(k_l + k_l' - 2\lambda), & u \neq 0. \end{cases} \tag{7.60}$$

The set of binary sequence pairs $\{(a_l, a_l')|0 \leqslant l \leqslant Q-1\}$ with two-valued property of Eq. (7.60), which is formed from the $(v.K, K', E, \lambda)$-DFP (S, S'), is a PCSP_Q^v if

$$Qv - 2\sum_{l=0}^{Q-1}(k_l + k_l' - 2\lambda) = 0. \tag{7.61}$$

Eq. (7.58) and Eq. (7.61) state the condition for the existence of a PCSP. The relationships between DFP and PCSP are given from Eqs. (7.59)–(7.61).

Therefore, the following construction methods for PCSP are also the construction methods for DFP.

Now we give the properties of PCSP. From the properties of Fourier transform spectrum of PCSP, we derive a condition for the existence of PCSP.

Theorem 7.3.25 *If* $\{(a_i, b_i)|0 \leqslant i \leqslant Q-1\}$ *is a* PCSP_Q^N, *then*

(1) $\{(-a_i, -b_i)|0 \leqslant i \leqslant Q-1\}$ *is also a* PCSP_Q^N;

(2) $\{(R(a_i), R(b_i))|0 \leqslant i \leqslant Q-1\}$ *is also a* PCSP_Q^N;

(3) $\{(F(a_i), F(b_i))|0 \leqslant i \leqslant Q-1\}$ *is also a* PCSP_Q^N;

(4) $\{(L(a_i), L(b_i))|0 \leqslant i \leqslant Q-1\}$ *is also a* PCSP_Q^N;

(5) $\{(D^{(q)}(a_i), D^{(q)}(b_i))|0 \leqslant i \leqslant Q-1\}$ *is also a* PCSP_Q^N;

(6) $\{(T^s(a_i), T^s(b_i))|0 \leqslant i \leqslant Q-1\}$ *is also a* PCSP_Q^N.

Proof From Definition 7.3.1, we have

(1) $P_{(-a_i, -b_i)}(u) = P_{(a_i, b_i)}(u)$;

(2) $P_{(R(a_i), R(b_i))}(u) = P_{(a_i, b_i)}(u)$;

(3) $P_{(F(a_i),F(b_i))}(u) = P_{(a_i,b_i)}(u)$;
(4) $P_{(L(a_i),L(b_i))}(u) = (-1)^u P_{(a_i,b_i)}(u)$;
(5) $P_{(D^{(q)}(a_i),D^{(q)}(b_i))}(u) = P_{(a_i,b_i)}(u)$;
(6) $P_{(T^s(a_i),T^s(b_i))}(u) = P_{(a_i,b_i)}(u)$.
Therefore, from Definition 7.3.10 the results of the theorem can be proved. □

From Theorem 7.3.25, it is known that the operators defined in Definition 7.3.1 are invariant operators of PCSP. Applying an invariant operator to a PCSP can yield another PCSP.

Theorem 7.3.26 *If* $\{(a_i,b_i)|0 \leqslant i \leqslant Q-1\}$ *is a* PCSP ^N_Q, *then* $\{(b_i,a_i)|0 \leqslant i \leqslant Q-1\}$ *is also a* PCSP ^N_Q.

Proof This theorem can be derived directly from

$$P_{(a_i,b_i)}(u) = P_{(b_i,a_i)}(N-u).$$ □

Theorem 7.3.27 *If* $\{(a_i,b_i)|0 \leqslant i \leqslant Q-1\}$ *is an* ACSP^N_Q, *then* $\{(a_i,b_i)|0 \leqslant i \leqslant Q-1\}$ *is also a* PCSP^N_Q.

Proof This theorem can be derived directly from

$$P_{(a_i,b_i)}(u) = A_{(a_i,b_i)}(u) + A_{(a_i,b_i)}(u-N).$$ □

Let $a = (a(0), a(1), \cdots, a(N-1))$ be a binary sequence with length N and entries ± 1. If

$$F_a(t) = \sum_{j=0}^{N-1} a(j)W^{tj}, \quad 0 \leqslant t \leqslant N-1, \tag{7.62}$$

then $F_a(t)$ is called Fourier transform spectra of the sequence a, where $W = \exp(2\pi i/N)$,$i = \sqrt{-1}$. Obviously $a(j) = 1/N \sum_{j=0}^{N-1} F_a(t)W^{-tj}, 0 \leqslant j \leqslant N-1$. Here $(a(0), a(1), \cdots, a(N-1))$ is called Fourier reverse transform of $(F_a(0), F_a(1), \cdots, F_a(N-1))$.

Theorem 7.3.28 *Let* $\{(a_i,b_i)|0 \leqslant i \leqslant Q-1\}$ *be a* PCSP^N_Q, *and let* $F_{a_i}(t)$ *and* $F_{b_i}(t)$ *be respectively Fourier transform spectra of* a_i *and* b_i, $i = 0, 1, \cdots, Q-1$, *then the equation:* $\sum_{i=0}^{Q-1}(F_{a_i}(t)F_{b_i}(-t)+2d(a_i,b_i)) = QN$ *holds, where* $d(a_i,b_i)$ *denotes the Hamming distance between* a_i *and* b_i, $i = 0, 1, \cdots, Q-1$.

Proof Let

$$\begin{aligned} c_i &= (c_i(0), c_i(1), \cdots, c_i(N-1)) \\ &= (a_i(0)b_i(0+u), a_i(1)b_i(1+u), \cdots, a_i(N-1)b_i(N-1+u)), \end{aligned} \tag{7.63}$$

then we have $F_{c_i}(t) = 1/N \sum_{x=0}^{N-1} W^{-u(t-x)} F_{a_i}(x)F_{b_i}(t-x)$. Therefore we have

$$F_{c_i}(0) = \frac{1}{N} \sum_{x=0}^{N-1} W^{ux} F_{a_i}(x)F_{b_i}(-x). \tag{7.64}$$

But from Eq. (7.62) and Eq. (7.63), we have

$$F_{c_i}(0) = \sum_{j=0}^{N-1} c_i(j) = \sum_{j=0}^{N-1} a_i(j)b_i(j+u) = P_{(a_i,b_i)}(u). \tag{7.65}$$

According to Eqs. (7.64), (7.65) and the fact that$\{(a_i, b_i)|0 \leqslant i \leqslant Q-1\}$ is a PCSP_Q^N, we have

$$\frac{1}{N}\sum_{i=0}^{Q-1}\sum_{x=0}^{N-1} W^{ux}F_{a_i}(x)F_{b_i}(-x) = \sum_{i=0}^{Q-1} R_{(a_i,b_i)}(u)$$
$$= \begin{cases} QN - 2\sum_{i=0}^{Q-1} d(a_i, b_i), & u = 0, \\ 0, & u \neq 0. \end{cases}$$

Let $F(u) = \sum_{i=0}^{Q-1}\sum_{x=0}^{N-1} W^{ux}F_{a_i}(x)F_{b_i}(-x)$, then

$$F(u) = \begin{cases} N\left(QN - 2\sum_{i=0}^{Q-1} d(a_i, b_i)\right), & u = 0, \\ 0, & u \neq 0. \end{cases}$$

Taking Fourier reverse transform to $F(u)$, we have

$$\sum_{i=0}^{Q-1} F_{a_i}(t)F_{b_i}(-t) = \frac{1}{N}\sum_{u=0}^{N-1} F(u)W^{-ut} = QN - 2\sum_{i=0}^{Q-1} d(a_i, b_i).$$ □

Using Theorem 7.3.28, we have the following existence condition for PCSP.

Theorem 7.3.29 *Let* $\{(a_i, b_i)|0 \leqslant i \leqslant Q-1\}$ *be a* PCSP_Q^N, *then the equation:*

$$\sum_{i=0}^{Q-1}((N - 2\text{pl}(a_i))(N - 2\text{pl}(b_i)) + 2d(a_i, b_i)) = QN$$

holds, where $\text{pl}(a_i)$ *denotes the number of elements of value* -1 *in the sequence* a_i, $i = 0, \cdots, Q-1$.

Proof Since $F_{a_i}(0) = \sum_{j=0}^{N-1} a_i(j) = N - 2\text{pl}(a_i)$ and $F_{b_i}(0) = \sum_{j=0}^{N-1} b_i(j) = N - 2\text{pl}(b_i)$ hold, by Theorem 7.3.28, the equation:

$$\sum_{i=0}^{Q-1}(F_{a_i}(0)F_{b_i}(0) + 2d(a_i, b_i))$$
$$= \sum_{i=0}^{Q-1}(N - 2\text{pl}(a_i))(N - 2\text{pl}(b_i) + 2d(a_i, b_i))$$
$$= QN$$

holds. □

Now we give recursive construction for PCSP.

Lemma 7.3.1 *Let* $a_i = (a_i(0), a_i(1), \cdots, a_i(N-1)), b_i = (b_i(0), b_i(1), \cdots, b_i(N-1)), a_i(j) = \pm 1, b_i(j) \pm 1, 0 \leqslant i \leqslant Q-1, 0 \leqslant j \leqslant N-1$, *and let*

$$A_i(x) = \sum_{j=0}^{N-1} a_i(j)x^j, \quad B_i(x) = \sum_{j=0}^{N-1} b_i(j)x^j \in Z[x],$$

then $\{(a_i, b_i)\}$ *is* PCSP_Q^N *if and only if*

$$\sum_{i=0}^{Q-1} A_i(x)B_i(x^{-1}) \equiv QN - 2\sum_{i=0}^{Q-1} d(a_i, b_i) \mod (x^N - 1).$$

Proof Since

$$\begin{aligned}\sum_{i=0}^{Q-1} A_i(x)B_i(x^{-1}) &= \sum_{i=0}^{Q-1}\sum_{j=0}^{N-1}\sum_{k=0}^{N-1} a_i(j)b_i(k)x^{j-k} \\ &= \sum_{i=0}^{Q-1}\left(\sum_{j=0}^{N-1} a_i(j)b_i(j) + \sum_{j,k=0,j\neq k}^{N-1} a_i(j)b_i(k)x^{j-k}\right) \\ &= QN - 2\sum_{i=0}^{Q-1} d(a_i, b_i) + \sum_{u=1}^{N-1} x^u \sum_{i=0}^{Q-1}\sum_{k=0}^{N-1} a_i(k+u)b_i(k) \\ &= QN - 2\sum_{i=0}^{Q-1} d(a_i, b_i) + \sum_{u=1}^{N-1} x^u \sum_{i=0}^{Q-1} P_{(a_i,b_i)}(u)\end{aligned}$$

holds, from Definition 7.3.10 and Theorem 7.3.11, we know that

$$\sum_{i=0}^{Q-1} A_i(x)B_i(x^{-1}) = QN - 2\sum_{i=0}^{Q-1} d(a_i, b_i)$$

if and only if $\{(a_i, b_i)\}$ is PCSP_Q^N. □

Theorem 7.3.30 *If there exist* $\mathrm{PCSP}_{Q_1}^{N_1}$, $\mathrm{PCSP}_{Q_2}^{N_2}, \cdots$, $\mathrm{PCSP}_{Q_t}^{N_t}$, *and there exists an* $S \times t$ *matrix with elements* ± 1, *whose columms are pairwise orthogonal, then there exists a* PCSP_{SQ}^{N}, *where* $N = N_1 + N_2 + \cdots + N_t, Q = \mathrm{L.C.M}(Q_1, Q_2, \cdots, Q_t)$, *i.e.,* Q *is the least common multiple of* $Q_1, Q_2, \cdots$ *and* Q_t.

Proof Assume that $[h(x, y)]$ is an $S \times t$ matrix with elements ± 1, and

$$\sum_{w=1}^{S} h(w, r)h(w, q) = 0, \quad 1 \leqslant r \neq q \leqslant t.$$

We also assume that $\{(a_i^{(r)}, b_i^{(r)},))|0 \leqslant i \leqslant Q_r - 1\}$ is a $\mathrm{PCSP}_{Q_r}^{N_r}$ for every r, $1 \leqslant r \leqslant t$. We use $A_i^{(r)}(x)$ and $B_i^{(r)}(x)$ to denote respectively the polynomials associated with $a_i^{(r)}$and $b_i^{(r)}$, and use $d(a_i^{(r)}, b_i^{(r)})$ to denote Hamming distance between $a_i^{(r)}$ and $b_i^{(r)}$, $0 \leqslant i \leqslant Q_r - 1$. From Lemma 7.3.1, we know that

$$\sum_{i=0}^{Q_r-1} (A_i^{(r)}(x)B_i^{(r)}(x^{-1}) + 2d(a_i^{(r)}, b_i^{(r)})) = N_r Q_r.$$

When $i \geqslant Q_r$, we define $a_i^{(r)} = a_{i\mathrm{mod}Q_r}^{(r)}$ and $b_i^{(r)} = b_{i\mathrm{mod}Q_r}^{(r)}$. Let c_{wi} and d_{wi} denote the following binary sequences with length $N = N_1 + N_2 + \cdots + N_t$,

$$c_{wi} = h(w, 1)a_i^{(1)} \oplus h(w, 2)a_i^{(2)} \oplus \cdots \oplus h(w, t)a_i^{(t)},$$

$$d_{wi} = h(w,1)b_i^{(1)} \oplus h(w,2)b_i^{(2)} \oplus \cdots \oplus h(w,t)b_i^{(t)},$$

$1 \leqslant w \leqslant S$, $0 \leqslant i \leqslant Q-1$. The polynomials associated with the sequences c_{wi} and d_{wi} are respectively

$$C_{wi}(x) = \sum_{r=1}^{t} h(w,r)A_i^{(r)}(x)x^{N(r)}$$

and

$$D_{wi}(x) = \sum_{r=1}^{t} h(w,r)B_i^{(r)}(x)x^{N(r)},$$

where $N(1)=0$, $N(2)=N_1,\cdots,N(r)=N_1+N_2+\cdots+N_{r-1}$, $r \geqslant 2$.

$$\begin{aligned}
&\sum_{w=1}^{S}\sum_{i=0}^{Q-1}(C_{wi}(x)D_{wi}(x^{-1}) + 2d(c_{wi},d_{wi}))\\
=&\sum_{w=1}^{S}\sum_{i=0}^{Q-1}\left(\left(\sum_{r=1}^{t} h(w,r)A_i^{(r)}(x)x^{N(r)}\right)\left(\sum_{r=1}^{t} h(w,r)B_i^{(r)}(x^{-1})x^{-N(r)}\right)\right.\\
&\left.+2\sum_{r=1}^{t} d(a_i^{(r)},b_i^{(r)})\right)\\
=&\sum_{i=0}^{Q-1}\left(\left(\sum_{r,q=1}^{t} A_i^{(r)}(x)B_i^{(q)}(x^{-1})x^{N(r)-N(q)}\right)\sum_{w=1}^{S} h(w,r)h(w,q)\right.\\
&\left.+2S\sum_{r=1}^{t} d(a_i^{(r)},b_i^{(r)})\right)\\
=&S\left(\sum_{r=1}^{t}\left(\sum_{i=0}^{Q-1} A_i^{(r)}(x)B_i^{(r)}(x^{-1}) + 2d(a_i^{(r)},b_i^{(r)})\right)\right) = S\sum_{r=1}^{t} N_rQ = SQN.
\end{aligned}$$

Therefore, $\{(c_{wi},d_{wi})|1 \leqslant w \leqslant S, 0 \leqslant i \leqslant Q-1\}$ form a PCSP_{SQ}^{N}. □

Theorem 7.3.31 *If there exist* $\mathrm{PCSP}_{Q_1}^{N_1}$, $\mathrm{PCSP}_{Q_2}^{N_2}$, $\cdots$, $\mathrm{PCSP}_{Q_t}^{N_t}$, *where* $N_1, N_2, \cdots, N_t$ *are pairwise relatively prime integers, and* $Q_1, Q_2, \cdots, Q_t$ *are also pairwise relatively prime integers, then there exists a* PCSP_{Q}^{N} *with* $N = N_1N_2\cdots N_t$, $Q = Q_1Q_2\cdots Q_t$.

Proof We only prove the theorem for the case $t=2$. The general case can be proved by induction on t.

Now assume $\{(a_i,b_i)\}$ is a $\mathrm{PCSP}_{Q_1}^{N_1}$, and $\{(c_i,d_i)\}$ is a $\mathrm{PCSP}_{Q_2}^{N_2}$. Let $f_{iQ_2+j}(lN_2+m) = a_i(l)c_j(m)$ and $g_{iQ_2+j}(lN_2+m) = b_i(l)d_j(m)$, where $0 \leqslant i \leqslant Q_1-1$, $0 \leqslant j \leqslant Q_2-1$, $0 \leqslant l \leqslant N_1-1$, and $0 \leqslant m \leqslant N_2-1$. For all $0 \leqslant u = u_1N_2+u_2 \leqslant N_1N_2-1$, $0 \leqslant u_1 \leqslant N_1-1$, $0 \leqslant u_2 \leqslant N_2-1$, we have

$$\sum_{i=0}^{Q_1-1}\sum_{j=0}^{Q_2-1} P_{(f_{iQ_2+j},g_{iQ_2+j})}(u)$$

$$= \sum_{i=0}^{Q_1-1} \sum_{j=0}^{Q_2-1} \sum_{l=0}^{N_1-1} \sum_{m=0}^{N_2-1} a_i(l)c_j(m)b_i(l+u_1)d_j(m+u_2)$$

$$= \sum_{i=0}^{Q_1-1} P_{(a_i,b_i)}(u_1) \sum_{j=0}^{Q_2-1} P_{(c_j,d_j)}(u_2)$$

$$= \begin{cases} F, & (u_1,u_2)=(0,0) \\ 0, & (u_1,u_2)\neq(0,0) \end{cases} = \begin{cases} F, & u=0, \\ 0, & u\neq 0. \end{cases}$$

This proves that $\{(f_i, g_i)|0 \leqslant r \leqslant Q_1Q_2 - 1\}$ form a $\mathrm{PCSP}_{Q_1Q_2}^{N_1N_2}$. □

In the following, we present a method to construct PCSP with DSP. This method not only indicates the importance of DSP study, but also provides a large number of PCSPs.

Theorem 7.3.32 *If there exists (v,k,k',e,λ)-DSP, then there exists* PCSP_2^v.

Proof Let (D, D') be a (v,k,k',e,λ)-DSP. We use $f_D(x)$ and $f_{D'}(x)$ to denote respectively $(0,1)$ Hall Polinominal of D and D', from Theorem 7.3.8, we have

$$f_D(x)f_{D'}(x^{-1}) = e - \lambda + \lambda T(x)(\mathrm{mod}\ x^v - 1).$$

Let $g_D(x)$ and $g_{D'}(x)$ denote respectively $(-1,1)$ Hall Polinominal of D and D', we have

$$g_D(x) = 2f_D(x) - T(x), g_{D'}(x) = 2f_{D'}(x) - T(x),$$

then

$$\begin{aligned} &g_D(x)g_{D'}(x^{-1}) \\ =&(2f_D(x) - T(x))(2f_{D'}(x^{-1}) - T(x^{-1})) \\ =&4f_D(x)f_{D'}(x^{-1}) - 2(f_D(x) + f_{D'}(x^{-1}))T(x) + T(x)T(x^{-1}) \\ =&(4\lambda + v - 2k - 2k')T(x) + 4(e-\lambda)(\mathrm{mod}\ x^v - 1). \end{aligned}$$

Next we construct a binary sequence pair (a, b) whose characteristic polinominal must satisfy the equation

$$f_a(x)f_b(x^{-1}) = -(4\lambda + v - 2k - 2k')T(x)(\mathrm{mod} x^v - 1).$$

From Theorem 7.3.10, we know $kk' = \lambda(v-1) + e$. Since $0 < k < v$, $0 < k' < v$, hence $k \geqslant \lambda$, $k' \geqslant \lambda$, then we can get $4\lambda + v - 2k - 2k' \leqslant v$, further $v \equiv (4\lambda + v - 2k - 2k')(\mathrm{mod} 2)$, let $f_a(x) = T(x)$ and let $\lfloor (v - 4\lambda + v - 2k - 2k')/2 \rfloor$ coefficients of $f_b(x)$ be "+1", and other $\lfloor (v - 4\lambda + v - 2k - 2k')/2 \rfloor$ coefficients of $f_b(x)$ be "−1", i.e., $f_b(1) = -(4\lambda + v - 2k - 2k')$, hence

$$f_a(x)f_b(x^{-1}) = -(4\lambda + v - 2k - 2k')T(x)(\mathrm{mod} x^v - 1).$$

Adding $g_D(x)g_{D'}(x^{-1})$ and $f_a(x)f_b(x^{-1})$, we have

$$g_D(x)g_{D'}(x^{-1}) + f_a(x)f_b(x^{-1}) = 4(e-\lambda)(\mathrm{mod} x^v - 1).$$

From Lemma 7.3.1, there exists PCSP_2^v. □

Example 7.3.8 It is known that there exists $(21, 11, 8, 8, 4)$-DSP,

$$D = \{1, 2, 3, 4, 6, 7, 8, 11, 12, 14, 16\},$$

$$D' = \{1, 2, 4, 7, 8, 11, 14, 16\}.$$

From Theorem 7.3.32, we can get $\text{PCSP}_2^{21}, \{(++++-+++--++-+-+-----, ++-+--++--+--+-+-----), (+++++++++++++++++++++, +++++++++++----------)\}$.

Example 7.3.9 It is known that there exists (31, 11, 6, 6, 2)-DSP,

$$D = \{0, 1, 2, 3, 4, 6, 8, 12, 16, 17, 24\},$$

$$D' = \{0, 3, 6, 12, 17, 24\}.$$

From Theorem 7.3.32, we can get PCSP_2^{31}, $\{(+++++-+-+---+---++------+------, +--+--+-----+----+------+------), (+++++++++++++++++++++++++++++++, ++++++++++++-------------------)\}$.

7.4 Sequence Unusual Analysis

7.4.1 Boolean Neural Network Design[28]

Implementing Boolean functions by using low-complexity neural networks is useful in many areas including digital logical design, coding and cryptography. Compared with Boolean functions implemented using digital logic circuits, those implemented using feedforward neural networks have characteristics of flexibility, simplicity, higher network performance and capability of generalization. In fact a three-layer feedforward neural network structure is available for implementation of arbitrary Boolean functions as long as the learning algorithm is effective. Since one neuron can memorize any linearly separable set, the complexity of a three-layer neural network is lower than that of digital circuits especially when the input dimension n is very large. In addition, the neural-based methods allow for generalization; in other words, they have the capability to classify (input) data that is not specified a priori (during training). Therefore, Boolean functions with neural networks have potential application to channel decoding of nonlinear code, nonlinear shift register synthesis and stream cipher analysis among others. Unfortunately, in application areas that require binary-to-binary mappings, the effectiveness of the back-propagation (BP)-type algorithm for the conventional analogue feedforward neural network comes into question because of factors such as poor convergence speed and local minimum among others. For instance, with the BP algorithm over 10^3 iterations become necessary to obtain two-binary-input XOR function with random initial weights, moreover the number of neurons in each layer is determined from experience before the algorithm is applied to train the net from exemplars.

In this subsection, we design a three-layer Boolean neural network (BNN) to implement an arbitrary Boolean function, where the hidden layer serves for characterizing or memorizing a certain pattern of set of patterns, and the output layer severs as the logic OR layer to obtain the desired Boolean function output. To reduce the number of hidden neurons for a given Boolean function, some set covering methods, such as the covering of a hypercube, Hamming sphere and set inhibition among others, are developed. Finally the availability of our approach is demonstrated by a typical example.

Let F_2^n be the n-dimensional vector space over binary field F_2. A Boolean function $f(X) = f(x_1, x_2, \cdots, x_n)$ is a mapping from F_2^n to F_2. The binary input vector $X = (x_1, x_2, \cdots, x_n), x_i \in F_2$ needs to be mapped into a bipolar sequence $C = (c_1, c_2, \cdots, c_n)^{\mathrm{T}}, c_i$

$\in \{-1, 1\}$ by a simple conversion: $c_i = 2x_i - 1$ for $i = 1, 2, \cdots, n$. The structure of the proposed three-layer binary neural network is illustrated by

$$h_j = U\left(\sum_{i=1}^{N} \omega_{1ij} c_i - T_j\right) = \begin{cases} 1, & \sum_{i=1}^{N} \omega_{1ij} \geqslant T_j, \\ 0, & \sum_{i=1}^{N} \omega_{1ij} < T_j, \end{cases} \tag{7.66}$$

$$y = U\left(\sum_{j=1}^{L} \omega_{2j} h_j - \theta\right) = \begin{cases} 1, & \sum_{j=1}^{L} \omega_{2j} h_j \geqslant \theta, \\ 0, & \sum_{j=1}^{L} \omega_{2j} h_j < \theta, \end{cases} \tag{7.67}$$

where $H = (h_1, h_2, \cdots, h_L)$ is the output vector of the hidden neurons; the output of the BNN $y \in \{0, 1\}$ corresponds to the output of the Boolean function $f(X)$, ω_{1ij} and ω_{2j} stand for feedforward connection weights of the hidden layer respectively, T_j and θ denote thresholds of the hidden layer and output layer respectively. It is assumed that all weights and thresholds are integers.

Some new concepts need to be defined before describing the learning strategies of the BNN.

Definition 7.4.1 *The set of all vectors in F_2^n is termed n-dimensional Hamming geometrical space, then elements $X^1, X^2, \cdots, X^{2^n}$ in F_2^n are called points of the n-dimensional Hamming geometrical space.*

Definition 7.4.2 *Subset $f^{-1}(1) = \{X \in F_2^n | f(X) = 1\}; f^{-1}(0) = \{X \in F_2^n | f(X) = 0\}$.*

Definition 7.4.3 *Assume the input and the output of a binary neuron t are denoted by X and y_t respectively. Thus, we propose that the neuron characterizes a set A as*

$$y_t(X) = \begin{cases} 1, & X \in A, \\ 0, & X \notin A. \end{cases} \tag{7.68}$$

For an arbitrary n-variable Boolean function, there are 2^n binary input patterns that can be associated with the 2^n points of an n-dimensional Hamming geometrical space, and be divided into two regions, $f^{-1}(1)$ and $f^{-1}(0)$. As a consequence, the implementation of the Boolean function by the BNN can be viewed as a covering set $f^{-1}(1)$ in the n-dimensional Hamming space by a combination of neurons that characterize different subsets. Since the proposed BNN has n inputs and one output, the complexity of the BNN depends on the number of the hidden neurons. The more patterns each neuron can characterize, the fewer the number of hidden neurons required for covering set $f^{-1}(1)$. Therefore, implementing a Boolean function with low complexity might be considered for enhancing the capability of characterizing sets by binary neurons.

Definition 7.4.4 *$n+1$ characterizing numbers denoted by $S = (s_0, s_1, \cdots, s_n)$, associated with a Boolean function $f(X)$ of n variables, are defined as*

$$S = \sum_{p=1}^{2^n} f(X^p) X^p = \sum_{p_1=1}^{|f^{-1}(1)|} X^{p1}, \tag{7.69}$$

where $X^p, 1 \leqslant p \leqslant 2^n$, *is a concatenation of the bias input,* $x_0^p = 1$, *with the pth vertex* $X'^p = (x_1^p, x_2^p, \cdots, x_n^p)$; $|f^{-1}(1)|$ *denotes the number of patterns in* $f^{-1}(1)$ *and* $X^{p_1} \in f^{-1}(1)$ *for* $1 \leqslant p_1 \leqslant |f^{-1}(1)|$.

It had been concluded that the linearly separable Boolean functions (those functions which can be realized by a binary neuron) are uniquely identified by their characterizing numbers, which will help to search for some specific sets, as will be demonstrated afterwards, to create corresponding neurons representing these sets.

Theorem 7.4.1 *If a binary neuron can characterize a set* A*, then it is also able to characterize* $\overline{A} = F_2^n \setminus A$*, where* $A \setminus B$ *shows the difference of sets* A *and* B*, thus* $\overline{A}$ *is the complementary set of* A.

Proof Assume the weights and threshold of a neuron characterizing set A are denoted by W, T respectively, the output is denoted by y. From Eq. (7.66) and Eq. (7.68), it evidently follows that

$$X \in A, \quad y = 1; \quad WX - T \geqslant 0;$$

$$X \notin A, \quad y = 0; \quad WX - T < 0.$$

Since both the weights and threshold are integers

$$X \in \overline{A}, \quad WX - T \leqslant -1.$$

If the weights and threshold are adjusted as: $W' = -W, T' = -T + 1$, we have

$$X \in A, \quad W'X - T' = -(WX - T) - 1 < 0,$$

then $y = 0$.

$$X \in \overline{A}, \quad W'X - T' = -(WX - T) - 1 \geqslant 0,$$

then $y = 1$.

Thus, the neuron is able to completely characterize the complementary set of A. □

Theorem 7.4.1 implies that the Boolean function could also be implemented by covering set $f^{-1}(0)$. In fact it is more efficient to select a smaller set as covering set because fewer neurons might be needed for a smaller covering set. With no loss of generality, let the logic of the covering set be 1 in the subsequent discussion.

Now, we explore the characterizing capability of binary neurons. To design a BNN for implementing the Boolean function $f(X)$, we create L hidden neurons and one output neuron to characterize the set $f^{-1}(1)$.

Theorem 7.4.2 *Assume the* j*th hidden neuron characterizes set* $A_j, 1 \leqslant j \leqslant L$*, where* $\bigcup_{j=1}^{L} A_j = f^{-1}(1)$*, then the* BNN *can characterize set* $\bigcup_{j=1}^{L} A_j$ *in terms of the following learning rule*

$$\begin{cases} \omega_{2j} = 1, \\ \theta = 1, \qquad j = 1, 2, \cdots, L. \end{cases} \tag{7.70}$$

Proof Given $X \in \bigcup_{j=1}^{L} A_j$, by Definition 7.4.3, we know that at least one hidden neuron is activated as "1" state. From Eq. (7.67), $o = \sum_{j=1}^{L} \omega_{2j} h_j = \sum_{j=1}^{L} h_j \geqslant 1 \Longrightarrow o - \theta \geqslant 0 \Rightarrow y = 1$.

Given $X \notin \bigcup_{j=1}^{L} A_j$, it follows that all L hidden neurons are inhibited as "0" state $o - \theta = \sum_{j=1}^{L} \omega_{2j} h_j - \theta = -1 \Longrightarrow y = 0$.

Hence the neuron in the output layer constrained by Eq. (7.70) can characterize set $\bigcup_{j=1}^{L} A_j$. □

From Theorem 7.4.2 we conclude that the output of the BNN is $y = 1$ if $X \in f^{-1}(1)$ as long as hidden neurons characterize the sets which just cover set $f^{-1}(1)$. Furthermore if $X \notin f^{-1}(1)$, $y = 0$, then the result that if $X \in f^{-1}(0), y = 0$ holds. Therefore, the fact that the output neuron characterizes the set $\bigcup_{j=1}^{L} A_j$ is consistent with the implementation of the Boolean function by the BNN. The remaining problem is how to construct hidden neurons so that set $f^{-1}(1)$ is efficiently covered with as few hidden neurons as possible.

Theorem 7.4.3 *The r-dimensional cube: $C_u = \{(x_1, x_2, \cdots, x_n) \in F_2^n | r$ components of X are not specified which are denoted as $*$; and the other $n - r$ components are explicitly specified 0 or 1$\} \in f^{-1}(1)$ can be characterized by one hidden neuron j, whose weights and threshold are determined in terms of the following rule*

$$\begin{cases} \omega_{1ij} = \begin{cases} 1, & if\ x_i = 1, \\ -1, & if\ x_i = 0, \\ 0, & if\ x_i = *, \quad i = 1, 2, \cdots, n, \end{cases} \\ T_j = n - r - 1. \end{cases} \tag{7.71}$$

Proof Suppose $X = (x_1, x_2, \cdots, x_n) \in C_u, c_i = 2x_i - 1$, we have

$$s_j = \sum_{i=1}^{n} \omega_{1ij} c_i = \sum_{i|x_i=1 \text{ or } 0} \omega_{1ij} c_i + \sum_{i|x_i=*} \omega_{1ij} c_i.$$

From Eq. (7.71), when $x_i = 1$ or 0, $\omega_{1ij} = c_i$ holds; otherwise $x_i = *, \omega_{1ij} = 0$. Thus, $s_j = n - r$.

From Eq. (7.66), we have $h_j = U(s_j - T_j) = U(1) = 1$.

On the other hand, if we suppose $X' = (x'_1, x'_2, \cdots, x'_n) \notin C_u$, then there exists at least one element in X' such that $x'_q \neq x_q$ and $x_q \neq *$, then $\omega_{1iq} \neq c'_q$. So

$$s_j = \sum_{i=1}^{n} \omega_{1ij} c'_i = \sum_{i|x_i=1 \text{ or } 0} \omega_{1ij} c'_i + \sum_{i|x_i=*} \omega_{1ij} c'_i = \sum_{i|x_i=1 \text{ or } 0} \omega_{1ij} c'_i \leqslant n - r - 2,$$

$$s_j - T_j \leqslant -1 \Rightarrow h_j = 0.$$ □

Theorem 7.4.3 ensures that if certain sets that are members of $f^{-1}(1)$ satisfy the definition of C_u, all 2^r patterns in any of them can be memorized by a single hidden neuron.

Theorem 7.4.4 *The Hamming sphere set*

$$R(X^c, d) = \{X \in F_2^n | d_H(X^c, X) \leqslant d\} \in f^{-1}(1)$$

can be characterized by one hidden neuron t, whose weights and threshold are determined in terms of the following rule:

$$\begin{cases} \omega_{1it} = c_i^c, \\ T_t = n - 2d - 1, \quad i = 1, 2, \cdots, L, \end{cases} \tag{7.72}$$

where $X^c = (x_1^c, x_2^c, \cdots, x_n^c)$ *and* d *are the center and radius of the Hamming sphere, respectively;* $d_H(,)$ *is the Hamming distance between two patterns,* $c_i^c = 2x_i^c - 1, i = 1, 2, \cdots, n$.

Proof Suppose $X = (x_1, x_2, \cdots, x_n) \in R(X^c, d), c_i = 2x_i - 1, i = 1, 2, \cdots, n$, we have $d_H(X, X^c) \leqslant d$,

$$s_t = \sum_{i=1}^{n} \omega_{1it} c_i = \sum_{x_i^t = x_i} c_i^c c_i + \sum_{x_i^t \neq x_i} c_i^c c_i,$$

if $x_i^c = x_i \Rightarrow c_i^c c_i = 1$; if $x_i^c \neq x_i \Rightarrow c_i^c c_i = -1$. Thus $s_t \geqslant n - d - d = n - 2d$.

From Eq. (7.66), $h_t = U(s_t - T_t) = 1$.

On the other hand, suppose $X' = (x_1', x_2', \cdots, x_n') \notin R(X^c, d)$, then we have $d_H(X, X^c) > d$,

$$s_t = \sum_{i=1}^{n} \omega_{1it} c_i' = \sum_{x_i^c = x_i} c_i^c c_i' + \sum_{x_i^c \neq x_i} c_i^c c_i' < n - 2d - 2 \Rightarrow s_t - T_t < 0.$$

From Eq. (7.66), $h_t = U(s_t - T_t) = 0$. □

Theorem 7.4.4 ensures that one single hidden neuron can characterize $\sum_{i=1}^{d} \begin{pmatrix} n \\ i \end{pmatrix} + 1, 1 \leqslant d \leqslant n$ patterns of the Hamming sphere set.

Corollary 7.4.1 *The characterizing numbers* $S_c = (s_0, s_1, \cdots, s_n)$ *of the* r*-dimensional cube* C_u *are given by*

$$s_i = \begin{cases} 2^r, & i = 0, \\ 0, & i \neq 0, x_i = 0, \\ 2^r, & i \neq 0, x_i = 1, \\ 2^{r-1}, & i \neq 0, x_i = *. \end{cases} \tag{7.73}$$

Corollary 7.4.2 *The characterizing numbers* $S_H = (s_0, s_1, \cdots, s_n)$ *of the Hamming sphere* $R(X^c, d)$ *are given by*

$$s_i = \begin{cases} \sum\limits_{p=0}^{d} \begin{pmatrix} n \\ p \end{pmatrix}, & i = 0, \\ \sum\limits_{p=0}^{d-1} \begin{pmatrix} n-1 \\ p \end{pmatrix}, & i \neq 0, x_i^c = 0, \\ \sum\limits_{p_1=0}^{d} \begin{pmatrix} n \\ p_1 \end{pmatrix} - \sum\limits_{p_2=0}^{d-1} \begin{pmatrix} n-1 \\ p_2 \end{pmatrix}, & i \neq 0, x_i^c = 1. \end{cases} \tag{7.74}$$

Corollary 7.4.1 and Corollary 7.4.2 can be easily proved from the definition of C_u, $R_{rb}(X^c, d)$ and the characterizing numbers. Since the characterizing numbers uniquely identify a linearly separable Boolean function, both C_u and $R(X^c, d)$ are linearly separable sets that coincide with linearly separable Boolean functions, thus C_u and $R(X^c, d)$ can be uniquely identified by their characterizing numbers. Therefor, Corollary 7.4.1 and 7.4.2 provide the corresponding learning algorithm with the means of judging whether the given sets are hypercubes or Hamming spheres during training, meanwhile determining all the properties of C_u and $R(X^c, d)$. We may start to search C_u and $R(X^c, d)$ from some smaller sets in

$f^{-1}(1)$, then increment the formed sets or search new sets among the remaining patterns. The search process continues until all patterns are included. Finally, the hidden layer is determined in terms of the proposed strategies.

Let set $A_1 = \{X^1, X^2, \cdots, X^p\}, X^i = (x_1^i, x_2^i, \cdots, x_n^i), i = 1, 2, \cdots, p$ be an ordered pair grouped from l coordinates of n-dimensional Hamming geometrical space; $A_2 = \{Y^1, Y^2, \cdots, Y^q\}, Y^j = (x_{l+1}^j, x_{l+2}^j, \cdots, x_n^j), j = 1, 2, \cdots, q$ is an ordered pair grouped from other remaining $n - l$ coordinates. The Cartesian product of subset A_1 with subset A_2 is defined as: $A_1 \times A_2 = \{< X, Y > | X \in A_1 \bigwedge Y \in A_2\}$, where $\bigwedge$ stands for logic AND.

Theorem 7.4.5 *Let A_1 be a complete l-dimensional Hamming space, A_2 is an $(n-l)$-dimensional sphere centered around $X^c = (x_{l+1}^c, x_{l+2}^c, \cdots, x_n^c)$ with radius d, then $A_1 \times A_2$ can be memorized by just one hidden neuron j:*

$$\left\{ \begin{array}{l} \omega_{1ij} = \left\{ \begin{array}{ll} 0, & i = 1, 2, \cdots, l, \\ c_i^c, & i = l+1, \cdots, n, \end{array} \right. \\ T_j = n - l - 2d - 1. \end{array} \right. \tag{7.75}$$

Proof Suppose $X^r = (x_1^r, x_2^r, \cdots, x_n^r) \in A_1 \times A_2$, from Eqs. (7.66) and (7.67) we have

$$s_j = \sum_{i=1}^{n} \omega_{1ij} c_i^r = \sum_{i=1}^{l} \omega_{1ij} c_i^r + \sum_{i=l+1}^{n} \omega_{1ij} c_i^r = \sum_{i=l+1}^{n} c_i^c c_i^r \leqslant n - l - 2d,$$

$$s_j - T_j = s_j - n + l + 2d + 1 \geqslant 1 \Rightarrow h_j = U(s_j - T_j) = 1.$$

Otherwise, if we suppose $X = X^q = (x_1^q, x_2^q, \cdots, x_n^q) \notin A_1 \times A_2$, there is only one possible case $(x_1^q, x_2^q, \cdots, x_l^q) \in A_1; (x_{l+1}^q, x_{l+2}^q, \cdots, x_n^q) \notin A_2$. It follows that

$$\sum_{i=1}^{l} \omega_{1ij} c_i^q = 0; \quad \sum_{i=l+1}^{n} \omega_{1ij} c_i^q \leqslant n - l - 2d - 2,$$

$$s_j - T_j = s_j - n + l + 2d + 1 \leqslant -1 \Rightarrow h_j = U(s_j - T_j) = 0. \qquad \square$$

Since $A_1 \times A_2$ consists of 2^l $(n-l)$-dimensional Hamming spheres, Theorem 7.4.5 significantly enhances the characterizing capability of a single hidden neuron. Therefore, the simplified implementation of Boolean functions is projected as the covering of set $f^{-1}(1)$ using cubes and/or hamming spheres corresponding to the hidden neurons. However, for a given Boolean function, it is not always easy to form a bigger set C_u or $R(X^c, d) \in f^{-1}(1)$. In order to further simplify the structure of the BNN, the constraint of covering set $A_i \in f^{-1}(1)$ may be loosened. In other words, we may choose very few patterns, which are members of $f^{-1}(0)$, to join the covering set together so that more vertices in $f^{-1}(1)$ are included in the set. So both the hidden neurons created for characterizing the covering set $A_i \in f^{-1}(1) \cup f^{-1}(0)$ and some additional hidden neurons called inhibition hidden neurons have to be created to ensure the correct mapping of the Boolean function.

Theorem 7.4.6 *Suppose the sets $A_1, A_2, \cdots, A_K, B_{K+1}, B_{K+2}, \cdots, B_L$ are characterized by L hidden neurons. Then the output neuron can characterize the set $\bigcup_{j=1}^{K} A_j \backslash \bigcup_{j=K+1}^{K} B_j$ in terms of rules as*

$$\omega_{2j} = \left\{ \begin{array}{ll} l, & j = 1, 2, \cdots, K, \\ -K, & j = K+1, \cdots, L, \end{array} \right. \qquad \theta = 1, \tag{7.76}$$

where $A_i \in f^{-1}(1) \cup f^{-1}(0), B_j \in f^{-1}(0)$ and $A \setminus B$ shows the difference between sets A and B.

Proof Three cases must be considered.

Case 1. $X \in \bigcup_{j=1}^{K} A_j$ and $X \notin \bigcup_{j=K+1}^{L} B_j$. From the proof of Theorem 7.4.1, it is concluded that at least one hidden neuron $j, 1 \leqslant j \leqslant K$ is activated.

From Eq. (7.66) $h_j = 0, j = K+1, \cdots, L$, then $y = 1$.

Case 2. $X \notin \bigcup_{j=1}^{K} A_j$, no hidden neuron $j, 1 \leqslant j \leqslant K$, is activated. So from Eqs. (7.67) and (7.76), irrespective of whether hidden neurons $k+1, \cdots, L$ are activated or inhibited, we have

$$o = \sum_{j=1}^{L} \omega_{2j} h_j \leqslant 0 \Rightarrow y = 0.$$

Case 3. $X \in \bigcup_{j=1}^{K} A_j$ and $X \in \bigcup_{j=K+1}^{L} B_j$. It follows that at least one inhibition neuron $j, K+1 \leqslant j \leqslant L$ is activated, and at most K hidden neurons $j, 1 \leqslant j \leqslant K$ are activated. from Eqs. (7.67) and (7.76),

$$o = \sum_{j=1}^{L} \omega_{2j} h_j = \sum_{j=1}^{K} \omega_{2j} h_j + \sum_{j=K+1}^{L} \omega_{2j} h_j \leqslant K - K = 0,$$

then $y = 0$. □

Remark 7.4.1 *Theorem 7.4.6 implies that some bigger set A_i might contain certain inhibition sets B_j characterized by corresponding inhibition hidden neurons. Then sets B'_j are inhibited in the output layer. The consideration of inhibition set increases the probability of constructing bigger covering sets, thus reducing the number of hidden neurons. Furthermore, Theorem 7.4.6 is also valid when some covering sets have a nonempty intersection. For instance, if $A_1 \cap A_2 \in f^{-1}(1)$, no change is needed in the network; otherwise, if $A_1 \cap A_2 \in f^{-1}(0)$, some hidden neurons $l, l > K > 2$ are created by Eq. (7.76) to characterize patterns in $A_1 \cap A_2$. If an input pattern is a member of $A_1 \cap A_2, h_1 = h_2 = 1$ and $h_l = 1$, then $\omega_{2l} = -K < \omega_{21} + \omega_{22}$. From Eqs. (7.76) and (7.67), if $A_1 \cap A_2 \in f^{-1}(0), y = 0$. Thus correct binary mapping is still guaranteed.*

Example 7.4.1 Consider the implementation of a 6-variable Boolean function

$$\begin{aligned} f(0-63) = (&1,1,1,0,1,0,0,1,1,0,0,0,0,0,0,0,1,0,0,0,0,0,0,0,0,0,0,0,\\ &0,0,0,1,1,1,0,0,1,1,1,1,1,0,0,1,1,1,1,1,0,0,0,0,0,0,0,1,1,0,\\ &0,0,0,0,0). \end{aligned}$$

Among the covering sets, there are two Hamming spheres centered around 000000 and 100000 with radius $d = 1$, which is the summed up Cartesian product of $A_1 = \{0, 1\}$ and $A_2 = \{00000, 00001, 00010, 00100, 01000, 10000\}$. Hence, from Eq. (7.75) we have $A_1 \times A_2 = \{000000, 000001, 000010, 000100, 001000, 010000, 100000, 100001, 100010, 100100, 101000, 110000\}$. A total of 12 patterns can be memorized by one hidden neuron: $W_{1i1}: 0, -1, -1, -1, -1, -1; T_1 = 3$. Meanwhile there exist two hypercubes: $B_1 = \{100100, 100101, 100110, 100111, 101100, 101101, 101110, 101111\} = \{10*1**\}$, $B_2 = \{101000, 101001, 111000, 111001\} = \{1*100*\}$. From Theorem 7.4.3, the two subsets can be characterized by hidden neuron 2 and hidden neuron 3, respectively. $W_{1i2}: 1, -1, 0, 1, 0, 0; T_2 = 2$; $W_{1i3}: 1, 0, 1, -1, -1, 0; T_3 = 3$. After covering sets $A_1 \times A_2$, B_1 and B_2 have been constructed, there is one vertex $B_3 = \{100110\}$ in $f^{-1}(1)$ that is not yet characterized by a hidden neuron. The hidden neuron 4 is created: $W_{1i4}: 1, -1, -1, 1, 1, -1; T_4 = 5$. As noted above, B_1

intersects $A_1 \times A_2$ at the pattern $X^{in1} = 100100$. Since $f(X^{in1}) = 0$, one inhibition hidden neuron 5 is created in terms of Eq. (7.76): $W_{1i5} : 1, -1, -1, 1, -1, -1; T_5 = 5$. In addition, B_2 intersects $A_1 \times A_2$ at the pattern $X^{in2} = 101000$. Since $f(X^{in1}) = 1$, no extra change to the neural network needs to be implemented. Finally, the output layer is determined according to Theorem 7.4.6: $\omega_{21} = \omega_{22} = \omega_{23} = \omega_{24} = 1; \omega_{25} = -4; \theta = 1$.

7.4.2 Linear Complexity and Random Sequences with Period 2^n[29]

In this subsection, we consider the expected variation of linear complexity when a random sequence with period 2^n is changed to its local complementation. The importance of the problem lies in that the linear complexity of a sequence is an important measure of its predictability.

Let $f(x) = f(x_0, x_1, \cdots, x_{n-1})$ be a Boolean function of n variables, where

$$x = (x_0, x_1, \cdots, x_{n-1}),$$

x_k takes values in the binary field GF(2). Its truth table, denoted by

$$a = (a_0, a_1, \cdots, a_{N-1}), \quad N = 2^n$$

can be regarded as a vector in $\mathrm{GF}(2)^N$, where

$$a_i = f(i_0, i_1, \cdots, i_{n-1}), \quad 0 \leqslant i \leqslant N-1, \tag{7.77}$$

$$i = i_{n-1} \cdots i_1 i_0 = \sum_{k=0}^{n-1} i_k 2^k \tag{7.78}$$

is the binary representation of i. It is well known that any Boolean function $f(x)$ can be expanded in polynomial form as

$$f(x) = \sum_{i=0}^{N-1} f_i x^i, \quad f_i = 0 \text{ or } 1, \tag{7.79}$$

where x^i stands for $x_0^{i_0}, x_1^{i_1} \cdots x_{n-1}^{i_{n-1}}$ with $x_k^0 = 1$ and $x_k^1 = x_k$. If $f(x) \not\equiv 0$, Equation (7.79) can be written as

$$f(x) = \sum_{j=1}^{L} x^{m_j}, \tag{7.80}$$

where $0 \leqslant m_1 < m_2 < \cdots < m_L \leqslant N-1$. The linear degree ($l$-degree) of the Boolean function $f(x)$, denoted by $l[f(x)]$, is defined by the largest integer m_L in Equation (7.80) while the nonlinear degree (degree) of the Boolean function $f(x)$, denoted by deg$[f(x)]$, is defined as the usual polynomial degree of $f(x)$. The l-degree of $f(x) \equiv 0$ is assumed to be $l[0] = -1$ for convenience.

A binary sequence

$$\underline{a} = (a, a, \cdots) = (a_0, a_1, \cdots, a_{N-1}, a_0, \cdots)$$

is said to be generated by the Boolean function $f(x)$ if the truth table of $f(x)$ is a. it is easy to see that the set of all Boolean functions of n variables, denoted by $\mathcal{B}$, is a linear algebra over GF(2) under the usual operations (addition, scalar product and product) for Boolean functions. Analogously, the set of all binary sequences with period 2^n, denoted by

$\mathcal{A}$, is a linear algebra over GF(2) under the usual operations (addition, scalar product and product) for sequences over GF(2). Furthermore, the mapping $M; f(x) \to \underline{a}$, where $\underline{a}$ is generated by $f(x)$, is a linear algebra isomorphism from $\mathcal{B}$ to $\mathcal{A}$.

Let T be the shift operator on $\mathcal{A}$, i.e.,

$$T(a_0, a_1, \cdots) = (a_1, a_2, \cdots).$$

The linear complexity of a periodic sequence $\underline{a}$, denoted by $c(\underline{a})$, is defined as

$$c(\underline{a}) = \min[\deg[p(T)]; p(T)\underline{a} = 0],$$

where $p(T)$ is a polynomial of T over GF(2).

Games and Chan have devised an algorithm for computing the linear complexity $c(\underline{a})$ of $\underline{a} \in \mathcal{A}$. If $\underline{a} \neq 0$, the linear complexity $c(\underline{a})$ can be computed recursively as follows. Initially, set $c_n = 0$ and $A_n = a$. At a typical step of the algorithm, the left half of $A_m, L(A_m) = (a_{m0}, \cdots, a_{m2^{m-1}-1})$, is added to the right half $R(A_m) = (a_{m2^{m-1}}, \cdots, a_{m2^m-1})$, the result being a sequence B_m of length 2^{m-1}. If $B_m = 0$, set $c_{m-1} = c_m$ and $A_{m-1} = L(A_m)$; otherwise, set $c_{m-1} = c_m + 2^{m-1}$ and $A_{m-1} = B_m$. Finally, the linear complexity of the sequence $\underline{a}$ is given by $c(\underline{a}) = c_0 + 1$.

Theorem 7.4.7 *If the sequence $\underline{a} = (a, a, \cdots)$ is generated by a Boolean function $f(x)$, then we have*

$$c(\underline{a}) = l[f(x)] + 1. \tag{7.81}$$

Proof Since $\underline{a} = \underline{0}$ is generated by $f(x) \equiv 0, l[0] = -1$ by our assumption, while $c(\underline{0}) = 0$ by definition. Hence Equation (7.81) is true in this case. Now we assume $\underline{a} \neq \underline{0}$, so $f(x) \not\equiv 0$. The proof relies on a similar algorithm for computing the l-degree of a Boolean function. If $f(x) \not\equiv 0$, its l-degree $l[f(x)]$ can be computed recursively as follows. Initially, set $l_n = 0$ and $f_n(x) = f(x)$. At a typical step of the algorithm, $f_m(x)$ is expressed as

$$f_m(x_0, \cdots, x_{m-1}) = x_{m-1}g_m(x_0, \cdots, x_{m-2}) + h_m(x_0, \cdots, x_{m-2}).$$

If $g_m \equiv 0$, set $l_{m-1} = l_m$ and $f_{m-1}(x) = h_m(x)$; otherwise, set $l_{m-1} = l_m + 2^{m-1}$ and $f_{m-1}(x) = g_m(x)$. Finally, the l-degree of $f(x)$ is given by $l[f(x)] = l_0$. According to the definition of $l[f(x)]$ $x^{l[f(x)]}$ is the highest l-degree term contained in the polynomial from of $f(x)$. Then, it is easy to check that the algorithm just given is correct. Because the truth table a of $f(x)$ is one period of $\underline{a}$, by contrasting this algorithm with Games and Chan's algorithm, we conclude that B_m and $L(A_m)$ are the truth tables of $g_m(x)$ and $h_m(x)$ respectively. Then, $g_m(x) \equiv 0$ if and only if $B_m = 0$. Hence, $c_0 = l_0$. This proves Equation (7.81). □

Lemma 7.4.1 *The linear complexity $c(\underline{a})$ possesses the following properties:*

(1) *$c(\underline{a}) \geqslant 0$, with equality of and only if $\underline{a} = \underline{0}$.*

(2) *$c(\underline{a}) \leqslant 2^n$, with equality if and only if $W(a)$ is odd, where $W(a)$ is the Hamming weight of one period of $\underline{a}$.*

(3) *$c(\underline{a} + \underline{b}) \leqslant max[c(\underline{a}), c(\underline{b})]$ with equality if and only if $c(\underline{a}) \neq c(\underline{b})$.*

(4) *$c(\underline{a} \cdot \underline{b}) \leqslant c(\underline{a}) + c(\underline{b}) - 1$, with equality if and only if $x^{l[f(x)]}$ and $x^{l[g(x)]}$ do not contain any common variable x_k, where $\underline{a}$ and $\underline{b}$ are generated by $f(x)$ and $g(x)$ respectively.*

Proof These properties can be proved easily by Theorem 7.4.7 and the corresponding properties of l-degree $l[f(x)]$ which are obviously true. □

Lemma 7.4.2 *Let n_k be the number of sequences in $\mathcal{A}$ with linear complexity k. Then we have*

$$n_0 = 1, \quad n_k = 2^{k-1}, \quad k = 1, 2, \cdots, 2^n.$$

Proof The conclusion can be verified by Theorem 7.4.7 and counting the number of Boolean functions in $\mathcal{B}$ with l-degree $k - 1$. □

Lemma 7.4.3 *For $0 \leqslant k \leqslant n$, we have*

$$min_{2^{k-1} < W(a) \leqslant 2^k} c(\underline{a}) = 2^{n-k}, \quad 0 \leqslant k \leqslant n. \tag{7.82}$$

Proof If $0 < c(\underline{a}) < 2^{n-k}$, then from Theorem 7.4.7, the sequence $\underline{a}$ is generated by a Boolean function $f(x)$ with $l[f(x)] < 2^{n-k} - 1$. This implies that $\deg[f(x)] < n - k$. Thus, a is a nonzero codeword of the $(n-k-1)$-th order Reed-Muller code. Hence, $W(a) > 2^k$. This means that Equation (7.82) is valid with $=$ replaced by $\geqslant$. On the other hand, for sequence $\underline{a}$ generated by x^{2n-k-1}, we have $W(a) = 2^k$ and $c(\underline{a}) = 2^{n-k}$. This proves Equation (7.82). □

Corollary 7.4.3 *For $0 \leqslant k \leqslant n - 1$, we have*

$$\min_{2^n - 2^k \leqslant W(a) < 2^n - 2^{k-1}} c(\underline{a}) = 2^{n-k}. \tag{7.83}$$

Proof Let $e = (1, 1, \cdots, 1)$ be the truth table of $f(x) \equiv 1$. Since $2^n - 2^k \leqslant W(a) < 2^n - 2^{k-1}$ is equivalent to $2^{k-1} < W(a + e) \leqslant 2^k$ and $l[f(x)] = l[f(x) + 1]$ if $l[f(x)] > 0$, equality; Equation (7.83) is proved by using Theorem 7.4.7 and Lemma 7.4.3. □

Let $\underline{S} = (S, S, \cdots) = (S_0, S_1, \cdots, S_{N-1}, S_0, \cdots)$ be a random sequence with period $N = 2^n$, where $S = (S_0, S_1, \cdots, S_{N-1})$ is its one period with N independent and uniformly distributed binary random variables. It means that the probability

$$P\{S = a\} = 2^{-N} \text{ for } a \in \text{GF}(2)^N.$$

Lemma 7.4.4 *For any integer N, we have*

$$\sum_{k=1}^{N} k 2^k = (N - 1)2^{N+1} + 2, \tag{7.84}$$

$$\sum_{k=1}^{N} k^2 2^k = (N^2 - 2N + 3)2^{N+1} - 6. \tag{7.85}$$

Proof It is easy to check that

$$\begin{aligned}\sum_{k=1}^{N} k 2^k &= \left[\frac{\mathrm{d}}{\mathrm{d}y}\left(\sum_{k=1}^{N} y^k 2^k\right)\right]_{y=1} = \left[\frac{\mathrm{d}}{\mathrm{d}y}\left(\frac{(2y)^{N+1} - 1}{2y - 1} - 1\right)\right]_{y=1} \\ &= (N - 1)2^{N+1} + 2,\end{aligned}$$

$$\begin{aligned}\sum_{k=1}^{N} k^2 2^k &= \left[\frac{\partial^2}{\partial y \partial z}\left(\sum_{k=1}^{N} y^k z^k 2^k\right)\right]_{y=1,z=1}\\ &= \left[\frac{\partial^2}{\partial y \partial z}\left(\frac{(2yz)^{N+1}-1}{2yz-1}-1\right)\right]_{y=1,z=1}\\ &= (N^2-2N+3)2^{N+1}-6.\end{aligned}$$

□

Theorem 7.4.8[32] *The expectation of $c(\underline{S})$ is given by*

$$Ec(\underline{S}) = N - 1 + 2^N, \quad N = 2^n. \tag{7.86}$$

Theorem 7.4.9 *The variance of $c(\underline{S})$ is given by*

$$\text{Var } c(\underline{S}) = Ec(\underline{S})^2 - [Ec(\underline{S})]^2 = 2 - (2N+1)2^{-N} - 2^{-2N}, \quad N = 2^n. \tag{7.87}$$

Proof Using Lemma 7.4.2 and Equation (7.85), we obtain

$$Ec(\underline{S})^2 = \sum_{k=1}^{N} \frac{2^{k-1}}{2^N} k^2 = \frac{1}{2^{N+1}} \sum_{k=1}^{N} k^2 2^k = N^2 - 2N + 3 - 3 \times 2^{-N}. \tag{7.88}$$

On the other hand, from Equation (7.86), we have

$$[Ec(\underline{S})]^2 = N^2 - 2N + 1 + 2(N-1)2^{-N} + 2^{-2N}. \tag{7.89}$$

Subtracting Equation (7.89) from Equation (7.88), we get Equation (7.87). □

The implication of Theorem 7.4.9 is that a large fraction of sequences in $\mathcal{A}$ have linear complexity near to $Ec(\underline{S}) = N - 1 + 2^{-N}$. Furthermore, the variance of $c(\underline{S})$ is virtually independent of n.

Let $\underline{a}, \underline{b}$ be two sequences in $\mathcal{A}$. $\underline{d} = \underline{a} + \underline{b}$ is called the local complementation of the sequence $\underline{a}$ at r places if $W(b) = r$ with $r < N$.

Theorem 7.4.10 *For any sequence $\underline{b} \in \mathcal{A}$, we have*

$$E|c(\underline{S}) - c(\underline{S} + \underline{b})| = [2^{c(\underline{b})+1} - 2]2^{-N}. \tag{7.90}$$

Proof Since $\underline{S}$ is uniformly distributed on $\mathcal{A}$, we have

$$E|c(\underline{S}) - c(\underline{S} + \underline{b})| = \sum_{\underline{a} \in \mathcal{A}} 2^{-N} |c(\underline{a}) - c(\underline{a} + \underline{b})|. \tag{7.91}$$

The summation in (7.91) can be divided into three parts. We treat each separately.

Case 1. $c(\underline{a}) > c(\underline{b})$. From (3) of Lemma 7.4.1, we have $c(\underline{a} + \underline{b}) = c(\underline{a})$. Hence, $|c(\underline{a}) - c(\underline{a} + \underline{b})| = 0$.

Case 2. $c(\underline{a}) < c(\underline{b})$. From (3) of Lemma 7.4.1, we have $c(\underline{a} + \underline{b}) = c(\underline{b})$. Hence, $|c(\underline{a}) - c(\underline{a} + \underline{b})| = c(\underline{b}) - c(\underline{a})$. From Lemma 7.4.2, $n_k = 2^{k-1}$ for $0 < k < c(\underline{b})$ and $n_0 = 1$.

Case 3. $c(\underline{a}) = c(\underline{b})$. From (3) of Lemma 7.4.1, we have $c(\underline{a} + \underline{b}) < c(\underline{b})$. Hence, $|c(\underline{a}) - c(\underline{a} + \underline{b})| = c(\underline{b}) - c(\underline{a} + \underline{b})$. Since for fixed $\underline{b} \in \mathcal{A}$, M: $\underline{a} \to \underline{a} + \underline{b}$ is a one-to-one mapping from $\mathcal{A}$ onto $\mathcal{A}$, the number of $\underline{a} \in \mathcal{A}$ with $c(\underline{a}) = c(\underline{b})$ and $c(\underline{a} + \underline{b}) = k$ is equal to n_k for $k < \underline{b}$. From Lemma 7.4.2, it is 2^{k-1} for $0 < k < c(\underline{b})$ and 1 for $k = 0$.

Substituting these results into Equation (7.91) and using Equation (7.84), we obtain

$$\begin{aligned} E|c(\underline{S}) - c(\underline{S}+\underline{b})| &= 2\left[\sum_{k=1}^{c(\underline{b})-1} \frac{2^{k-1}}{2^N}(c(\underline{b})-k) + \frac{c(\underline{b})}{2^N}\right] \\ &= \frac{1}{2^N}\left[\sum_{k=1}^{c(\underline{b})-1} (c(\underline{b})-k)2^k + 2c(\underline{b})\right] \\ &= (2^{c(\underline{b})+1} - 2)2^{-N}. \end{aligned}$$

□

The implication of Theorem 7.4.10 is that for a large fraction of sequences in $\mathcal{A}$, the variation of linear complexity is small when the sequence is changed to its local complementation at any given place.

An interesting problem is to calculate

$$E[\max_{W(b)\leqslant W} c(\underline{S}+\underline{b})] \tag{7.92}$$

and

$$E[\max_{W(b)\leqslant W} |c(\underline{S}) - c(\underline{S}+\underline{b})|]. \tag{7.93}$$

Since it is difficult to get exact results even for $W = 1$, we turn to estimating their bounds.

Theorem 7.4.11 *For $W = 1$, we have*

$$\begin{aligned} &N - 1 + 2^{-N} - \left(\frac{n}{2} + 1 - 2^{-(N-n)}\right) \\ &\leqslant E[\min_{W(b)\leqslant 1} c(\underline{S}+\underline{b})] \\ &\leqslant N - 1 + 2^{-N}, \quad N = 2^n. \end{aligned} \tag{7.94}$$

Proof From the definition of $\underline{S}$, we have

$$E[\min_{W(b)\leqslant 1} c(\underline{S}+\underline{b})] = \sum_{\underline{a}\in\mathcal{A}} 2^{-N} \min_{W(b)\leqslant 1} c(\underline{a}+\underline{b}). \tag{7.95}$$

From (2) of Lemma 7.4.1, $c(\underline{b}) = 2^n$ for any $\underline{b}$ with $W(b) = 1$. Then, the summation in Equation (7.95) can be divided into two parts. We treat them separately.

Case 1. $c(\underline{a}) < 2^n$. From (3) of Lemma 7.4.1, we have

$$\min_{W(b)\leqslant 1} c(\underline{a}+\underline{b}) = c(\underline{a}).$$

From Lemma 7.4.2, $n_k = 2^{k-1}$ for $0 < k < 2^n$ and $n_0 = 1$.

Case 2. $c(\underline{a}) = 2^n$. From (3) of Lemma 7.4.1, we have

$$\min_{W(b)\leqslant 1} c(\underline{a}+\underline{b}) = \min_{W(b)=1} c(\underline{a}+\underline{b}).$$

Let

$$\mathcal{A}_k = \{\underline{a} = \underline{a}' + \underline{b};\ c(\underline{a}') = k, W(b) = 1\}, \quad 0 \leqslant k < 2^n.$$

From Lemma 7.4.2, the number of different sequences contained in $\mathcal{A}_k$ is at most 2^{n+k-1} for $0 < k < 2^n$ and 2^n for $k = 0$. It is easy to see that $\min_{W(b)=1} c(\underline{a}+\underline{b}) = k$ implies that

$\underline{a} \in \mathcal{A}_k$. But the converse is not always true; $\underline{a} \in \mathcal{A}_k$ only implies $\min_{W(b)=0} c(\underline{a}+\underline{b}) \leqslant k$. Hence, the number of sequences $\underline{a} \in \mathcal{A}$ such that $\min_{W(b)=1} c(\underline{a}+\underline{b}) = k$ is less than or equal to 2^{n+k-1} for $0 < k < 2^n$ and 2^n for $k = 0$.

Substituting these result into Equation (7.95), we get the lower bound

$$\begin{aligned} E[\min_{W(b)\leqslant 1} c(\underline{S}+\underline{b})] &\geqslant \frac{1}{2^N}\left[\sum_{k=1}^{N-1} k2^{k-1} + \sum_{k=1}^{K} k2^{n+k-1}\right] \\ &= \frac{N-2}{2} + 2^{-N} + [(K-1)2^{K+1}+2]2^{-(N-n+1)}, \end{aligned} \tag{7.96}$$

where K is the solution for the equation

$$2^n + \sum_{k=1}^{K} 2^{n+k-1} = 2^{n+K} = 2^{N-1}.$$

So $K = N - n - 1$. Substituting it into Equation (7.96), we obtain the lower bound in Equation (7.94). The upper bound in Equation (7.94) is obtained by $E[\min_{W(b)\leqslant 1} c(\underline{S}+\underline{b})] \leqslant Ec(\underline{S})$ and Theorem 7.4.8. □

Theorem 7.4.12 *For $W = 1$, we have*

$$2 - 2^{-(N-1)} \leqslant E[\max_{W(b)\leqslant 1} |c(\underline{S}) - c(\underline{S}+\underline{b})|] \leqslant 2 - 2^{-N} + \frac{n}{2} - 2^{-(N-1)}, \quad N = 2^n. \tag{7.97}$$

Proof Similar to the proof of Theorem 7.4.11, we have

$$\begin{aligned} &E[\max_{W(b)\leqslant 1} |c(\underline{S}) - c(\underline{S}+\underline{b})|] \\ &\leqslant \frac{1}{2^N}\left[\sum_{k=1}^{N-1}(N-k)2^{k-1} + N + \sum_{k=1}^{K}(N-k)2^{n+k-1}N2^n\right] \\ &= 1 - 2^{-N} + N2^{-(N-K-n)} - (K-1)2^{-(N-K-n)} - 2^{-(N-n)} \\ &= 2 - 2^{-N} + \frac{n}{2} - 2^{-(N-n)}. \end{aligned} \tag{7.98}$$

This proves the upper bound in Equation (7.97). The lower bound in Equation (7.97) is proved by

$$E[\max_{W(b)\leqslant 1} |c(\underline{S}) - c(\underline{S}+\underline{b})|] \geqslant E|c(\underline{S}) - c(\underline{S}+\underline{b})|$$

and Theorem 7.4.10, where $\underline{b}$ is a fixed sequence with $W(b) = 1$. □

The implication of Theorem 7.4.12 is that for a large fraction of sequences in $\mathcal{A}$, the variation of linear complexity is relatively small when the sequence is changed to its local complementation at any one place.

Using the same method, we can get bounds of Equations (7.92) and (7.93) for $W > 1$. But we think that the larger the integer W is, the looser the bounds are.

7.4.3 Periodic Ambiguity Functions of EQC-Based TFHC

Coherent active radar and sonar echolocation systems often use time-frequency hop pulse train signals. Since range measurement or high target resolution is virtually always desired, these signals must be chosen in such a way that their autocorrelation functions exhibit a narrow mainlobe and adequately small sidelobes; these pulse compression characteristics are necessary to determine precisely the time of arrival of the received signal. On the other hand, if several active echolocation systems view the same target complex, signals from one system may be interpreted as echoes or outputs from the other system(s). A similar situation arises in asynchronous spread spectrum communications if crosstalk occurs between two or more of the signals considered. These interference problems are typical of such multiuser environments. To achieve jamming resistance or low probability of intercept, it is necessary to use a sequence of tome-frequency hop codes with small cross-correlation functions between any two elements of the sequence. As the exact time of arrival of the received signal is unknown a priori, this property is required to minimize the output of the matched filter for the correct signal in cases where spurious codes are present within the received signal. The need for finding time-frequency hop codes which have simultaneously good autocorrelation functions and small mutual cross-correlation functions is therefore well motivated.

A time-frequency hop code can be defined as a mapping from a set of time slots, represented by a finite set of integers, onto a set of equally spaced discrete frequencies, also represented by a finite set of integers. A convenient way of representing such a mapping is with a (0,1)-valued array. The rows in the array correspond to frequency channels, while the columns correspond to the time slots. The entire array constitutes one codeword. Thus every time-frequency hop codeword can be equivalently defined by a (0,1)-valued array $A = [A(i,j)], 0 \leqslant i,j \leqslant N-1$, of size $N \times N$ such that there is one and only one non-zero element in each column.

Another significant description of a time-frequency hop codeword is by the well-known placement operator, which is, in fact, mapping $y(k)$ from the set $\{0,1,\cdots,N-1\}$ to itself, where $y(k)$ stands for the position of the non-zero element in the kth column of the codeword array. For example, each codeword of the time-frequency hop codes based upon extended quadratic congruences (TFHC-EQC), proposed by Bellegarda and Titlebaum, is represented by the following placement operator.

$$y(k) = \begin{cases} \left[\dfrac{ak(k+1)}{2}\right] \pmod{N}, & \text{if } 0 \leqslant k \leqslant \dfrac{N-1}{2}, \\ \left[\dfrac{bk(k+1)}{2} + \dfrac{(a-b)(N^2-1)}{8}\right] \pmod{N}, & \text{if } \dfrac{N-1}{2} \leqslant k \leqslant N-1, \end{cases} \tag{7.99}$$

where N is an odd prime integer, $a, b, 1 \leqslant a, b \leqslant N-1$, are integers such that a and b are not both quadratic residue (OR) or quadratic nonresidue (NQR).

Let $\{a_1, a_2, \cdots, a_{(N-1)/2}\}$ and $\{b_1, b_2, \cdots, b_{(N-1)/2}\}$ be the sets of QR and NQR mod N, respectively. Each TFHC-EQC family consists of $N-1$ member codewords, which are defined by the placement operators of Equation (7.99) by setting $(a,b) = (a_1,b_1), (b_1,a_2), (a_2,b_2), (b_2,a_3), \cdots, (a_{(N-1)/2}, b_{(N-1)/2}), (b_{(N-1)/2}, a_1)$, respectively.

From the theoretical points of view, there are two main approaches to investigate the correlations of signals. One is the aperiodic ambiguity function and the other is the periodic ambiguity function.

Definition 7.4.5 *Let $B = [B(i,j)]$ and $D = [D(i,j)], 0 \leqslant i,j \leqslant N-1$, be two 2-dimensional signals of size $N \times N$. Then for each $0 \leqslant s,t \leqslant N-1$, the respectively aperiodic $A_{BD}(s,t)$ and periodic $P_{BD}(s,t)$ cross-ambiguity function between B and D is* *defined by*

$$A_{BD}(s,t)=\sum_{i=0}^{N-1-s}\sum_{j=0}^{N-1-t}B(i,j)D(i+s,j+t) \tag{7.100}$$

and

$$P_{BD}(s,t)=\sum_{i=0}^{N-1}\sum_{j=0}^{N-1}B(i,j)D(i\oplus s,j\oplus t), \tag{7.101}$$

here and hereafter $a\oplus b=:(a+b)\bmod N$*. If* $B=D$*, then* $A_{BD}(\cdot)=:A_B$ *and* $P_{BD}(\cdot)=:P_B(\cdot)$ *is called, respectively, the aperiodic and periodic auto-ambiguity function.*

A direct consequence of Definition 7.4.5 is the following lemma.

Lemma 7.4.5 *Let* $y'(k)$ *and* $y(k), 0\leqslant k\leqslant N-1$*, be the placement operators of two time-frequency hop codewords. Then for each (2D) periodic shift pair* $(s,t), 0\leqslant s,t\leqslant N-1$*, the corresponding value of periodic cross-ambiguity function between these two codewords is equal to the number of zeros (in the variable* k*) of the following cyclic placement difference equation:*

$$(y'\Delta y)_{k,s,t}=[y'(k\oplus s)+t-y(k)](\text{mod }N), \tag{7.102}$$

where $0\leqslant k,s,t\leqslant N-1$*, and* s *and* t *correspond to the horizontal and vertical cyclic shift, respectively.*

Proof In fact, the placement operator of the cyclicly shifted codeword is exactly $[y'(k\oplus s)+t]$ (mod N). □

There exist little relationships between the periodic and aperiodic ambiguity functions for the general (2D) signals, while for the case of time-frequency hop codes we have the following.

Lemma 7.4.6 *Let* $B=[B(i,j)]$ *and* $D=[D(i,j)], 0\leqslant i,j\leqslant N-1$*, be two given time-frequency hop codewords. Then*

$$\max_{s,t}A_{B,D}(s,t)\leqslant \max_{s,t}P_{BD}(s,t)\leqslant 4\max_{s,t}A_{B,D}(s,t). \tag{7.103}$$

Proof On one hand, the fact of $B(i,j),D(i,j)\in\{0,1\}$ and Definition 7.4.5 simply imply the left inequality. On the other hand,

$$\begin{aligned}P_{BD}(s,t)&=\sum_{i=0}^{N-1}\sum_{j=0}^{N-1}B(i,j)D(i\oplus s,j\oplus t)\\&=\sum_{i=0}^{N-1-s}\sum_{j=0}^{N-1-t}B(i,j)D(i+s,j+t)\\&\quad+\sum_{i=0}^{N-1-s}\sum_{j=N-t}^{N-1}B(i,j)D(i+s,j+t-N)\\&\quad+\sum_{i=N-s}^{N-1}\sum_{j=0}^{N-1-t}B(i,j)D(i+s-N,j+t)\end{aligned}$$

$$\sum_{i=N-s}^{N-1}\sum_{j=N-t}^{N-1} B(i,j)D(i+s-N, j+t-N)$$

$$= A_{BD}(s,t) + \sum_{i=0}^{N-1-s}\sum_{j=0}^{t-1} B(i,j)D(i+s,j)$$

$$+\sum_{i=0}^{s-1}\sum_{j=0}^{N-t-1} B(i,j)D(i,j+t) + \sum_{i=0}^{s-1}\sum_{j=0}^{t-1} B(i,j)D(i,j)$$

$$\leqslant A_{BD}(s,t) + A_{BD}(s,0) + A_{BD}(0,t) + A_{BD}(0,0)$$

$$\leqslant 4\max_{s,t} A_{BD}(s,t), \tag{7.104}$$

where the last to second inequality is also due to the fact of $B(i,j), D(i,j) \in \{0,1\}$.

Thus the right-hand side inequality in Equation (7.103) is proved. □

This lemma asserts that time-frequency hop codes having good periodic ambiguity functions have also good aperiodic ambiguity functions. In the following text, it is found that some subset of TFHC-EQC has simultaneously good periodic and aperiodic ambiguity functions.

It has been proved that the aperiodic auto-ambiguity function of the TFHC-EQC is upper bounded by $4 \times 4 = 16$. The following theorem improves this bound from 16 to 6.

Theorem 7.4.13 *The periodic auto-ambiguity function of* TFHC-EQC *is tightly upper bounded by* 6.

Assume that $y'(k) = y(k)$. It is easy to see that Eq. (7.102) has no zeros, if $s = 0$ and $t \neq 0$, or $t = 0$ and $s \neq 0$. In the following, we assume that $s \neq 0$ and $t \neq 0$. In order to prove that the periodic auto-ambiguity of TFHC-EQC is upper bounded by 6, we divide Eq. (7.102) into the following four equations.

Case 1. $0 \leqslant (k+s) \bmod N \leqslant (N-1)/2$ and $0 \leqslant k \leqslant (N-1)/2$.

$$\begin{aligned}(y\Delta y)_{k,s,t} &= [y(k+s) + t - y(k)] (\text{mod } N)\\ &= \left[\frac{a(k+s)(k+s+1)}{2} + t - \frac{ak(k+1)}{2}\right] (\text{mod } N)\\ &= \left[\frac{as(2k+s+1)}{2} + t\right] (\text{mod } N)\end{aligned} \tag{7.105}$$

which is a linear polynomial (of the variable k) over the field GF(N). Thus Eq. (7.105) has at most one zero.

Case 2. $0 \leqslant (k+s) \bmod N \leqslant (N-1)/2$ and $(N-1)/2 \leqslant k \leqslant N-1$.

$$\begin{aligned}(y\Delta y)_{k,s,t} &= [y(k+s) + t - y(k)] (\text{mod } N)\\ &= \left[\frac{a(k+s)(k+s+1)}{2} + t - \frac{bk(k+1)}{2} - \frac{(a-b)(N^2-1)}{8}\right] (\text{mod } N)\end{aligned} \tag{7.106}$$

which is a polynomial of order 2 in the field GF(N). Thus Eq. (7.106) has at most two zeros.

Case 3. $(N-1)/2 \leqslant (k+s) \bmod N \leqslant N-1$ and $(N-1)/2 \leqslant k \leqslant N-1$.

$$\begin{aligned}(y\Delta y)_{k,s,t} &= [y(k+s)+t-y(k)] \pmod N \\ &= \left[\frac{b(k+s)(k+s+1)}{2}+\frac{(a-b)(N^2-1)}{8}\right. \\ &\quad \left. -\frac{bk(k+1)}{2}-\frac{(a-b)(N^2-1)}{8}+t\right] \pmod N \\ &= \left[\frac{bi(2k+s+1)}{2}+t\right] \pmod N.\end{aligned} \tag{7.107}$$

By the same reason as that in Case 1, we know that Eq. (7.107) has at most one zero.

Case 4. $(N-1)/2 \leqslant (k+s) \bmod N \leqslant N-1$ and $0 \leqslant k \leqslant (N-1)/2$.

$$\begin{aligned}(y\Delta y)_{k,s,t} &= [y(k+s)+t-y(k)] \pmod N \\ &= \left[\frac{b(k+s)(k+s+1)}{2}+\frac{(a-b)(N^2-1)}{8}-\frac{ak(k+1)}{2}+t\right] \pmod N.\end{aligned} \tag{7.108}$$

From the same reason as that in Case 2, we know that Eq. (7.108) has at most two zeros.

Therefore from Cases 1-4, we know that if $y'(k) = y(k)$, Eq. (7.102) has at most $1+2+1+2=6$ zeros, thus by Lemma 7.4.5, the periodic auto-ambiguity function of TFHC-EQC is upper bounded by 6.

The following Example shows that the value 6 is a tight upper bound.

Example 7.4.2 Consider the TFHC-EQC with $N=23, a=1$ (OR), and $b=11$ (NQR). From Eq. (7.99),

$$y(k)=\begin{cases}\left[\dfrac{k(k+1)}{2}\right] \pmod{23}, & \text{if } 0 \leqslant k \leqslant 11, \\ \left[\dfrac{11k(k+1)}{2}-\dfrac{10(23^2-1)}{8}\right] \pmod{23}, & \text{if } 11 \leqslant k \leqslant 22.\end{cases} \tag{7.109}$$

Or, equivalently, $y(k) = \{0\ 1\ 3\ 6\ 10\ 15\ 21\ 5\ 13\ 22\ 9\ 20\ 14\ 19\ 12\ 16\ 8\ 11\ 2\ 4\ 17\ 18\ 7\}$.

If $(s,t)=(6,1)$, its cyclic placement difference function $(y\Delta y)_{k,6,1} = [y(k+6)+1-y(k)]$ (mod 23) has 6 zeros $k = 4, 8, 10, 16, 18$, and 19. Thus the upper bound 6 is achievable.

The $N-1$ codewords of TFHC-EQC have their placement operators determined by Eq. (7.99) by setting $(a,b) = (a_1,b_1),(b_1,a_2),(a_2,b_2),(b_2,a_3),\cdots,(a_{(N-1)/2},b_{(N-1)/2}),$ $(b_{(N-1)/2},a_1)$, respectively.

Particularly, the placement operators of the first $((a,b)=(a_1,b_1))$ and the second $((a,b)=(b_1,a_2))$ codewords are, respectively,

$$y'(k)=\begin{cases}\left[\dfrac{a_1k(k+1)}{2}\right] \pmod N, & \text{if } 0 \leqslant k \leqslant \dfrac{N-1}{2}, \\ \left[\dfrac{b_1k(k+1)}{2}-\dfrac{(a_1-b_1)(N^2-1)}{8}\right] \pmod N, & \text{if } \dfrac{N-1}{2} \leqslant k \leqslant N-1\end{cases} \tag{7.110}$$

and

$$y(k) = \begin{cases} \left[\dfrac{b_1 k(k+1)}{2}\right] \pmod N, & \text{if } 0 \leqslant k \leqslant \dfrac{N-1}{2}, \\ \left[\dfrac{a_1 k(k+1)}{2} + \dfrac{(b_1 - a_2)(N^2-1)}{8}\right] \pmod N, & \text{if } \dfrac{N-1}{2} \leqslant k \leqslant N-1. \end{cases} \tag{7.111}$$

Let $s = (N-1)/2$ and $0 \leqslant k \leqslant (N-1)/2$, then $(N-1)/2 \leqslant k+s \leqslant N-1$. In this case, the cyclic placement difference function in Eq. (7.102) becomes

$$\begin{aligned} (y'\Delta y)_{k,s,t} &= [y'(k+s) + t - y(k)] \pmod N \\ &= \left[\frac{b_1 k(k+1)}{2} + \frac{(a_1 - b_1)(N^2-1)}{8} + t - \frac{b_1 k(k+1)}{2}\right] \pmod N \\ &= \left[\frac{(a_1 - b_1)(N^2-1)}{8} + t\right] \pmod N. \end{aligned} \tag{7.112}$$

It is clear that if

$$t = \left[N - \frac{(a_1 - b_1)(N^2-1)}{8}\right] \pmod N,$$

then Eq. (7.112) is identically zero, i.e., every integer k, $0 \leqslant k \leqslant (N-1)/2$, is a zero. Therefore at the shift $s = (N-1)/2$ and

$$t = \left[N - \frac{(a_1 - b_1)(N^2-1)}{8}\right] \pmod N, \tag{7.113}$$

the cyclic placement difference function between $y'(k)$ and $y(k)$ has at least $(N-1)/2$ zeros. Thus we have proved the following result.

Theorem 7.4.14 *The upper bound of the periodic cross-ambiguity functions of* TFHC-EQC *is larger than or equal to* $(N-1)/2$.

From Theorem 7.4.14, it is known that the periodic cross-ambiguity functions of TFHC-EQC are too large to be used. Now, we show that some subsets of TFHC-EQC have much smaller periodic and hence aperiodic cross-ambiguity functions.

For example, let $\{a_1, a_2, \cdots, a_{(N-1)/2}\}$ and $\{b_1, b_2, \cdots, b_{(N-1)/2}\}$ be the set of QR and NQR mod N, respectively. Then 8 is the tight periodic cross-ambiguity upper bound of the subset TFHC-EQC consisting of $(N-1)/2$ (not $(N-1)$ as in the original TFHC-EQC!) member codewords having their placement operators defined by Eq. (7.99) by setting $(a, b) = (a_1, b_1), (a_2, b_2), \cdots, (a_{(N-1)/2}, b_{(N-1)/2})$, respectively (note that those placement operators corresponding to $(a, b) = (b_1, a_2), (b_2, a_3), \cdots, (b_{(N-1)/2}, a_1)$ have been deleted from the original TFHC-EQC).

In general, we have the following theorem.

Theorem 7.4.15 *Let $y'(k)$ and $y(k)$ be two placement operators defined by Eq.* (7.99) *by (a, b) and (c, d), respectively. If a, b, c, and d are different from each other, then the cyclic placement difference function between $y'(k)$ and $y(k)$ has at most 8 zeros, i.e., the corresponding periodic cross-ambiguity function is upper bounded by 8. Moreover 8 is the tight upper bound.*

In order to prove Theorem 7.4.15, we divide the cyclic placement difference function in Eq. (7.102) into the following four equations.

Case 1. $0 \leqslant (k+s) \bmod N \leqslant (N-1)/2$ and $0 \leqslant k \leqslant (N-1)/2$.

$$\begin{aligned}(y'\Delta y)_{k,s,t} &= [y'(k+s)+t-y(k)] (\text{mod } N)\\ &= \left[\frac{a(k+s)(k+s+1)}{2}+t-\frac{ck(k+1)}{2}\right] (\text{mod } N)\\ &= \left[\frac{a-c}{2}k^2+\frac{a(2s+1)-c}{2}+\frac{as(s+1)}{2}+t\right] (\text{mod } N),\end{aligned} \tag{7.114}$$

which is a polynomial of order 2, for $a \neq c$, in the field GF(N). Thus Eq. (7.114) has at most two zeros.

Case 2. $0 \leqslant (k+s) \bmod N \leqslant (N-1)/2$ and $(N-1)/2 \leqslant k \leqslant N-1$.

$$\begin{aligned}(y'\Delta y)_{k,s,t} &= [y'(k+s)+t-y(k)] (\text{mod } N)\\ &= \left[\frac{a(k+s)(k+s+1)}{2}+t-\frac{dk(k+1)}{2}-\frac{(c-d)(N^2-1)}{8}\right] (\text{mod } N)\\ &= \left[\frac{a-d}{2}k^2+\frac{a(2s+1)-d}{2}k+\frac{as(s+1)}{2}+t-\frac{(c-d)(N^2-1)}{8}\right] (\text{mod } N)\end{aligned} \tag{7.115}$$

which is a polynomial of order 2, for $a \neq d$, in the field GF(N). Thus Eq. (7.115) has at most two zeros.

Case 3. $(N-1)/2 \leqslant (k+s) \bmod N \leqslant N-1$ and $(N-1)/2 \leqslant k \leqslant N-1$.

$$\begin{aligned}(y'\Delta y)_{k,s,t} &= [y'(k+s)+m-y(k)] (\text{mod } N)\\ &= \left[\frac{b(k+s)(k+s+1)}{2}+t-\frac{dk(k+1)}{2}-\frac{(c-d)(N^2-1)}{8}\right] (\text{mod } N)\\ &= \left[\frac{b(k+s)(k+s+1)}{2}+\frac{(a-b)(N^2-1)}{8}\right.\\ &\quad \left. -\frac{dk(k+1)}{2}-\frac{(c-d)(N^2-1)}{8}+t\right] (\text{mod } N)\\ &= \left[\frac{(b-d)}{2}k^2+\frac{b(2s+1)-d}{2}k+\frac{bs(s+1)}{2}+m\right] (\text{mod } N)\end{aligned} \tag{7.116}$$

which is a polynomial of order 2, for $d \neq c$, in the field GF(N). Thus Eq. (7.116) has at most two zeros.

Case 4. $(N-1)/2 \leqslant (k+s) \bmod N \leqslant N-1$ and $0 \leqslant k \leqslant (N-1)/2$.

$$\begin{aligned}(y'\Delta y)_{k,s,t} &= [y'(k+s)+t-y(k)] (\text{mod } N)\\ &= \left[\frac{b(k+s)(k+s+1)}{2}+\frac{(a-b)(N^2-1)}{8}-\frac{ck(k+1)}{2}+t\right] (\text{mod } N)\\ &= \left[\frac{b-c}{2}k^2+\frac{b(2s+1)-c}{2}k+\frac{bs(s+1)}{2}+t+\frac{(a-b)(N^2-1)}{8}\right]\end{aligned} \tag{7.117}$$

which is a polynomial of order 2, for $b \neq c$, in the field GF(N). Thus Eq. (7.117) has at most two zeros.

Thus from Cases 1–4, the cyclic placement difference function between $y'(k)$ and $y(k)$ has at most $2+2+2+2=8$ zeros.

The following Example 7.4.3 shows that the upper bound 8 is achievable.

Example 7.4.3 Let $N=23, a=18$ (QR), $b=14$ (NQR), $c=2$ (QR), and $d=11$ (NQR). The placement operators of $(a,b)=(18,14)$ and $(c,d)=(2,11)$ defined by Eq. (7.99) are $y'(k)=\{0\ 18\ 8\ 16\ 19\ 17\ 10\ 21\ 4\ 5\ 1\ \ 15\ 22\ 20\ 9\ 12\ 6\ 14\ 13\ 3\ 7\ 2\ 11\}$ and $y(k)=\{0\ 2\ 6\ 12\ 20\ 7\ 19\ 10\ 3\ 21\ 18\ 17\ 11\ 16\ 9\ 13\ 5\ 8\ 22\ 1\ 14\ 15\ 4\}$, respectively. If $(s,t)=(5,6)$, the cyclic placement difference function between $y'(k)$ and $y(k)$, $(y'\triangle y)_{k,5,6}=[y'(k+5)+6-y(k)] \pmod{23}$ has eight zeros $k=0,5,8,10,14,15,19$, and 20.

Thus we have found a subset of TFHC-EQC that has simultaneously good periodic and aperiodic auto- and cross-ambiguity functions. The volume of this subset is $(N-1)/2$.

7.4.4 Auto-, Cross-, and Triple Correlations of Sequences[31]

Code-Division Multiple-Access (CDMA) allows serval users simultaneous access to a common channel by assigning a distinct code sequence to the user, thus enabling the user to distinguish his signal from that of other users. When Phase-Shift Keying (PSK) is the method of modulation employed in such a communication system, the code symbols are required to be complex roots of unity. In other words, the sequences should be polyphase. Besides in CDMA, polyphase sequences with good periodic auto- and/or crosscorrelation functions have been widely used in spread-spectrum communications, multiple-access communications, radar, sonar systems, etc. In recent years, numerous papers have been published on the design and analysis of such sequences.

Let $b(0), b(1), \cdots, b(L-1)$ be a sequence of period L, where $b(i)=-1$ or 1. The triple correlation $T(r,s), 0 \leqslant r,s \leqslant L-1$, of the sequence is defined by

$$T(r,s)=\frac{1}{L}\sum_{i=0}^{L-1} b(i)b(i+r)b(i+s), \tag{7.118}$$

where $i+r \equiv (i+r)\text{mod}L$ and $i+s \equiv (i+s)\text{mod}L$.

Triple correlation is less popular than the well known standard auto- and cross- correlations but in general, the triple correlation knows more information about the original sequence than do the ordinary correlations. In fact, it has been proved by A. Lohmann that: "If the signal $h(t)$ is real and of finite extent, then it is possible to reconstruct $h(t)$ (apart from a shift) from its triple correlation." In particular, Marner and coworkers proved that the feedback connection of an m-sequence can be uniquely determined by the triple correlation of the given sequence. In a word, there are indeed situations where it is quite favorable to evaluate the signals by means of their triple correlations.

The universal form of polyphase sequence is

$$a=(a_0,a_1,\cdots,a_{L_1})=(\alpha^{h(0)},\alpha^{h(1)},\cdots,\alpha^{h(L-1)}),$$

where $h(n)$ is a function of n and α, a primitive root of unity. Up to now, most of the known good polyphase sequences are designed by taking $h(n)$ as the trace function of their combinations. However it is well known that polynomials are the simplest functions. What will happen, if we take $h(n)$ as polynomials? This is the question that we try to answer. For simplicity, we assume that $L(>3)$, the length of sequences, is the prime integer and α is the complex primitive L-th root of unity.

In order to evaluate our new sequences, we restate the following known lemma.

Lemma 7.4.7[34] *For a family of M uniform sequences, each of period L, the maximal magnitudes of the sidelobes θ_α, θ_c of auto- and cross-correlation are lower bounded by* $\frac{\theta_c^2}{L} + \frac{(L)\theta_\alpha^2}{L(M-1)L} \geqslant 1.$

We concentrate on polyphase sequence family of the form

$$A = \{a^{(r)} = (a_0^{(r)}, a_1^{(r)}, \cdots, a_{L-1}^{(r)})\},$$
$$\text{with } a_n^{(r)} = \alpha^{f(n)+rg(n)}, \quad 0 \leqslant n \leqslant L-1, \tag{7.119}$$

where we assume that $L(> 3)$ is a prime integer, $\alpha = \exp(2\pi j/L)$ is the complex primitive L-th root of unity, both $f(n)$ and $g(n)$ are polynomials in variable n. In Eq. (7.119), the integer index r is used to identify the r-th member sequence in the family. Therefore, the periodic correlation between the r-th and s-th member sequences $a^{(r)}$ and $a^{(s)}$ can be uniformly defined by

$$R_{rs}(r) = \sum_{n=0}^{L-1} a_n^{(r)}(a_{n+r}^{(s)})^* = \sum_{n=0}^{L-1} a^{f(n)+rg(n)-f(n+r)-sg(n+r)}. \tag{7.120}$$

Equivalently, we have

$$|R_{rs}(\tau)|^2 = \sum_{n=0}^{L-1} a_n^{(r)}(a_{n+r}^{(s)})^* \sum_{m=0}^{L-1} a_m^{(r)}(a_{m+r}^{(s)})^*$$
$$= \sum_{m,n=0}^{L-1} a^{H(n,m,\tau^*,r,s)}, \tag{7.121}$$

where

$$H(n,m,\tau;r,s) = [f(n) + f(m+\tau) - (f(n+\tau) + f(m)))]$$
$$+ r[g(n) - g(m)] - s[g(n+\tau) - g(m+\tau)]$$
$$= F(n,m,\tau) + G(n,m,\tau;r,s), \tag{7.122}$$

$$F(n,m,\tau) = [(f(n) + f(m+\tau)] - [f(n+\tau) + f(m)], \tag{7.123}$$

$$G(n,m,\tau;r,s) = r[g(n) - g(m)] - s[g(n+\tau) - g(m+\tau)]. \tag{7.124}$$

Now we begin to analyze the correlations of the sequence family in Eq. (7.119), according to the degrees of polynomials $f(n)$ and $g(n)$. At first we restate a popular lemma.

Lemma 7.4.8 $\sum_{n=0}^{L-1} \beta^n = 0$, *if β is a primitive L-th root of unit.*

Case 1. $\deg f(n) = 0$, e.g., $f(n) = 0$. In this case, the degree of $g(n)$ must be positive and $F(n,m,\tau) = 0$.

Case 1.1. $\deg g(n) = 1$, e.g., $g(n) = an + b, a \neq 0$. Therefore

$$
\begin{aligned}
R_{rs}(\tau) &= \sum_{n=0}^{L-1} a^{rg(n)-sg(n+\tau)} \\
&= \sum_{n=0}^{L-1} a^{n(ar-as)+rb-as\tau-sb} \\
&= a^{rb-as\tau-sb} \sum_{n=0}^{L-1} a^{n(a\tau-as)} = 0, \quad \text{for } r \neq s.
\end{aligned}
$$

Thus we have

Observation 7.4.1 If $\deg f(n) = 0$ and $\deg g(n) = 1$, the family $\{a^{(r)} : 0 \leqslant r \leqslant L-1\}$ consists of L polyphase sequences with identical zero cross-correlation (their autocorrelations have the same magnitude L, which is not good.), where the member sequence $a^{(r)}$ is defined by Eq. (7.119).

Case 1.2. $\deg g(n) = 2$, e.g. $g(n) = an^2 + bn + c, a \neq 0$. Therefore

$$
\begin{aligned}
G(n,m,\tau;r,s) &= r\left[an^2 + bn + c - am^2 - bm - c\right] \\
&\quad - s\left[a(n+\tau)^2 + b(n+\tau) + c - a(m+\tau)^2 - b(m+\tau) - c\right] \\
&= r\left[a(n-m)(n+m) + b(n-m)\right] \\
&\quad - s\left[a(n-m)(n+m+2\tau) + b(n-m)\right] \\
&= (n-m)\left[(r-s)a(n+m) + rb - sb - 2a\tau s\right].
\end{aligned}
$$

From Eq. (7.121), we have

$$
\begin{aligned}
|R_{rs}(\tau)|^2 &= \sum_{m,n=0}^{L-1} a^{(n-m)[(r-s)a(n+m)+rb-sb-2a\tau s]} \\
&= \sum_{n,q=0}^{L-1} a^{q[(r-s)a(2n-q)+rb-sb-2a\tau s]}, \quad \text{where } q = n-m \\
&= \sum_{q=0}^{L-1} a^{q[(r-s)aq+rb-sb-2a\tau s]} \sum_{n=0}^{L-1} a^{2q(r-s)an}.
\end{aligned}
$$

Thus, from Lemma 7.4.8, the auto- and cross-correlations of this sequence family are

$$
|R_{rs}(\tau)|^2 = L, r \neq s, \text{ and } |R_{rs}(\tau)|^2 = 0, \tau \neq 0, \text{ and } r \neq 0. \tag{7.125}
$$

Observation 7.4.2 If $\deg f(n) = 0$, $\deg g(n) = 2$, then the family $\{a^{(r)} : 1 \leqslant r \leqslant L-1\}$ consists of $M = L-1$ polyphase sequences with their auto- and cross-correlations determined by Eq. (7.125), where the member sequence $a^{(r)}$ is defined by Eq. (7.119). Thus, for this family, $\theta_a = 0$ and $\theta_c = \sqrt{L}$, and $\frac{\theta_c^2}{L} + \frac{(L-1)\theta_a^2}{L(M-1)L} = 1$, which meets the lower bound in Lemma 7.4.7. In other words, we have found an optimal sequence family with respect to Sarwate's bound. It should be pointed out that the known Alltop's quadric sequences and Chu's sequences are very similar to our special subcase of $f(n) = 0$ and $g(n) = n^2$.

Case 1.3. $\deg g(n) = 3$, e.g., $g(n) = an^3 + bn^2 + cn + d, a \neq 0$. Then

$$
\begin{aligned}
H(n,m,\tau;r,s) &= r\left[g(n) - g(m) + g(m+\tau) - g(n+\tau)\right] \\
&= r(m-n)\left[3a\tau(m+n) + 3a\tau^3 + 2b\tau\right].
\end{aligned}
$$

For $r \neq 0, \tau \neq 0$, we have

$$
\begin{aligned}
|R_{rr}(\tau)|^2 &= \sum_{m,n=0}^{L-1} a^{H(n,m,\tau;r,s)} = \sum_{m,n=0}^{L-1} a^{r(m-n)\left[3a\tau(m+n)+3a\tau^3+2b\tau\right]} \\
&= \sum_{n,q=0}^{L-1} a^{rq\left[3a\tau(2n+q)+3a\tau^3+2b\tau\right]} \text{ (where } q = m - n) \\
&= \sum_{q=0}^{L-1} a^{rq\left[3a\tau q+3a\tau^2+2b\tau\right]} \sum_{n=0}^{L-1} a^{6a\tau rqn} = L
\end{aligned}
$$

The last equation results from Lemma 7.4.8.

Observation 7.4.3 If deg $f(n) = 0$, and deg $g(n) = 3$, then every member sequence in the family $\left\{a^{(r)} : 1 \leqslant r \leqslant L-1\right\}$ has an identical out-phase autocorrelation magnitude of $\sqrt{L}$ (the cross correlations of this family are not good), where the member sequence $a^{(r)}$ is defined by Eq. (7.119).

By the same approach, the following observations can be proved, Their proof details are omitted.

Observation 7.4.4 If deg $f(n) = 1$, and deg $g(n) = 0$, then the sequence family $\left\{a^{(r)} : 1 \leqslant r \leqslant L-1\right\}$ defined by Eq.(7.119), is of the identical auto- and cross-correlation magnitude of L, i.e., $|R_{rs}(\tau)| = L$, for all r, s and τ (this family is not good).

Observation 7.4.5 If deg $f(n) = 1$, deg $g(n) = 1$, then the family $\{a^{(r)} : 1 \leqslant r \leqslant L-1\}$ consists of L polyphase sequences with identical zero cross-correlation (their autocorrelations have the same magnitude L, which is not good).

Observation 7.4.6 If deg $f(n) = 1$, deg $g(n) = 2$, then the family $\{a^{(r)} : 1 \leqslant r \leqslant L-1\}$ consists of $M = L-1$ polyphase sequences with their auto- and cross-correlations determined by Eq. (7.125), where the member sequence $a^{(r)}$ is defined by Eq. (7.119). Thus, for this family, $\theta_a = 0$ and $\theta_c = \sqrt{L}$, and $\dfrac{\theta_c^2}{L} + \dfrac{(L-1)\theta_a^2}{L(M-1)L} = 1$, which meets the lower bound in Lemma 7.4.7. In other words, we have found another optimal sequence family with respect to Sarwate's bound.

Observation 7.4.7 If deg $f(n) = 1$, and deg $g(n) = 3$, then every member sequence in the family $\left\{a^{(r)} : 1 \leqslant r \leqslant L-1\right\}$ has an identical out-phase autocorrelation magnitude of $\sqrt{L}$ (The cross-correlations of this family are not good).

Observation 7.4.8 If deg $f(n) = 2$, and deg $g(n) = 0$, then the family $\{a^{(r)} : 1 \leqslant r \leqslant L-1\}$ is of zero out-phase and L in-phase auto- and cross-correlation, i.e., for any r and s $|R_{rs}(\tau)| = 0$, $\tau \neq 0$, and $|R_{rs}(0)| = L$.

Observation 7.4.9 If deg $f(n) = 2$, e.g., $f(n) = un^2 + vn + \omega$, $u \neq 0$ and deg $g(n) = 1$, e.g., $g(n) = an + b$, $a \neq 0$, then the family $\left\{a^{(r)} : 1 \leqslant r \leqslant L-1\right\}$ is of two-valued auto- and cross-correlation magnitude. Exactly, $|R_{rs}(\tau)| = 0$, for $r = s$ and $\tau \neq 0$ or for $r \neq s$ and $\tau \neq \dfrac{a(r-s)}{2u}$; otherwise $|R_{rs}(\tau)| = L$.

Observation 7.4.10 If deg $f(n) = 2$, e.g., $f(n) = un^2 + vn + \omega$, $u \neq 0$ and deg $g(n) = 2$, e.g. $g(n) = an^2 + bn + c$, $a \neq 0$, then the family $\{a^{(r)} : 1 \leqslant r \leqslant L-1$, and r $\neq$ −u/a$\}$ consists of $M = L-1$ sequences with their auto- and cross-correlations $|R_{rr}(\tau)| = 0$, for $r \neq s$. Thus $\theta_a = 0$ and $\theta_c = \sqrt{L}$, and $\frac{\theta_c^2}{L} + \frac{(L-1)\theta_a^2}{L(M-1)L} = 1$. In other words, we have found the third optimal polyphase sequence family with respect to Sarwate's bound.

Observation 7.4.11 If deg $f(n) = 2$, and deg $g(n) = 3$, then every member sequence in the family $\left\{a^{(r)} : 1 \leqslant r \leqslant L-1\right\}$ has the same out-phase autocorrelation magnitude of $\sqrt{L}$.

Observation 7.4.12 If deg $f(n) = 3$, deg $g(n) = 0$, then the family $\{a^{(r)} : 1 \leqslant r \leqslant L-1\}$ is of two-valued auto- and cross-correlation magnitude. Exactly, for any r and s, $|R_{rs}(\tau)| = \sqrt{L}$, if $\tau \neq 0$ and $|R_{rs}(0)| = L$.

Observation 7.4.13 If deg $f(n) = 3$, deg $g(n) = 1$, then the family $\{a^{(r)} : 1 \leqslant r \leqslant L-1\}$ satisfies $|R_{rr}(0)| = L$, $|R_{rr}(\tau)| = \sqrt{L}$, for $\tau \neq 0$ and for $r \neq s$ $|R_{rs}(0)| = 0$ and $|R_{rs}(\tau)| = \sqrt{L}$, if $\tau \neq 0$. In other words, this family consists of $M = L$ uniform polyphase sequences so that $\theta_a = \theta_c = \sqrt{L}$, thus $\frac{\theta_c^2}{L} + \frac{(L-1)\theta_a^2}{L(M-1)L} = 1 + \frac{1}{L} \to 1$, i.e., this sequence family is asymptotically optimal with respect to Sarwate's bound.

Observation 7.4.14 If deg $f(n) = 3$, e.g., $f(n) = un^3 + vn^2 + \omega n + x$, $u \neq 0$ and deg $g(n) = 2$, e.g. $g(n) = an^2 + bn + c$, $a \neq 0$. Then, the family $\left\{a^{(r)} : 1 \leqslant r \leqslant L-1\right\}$ is of two-valued auto- and cross-correlations magnitude. Exactly, $|R_{rr}(\tau)| = \sqrt{L}$, for $\tau \neq 0$; and $|R_{rs}(\tau)| = \sqrt{L}$, for $\tau \neq a(r-s)/(6u)$.

Observation 7.4.15 If deg $f(n) = 3$, deg $g(n) = 3$, then every member sequence in the family $\left\{a^{(r)} : 1 \leqslant r \leqslant L-1\right\}$ is of two-valued autocorrelation magnitude. Exactly, for every r, $|R_{rr}(\tau)| = \sqrt{L}$, for $\tau \neq 0$.

After the analysis of polyphase sequence families $\left\{a^{(r)} = \left(a_0^{(r)}, a_1^{(r)}, \cdots, a_{L-1}^{(r)}\right)\right\}$, $a_n^{(r)} = a^{f(n)+rg(n)}$, for all possible cases of $0 \leqslant$ deg $f(n) \leqslant 3$ and $0 \leqslant$ deg $g(n) \leqslant 3$, we find three families of optimal sequences (Observations 7.4.2, 7.4.6 and 7.4.10); one family of asymptotically optimal sequences (Observation 7.4.13); four families of sequences with both good auto- and cross-correlations (Observations 7.4.8, 7.4.9, 7.4.12, and 7.4.14); four families of sequences with good autocorrelations (Observations 7.4.3, 7.4.7, 7.4.11). In a word, except for the family in Observation 7.4.4, all the families have good auto- and/or cross-correlation functions.

Luke sequences are a family of polyphase sequences consisting of N member sequences of period N, where $N = p^r - 1$, p =prime and $r > 1$, the i-th member sequence $(S_i(0), S_i(1), \cdots, S_i(N-1))$ is generated by a given p-ary m-sequence $(S(0), S(1), \cdots, S(N-1))$ is the form of

$$S_i(n) = \exp[j2\pi(S(n)/p + in/N)], \quad 0 \leqslant n, i \leqslant N-1. \tag{7.126}$$

Theorem 7.4.16 *If $p = 2$ and $\gcd(N, 3) = 1$, the corresponding Luke sequences can be recovered by only a very small part of their triple correlations.*

Proof Let $y = (y(0), y(1), \cdots, y(N-1))$ be a vector of length N with each $y(n)$ defined by

$$y(n) = \exp[j2\pi(S(n)/p], \quad 0 \leqslant n \leqslant N-1. \tag{7.127}$$

The Fourier transformation vector $F_y = (F_y(0), \cdots, F_y(N-1))$ of the vector $y = (y(0), \cdots, y(N-1))$ is

$$F_y(m) = \sum_{n=0}^{N-1} y(n)\exp[j2\pi mn/N] = \sum_{n=0}^{N-1} \exp[j2\pi S(n)/p + mn/N]. \tag{7.128}$$

Let the triple correlation of the i-th member sequence be denoted by $T_i(\tau_1, \tau_2)$, by Eq. (7.126),

$$\begin{aligned} T_i(\tau_1,\tau_2) &= \sum_{n=0}^{N-1} S_i(n)S_i(n+\tau_1)S_i(n+\tau_2) \\ &= \sum_{n=0}^{N-1} \exp\Big\{j2\pi\Big[\frac{S(n)S(n+\tau_1)S(n+\tau_2)}{p} \\ &\quad + \frac{i(n+n+\tau_1+n+\tau_2)}{N}\Big]\Big\} \end{aligned}$$

If $p = 2$ and $\tau_1 = \tau_2 = \tau$, we have

$$T_i(\tau,\tau) = \sum_{n=0}^{N-1} \exp\left[j2\pi\left[\frac{S(n)}{2} + \frac{i(3n+2\tau)}{N}\right]\right].$$

In particular, for $\tau = 0$, we have

$$T_i(0,0) = \sum_{n=0}^{N-1} \exp\left[j2\pi\left[\frac{S(n)}{2} + \frac{i(3n)}{N}\right]\right] = F_y(3i), \quad 0 \leqslant i \leqslant N-1. \tag{7.129}$$

Because $\gcd(N,3) = 1$, the binary matrix $A = [A_{ij}]$, $A_{ij} = 1$ if $i = 3j$, otherwise $A_{ij} = 0$, which is a permutation matrix. Thus Eq. (7.129) can be written as

$$(T_0(0,0), \cdots, T_{N-1}(0,0)) = (F_y(0), \cdots, F_y(N-1))A$$

or equivalently

$$(F_y(0), \cdots, F_y(N-1)) = (T_0(0,0), \cdots, T_{N-1}(0,0))A^{-1}.$$

Therefore, by only N triple correlation coefficients $T_0(0,0), \cdots, T_{N-1}(0,0)$, we can uniquely recover the Fourier transformation vector of the sequence

$$(y(0), \cdots, y(N-1)) = (\exp(j\pi S(0)), \cdots, \exp(j\pi S(N-1))).$$

In other words, the mother m-sequences $(S(0), \cdots, S(N-1))$ of the Luke sequences or equivalently the Luke sequences themselves have been recovered. □

The i-th member sequence in Observation 7.4.5 is rewritten as

$$a^{(i)} = (a_0^{(i)}, \cdots, a_{L-1}^{(i)}), \quad \text{where } a_n^{(i)} = \exp\left[i2\pi\frac{f(n)+in}{L}\right], \tag{7.130}$$

where $L(>3)$ a prime integer, and $f(n)=an^3+bn^2+cn+d$ a polynomial of degree 3 in the field GF(L).

Theorem 7.4.17 *The sequence defined by Eq.* (7.130) *can also be recovered by only a small part of its triple correlations.*

Proof In the same way as used in Theorem 7.4.16, it can be proved that the vector $(\exp(i2\pi f(0)/L),\cdots,\exp(i2\pi f(L-1)/L))$, can be uniquely recovered by only $(T_0(0,0),\cdots,T_{N-1}(0,0))$. In other words, the mother polynomial $f(n)$, or equivalently the sequence family itself, can be uniquely recovered by its triple correlations. □

In the following, we will prove that the peaks of the triple correlation of m-sequences grow up from a Costas array. For the self-contained, we restate the definition of Costas arrays[36].

A permutation matrix $A=[A_{ij}], 1\leqslant i,j\leqslant n$, of order n is defined as a Costas array if for every nonzero integer pair $(r,s)\neq(0,0), |r|\leqslant n, |s|\leqslant n$, the ordinary non-cyclic correlation function of the matrix A satisfies

$$C(r,s)=\sum_{i,j=1}^{n}A(i,j)A(i+r,j+s)\leqslant 1, \tag{7.131}$$

where "non-cyclic" means that $A(i+r,j+s)=0$ if $i+r$ or $j+s$ is beyond the range of $[1,n]$.

At the first glance, it seems that there is little relationship between Costas arrays and triple correlation. Fortunately we find that the peaks of triple correlation of m-sequence grow up perfectly from a Costas array. Exactly, we can prove the following theorem.

Theorem 7.4.18 *Let $b(0),b(1),\cdots,b(2^n-2)$ be an m sequence of length $L=2^n-1$ and $T(r,s), 0\leqslant r,s\leqslant L-1$, be the triple correlation of the m sequence, then the binary matrix $A=[A(i,j)], 1\leqslant i,j\leqslant L-1$, defined by $A(i,j)=1$ if $T(i,j)$ a peak of the triple correlation, otherwise $A(i,j)=0$, is a Costas array of order $L-1$. Consequently, the number of peaks is exactly $L-1$.*

Proof By the trace function expression of m sequence, it is known that there exists a primitive element $a\in\mathrm{GF}(2^n)$ and an non-zero element $\theta\in\mathrm{GF}(2^n)$ so that $b(i)=Tr_1^n(\theta a^i), 0\leqslant i\leqslant L-1$, where $Tr_1^n(x)=\sum_{k=0}^{n-1}x^{2^k}$ is the so-called trace function, which is a linear map from GF(2^n) to GF(2). Thus, the triple correlation $T(r,s)$ of the m-sequence can be formulated by

$$\begin{aligned}T(r,s)&=\frac{1}{L}\sum_{i=0}^{L-1}(-1)^{b(i)+b(i+r)b(i+s)}\\&=\frac{1}{L}\sum_{i=0}^{L-1}(-1)^{\mathrm{Tr}_1^n(\theta\alpha^i)+\mathrm{Tr}_1^n(\theta\alpha^{i+r})+\mathrm{Tr}_1^n(\theta\alpha^{i+s})}\\&=\frac{1}{L}\sum_{i=0}^{L-1}(-1)^{\mathrm{Tr}_1^n(\theta(1+\alpha^r+\alpha^s)\alpha^i)}.\end{aligned} \tag{7.132}$$

Therefore, if the integers s and r satisfy $1+\alpha^r+\alpha^s=0$, then $T(r,s)=1$, i.e., a peak occurs; otherwise if $1+\alpha^r+\alpha^s\neq 0$, then the sequence $\{\mathrm{Tr}_1^n(\theta(1+\alpha^r+\alpha^s)\alpha^i)\}$ is another

m-sequence, then $T(r,s) = -1/L$, which is due to the balance property of m-sequence and Eq. (7.132).

Now it is clear that $T(r,s)$ is a peak if $\alpha^r + \alpha^s = 1$. Thus, the binary matrix defined in the theorem can be equivalently defined as $A(i,j) = 1$ iff $\alpha^r + \alpha^s = 1$, which is clearly a Costas array of order $L-1$. In fact, at first $A = [A(i,j)]$ is a permutation matrix, since every non-zero element of the form $1+\alpha^r$ can be uniquely expressed by α^s; secondly the ordinary non-cyclic correlation of A is $C(r,s) = \sum_{i,j=1}^{n} A(i,j)A(i+r,j+s) \leqslant 1$, since the simultaneous equations $\alpha^x + \alpha^y = 1$ and $\alpha^{x+r} + \alpha^{y+s} = 1$ have at most one solution of the form $y = log_\alpha(1+\alpha^r)$ and $y = \log_\alpha(1+\alpha^s)$. The theorem follows. □

Remark 7.4.2 *Marner and coworkers proved that the feedback connection of an m sequence can be recovered from its triple correlations. Yet Marner's reconstruction is mainly based on decomposing a polynomial form $x^r + x^s + 1$ into the multiplication of a polynomial $f(x)$ of degree n with $f(\alpha) = 0$ and the other polynomial $g(x)$ with $g(\alpha) = 1$. Then the polynomial $f(x)$ is claimed to be the feedback connection of this m-sequence. Yet in general, it is very difficult to decompose a polynomial, especially when the degree becomes higher and higher. Therefore, Marner's decomposition reconstruction is limited in engineering applications. In the following, we will show an alternative reconstruction method, which is, in general, much easier than the aforementioned direct-decomposition approach.*

Let $(r_1,s_1),\cdots,(r_{L-1},s_{L-1})$ be all the peak positions of the triple correlation of some m-sequence. From the number $L-1$ of peaks, the length or equivalently the degree n of the feedback connection of the m-sequence can be recovered by the equation $L = 2^n - 1$. Every peak position (r_i,s_i) corresponds to a polynomial $f_i(x) = x^{r_i} + x^{s_i} + 1$ satisfying $f_i(\alpha) = 0$. Assume that $f(x)$ is the feedback connection of the m-sequence, then $f(\alpha) = 0$ and $f(x)$ divides $\gcd\{f_i(x),\cdots,f_z(x)\}$ for all possible $\{i,j,\cdots,z\} \subset \{1,\cdots,L-1\}$, where $\gcd[a(x),b(x)]$ is the popular great-common-divisor, which has a smaller degree and can be easily calculated. Because of the identity $\gcd[a(x),b(x),c(x)] = \gcd\{\gcd[a(x),b(x)],c(x)\}$, the polynomial $\gcd\{f_i(x),f_j(x),\cdots,f_z(x)\}$ can be easily determined in recursion. If we are lucky to find some $h(x) = \gcd\{f_i(x),f_j(x),\cdots,f_z(x)\}$ of degree n, then this polynomial $h(x)$ is just the desiced feedback connection of the m-sequence. If, unfortunately, every possible $\gcd\{f_i(x),f_j(x),\cdots,f_z(x)\}$ has its degree larger than n, we now choose, among such polynomials, one of the smallest degree and then apply Marner's decomposition to this polynomial of much smaller degree.

Example 7.4.4 In order to show the superiority of our new approach, we consider the same example as that used by Marner. Let $\{b(i), 0 \leqslant i \leqslant 30\}$ be the m-sequence with its feedback connection $f(x) = 1+x+x^2+x^3+x^5$. The peak positions of this m-sequence are (1,12), (12,1), (2,24), (24,2), (3,8), (8,3), (4,17), (17,4), (5,28), (28,5), (6,16), (16,6), (7,9), (9,7), (10,25), (25,10), (11,30), (30,11), (13,27), (27,13), (14,18), (18,14), (15,21),(21,15), (19,20), (20,19), (22,29), (29,22), (23,26), (26,23). Now we will use the new approach to reconstruct the feedback connection $f(x)$ based only on these 30 peak positions.

At first, from the number 30 of peak positions and the equation $L-1 = 30$, we get the length $L = 31$ of the m-sequence, or equivalently $\deg f(x) = 5$. On the other hand, among the 30 polynomials $f_i(x) = x^{r_i} + x^{s_i} + 1, 1 \leqslant i \leqslant 30$, corresponding to the peak position $\{(r_i,s_i) : 1 \leqslant i \leqslant 30\}$, the polynomials $a(x) = 1+x^3+x^8$ and $b(x) = 1+x^7+x^9$, corresponding to the peak positions (3,8) and (7,9) respectively, are of the smallest degrees.

It is lucky that $\gcd[a(x), b(x)] = 1 + x + x^2 + x^3 + x^5$ is of degree 5, the expected degree of $f(x)$, thus this polynomial is exactly the feedback connection of the given m-sequence.

Remark 7.4.3 *Besides the m-sequence, Warner's approach can also be applied to reconstruction of the feedback connection of the mother m-sequences used in Gold sequences, Gold-like sequences, Dual-BCH sequences and Kasami sequences*[34]. *Some other known sequences may also be recovered by their triple correlations. It is a very interesting exercise to find as many such sequences as possible.*

Remark 7.4.4 *In general, triple corrections contain much more information about the original signal than the ordinary autocorrelations, but we can find extreme cases in which both triple correlations and autocorrelations know the same information. For example, let* $x = (x_0, x_1, \cdots, x_{N-1})$ *be a codeword of a* $(N = 2^n - 1, k)$ *Reed-Solomon code with* $N = 2^n - 1 =$ *prime and* $k < (N+2)/3$, *then both triple and autocorrelations know only the first coordinate of the information vector* $a = (a_0, a_1, \cdots, a_{k-1})$ *corresponding to the codeword* x. *In fact, the codeword coordinate* $x_j, 0 \leqslant j \leqslant N-1$, *is defined by*

$$x_j = \sum_{i=0}^{k-1} \alpha_i \beta^{ij}, \quad 0 \leqslant j \leqslant N-1, \tag{7.133}$$

where β *is a primitive element in* GF(2^n). *Thus, the triple correlation* $T(r, s), 0 \leqslant r, s \leqslant N-1$, *of this codeword is*

$$\begin{aligned} T(r,s) &= \sum_{i=0}^{N-1} x_i x_{i+r} x_{i+s} \\ &= \sum_{i=0}^{N-1} \left(\sum_{p=0}^{k-1} \alpha_p \beta^{pi}\right) \left(\sum_{q=0}^{k-1} \alpha_q \beta^{q(i+r)}\right) \left(\sum_{t=0}^{k-1} \alpha_t \beta^{t(i+s)}\right) \\ &= \sum_{p,q,t=0}^{k-1} \alpha_p \alpha_q \alpha_t \beta^{qr} \beta^{ts} \sum_{i=0}^{N-1} \beta^{(p+q+t)i} \\ &= \alpha_0^3 \sum_{i=0}^{N-1} 1 + \sum_{p,q,t=0, p+q+t \neq 0}^{k-1} \alpha_p \alpha_q \alpha_t \beta^{qr} \beta^{ts} \sum_{i=0}^{N-1} \alpha^i \\ &= \alpha_0^3 + \sum_{p,q,t=0, p+q+t \neq 0}^{k-1} \alpha_p \alpha_q \alpha_t \beta^{qr} \beta^{ts} 0 \\ &= \alpha_0^3, \end{aligned} \tag{7.134}$$

where in the last to second equation $\alpha = \beta^{p+q+t}$ *is another primitive element in* GF(2^n) *since* $0 < p+q+t < N-1$ *and* N *is prime.*

In the same way, it can be proved that

$$R(\tau) = \sum_{i=0}^{N-1} x_i x_{i+r} = \alpha_0^2, \quad 0 \leqslant \tau \leqslant N-1. \tag{7.135}$$

From Eqs. (7.134) and (7.135), it is clear that both the triple and autocorrelations of x know only a_0, consequently, they contain the same information about the original signal x.

Remark 7.4.5 *Contrary to the m-sequences, in some other cases, triple correlations know little information about the original signal. For example, let* $y^{(i)} = (y_0^{(i)}, y_1^{(i)}, \cdots, y_{p-1}^{(i)})$ *be a member sequence in the family* $\{y^{(i)} : 1 \leqslant i \leqslant p-1\}$ *defined by* $y_n^{(i)} = \exp(j2\pi in/p)$, *where* $p = $ *prime. Thus, the triple correlation* $T_i(r, s)$ *of* $y^{(i)}$ *is* $T_i(r,s) = \sum_{m=0}^{p-1} \exp(j2\pi i(3n+r+s)/p) = \sum_{m=0}^{p-1} \exp(j2\pi im/p) = 0$, *which is independent of the integer* $i, 1 \leqslant i \leqslant p-1$. *In other word,* $T_i(r,s)$ *contains little information about* $y^{(i)}$.

Bibliography

[1] Yang Yixian and Guo Baoan. Further enumerating Boolean functions of cryptographic significance. *J. of Cryptology*, 1995, 8: 115–122.

[2] Mitchell C J. Enumerating Boolean functions of cryptographic significance. *J. of Cryptology*, 1990, 2: 155–170.

[3] Tian Haijian, Yang Yixian, Wang Jianyu. Enumerating correlation-immune Boolean functions of order one. *J. of Electronics*, 1998, 15(1): 50–57.

[4] Luke O'Connor, Andrew Klapper. Algebraic nonlinearity and it's applications to cryptography. *J. of Cryptology*, 1994, 7(4): 213–227.

[5] Wen Qiaoyan and Yang Yixian. Application of errorcorrecting codes to the construction of Boolean functions of cryptographic significance. *Chinese Journal of Electronics*, 1999, 8(4): 396–398.

[6] Wen Qiaoyan, Yang Yixian. Construction and enumerating of resilient-functions. *Chinese Journal of Electronics*, 2003, 12(1): 15–19.

[7] Carlet C. Partially-bent functions. *Proc. Of Crypto' 92*, 1992: 280–291.

[8] Preneel B. et al. Boolean functions satisfying higher order propagation criteria. *Advances in Cryptology-Eurocrypt' 91. LNCS 547.* Springer-Verlag, 1991: 141–152.

[9] Siegenthaler T. Correlation-Immunity of nonlinear combining functions for cryptographic applications. *IEEE Trans. On Inform. Theory*, 1984, IT30(5): 776–780.

[10] Bierbrauer J. et al. Bounds for resilient functions and orthogonal arrays. *Advance in Cryptology CRYPTO'94, LNCS.* Springer-Verlag, 1995, 839: 247–256.

[11] Xiaomo Zhang, et al. Cryptographically resilient functions. *IEEE Trans. Inform. Theory*, 1997, 43: 1740–1747.

[12] Xiaomo Zhang, et al. On nonlinear resilient functions. *Advance in Cryptology Eurocrypt'95.* Berlin: Springer-Verlag, 1996: 274–290.

[13] Lusheng Chen, et al. On the constructions of new resilient functions from old ones. *IEEE Trans. On Inform. Theory*, 1999, 45(6): 2077–2082.

[14] Yang Yixian. Correlation-Immunity of Boolean functions. *IEEE Electronics Letters*, 1987, 23(25): 1335–1336.

[15] Yang Yixian. Correlation-immunity of Boolean functions. *Selected Papers for the Journal of BUPT*, 1990, 1(1): 50–57.

[16] Yang Yixian. On entropy immunity of feedforward networks. *J. of Electronics Letters*, 1991, 8(5): 297–306.

[17] Rueppel R. *Analysis and Design of Stream Ciphers.* New York: Springer-Verlag, 1986.

[18] Sun Wei, Yang Yixian. Correlations of pseudo-generalized geometric sequences. *1997 IEEE Int. Symposium on Information Theory*, June 29-July 4, 1997: 44. Ulm, Germany.

[19] Sun Wei and Yang Yixian. Correlation functions of a family of generalized geometric sequences. *Discrete Applied Mathematics*, 1997, 80: 193–201.

[20] Wei Sun, Andrew Klapper, and Yixian Yang. On correlations of a family of generalized geometric sequences. *IEEE Transactions on Information Theory*, 2001, 47(6): 2609–2618.

[21] A. Klapper, et al. Cross-correlations of linearly and quadratically related geometric sequences and GMW sequences. *Discrete Appl. Math.*, 1993, 46: 1–20.

[22] Lidl R and Niederreiter H. Finite Fields. *Encyclopedia of Mathematics and Its Applications.* Reading, MA:Addision Wesley, 1983, 20.

[23] Carlitz L. Explicit evaluation of certain exponential sums. *Math. Scand*, 1979, 44: 5–16.

[24] Klapper A. d-form sequences: families of sequences with low correlation values and large linear span. *IEEE Trans. On Inform. Theory*, 1995, IT-41: 423–431.

[25] Baumert L. D. Cyclic difference sets. *Lecture Notes in Mathematics*, New York: Springer-Verlag, 1971, 182.

[26] Feng K, Shiue P. J., et al. On aperiodic and periodic complementary binary sequences. *IEEE Trans. On Inform. Theory*, 1999, 45(1): 296–202.

[27] Bomer L. Antweiler M. Periodic complementary binary sequences. *IEEE Trans. On Inform. Theory*, 1990, 36(6): 1478–1494.

[28] Ma Xiaomin, Yang Yixian, and Zhan Zhaozhi. Boolean neural network design using set covering in Hamming geometrical space. *IEICE Trans. On Fundamentals of Electronics, Communications and Computer Sciences*, 1999, E82-A: 2285–2290.

[29] Zhang Zhaozhi and Yang Yixian. Linear complexity and random sequences with period 2^n. *Systems Science and Mathematical Sciences*, 1990, 3(2): 136–142.

[30] Yi xian Yang and Xin Xinniu. Periodic ambiguity functions of EQC-Based TFHC. *IEEE Trans. on Aerospace and Electronic Systems*, 1998, 34(1): 194–199.

[31] Yang Yixian. Auto-, cross-, and triple-correlations of sequences. *J. of Univ. Of Posts and Telecomm.*, 1996, 3(2): 1–8.

[32] Rueppel R A. Linear Complexity and Random Sequences. *Proc. of Eurocrypt'85.* 167–188.

[33] Yixian Yang. The calculation of linear complexity for sequences with length 2^n and q^n. *Proc. Of BIWIT'88*, DII2: 1–3.

[34] Sarwate D V. Bounds on cross-correlation and autocorrelation of sequences. *IEEE Trans. Inform. Theory*, 1979, IT-25: 720–724.

[35] Luke H D. Families of polyphase sequences with near-optimal two-valued auto-and cross-correlation functions. *Electron. Lett.*, 1992, 28(1): 1–2.

[36] Golomb S. W. and Taylor H. Constructions and properties of Costas arrays. *Proc. Of the IEEE*, 72(9): 1143-1163.

Chapter 8

Design and Analysis of Arrays

There are a large number of problems in system engineering that require families of signal arrays which have one or both of the following two properties:

(1) Each signal array in the family is easily distinguished from a time-frequency shifted version of itself;

(2) Each signal array in the family is easily distinguished from (possibly time-frequency shifted version of) every other signal array in the family.

The first property is important for such applications as ranging systems, radar systems, and spread-spectrum communications systems. The second is important for simultaneous ranging to several targets, multiple-terminal system identification, and code-division multiple-access communications systems.

One of the most widely used approaches to distinguish a signal with its shifted version is the correlation function approach. Thus many kinds of correlation functions have been defined for the aim of signal distinguishing. Examples of correlation functions include ambiguity functions, periodic correlation functions, non-periodic correlation functions, dyadic correlation functions, triple-correlation functions, and Hamming correlation functions, etc. Two signals can be easily distinguished from each other if at least one of their correlation functions looks like a "thumbtack." Thus it becomes very interesting to search for signals (arrays) with ideal auto- or cross-correlation functions.

In 1984, Costas introduced a new two-dimensional pattern, called Costas arrays (see [1]), in the time-frequency plane. A Costas array is a permutation matrix with the property that vectors connecting two 1s of the matrix are all distinct vectors (that is, no two vectors are equal in both amplitude and slope). Costas arrays can be used to provide within time-bandwidth limitations uncoupled parameter estimates of any degree of resolution, because of their "thumbtack" auto-ambiguity functions. Thus, a Costas array is a very important kind of radar and sonar signal.

In many spread-spectrum (SS) and code-division multiple-access (CDMA) systems implemented in fiber-optics media with incoherent signaling, the signature sequences are represented as n-tuple with elements from the set $\{0,1\}$. In order for a receiver, in this case an optical correlator, to correctly recognize the active users, the employed signature sequences must be quasi-orthogonal to one another. The so-called optical orthogonal codes (OOC) are examples of such codes.

This chapter concentrates on the designs and analysis of Costas arrays and OOCs. Correlation properties, constructions, enumerations and parameters bounds will be presented.

8.1 Costas Arrays

8.1.1 Correlations of Costas Arrays

Definition 8.1.1 *Let $A = [A(i, j)]$ and $B = [B(i, j)]$, $0 \leqslant i \leqslant N-1$, $0 \leqslant j \leqslant M-1$, be two arrays of the same size $N \times M$. The following two arrays $R_{AB} = [R_{AB}(s, t)]$, $0 \leqslant s \leqslant N-1$, $0 \leqslant t \leqslant M-1$, and $T_{AB} = [T_{AB}(u, v)]$, $-N+1 \leqslant u \leqslant N-1$, $-M+1 \leqslant v \leqslant M-1$, are called periodic and non-periodic, respectively, cross-correlation arrays between A and B, if*

$$
\begin{aligned}
R_{AB}(s,t) &= \sum_{i=0}^{N-1}\sum_{j=0}^{M-1} A(i,j)B(i \oplus s, j \oplus t),\\
&\quad 0 \leqslant s \leqslant N-1, 0 \leqslant t \leqslant M-1, \qquad (8.1)\\
T_{AB}(u,v) &= \sum_{i=0}^{N-1-u}\sum_{j=0}^{M-1-v} A(i,j)B(i+u, j+v),\\
&\quad -N+1 \leqslant u \leqslant N-1, -M+1 \leqslant v \leqslant M-1, \qquad (8.2)
\end{aligned}
$$

where $N-1 \geqslant i \oplus s \equiv (i+s)\text{mod}N \geqslant 0$, $M-1 \geqslant j \oplus t \equiv (j+t)\text{mod}M \geqslant 0$, and $B(i+u, j+v) = 0$, if $i+u$ exceeds the range $[-N+1, N-1]$ or $j+v$ exceeds $[-M+1, M-1]$. If $A = B$, then $R_A := R_{AA} = [R_{AA}(s,t)]$ and $T_A := T_{AA} = [T_{AA}(u,v)]$ are called the periodic and non-periodic, respectively, autocorrelation arrays of A. If $(s,t) \neq (0,0)$ (resp. $(u,v) \neq (0,0)$), then $R_{AB}(s,t)$ (resp. $T_{AB}(u,v)$) is called the out-of-phase periodic (resp. non-periodic) cross-correlation between A and B. $R_{AB}(0,0)$ (resp. $T_{AB}(0,0)$) is called the in-phase periodic (resp. non-periodic) cross-correlation between A and B.

Sometimes the periodic cross- (resp. auto-)correlation arrays are shortly called cross- (resp. auto-)correlation arrays; the non-periodic cross- (resp. auto-)correlation arrays are called cross- (resp. auto-) ambiguity arrays. While we prefer to use the more intuitive periodic and non-periodic terms in this chapter to avoid misunderstanding.

Definition 8.1.2[1] *A permutation matrix $A = [A(i, j)]$, $0 \leqslant i, j \leqslant N-1$, of order $N \times N$ is called a Costas array if and only if its non-periodic autocorrelation is upper bounded by 1, i.e., $T_{AA}(u, v) \leqslant 1$ for all $-N+1 \leqslant u, v \leqslant N-1$ and $(u, v) \neq (0, 0)$.*

The following classical Lagrange's Theorem plays a very important role in this chapter.

Lemma 8.1.1 *Let F be a field (finite or infinite), then every polynomial $f(x)$ in F has at most* $\deg(f(x))$ *zeros in F. Where* $\deg(f(x))$ *is the degree of $f(x)$.*

The aim of restating the classical Lagrange's Theorem is to address the fact that this theorem works in only fields. Any abuse of this theorem by applying it to the other algebraic structures will possibly result in mistakes. Particuly, Lagrange's Theorem can't be used to estimate the zeros of polynomials in a non-field structure. For example, in the ring Z_8 the polynomial $f(x) = x^2 - 1$ of degree 2 has 4 zeros: 1,3,5, and 7.

Lemma 8.1.2 *Let $A = [A(i, j)]$ and $B = [B(i, j)]$, $0 \leqslant i \leqslant N-1$, $0 \leqslant j \leqslant M-1$, be two arrays with non-negative elements, i.e., $A(i, j), B(i, j) \geqslant 0$, then their periodic and non-periodic cross-correlation arrays satisfy the following inequality*

$$\max_{u,v} T_{AB}(u,v) \leqslant \max_{s,t} R_{AB}(s,t) \leqslant 4 \max_{u,v} T_{AB}(u,v), \tag{8.3}$$

where the first and third "max" are over the area $-N+1 \leqslant u \leqslant N-1$, $-M+1 \leqslant v \leqslant M-1$, *and the second "max" is over* $0 \leqslant s \leqslant N-1$, $0 \leqslant t \leqslant M-1$.

Proof The area $\{0 \leqslant i \leqslant N-1, 0 \leqslant j \leqslant M-1\}$ can be divided into the following four disjointed sub-areas: (1) $\{0 \leqslant i \leqslant N-1-s, 0 \leqslant j \leqslant M-1-t\}$; (2) $\{0 \leqslant i \leqslant N-1-s, M-t \leqslant j \leqslant M-1\}$; (3) $\{N-s \leqslant i \leqslant N-1, 0 \leqslant j \leqslant M-1-t\}$; (4) $\{N-s \leqslant i \leqslant N-1, M-t \leqslant j \leqslant M-1\}$. According to this division, the lemma immediately follows from the next identity, for $0 \leqslant s \leqslant N-1$, $0 \leqslant t \leqslant M-1$,

$$\begin{aligned} R_{AB}(s,t) &= \sum_{i=0}^{N-1}\sum_{j=0}^{M-1} A(i,j)B(i \oplus s, j \oplus t) \\ &= \sum_{i=0}^{N-1-s}\sum_{j=0}^{M-1-t} A(i,j)B(i+s, j+t) \\ &\quad + \sum_{i=0}^{N-1-s}\sum_{j=M-t}^{M-1} A(i,j)B(i+s, j+t-M) \\ &\quad + \sum_{i=N-s}^{N-1}\sum_{j=0}^{M-1-t} A(i,j)B(i+s-N, j+t) \\ &\quad + \sum_{i=N-s}^{N-1}\sum_{j=M-t}^{M-1} A(i,j)B(i+s-N, j+t-M) \\ &= T_{AB}(s,t) + T_{AB}(s,t-M) + T_{AB}(s-N,t) + T_{AB}(s-N,t-M). \end{aligned}$$ □

This lemma will be frequently used in the following text to upper bound the periodic correlations by the non-periodic ones. The proof of this lemma is described in detail in the appendix section of the paper.

From Definition 8.1.2 we know that the out-of-phase non-periodic autocorrelation of every Costas array is upper bounded by 1, thus the following theorem is straightforward from Lemma 8.1.2.

Theorem 8.1.1 *The out-of-phase* periodic *autocorrelation of every Costas array,* $A = [A(i,j)]$, *is upper bounded by* 4, *i.e.,* $\max_{s,t}[R_{AA}(s,t)] \leqslant 4$.

The following example shows that the upper bound in Theorem 8.1.1 is tight.

Example 8.1.1 Let $A = [A(i,j)]$ be a 15×15 Golomb-Costas array generated by the two primitive elements $\alpha = \beta = 10$ in the field GF(17), i.e., $A(i,j) = 1$ if $\alpha^i + \beta^j \equiv 1 \bmod 17$, $A(i,j) = 0$ otherwise. Then the periodic autocorrelation array of A is

$$R_{AA} = \begin{bmatrix} 15 & 0 & 0 & 0 & 0 & 0 & 0 & 0 & 0 & 0 & 0 & 0 & 0 & 0 & 0 \\ 0 & 0 & 1 & 1 & 1 & 1 & 2 & 1 & 1 & 2 & 1 & 0 & 2 & 1 & 1 \\ 0 & 1 & 0 & 1 & 2 & 1 & 1 & 2 & 0 & 1 & 0 & 2 & 1 & 2 & 1 \\ 0 & 1 & 1 & 0 & 2 & 1 & 0 & 2 & 0 & 2 & 2 & 1 & 0 & 1 & 2 \\ 0 & 1 & 2 & 2 & 0 & 1 & 0 & 2 & 2 & 0 & 1 & 1 & 1 & 2 & 0 \\ 0 & 1 & 1 & 1 & 1 & 0 & 1 & 1 & 1 & 0 & 4 & 1 & 2 & 0 & 1 \\ 0 & 2 & 1 & 0 & 0 & 1 & 2 & 1 & 0 & 3 & 0 & 0 & 2 & 1 & 2 \\ 0 & 1 & 2 & 2 & 2 & 1 & 1 & 2 & 0 & 0 & 1 & 2 & 0 & 0 & 1 \\ 0 & 1 & 0 & 0 & 2 & 1 & 0 & 0 & 2 & 1 & 1 & 2 & 2 & 2 & 1 \\ 0 & 2 & 1 & 2 & 0 & 0 & 3 & 0 & 1 & 2 & 1 & 0 & 0 & 1 & 2 \\ 0 & 1 & 0 & 2 & 1 & 4 & 0 & 1 & 1 & 1 & 0 & 1 & 1 & 1 & 1 \\ 0 & 0 & 2 & 1 & 1 & 1 & 0 & 2 & 2 & 0 & 1 & 0 & 2 & 2 & 1 \\ 0 & 2 & 1 & 0 & 1 & 2 & 2 & 0 & 2 & 0 & 1 & 2 & 0 & 1 & 1 \\ 0 & 1 & 2 & 1 & 2 & 0 & 1 & 0 & 2 & 1 & 1 & 2 & 1 & 0 & 1 \\ 0 & 1 & 1 & 2 & 0 & 1 & 2 & 1 & 1 & 2 & 1 & 1 & 1 & 1 & 0 \end{bmatrix}.$$

Clearly, if $(s,t) \neq (0,0)$, then $\max R_{AA}(s,t) = 4$, the upper bound in Theorem 8.1.1 is reached.

Lemma 8.1.3 *Let $A = [A(i,j)]$, $B = [B(i,j)]$, $0 \leqslant i \leqslant N-1$, $0 \leqslant j \leqslant M-1$, be two arrays with non-negative elements. Let C and D be the sub-arrays of A and B, respectively, consisiting of the last $N-1$ rows (or $M-1$ columns). Then* $\max T_{CD}(u,v) \leqslant \max T_{AB}(u,v)$.

This lemma will be frequently used to estimate the non-periodic cross-correlations for the variants of Welch-Costas and Golomb-Costas arrays.

Proof Without loss in generality, we assume that $C = [C(i,j)]$ and $D = [D(i,j)]$ consist of the last $N-1$ rows of A and B, respectively, i.e., $C(i,j) = A(i+1,j)$ and $D(i,j) = B(i+1,j)$, $0 \leqslant i \leqslant N-2$, $0 \leqslant j \leqslant M-1$. The proof is finished by the following inequality, for every $-N+2 \leqslant u \leqslant N-2$, $-M+1 \leqslant v \leqslant M-1$,

$$\begin{aligned} T_{CD}(u,v) &= \sum_{i=0}^{N-2-u} \sum_{j=0}^{M-1-v} C(i,j)D(i+u,j+v) \\ &= \sum_{i=0}^{N-2-u} \sum_{j=0}^{M-1-v} A(i+1,j)B(i+1+u,j+v) \\ &= \sum_{i=1}^{N-1-u} \sum_{j=0}^{M-1-v} A(i,j)B(i+u,j+v) \\ &\leqslant \sum_{i=0}^{N-1-u} \sum_{j=0}^{M-1-v} A(i,j)B(i+u,j+v) = T_{AB}(u,v). \end{aligned}$$

The last inequality is due to the non-negative assumption of $A(i,j)$ and $B(i,j)$. □

Definition 8.1.3[1] *Let p be an odd prime integer, α a primitive element in the finite field* GF(p), *and η a non-zero element in* GF(p). *An array $A = [A(i,j)]$, $0 \leqslant i \leqslant p-2$, $1 \leqslant j \leqslant p-1$, of size $(p-1) \times (p-1)$ is called a Welch-Costas* (WC) *array generated by (η, α) if*

$$A(i,j) = \begin{cases} 1, & \text{if } j = \eta\alpha^i \bmod p, \\ 0, & \text{otherwise}. \end{cases} \tag{8.4}$$

There are two kinds of variants of WC arrays which are defined as follows in Ref.[1]:

The first variant WC, denoted by WCI, is a $(p-2) \times (p-2)$ array produced by simply deleting the first row $(i=0)$ and column $(j=1)$ of a WC array.

The second variant WC, denoted by WCII, works only when 2 is a primitive element in GF(p). A WCII array is of size $(p-3) \times (p-3)$ which is produced by simply deleting the first two rows $(i=0,1)$ and columns $(j=1,2)$ of a WC array.

In the following we will show some more general results about both periodic and non-periodic cross-correlations of WC, WCI, and WCII arrays. First, we show an example.

Example 8.1.2 Let $p=11$, A and B be the Welch-Costas arrays defined by $(\eta=1, \alpha=2)$ and $(\mu=1, \beta=8=\alpha^3)$, respectively. Then

$$A = \begin{bmatrix} 0&1&0&0&0&0&0&0&0&0\\ 0&0&0&1&0&0&0&0&0&0\\ 0&0&0&0&0&0&0&1&0&0\\ 0&0&0&0&1&0&0&0&0&0\\ 0&0&0&0&0&0&0&0&0&1\\ 0&0&0&0&0&0&0&0&1&0\\ 0&0&0&0&0&0&1&0&0&0\\ 0&0&1&0&0&0&0&0&0&0\\ 0&0&0&0&0&1&0&0&0&0\\ 1&0&0&0&0&0&0&0&0&0 \end{bmatrix} \tag{8.5}$$

and

$$B = \begin{bmatrix} 0&0&0&0&0&0&0&1&0&0\\ 0&0&0&0&0&0&0&0&1&0\\ 0&0&0&0&0&1&0&0&0&0\\ 0&0&0&1&0&0&0&0&0&0\\ 0&0&0&0&0&0&0&0&0&1\\ 0&0&1&0&0&0&0&0&0&0\\ 0&1&0&0&0&0&0&0&0&0\\ 0&0&0&0&1&0&0&0&0&0\\ 0&0&0&0&0&0&1&0&0&0\\ 1&0&0&0&0&0&0&0&0&0 \end{bmatrix}, \tag{8.6}$$

The periodic cross-correlation matrix R_{AB} between A and B is

$$R_{AB} = \begin{bmatrix} 0&1&2&1&2&1&2&1&0&0\\ 0&3&0&1&0&1&0&3&0&2\\ 0&1&1&1&4&1&1&1&0&0\\ 1&1&1&1&0&1&1&1&1&2\\ 2&2&1&0&0&0&1&2&2&0\\ 2&1&0&1&0&1&0&1&2&2\\ 1&0&4&0&0&0&4&0&1&0\\ 0&0&1&2&2&2&1&0&0&2\\ 3&0&0&2&0&2&0&0&3&0\\ 1&1&0&1&2&1&0&1&1&2 \end{bmatrix}. \tag{8.7}$$

There are three 4's in this periodic cross-correlation matrix R_{AB}, thus the out-of-phase maximum of R_{AB} is 4, i.e., $\max(R_{AB}(s,t)) = 4$. The relative power between β and α is 3, i.e., $\beta = 8 = \alpha^3 = 2^3$, thus $\max(R_{AB}(s,t))$ is not upper bounded by the relative power.

The periodic autocorrelation of the matrix A in Equation (8.5) is

$$R_{AA} = \begin{bmatrix} 1 & 1 & 1 & 1 & 2 & 1 & 1 & 1 & 1 & 0 \\ 1 & 1 & 1 & 2 & 0 & 2 & 1 & 1 & 1 & 0 \\ 1 & 1 & 2 & 1 & 0 & 1 & 2 & 1 & 1 & 0 \\ 1 & 2 & 0 & 1 & 2 & 1 & 0 & 2 & 1 & 0 \\ 2 & 0 & 2 & 0 & 2 & 0 & 2 & 0 & 2 & 0 \\ 1 & 2 & 0 & 1 & 2 & 1 & 0 & 2 & 1 & 0 \\ 1 & 1 & 2 & 1 & 0 & 1 & 2 & 1 & 1 & 0 \\ 1 & 1 & 1 & 2 & 0 & 2 & 1 & 1 & 1 & 0 \\ 1 & 1 & 1 & 1 & 2 & 1 & 1 & 1 & 1 & 0 \\ 0 & 0 & 0 & 0 & 0 & 0 & 0 & 0 & 0 & 10 \end{bmatrix}, \tag{8.8}$$

which takes 2 as its maximum out-of-phase periodic autocorrelation.

Example 8.1.3 In general, the maximum non-periodic correlation is upper bounded by that of periodic correlation, if the arrays are binary, i.e., with elements equal 0 or 1. In this example, we will show another two Welch-Costas arrays such that their non-periodic cross-correlation is larger than the relative power too. Let $p = 7$, $\alpha = 3 = \beta$ (thus the relative power $l = 1$), $\eta = 1$, and $\mu = 2$. Let A and B be Welch-Costas arrays defined by (η, α) and (μ, β), respectively, i.e.,

$$A = \begin{bmatrix} 0 & 0 & 1 & 0 & 0 & 0 \\ 0 & 1 & 0 & 0 & 0 & 0 \\ 0 & 0 & 0 & 0 & 0 & 1 \\ 0 & 0 & 0 & 1 & 0 & 0 \\ 0 & 0 & 0 & 0 & 1 & 0 \\ 1 & 0 & 0 & 0 & 0 & 0 \end{bmatrix}, \tag{8.9}$$

$$B = \begin{bmatrix} 0 & 0 & 0 & 0 & 0 & 1 \\ 0 & 0 & 0 & 1 & 0 & 0 \\ 0 & 0 & 0 & 0 & 1 & 0 \\ 1 & 0 & 0 & 0 & 0 & 0 \\ 0 & 0 & 1 & 0 & 0 & 0 \\ 0 & 1 & 0 & 0 & 0 & 0 \end{bmatrix}. \tag{8.10}$$

The non-periodic cross-correlation between A and B is

$$T_{AB} = \begin{bmatrix} 0 & 0 & 0 & 0 & 0 & 0 & 0 & 0 & 0 & 0 & 1 \\ 0 & 0 & 0 & 0 & 0 & 0 & 1 & 0 & 1 & 0 & 0 \\ 0 & 0 & 0 & 0 & 1 & 0 & 0 & 1 & 0 & 1 & 0 \\ 0 & 0 & 0 & 0 & 0 & 4 & 0 & 0 & 0 & 0 & 0 \\ 0 & 1 & 0 & 1 & 0 & 0 & 1 & 1 & 0 & 1 & 0 \\ 0 & 0 & 1 & 1 & 1 & 0 & 1 & 1 & 1 & 0 & 0 \\ 1 & 0 & 1 & 0 & 1 & 0 & 1 & 0 & 1 & 0 & 0 \\ 0 & 0 & 1 & 1 & 1 & 0 & 0 & 1 & 0 & 0 & 0 \\ 0 & 1 & 0 & 1 & 0 & 0 & 1 & 0 & 0 & 0 & 0 \\ 0 & 0 & 0 & 0 & 0 & 2 & 0 & 0 & 0 & 0 & 0 \\ 0 & 0 & 0 & 0 & 1 & 0 & 0 & 0 & 0 & 0 & 0 \end{bmatrix}. \tag{8.11}$$

Thus the maximum out-of-phase non-periodic cross-correlation $\max(T_{AB}(s,t)) = 4$, which is larger than the relative power 1.

In general, we have the following results about the upper bounds of periodic and non-periodic correlations between Welch-Costas arrays:

Theorem 8.1.2 *Let p be a prime, $\alpha = \beta$ primitive elements in $\mathrm{GF}(p)$, $\eta = \alpha^s$ and $\mu = \beta^t$ any two non-zero elements in $\mathrm{GF}(p)$. Then the maximum non-periodic cross-correlation between the Welch-Costas arrays A and B defined by (η, α) and (μ, β), respectively, is $\max(|s-t|, p-1-|s-t|)$, which is clearly larger than the relative power $l = 1$.*

Proof This theorem follows from the fact that B is from A by periodically shift $|s-t|$ units in horizontal direction. □

Can Theorem 8.1.2 be generalized to the cases of $\alpha \neq \beta$? In order to find the answer, we used 3 workstations to simultaneously search for more than 100 hours. But, we failed to find any pair of Welch-Costas arrays such that their maximum non-periodic cross-correlations is larger than the relative powers. Thus, we are motivated to conjecture that: if $\alpha \neq \beta$, then the maximum non-periodic cross-correlation between every pair of Welch-Costas arrays A and B, defined by (α, η) and (β, μ), respectively, for any non-zeros η and μ, is upper bounded by the relative power l, where l is determined by $\alpha^l = \beta$. The following Theorem 8.1.3 will prove that this conjecture is indeed true.

Theorem 8.1.3 *Let $A = [A(i,j)]$ and $B = [B(i,j)]$ be two WC arrays generated by (η, α) and (μ, β), respectively. Where α and β are primitive elements in $\mathrm{GF}(p)$, p prime, η and μ are two non-zero elements. If $\alpha\beta \equiv 1 \mathrm{mod} p$, then the non-periodic and periodic cross-correlations between A and B are upper bounded by 2 and 8, respectively, i.e., $\max T_{AB}(u,v) \leqslant 2$ and $\max R_{AB}(s,t) \leqslant 8$.*

Proof By Definition 8.1.1 and Equation (8.4), the non-periodic cross-correlation between A and B is

$$
\begin{aligned}
T_{AB}(u,v) &= \sum_{i=0}^{p-2-u} \sum_{j=1}^{p-1-v} A(i,j)B(i+u, j+v) \\
&= \sum_{i=0}^{p-2-u} A(i, (\eta\alpha^i)\mathrm{mod}p)B(i+u, (\eta\alpha^i)\mathrm{mod}p + v) \\
&= \sum_{i=0}^{p-2-u} B(i+u, (\eta\alpha^i)\mathrm{mod}p + v).
\end{aligned}
$$

Thus, $T_{AB}(u,v)$ is equivalent to the number of i, $0 \leqslant i \leqslant p-2-u$, such that $B(i+u, (\eta\alpha^i)\mathrm{mod}p + v) = 1$, which is

$$(\eta\alpha^i)\mathrm{mod}p + v = (\mu\beta^{(i+u)\mathrm{mod}(p-1)})\mathrm{mod}p \tag{8.12}$$

$$\Longrightarrow \eta\alpha^i + v \equiv \mu\beta^{(i+u)}\mathrm{mod}p. \tag{8.13}$$

It is worthwhile to note that Equation 8.12 and Equation 8.13 are not equivalent to each other. Precisely, a solution i of Equation 8.12 is surely a solution of Equation 8.13, but the reverse is not true, because the plus "+" in the left of Equation 8.12 is operated in the real field, while Equation 8.13 is operated in the finite field GF(p). If i_1 and i_2 are two different solutions of Equation 8.12, i.e., $i_1 \not\equiv i_2 \mathrm{mod}(p-1)$, they are clearly different for Equation 8.13 as well. Thus $T_{AB}(u,v)$ is upper bounded by the number of solutions i in Equation 8.13. Because of $\alpha\beta \equiv 1(\mathrm{mod}p)$, Equation 8.13 becomes $\eta\alpha^i + v \equiv \mu\alpha^{-i-u}(\mathrm{mod}p) \Longleftrightarrow \eta(\alpha^i)^2 + v\alpha^i - \mu\alpha^{-u} \equiv 0(\mathrm{mod}p)$. The left hand of this equation is a polynomial, in the variable α^i, of degree 2 in GF(p). Hence by Lagrange's Theorem, $T_{AB}(u,v) \leqslant 2$, the first assertion follows. The second assertion of the theorem is due to Lemma 8.1.2 and the first assertion. □

Theorem 8.1.4 *Let $A = [A(i,j)]$ and $B = [B(i,j)]$ be two* WC *arrays generated by (η,α) and (μ,β), respectively. Let $\beta = \alpha^l$, $l \geqslant 2$, then the non-periodic cross-correlation between A and B is upper bounded by* $\min(l, p-1-l) \leqslant (p-1)/2$. *Consequently, their periodic cross-correlation is upper bounded by* $4\min(l, p-1-l) \leqslant (p-1)/2$.

Proof By the same argument as that used in the proof of Theorem 8.1.3, $T_{AB}(u,v)$ is upper bounded by the number of solutions i in Equation 8.13. Because of $\beta = \alpha^l$, Equation 8.13 becomes $\eta\alpha^i + v \equiv \mu\alpha^{(i+u)l}(\mathrm{mod}p)$

$$\Longleftrightarrow (\mu\alpha^{ul})(\alpha^i)^l - \eta\alpha^i - v \equiv 0 \ \mathrm{mod}p. \tag{8.14}$$

The left hand of Equation 8.14 is clearly a polynomial, in the variable α^i, of degree l, in GF(p), thus $\max T_{AB}(u,v) \leqslant l$.

Equation 8.14 is also equivalent to

$$(\mu\alpha^{ul}) - \eta(\alpha^i)^{-l+1} - v(\alpha^i)^{-l} \equiv 0 \ \mathrm{mod}p. \tag{8.15}$$

Note that $x^{p-1} \equiv 1\mathrm{mod}p$ for every non-zero element x in GF(p), thus the left hand of Equation 8.15 is $(\mu\alpha^{ul}) - \eta(\alpha^i)^{-l+1} - v(\alpha^i)^{-l} = (\mu\alpha^{ul}) - \eta(\alpha^i)^{p-l} - v(\alpha^i)^{p-l-1}$, which is a polynomial, in the variable α^i of degree $p-l$ in GF(p), thus $\max T_{AB}(u,v) \leqslant p-l$. □

Theorem 8.1.5 *Let $A = [A(i,j)]$, $B = [B(i,j)]$ be two* WC *arrays generated by (η,α) and (μ,β), respectively. Let $A' = [A'(i,j)]$, $B' = [B'(i,j)]$ (resp. $A'' = [A''(i,j)]$, $B'' = [B''(i,j)]$) are the* WCI *(resp.* WCII, *if 2 is a primitive element in* GF(p)) *arrays corresponding to A and B. Then the non-periodic and periodic cross-correlations between A', B' (and A'', B'') are respectively upper bounded:*

(1) 2 *and* 8, *if* $\alpha\beta \equiv 1\mathrm{mod}p$.

(2) $\min(l, m, p-1-l, p-1-m)$ *and* $4\min(l, m, p-1-l, p-1-m)$, *if* $\beta = \alpha^l$, $l \geqslant 2$, *where* $ml \equiv 1\mathrm{mod}(p-1)$.

(3) $\max(|s-t|, p-1-|s-t|)$ *and* $4\max(|s-t|, p-1-|s-t|)$, *if* $\alpha = \beta$, $\eta = \alpha^s$, $\mu = \alpha^t$.

Proof The first assertion is due to Theorem 8.1.3, Lemma 8.1.2, Lemma 8.1.3, and Definition 8.1.3.

The second assertion is due to Theorem 8.1.4, Lemma 8.1.2, Lemma 8.1.3, and Definition 8.1.3.

The third assertion is due to Theorem 8.1.2, Lemma 8.1.2, Lemma 8.1.3, and Definition 8.1.3. □

We have presented a few upper bounds for the maximum of the correlation values. In the following, we turn to the exact values of the in-phase correlations. These exact values are helpful in some cases, e.g., they provide us some information about the lower bounds of the maximum of the correlation values.

Theorem 8.1.6 *Let $A = [A(i,j)]$ and $B = [B(i,j)]$ be two* WC *arrays in Theorem* 8.1.4 *with $\eta = \mu$, then their non-periodic (and periodic) cross-correlation at the shift $(0,0)$ is $T_{AB}(0,0) = \gcd(l-1, p-1)$, where $\gcd(x,y)$ denotes the greatest common divisor between x and y.*

Proof In the same argument as in the proof of Theorem 8.1.3, $T_{AB}(0,0)$ is equivalent to the number of i, $0 \leqslant i \leqslant p-2$, such that

$$(\eta\alpha^i)\text{mod}p = (\mu\beta^i)\text{mod}p. \tag{8.16}$$

Equation 8.16 is equivalent to the following equation in the field GF(p): $\eta\alpha^i \equiv \mu\beta^i(\text{mod}p)$ $\Longleftrightarrow \eta\alpha^i - \mu\alpha^{li} \equiv 0(\text{mod}p) \Longleftrightarrow (\alpha^i)^{l-1} \equiv \eta\mu^{-1} \equiv 1(\text{mod}p) \Longleftrightarrow i(l-1) \equiv 0\text{mod}(p-1)$ $\Longleftrightarrow i\left(\dfrac{l-1}{\gcd(l-1,p-1)}\right) \equiv 0\text{mod}\left(\dfrac{p-1}{\gcd(l-1,p-1)}\right) \Longleftrightarrow w := \dfrac{p-1}{\gcd(l-1,p-1)}\,|i$ and $0 \leqslant i \leqslant p-2$, where $x\,|y$ means y is a multiple of x. While there are exactly $\left\lfloor \dfrac{p-1}{w} \right\rfloor =$ $\gcd(l-1,p-1)$ integers i satisfying $0 \leqslant i \leqslant p-2$ and $w\,|i$, the theorem follows ($\lfloor x \rfloor$ is the floor of x, i.e., the largest integer smaller than or equal to x). □

In the same way as used in the proof of Theorem 8.1.2, it is easy to prove the following corollary.

Corollary 8.1.1 *Let $A = [A(i,j)]$, $B = [B(i,j)]$ be the arrays in Theorem* 8.1.2, *then $T_{AB}(0,0) = 0$, if $\eta \neq \mu$, and $T_{AB}(0,0) = p-1$, if $\eta = \mu$.*

Theorem 8.1.7 *Let $A = [A(i,j)]$, $B = [B(i,j)]$ be two* WC *arrays generated by (η,α) and (μ,β), respectively. Let $A' = [A'(i,j)]$, $B' = [B'(i,j)]$ (resp. $A'' = [A''(i,j)]$, $B'' = [B''(i,j)]$) are the WCI (resp.* WCII, *if 2 is a primitive element in* GF(p)) *arrays corresponding to A and B. Then*

(1) $T_{A'B'}(0,0) = \gcd(l-1,p-1) - 1$ *and* $T_{A''B''}(0,0) = \gcd(l-1,p-1) - 2$, *if* $\beta = \alpha^l$, $l \geqslant 2$, $\eta = \mu$.

(2) $T_{A'B'}(0,0) = T_{A''B''}(0,0) = 0$, *if* $\beta = \alpha$, *and* $\eta \neq \mu$.

Proof The first assertion is due to Theorem 8.1.6, Lemma 8.1.2, Lemma 8.1.3, and Definition 8.1.3. The second assertion is due to Corollary 8.1.1, Lemma 8.1.2, Lemma 8.1.3, and Definition 8.1.3. □

Definition 8.1.4[1] *Let $q = p^r$ be a prime power, α and β two primitive elements in the field* GF(q).

(1) *The array $A = [A(i,j)]$, $1 \leqslant i,j \leqslant q-2$, with*

$$A(i,j) = \begin{cases} 1, & \text{if } \alpha^i + \beta^j = 1, \\ 0, & \text{otherwise} \end{cases} \tag{8.17}$$

is called a Golomb-Costas array, or for short, GC *array, generated by (α,β).*

(2) *If* $\alpha+\beta = 1$, *deleting the first row* ($i = 1$) *and column* ($j = 1$) *from the above* A, *we get the first variant* GC *array, denoted* GCI, *generated by* (α,β) *which is of size* $(q-3)\times(q-3)$.
(3) *If* $\alpha + \beta = 1$ *and* $q = 2^r$, *deleting the first two rows* ($i = 1,2$) *and columns* ($j = 1, 2$) *from the above* A, *we get the second variant* GC *array, denoted* GCII, *generated by* (α,β) *which is of size* $(q-4)\times(q-4)$.
(4) *If* $\alpha+\beta = 1$ *and* $\alpha^2+\beta^{-1} = 1$ *in* GF(q), *deleting the first two rows* ($i = 1,2$) *and the first and the* $(q-2)$*-th columns* ($j = 1, q-2$) *from the above* A, *we get the third variant* GC *array, denoted* GCIII, *generated by* (α,β) *which is of size* $(q-4)\times(q-4)$.
(5) *If* $\alpha+\beta = 1$ *and* $\alpha^2+\beta^{-1} = 1$ (*thus* $\alpha^{-1}+\beta^2 = 1$), *deleting the first, the second and the* $(q-2)$*-th rows and columns* ($i = 1,2,q-2$, $j = 1,2,q-2$) *from the above* A, *we get the fourth variant* GC *array, denoted* GCIV, *generated by* (α,β) *which is of size* $(q-5)\times(q-5)$.

When $\alpha = \beta$, *a* GC *array is also called a Lemple-Costas array*[1].

Example 8.1.4 Let $q = p = 11$, $C = [C(i,j)]$ and $D = [D(i,j)]$ be the Golomb-Costas arrays defined by $(\alpha = 8, \beta = 6)$ and $(\sigma = 2, \rho = 8)$, respectively, thus

$$C = \begin{bmatrix} 0&0&0&0&0&0&0&1&0\\ 0&1&0&0&0&0&0&0&0\\ 1&0&0&0&0&0&0&0&0\\ 0&0&0&0&0&0&1&0&0\\ 0&0&0&0&0&0&0&0&1\\ 0&0&0&1&0&0&0&0&0\\ 0&0&0&0&1&0&0&0&0\\ 0&0&1&0&0&0&0&0&0\\ 0&0&0&0&0&1&0&0&0 \end{bmatrix} \tag{8.18}$$

and

$$D = \begin{bmatrix} 0&0&0&0&1&0&0&0&0\\ 1&0&0&0&0&0&0&0&0\\ 0&0&0&1&0&0&0&0&0\\ 0&0&0&0&0&0&0&0&1\\ 0&0&0&0&0&0&1&0&0\\ 0&0&0&0&0&1&0&0&0\\ 0&0&0&0&0&0&0&1&0\\ 0&1&0&0&0&0&0&0&0\\ 0&0&1&0&0&0&0&0&0 \end{bmatrix}, \tag{8.19}$$

The periodic cross-correlation between C and D is

$$R_{CD} = \begin{bmatrix} 0&0&1&2&1&2&0&1&2\\ 0&0&0&1&2&3&2&1&0\\ 0&1&0&0&2&0&1&1&\mathbf{4}\\ 1&1&3&2&0&0&0&2&0\\ 1&1&0&1&2&2&0&2&0\\ 1&2&0&2&0&0&1&2&1\\ 2&3&1&1&1&0&1&0&0\\ 2&0&2&0&1&0&2&0&2\\ 2&1&2&0&0&2&2&0&0 \end{bmatrix}. \tag{8.20}$$

From this matrix, we know that $\max(R_{CD}(s,t)) = 4$. Note that $\alpha = 8 = 2^3 = \sigma^3$ and $6 = \beta = 8^3 = \rho^3$, thus the relative power between α and σ (resp. β and ρ) is 3 (resp. 3). Hence the periodic cross-correlation between two Golomb-Costas arrays is larger than the maximum relative powers, in some cases. The following theorem answers the relationships between the maximum correlations and the relative powers.

Let α be a primitive element, and b a non-zero element in GF(q), then we define $\log_\alpha b$ to be an integer k such that $0 \leqslant k \leqslant q-2$ and $\alpha^k = b$ in GF(q), i.e., $\log_\alpha b$ is the discrete logarithm of b to the base α. The notation $\log_\alpha b$ will be frequently used in the proofs of the following theorems and corollaries.

Theorem 8.1.8 *Let $A = [A(i,j)]$ and $B = [B(i,j)]$, $1 \leqslant i,j \leqslant q-2$, be two* GC *arrays generated by (α,β) and (ρ,σ), respectively. If $\beta = \rho^m$ and $\alpha = \sigma^l$, $l \geqslant 2$ or $m \geqslant 2$, then the non-periodic and periodic cross-correlation between A and B are upper bounded by $\max(l,m)$ and $4\max(l,m)$, respectively, i.e., $\max T_{AB}(u,v) \leqslant \max(l,m)$ and $\max R_{AB}(s,t) \leqslant 4\max(l,m)$*

This theorem can be slightly improved. In fact, if $\beta = \rho^m$ and $\alpha = \sigma^l$, then $\rho = \beta^{m'}$ and $\sigma = \alpha^{l'}$, where $mm' \equiv 1 \bmod (q-1)$ and $ll' \equiv 1 \bmod (q-1)$. Thus Theorem 8.1.8 is, in fact, $\max T_{AB}(u,v) \leqslant \min[\max(l,m), \max(l',m')]$ and $\max R_{AB}(s,t) \leqslant 4\min[\max(l,m), \max(l',m')]$.

Proof From Definition 8.1.4 and Definition 8.1.1, for $-q+3 \leqslant u,v \leqslant q-3$,

$$
\begin{aligned}
T_{AB}(u,v) &= \sum_{i=1}^{q-2-u} \sum_{j=1}^{q-2-v} A(i,j)B(i+u,j+v) \\
&= \sum_{i=1}^{q-2-u} A(i,\log_\beta(1-\alpha^i))B(i+u,\log_\beta(1-\alpha^i)+v) \\
&= \sum_{i=1}^{q-2-u} B(i+u,\log_\beta(1-\alpha^i)+v).
\end{aligned}
$$

Thus, $T_{AB}(u,v)$ is equivalent to the number i, $1 \leqslant i \leqslant q-2-u$, such that $B(i+u,\log_\beta(1-\alpha^i)+v) = 1$

$$\Longleftrightarrow \log_\beta(1-\alpha^i)+v = \log_\rho(1-\sigma^{i+u}) \tag{8.21}$$

$$\Longrightarrow \log_\beta(1-\alpha^i)+v \equiv \log_\rho(1-\sigma^{i+u}) \bmod (q-1). \tag{8.22}$$

It is worthwhile to point out that Equation 8.21 is not equivalent to Equation 8.22. Precisely, a solution i of Equation 8.21 is a solution of Equation 8.22, but the reverse is not true, because the plus "+" in Equation 8.21 is operated in the real field, while the plus "+" in Equation 8.22 is operated under mod$(q-1)$. If i_1 and i_2 are two different solutions of Equation 8.21, i.e., $i_1 \not\equiv i_2 \bmod (q-1)$, they are clearly different in Equation 8.22 as well. Thus $T_{AB}(u,v)$ is upper bounded by the number of solutions i in Equation 8.22. Because of $x^{q-1} = 1$, for every non-zero element x in GF(q), thus Equation 8.22 is equivalent to

$$(1-\alpha^i)\beta^v = \beta^{\log_\rho(1-\sigma^{i+u})} \text{ in GF(q)}. \tag{8.23}$$

Because of $\beta = \rho^m$ and $[m \log_\rho(1-\sigma^{i+u})] \bmod (q-1) = \log_\rho(1-\sigma^{i+u})^m$, thus Equation 8.23 is equivalent to

$$(1-\alpha^i)\beta^v = (1-\sigma^{i+u})^m \text{ in GF(q)}. \tag{8.24}$$

Setting $\alpha = \sigma^l$ into Equation 8.24, we have

$$(1-\sigma^{il})\beta^v - (1-\sigma^i\sigma^u)^m = 0 \text{ in GF(q)}. \tag{8.25}$$

The left hand of Equation 8.25 is clearly a polynomial, in variable σ^i, of degree less than or equal to $\max(m,l)$, thus it has at most $\max(m,l)$ solutions σ^i in GF(q). Consequently $T_{AB}(u,v) \leqslant \max(m,l)$. □

Theorem 8.1.9 *Let $A = [A(i,j)]$, $B = [B(i,j)]$ be the* GC *arrays in Theorem* 8.1.8. *If $\beta\rho = 1$ and $\alpha\sigma = 1$ in* GF(q), *then the non-periodic and periodic cross-correlations between A and B are upper bounded by* 2 *and* 8, *respectively.*

Proof Using the same argument as that in the proof of Theorem 8.1.8, $T_{AB}(u,v)$ is upper bounded by the number of solutions i of the following equation in GF(q): $(1-\sigma^{-i})\beta^v = (1-\sigma^{i+u})^{-1}$, which is, in fact, Equation 8.24 for $\alpha = \sigma^{-1}$ and $m = -1 \Longleftrightarrow (1-\sigma^u\sigma^i)(1-\sigma^{-i})\beta^v - 1 = 0$ in GF$(q) \Longleftrightarrow (1-\sigma^u\sigma^i)(\sigma^i - 1)\beta^v - \sigma^i = 0$ in GF(q), which has, clearly, at most two solutions σ^i in GF(q). □

From the same motivation as that of the Welch-Costas arrays, we present here a few results about the exact values of the in-phased correlation between two Golomb-Costas arrays.

Theorem 8.1.10 *Let $A = [A(i,j)]$, $B = [B(i,j)]$ be the* GC *arrays in Theorem* 8.1.8. *If $\beta = \rho^m$, $m \geqslant 2$, and $\alpha = \sigma$, then $T_{AB}(0,0) = \gcd(m-1, q-1) - 1$.*

Proof Using the same argument as that in the proof of Theorem 8.1.8, $T_{AB}(0,0)$ is equivalent to the number i, $1 \leqslant i \leqslant q-2$ such that $\log_\beta(1-\alpha^i) = \log_\rho(1-\alpha^i)$, which is, in fact, Equation 8.22 for $u = 0$ and $v = 0 \Longleftrightarrow (1-\alpha^i) = \beta^{\log_\rho(1-\alpha^i)} = (1-\alpha^i)^m$ in GF(q)

$$\Longleftrightarrow (1-\alpha^i)^{m-1} = 1, \quad \text{in GF}(q) \tag{8.26}$$

Because i is a solution of Equation 8.26 if and only if $1-\alpha^i = \alpha^k (1 \leqslant k \leqslant q-2)$ and $\alpha^{k(m-1)} = 1$, i.e.,

$$k(m-1) \equiv 0 \bmod (q-1), \quad 1 \leqslant k \leqslant q-2. \tag{8.27}$$

From the proof of Theorem 8.1.6, it is known that Equation 8.27 has exactly $\gcd(m-1, q-1)$ solutions, in which $\gcd(m-1, q-1) - 1$ of them are non-zero. □

A direct corollary of Theorem 8.1.10 is that if $A = [A(i,j)]$ and $B = [B(i,j)]$ are two GC arrays in Theorem 8.1.10, then $\max T_{AB}(u,v) \geqslant \gcd(m-1, q-1) - 1$.

Theorem 8.1.11 *Let $A = [A(i,j)]$, $B = [B(i,j)]$ be two* GC *arrays generated by (α,β) and (ρ,σ), respectively. Let $\beta = \rho^m$ and $\alpha = \sigma^l$. And let $A_k = [A_k(i,j)]$, $B_k = [B_k(i,j)]$ be the k-th, $1 \leqslant k \leqslant 4$, variant* GC *arrays corresponding to A and B.*

(1) *If $l \geqslant 2$ or $m \geqslant 2$, then the non-periodic and periodic cross-correlations between A_k and B_k, $1 \leqslant k \leqslant 4$, are upper bounded by* $\max(m,l)$ *and* $4\max(m,l)$, *respectively.*

(2) *If* $\beta\rho = 1$ *and* $\alpha\sigma = 1$, *then the non-periodic and periodic cross-correlations between* A_k *and* B_k, $1 \leqslant k \leqslant 4$, *are upper bounded by* 2 *and* 8, *respectively.*

(3) *If* $m \geqslant 2$ *and* $l = 1$, *then* $T_{A_1B_1}(0,0) = \gcd(m-1, q-1) - 2$, $T_{A_2B_2}(0,0) = \gcd(m-1, q-1) - 3 = T_{A_3B_3}(0,0)$, *and* $T_{A_4B_4}(0,0) = \gcd(m-1, q-1) - 4$.

Proof The first assertion is due to Theorem 8.1.8, Lemma 8.1.3, Definition 8.1.4, and Lemma 8.1.2.

The second assertion is due to Theorem 8.1.9, Lemma 8.1.3, Definition 8.1.4, and Lemma 8.1.2.

The third assertion is due to Theorem 8.1.10, Lemma 8.1.3, Definition 8.1.4, and Lemma 8.1.2. □

8.1.2 Algebraically Constructed Costas Arrays

We know that it is an important problem to find two or more Costas arrays such that their non-periodic cross-correlation is ideal ($\leqslant 1$) when the allowable Doppler shift is constrained. Mathematically, this problem is equivalent to finding a family of Costas arrays of size $N \times N$ such that for each pair members A and B:

$$T_{AB}(u,v) \leqslant 1, \quad |u| \leqslant N-1, \quad |v| \leqslant m \leqslant N-1 \tag{8.28}$$

or equivalently,

$$T_{AB}(u,v) \leqslant 1, \quad |u| \leqslant n \leqslant N-1, \quad |v| \leqslant N-1 \tag{8.29}$$

We now will present eight new families of such good Costas arrays.

Note that every $M \times M$ permutation matrix can be represented by an integer sequence $a_0, a_1, \cdots, a_{M-1}$, $0 \leqslant a_i \leqslant M-1$, such that the entry of the j-th column and a_j-th row of the matrix is "1". If two $M \times M$ arrays A and B are represented by $a_0, a_1, \cdots, a_{M-1}$ and $b_0, b_1, \cdots, b_{M-1}$, respectively, then the condition

$$T_{AB}(u,v) \geqslant 2 \tag{8.30}$$

implies that for some integers k and h, $0 \leqslant k \leqslant M-1$, $1 \leqslant h \leqslant M-1$,

$$a_k - a_{k+h} = b_{k-u} - b_{k+h-u}. \tag{8.31}$$

In fact, Equation (8.31) means that the "1"s of array A at the k-th and $(k+h)$-th columns overlap with the "1" s of array B on the $(k-u)$-th and $(k+h-u)$-th columns, respectively, when B is horizontally shifted by u and vertically shifted by some v.

The following algorithm will be frequently used in the sequel.

Algorithm A

Initial. Let A_0 be an $M \times M$ matrix.

Step 1. Extend the above matrix A_0 horizontally to the right by periodic extension with period M.

Step 2. From this periodically extended matrix, take an $M \times M$ subarray starting at the N-th column and call it A_1. Similarly, A_j is defined to be the $M \times M$ subarray starting at the Nj-th column, for $0 \leqslant j \leqslant Q-1$.

Resultant Arrays The matrice $A_0, A_1, \cdots, A_{Q-1}$ are the arrays we wanted.

The first family of Costas arrays with small cross-coincidences are heavily dependent on the so-called discrete logarithm $\log_\alpha(b)$, which is a function over GF(q), q a prime power. Let α be a primitive element, b non-zero element of GF(q). Then $\log_\alpha(b) = k$ if and only if $0 \leqslant k \leqslant q-2$ and $\alpha^k = b$ in GF(q).

Let $M = q-3$, α and β two primitive elements in GF(q) and $\alpha + \beta = 1$, and $q = p^r$ a prime power.

Then by the definition the integers $\log_\alpha(1-\beta^2)$, $\log_\alpha(1-\beta^3)$, $\cdots$, $\log_\alpha(1-\beta^{q-2})$ are all distinct and exhaust all elements in the set $\{2, 3, \cdots, q-2\}$. Define positive integers $N(>1)$ and Q satisfying

$$NQ \leqslant M = q-3. \tag{8.32}$$

Let $d_j = Nj+2$, $0 \leqslant j \leqslant Q-1$. Then for each j, $0 \leqslant j \leqslant Q-1$, the following sequence

$$A_j \sim [\log_\alpha(1-\beta^{d_j}) - 2, \log_\alpha(1-\beta^{d_j+1}) - 2, \cdots, \log_\alpha(1-\beta^{d_j+M-1}) - 2] \tag{8.33}$$

represents a permutation array, denoted by A_j. Moreover, we have the following theorem.

Theorem 8.1.12 *The set of arrays $\{A_0, A_1, \cdots, A_{Q-1}\}$ defined by Equation (8.33) is a family of Costas arrays satisfying*

$$T_{A_iA_j}(u,v) \leqslant 1, \quad |u| \leqslant N-1, \quad |v| \leqslant M-1 \tag{8.34}$$

for all $0 \leqslant i, j \leqslant Q-1$, where N and Q are determined by Equation (8.32).

Proof For each fixed j, $0 \leqslant j \leqslant Q-1$, the array A_j is clearly a permutation matrix. Thus it is necessary to prove that (1) A_j is a Costas array, and (2) Equation (8.34) is satisfied by each pair A_i and A_j, $0 \leqslant i \neq j \leqslant Q-1$.

In the following we use "$X \Longleftrightarrow Y$" and "$X \Longrightarrow Y$" to stand for the statements "X if and only if Y" and "if X, then Y", respectively.

(1) Let $A_j = [A_j(n,m)]$, $0 \leqslant n, m \leqslant M-1$, by the definition, $A_j(n,m) = 1$ iff $n = \log_\alpha(1-\beta^{d_j+m}) - 2$. Thus

$$\begin{aligned}
T_{A_j}(u,v) &= \sum_{m,n=0}^{M-1} A_j(n,m)A_j(n+u, m+v) \\
&= \sum_{m=0}^{M-1} A_j(\log_\alpha(1-\beta^{d_j+m}) - 2, m)_j(\log_\alpha(1-\beta^{d_j+m}) - 2 + u, m+v) \\
&= \sum_{m=0}^{M-1} A_j(\log_\alpha(1-\beta^{d_j+m}) - 2 + u, m+v).
\end{aligned} \tag{8.35}$$

Because of that

$$\begin{aligned}
&A_j(\log_\alpha(1-\beta^{d_j+m}) - 2 + u, m+v) = 1 \\
\Longleftrightarrow\ &\log_\alpha(1-\beta^{d_j+m}) - 2 + u = \log_\alpha(1-\beta^{d_j+m+v}) - 2 \\
\Longrightarrow\ &\alpha^u(1-\beta^{d_j+m}) = 1 - \beta^{d_j+m+v} \\
\Longleftrightarrow\ &\beta^m(\alpha^u\beta^{d_j} - \beta^{d_j+v}) = \alpha^u - 1,
\end{aligned}$$

which has exactly one solution m if $\alpha^u \neq \beta^v$, or has no solution m, if $\alpha^u = \beta^v$. Hence $T_{A_j}(u,v) \leqslant 1$, i.e., A_j is indeed a Costas array.

(2) In order to prove Equation (8.34), for $0 \leqslant i \neq j \leqslant Q-1$, we assume on the contrary that $T_{A_iA_j}(u,v) \geqslant 2$ for some $|u| \leqslant N-1$, $|v| \leqslant M-1$. Then from Equation (8.30) and Equation (8.31), there exist integers k and h, $0 \leqslant k \leqslant M-1$, $1 \leqslant h \leqslant M-1$, such that the following equality holds

$$\begin{aligned}
&\log_\alpha(1-\beta^{d_j+k}) - \log_\alpha(1-\beta^{d_j+k+h}) = \log_\alpha(1-\beta^{d_i+k-u}) - \log_\alpha(1-\beta^{d_i+k+h-u})\\
&\Longrightarrow (1-\beta^{d_j+k})(1-\beta^{d_i+k+h-u}) = (1-\beta^{d_i+k-u})(1-\beta^{d_j+k+h})\\
&\Longleftrightarrow \beta^{d_i+k-u}(\beta^h-1) = \beta^{d_j+k}(\beta^h-1)\\
&\Longleftrightarrow \beta^{d_i+k-u} = \beta^{d_j+k}, \qquad \text{for } h \geqslant 1\\
&\Longleftrightarrow d_i - u \equiv d_j(\bmod(M+2)), \qquad \text{for } \beta^{M+2} = \beta^{q-1} = 1,
\end{aligned}$$

which is impossible for $i \neq j$ and $|u| \leqslant N-1$. In fact, $d_i - d_j = N(i-j)$. Using the condition $i \neq j$ and $0 \leqslant i,j \leqslant Q-1$, one obtains $N \leqslant |d_i - d_j| \leqslant N(Q-1) \leqslant M-N$. If $|u| \leqslant N-1$, it follows that $1 \leqslant |d_i - d_j - u| \leqslant M$. This implies that $d_i - d_j - u \not\equiv 0 \bmod (M+2)$, i.e., the equation $d_i - u \equiv d_j \bmod (M+2)$ is impossible. □

The second family of Costas arrays with small cross-coincidences can be constructed as follows.

Let $q = 2^r$, $M = q-4$, α and β two primitive elements in the finite field GF(q) such that $\alpha + \beta = 1$.

Then the integers $\log_\alpha(1-\beta^3)$, $\log_\alpha(1-\beta^4)$, $\cdots$, $\log_\alpha(1-\beta^{q-2})$ are all distinct and exhaust all elements in the set $\{3, 4, \cdots, q-2\}$. Define positive integers $N(>1)$ and Q satisfying

$$NQ \leqslant M = q - 4 = 2^r - 4. \tag{8.36}$$

Let $d_j = Nj + 3$, $0 \leqslant j \leqslant Q-1$, then for each j, $0 \leqslant j \leqslant Q-1$, the following sequence

$$A_j \sim [\log_\alpha(1-\beta^{d_j}) - 3, \log_\alpha(1-\beta^{d_j+1}) - 3, \cdots, \log_\alpha(1-\beta^{d_j+M-1}) - 3] \tag{8.37}$$

represents a permutation array, denoted by A_j. Moreover, we have the following theorem.

Theorem 8.1.13 *The arrays set $\{A_0, A_1, \cdots, A_{Q-1}\}$ defined by Equation (8.37) is a family of Costas arrays satisfying*

$$T_{A_iA_j}(u,v) \leqslant 1, \qquad |u| \leqslant N-1, \quad |v| \leqslant M-1 \tag{8.38}$$

for all $0 \leqslant i,j \leqslant Q-1$. Where N and Q are determined by Equation (8.36).

This theorem can be proved by using the same method as that for Theorem 8.1.12.

The third family of Costas arrays with small cross-coincidences can be constructed as follows.

Let $M = q-4$, α and β two primitive elements in the finite field GF(q) such that $\alpha + \beta = 1$ and $\alpha^2 + \beta^{-1} = 1$, q a prime power.

Then the integers $\log_\alpha(1-\beta^2)$, $\log_\alpha(1-\beta^3)$, $\cdots$, $\log_\alpha(1-\beta^{q-3})$ are all distinct and exhaust all elements in the set $\{3, 4, \cdots, q-2\}$. Define positive integers $N(>1)$ and Q satisfying

$$NQ \leqslant M = q - 4. \tag{8.39}$$

Let $d_j = Nj + 2$, $0 \leqslant j \leqslant Q-1$, then for each j, $0 \leqslant j \leqslant Q-1$, the following sequence

$$A_j \sim [\log_\alpha(1-\beta^{d_j}) - 3, \log_\alpha(1-\beta^{d_j+1}) - 3, \cdots, \log_\alpha(1-\beta^{d_j+M-1}) - 3] \tag{8.40}$$

represents a permutation array, denoted by A_j. Moreover, we have the following theorem.

Theorem 8.1.14 *The arrays set $\{A_0, A_1, \cdots, A_{Q-1}\}$ defined by Equation (8.40) is a family of Costas arrays satisfying*

$$T_{A_iA_j}(u,v) \leqslant 1, \quad |u| \leqslant N-1, \quad |v| \leqslant M-1 \tag{8.41}$$

for all $0 \leqslant i,j \leqslant Q-1$. Where N and Q are determined by Equation (8.39).

This theorem can be proved in the same way as Theorem 8.1.12.

The fourth family of Costas arrays with small cross-coincidences can be constructed as follows.

Let $M = q-5$, α and β two primitive elements in the finite field GF(q) such that $\alpha+\beta = 1$ and $\alpha^2+\beta^{-1} = 1$ and $\alpha^{-1}+\beta^2 = 1$, q a prime power.

Then the integers $\log_\alpha(1-\beta^3)$, $\log_\alpha(1-\beta^4)$, $\cdots$, $\log_\alpha(1-\beta^{q-3})$ are all distinct and exhaust all elements in the set $\{3, 4, \cdots, q-3\}$.

Define positive integers $N(>1)$ and Q satisfying

$$NQ \leqslant M = q-5. \tag{8.42}$$

Let $d_j = Nj+3$, $0 \leqslant j \leqslant Q-1$, then for each j, $0 \leqslant j \leqslant Q-1$, the following sequence

$$A_j \sim [\log_\alpha(1-\beta^{d_j})-3, \log_\alpha(1-\beta^{d_j+1})-3, \cdots, \log_\alpha(1-\beta^{d_j+M-1})-3] \tag{8.43}$$

represents a permutation array, denoted by A_j. Moreover, we have the following theorem.

Theorem 8.1.15 *The arrays set $\{A_0, A_1, \cdots, A_{Q-1}\}$ defined by Equation (8.43) is a family of Costas arrays satisfying*

$$T_{A_iA_j}(u,v) \leqslant 1, \quad |u| \leqslant N-1, \quad |v| \leqslant M-1 \tag{8.44}$$

for all $0 \leqslant i,j \leqslant Q-1$, where N and Q are determined by Equation (8.42).

The fifth family of Costas arrays with small cross-coincidences can be constructed as follows.

Let $M = q-4$, α a primitive element in the finite field GF(q) such that $\alpha^2+\alpha = 1$, q a prime power.

Then the integers $\log_\alpha(1-\beta^3)$, $\log_\alpha(1-\beta^4)$, $\cdots$, $\log_\alpha(1-\beta^{q-2})$ are all distinct and exhaust all elements in the set $\{3, 4, \cdots, q-2\}$.

Define positive integers $N(>1)$ and Q satisfying

$$NQ \leqslant M = q-4. \tag{8.45}$$

Let $d_j = Nj+3$, $0 \leqslant j \leqslant Q-1$, then for each j, $0 \leqslant j \leqslant Q-1$, the following sequence

$$A_j \sim [\log_\alpha(1-\alpha^{d_j})-3, \log_\alpha(1-\alpha^{d_j+1})-3, \cdots, \log_\alpha(1-\alpha^{d_j+M-1})-3] \tag{8.46}$$

represents a permutation array, denoted by A_j. Moreover, we have the following theorem.

Theorem 8.1.16 *The arrays set $\{A_0, A_1, \cdots, A_{Q-1}\}$ defined by Equation (8.46) is a family of Costas arrays satisfying*

$$T_{A_iA_j}(u,v) \leqslant 1, \quad |u| \leqslant N-1, \quad |v| \leqslant M-1 \tag{8.47}$$

for all $0 \leqslant i, j \leqslant Q-1$, *where* N *and* Q *are determined by Equation* (8.45).

The sixth family of Costas arrays with small cross-coincidences can be constructed as follows.

Let p be such an odd prime that 2 is a primitive element in GF(p), and $M = p - 3$.

If $\alpha = 2^{-1}$, then $\alpha^1 + \alpha^1 = 1$ thus the integers $\log_\alpha(1-\alpha^2)$, $\log_\alpha(1-\alpha^3)$, $\cdots$, $\log_\alpha(1-\alpha^{q-2})$ are all distinct and exhaust all elements in the set $\{2, 3, \cdots, q-2\}$.

Define positive integers $N(>1)$ and Q satisfying

$$NQ \leqslant M = p - 3. \tag{8.48}$$

Let $d_j = Nj + 2$, $0 \leqslant j \leqslant Q-1$, then for each j, $0 \leqslant j \leqslant Q-1$, the following sequence

$$A_j \sim [\log_\alpha(1-\alpha^{d_j}) - 2, \log_\alpha(1-\alpha^{d_j+1}) - 2, \cdots, \log_\alpha(1-\alpha^{d_j+M-1}) - 2] \tag{8.49}$$

represents a permutation array, denoted by A_j. Moreover, we have the following theorem.

Theorem 8.1.17 *The arrays set* $\{A_0, A_1, \cdots, A_{Q-1}\}$ *defined by Equation* (8.49) *is a family of Costas arrays satisfying*

$$T_{A_iA_j}(u, v) \leqslant 1, \quad |u| \leqslant N-1, \quad |v| \leqslant M-1 \tag{8.50}$$

for all $0 \leqslant i, j \leqslant Q-1$, *where* N *and* Q *are determined by Equation* (8.48).

The seventh family of Costas arrays with small cross-coincidences can be constructed as follows.

Let p be an odd prime number, $M = p - 2$, α a primitive element in GF(p). Because of that the integers α, α^2 , $\cdots$, α^{p-2} are all distinct and exhaust all elements in the set $\{2, 3, \cdots, p-1\}$.

Define positive integers $N(>1)$ and Q satisfying

$$NQ \leqslant M = p - 2. \tag{8.51}$$

Let $d_j = Nj + 1$, $0 \leqslant j \leqslant Q-1$, then for each j, $0 \leqslant j \leqslant Q-1$, the following sequence

$$A_j \sim [\alpha^{d_j} - 2, \alpha^{d_j+1} - 2, \cdots, \alpha^{d_j+M-1} - 2] \tag{8.52}$$

represents a permutation array, denoted by A_j. Moreover, we have the following theorem.

Theorem 8.1.18 *The arrays set* $\{A_0, A_1, \cdots, A_{Q-1}\}$ *defined by Equation* (8.52) *is a family of Costas arrays satisfying*

$$T_{A_iA_j}(u, v) \leqslant 1, \quad |u| \leqslant N-1, \quad |v| \leqslant M-1 \tag{8.53}$$

for all $0 \leqslant i, j \leqslant Q-1$, *where* N *and* Q *are determined by Equation* (8.51).

Proof For each fixed j, $0 \leqslant j \leqslant Q-1$, the array A_j is clearly a permutation matrix. Thus it is necessary to prove that (1) A_j is a Costas array, and (2) Equation (8.53) is satisfied by each pair A_i and A_j, $0 \leqslant i \neq j \leqslant Q-1$.

(1) Let $A_j = [A_j(n,m)]$, $0 \leqslant n, m \leqslant M-1$, by the definition, $A_j(n,m) = 1$ iff $n = \alpha^{m+d_j}(\text{mod}p) - 2$, $0 \leqslant i, j \leqslant p-3$. Thus

$$\begin{aligned} T_{A_j}(u,v) &= \sum_{m,n=0}^{M-1} A_j(n,m)A_j(n+u, m+v) \\ &= \sum_{m=0}^{M-1} A_j(\alpha^{m+d_j} - 2, m)A_j(\alpha^{m+d_j} - 2 + u, m+v) \\ &= \sum_{m=0}^{M-1} A_j(\alpha^{m+d_j} - 2 + u, m+v). \end{aligned} \tag{8.54}$$

Because of that

$$\begin{aligned} & A_j(\alpha^{m+d_j} - 2 + u, m+v) = 1 \\ \Longleftrightarrow & \alpha^{m+d_j} - 2 + u = \alpha^{m+d_j+v} - 2 \\ \Longrightarrow & \alpha^{m+d_j}(\alpha^v - 1) = u \end{aligned}$$

which has exactly one solution m if $v \neq 0$, or has no solution m, if $v = 0$ (thus $u \neq 0$). Hence $T_{A_j}(u,v) \leqslant 1$, i.e., A_j is indeed a Costas array.

(2) In order to prove Equation (8.34), for $0 \leqslant i \neq j \leqslant Q-1$, we assume on the contrary that $T_{A_iA_j}(u,v) \geqslant 2$ for some $|u| \leqslant N-1$, $|v| \leqslant M-1$. Then from Equation (8.30) and Equation (8.31), there exist integers k and h, $0 \leqslant k \leqslant M-1$, $1 \leqslant h \leqslant M-1$, such that the following equality holds

$$\begin{aligned} & \alpha^{d_j+k} - \alpha^{d_j+k+h} \equiv \alpha^{d_i+k-u} - \alpha^{d_i+k+h-u}(\text{mod}p) \\ \Longrightarrow & \alpha^{d_i+k}(1-\alpha^h) \equiv \alpha^{d_j+k-u}(1-\alpha^h)(\text{mod}p) \\ \Longleftrightarrow & \alpha^{d_i} \equiv \alpha^{d_j-u}(\text{mod}p) \\ \Longleftrightarrow & d_i \equiv d_j - u(\text{mod}(p-1)) \end{aligned}$$

which is impossible for $i \neq j$ and $|u| \leqslant N-1$. In fact, $d_i - d_j = N(i-j)$. Using the condition $i \neq j$ and $0 \leqslant i, j \leqslant Q-1$, one obtains $N \leqslant |d_i - d_j| \leqslant N(Q-1) \leqslant M-N$. If $|u| \leqslant N-1$, it follows that $1 \leqslant |d_i - d_j - u| \leqslant M$. This implies that $d_i - d_j - u \not\equiv 0\text{mod}(M+1)$, i.e., the equation $d_i - u \equiv d_j\text{mod}(M+1)$ is impossible.

□

The eighth family of Costas arrays with small cross-coincidences can be constructed as follows.

Let p be an odd prime number such that $\alpha = 2$ is a primitive element in GF(p), and let $M = p-3$. Because of that the integers α^2, α^3, $\cdots$, α^{p-2} are all distinct and exhaust all elements in the set $\{3, 4, \cdots, p-1\}$.

Define two positive integers $N(>1)$ and Q satisfying

$$NQ \leqslant M = p-3. \tag{8.55}$$

Let $d_j = Nj + 2$, $0 \leqslant j \leqslant Q-1$, then for each j, $0 \leqslant j \leqslant Q-1$, the following sequence

$$A_j \sim [\alpha^{d_j} - 3, \alpha^{d_j+1} - 3, \cdots, \alpha^{d_j+M-1} - 3] \tag{8.56}$$

represents a permutation array, denoted by A_j. Moreover, we have the following theorem.

Theorem 8.1.19 *The set of arrays $\{A_0, A_1, \cdots, A_{Q-1}\}$ defined by Equation (8.56) is a family of Costas arrays satisfying*

$$T_{A_iA_j}(u, v) \leqslant 1, \quad |u| \leqslant N-1, \quad |v| \leqslant M-1 \tag{8.57}$$

for all $0 \leqslant i, j \leqslant Q-1$, where N and Q are determined by Equation (8.55).

This theorem can be proved in the same way as Theorem 8.1.18.

The main purpose of this subsection is to study the construction of waveforms for radar and sonar applications that have ideal range-Doppler auto- and cross-ambiguity properties. For frequency-hopping signals, this problem leads one to search for multiple arrays (hopping patterns) with a small number of self and mutal hits. The main results, Theorems 8.1.11-8.1.19, provide eight families of Costas arrays having ideal cross-correlation for constraint Doppler shift. It is clear that there is a trade off between the number Q of Costas arrays contained in the new families and the Doppler constraint N.

8.1.3 Enumeration Limitation of Costas Arrays[2]

Recall that a permutation matrix $A = [A_{ij}](1 \leqslant i, j \leqslant n)$ of order $n \times n$ is said to be a Costas array iff for any non-zero pair integers $(r, s) \neq (0, 0), |r| \leqslant n, |s| \leqslant n$, the non-cyclic correction function of the matrix A satisfies

$$c(r, s) = \sum_{i,j=1}^{n} A_{ij}A_{(i+r)(j+s)} \leqslant 1. \tag{8.58}$$

Remark 8.1.1 (1) *A is a permutation matrix, therefore $A_{ij} = 0$ or 1.* (2) *$A_{(i+r)(j+s)}$ is defined to be zero when $i + r$ or $j + s$ is out of the range of $[1, n]$.* (3) *The definition of Costas arrays can also be further simplified.*

Let the diagonal matrices

$$B_1 = \begin{bmatrix} 1 & 0 \\ 0 & 1 \end{bmatrix} = \text{diag } (1, 1), \quad B_2 = \begin{bmatrix} 1 & 0 & 0 \\ 0 & 0 & 0 \\ 0 & 0 & 1 \end{bmatrix} = \text{diag } (1, 0, 1),$$

in general let

$$B_i = \text{diag } (1, 0, \cdots, 0, 1), \quad i = 1, 2, \cdots.$$

Lemma 8.1.4 *For any fixed positive integer i, if the permutation matrix A contains at least two sub-matrices of the form B_i then A is not a Costas array.*

Illustration *"A contains as least two sub-matrices of the form B_i" means that A contains at least two sub-matrices $A_1 = [a_{uv}]$ and $A_2 = [b_{uv}](1 \leqslant u, v \leqslant i+1)$ of order $i+1$ by $i+1$ such that*

$$a_{11} = a_{(i+1)(i+1)} = b_{11} = b_{(i+1)(i+1)} = 1$$

and $a_{(i+1)(i+1)}$ and b_{11} respectively stay at different places in the permutation matrix A.

This lemma can be directly verified by the definition of Costas arrays.

Let $C(n)$ is the number of Costas arrays of order n by n. In their fifth conjecture Golomb and Taylor conjectured that $\lim_{n \to \infty} C(n)/n! = 0$. This means that if the matrix

A is arbitrarily taken out from the set of permutation matrices with order n by n, then A is a Costas array with an arbitrarily small probability provided that n is large enough. Now in this subsection, let us show a proof for this conjecture.

After arbitrarily fixing the positive integer k, all the following inferences are based on the basic assumption of $n > k$.

Let P_n be the set of all the permutation matrices of order n by n; C_n be the set of all the Costas arrays of order n by n; B_{1n} be the set of all n by n permutation matrices containing at least two sub-matrices of the form B_1; B_{2n} be the set of all n by n permutation matrices containing at least two sub-matrices of the form B_2; $\cdots$, in general let $B_{mn}(1 \leqslant m \leqslant k)$ be the set of all n by n permutation matrices containing at least two sub-matrices of the form $B_m = \text{diag}(1, 0, \cdots, 0, 1)$ (where the meaning of "Containing at least two sub-matrices of the form B_i" has been explained under the Lemma 8.1.4). Again assume that $N(X)$ is the number of elements contained in the set X (i.e., the volume of set X). Then it is clear that $N(P_n) = n!$ and $N(C_n) = C(n)$.

In addition, from the Lemma 8.1.4 we know that

$$C_n \subseteq P_n - [B_{1n} \cup B_{2n} \cup \cdots \cup B_{kn}].$$

This formula is in fact the equivalent restatement of Lemma 8.1.4. It is a very important formula to the present subsection although it is very simple. In this formula we have

$$\begin{aligned} C(n) &= N(C_n) \leqslant N(P_n - (B_{1n} \cup B_{2n} \cup \cdots \cup B_{kn})) \\ &= n! - N(B_{1n} \cup B_{2n} \cup \cdots \cup B_{kn}), \end{aligned}$$

therefore

$$C(n)/n! \leqslant 1 - N(B_{1n} \cup B_{2n} \cup \cdots \cup B_{kn})/n! \tag{8.59}$$

by the principle of inclusion and exclusion we know

$$\begin{aligned} N(B_{1n} \cup \cdots \cup B_{kn}) = &\sum_{i=1}^{k} N(B_{in}) - \sum_{1 \leqslant i<j \leqslant k} N(B_{in} \cap B_{jn}) + \cdots \\ &+(-1)^{r-1} \sum_{1 \leqslant i_1<i_2<\cdots<i_r \leqslant k} N(B_{i_1 n} \cap B_{i_2 n} \cap \cdots \cap Bi_r n) \\ &+ \cdots +(-1)^{k-1} N(B_{1n} \cap B_{2n} \cap \cdots \cap B_{kn}). \end{aligned} \tag{8.60}$$

Now let us first estimate the upper and down bounds of $N(B_{1n}), N(B_{2n}), \cdots, N(B_{kn})$.

First we consider $N(B_{1n})$, i.e., the number of n by n permutation matrices containing at least two sub-matrices $D_1 = \begin{bmatrix} 1 & 0 \\ 0 & 1 \end{bmatrix}$ and $D_2 = \begin{bmatrix} d_{11} & d_{12} \\ d_{21} & d_{22} \end{bmatrix} = \begin{bmatrix} 1 & 0 \\ 0 & 1 \end{bmatrix}$.

There are $(n-1)^2$ possible methods to place the $D_1 = \begin{bmatrix} 1 & 0 \\ 0 & 1 \end{bmatrix}$ into a n by n empty matrix $E = [E_{ij}], (1 \leqslant i, j \leqslant n)$. If the order is in our consideration (i.e., when the places of D_1 and D_2 are exchanged we think that a new matrix is formed), then after D_1 has been put into E, we should consider the following three cases to enumerate the possible methods of placing D_2 into the matrix E:

Case 1. If the sub-matrix D_1 is placed at one of the four corners of the matrix E, i.e.,

$$D_1 = \begin{bmatrix} E_{11} & E_{12} \\ E_{21} & E_{22} \end{bmatrix}, \quad \text{or } D_1 = \begin{bmatrix} E_{1(n-1)} & E_{2(n-1)} \\ E_{1n} & E_{2n} \end{bmatrix},$$

$$\text{or } D_1 = \begin{bmatrix} E_{(n-1)1} & E_{n1} \\ E_{(n-1)2} & E_{n2} \end{bmatrix}, \quad \text{or } D_1 = \begin{bmatrix} E_{(n-1)(n-1)} & E_{(n-1)n} \\ E_{n(n-1)} & E_{nn} \end{bmatrix},$$

then there are $(n-1)^2$ possible methods to place the sub-matrix D_2 into E.

In order to prove this enumeration result, let us assume, without losing the generality, that D_1 has been placed at the left upper corner of E. Because of

$$D_2 = \begin{bmatrix} d_{11} & d_{12} \\ d_{21} & d_{22} \end{bmatrix},$$

the place of D_2 is completely determined by the place of its element d_{11}.

If the left upper corner element d_{11} of D_2 is put in the place of E_{11} then D_1 is covered by D_2, this is a prohibitive case (because we have arranged in advance that E contains at least two different sub-matrices of the form B_1). In the same way it can be proved that d_{11} cannot be put in the place of E_{22} too.

If d_{11} is put in the place of $E_{1i}(i > 1)$ then there are at least two "1"s contained in the first row of E. This is also a prohibitive case (because E has been asked to be a permutation matrix). In the same way it can be proved that d_{11} cannot be put in the first row, second row, first column and the second column.

If d_{11} is put in the last row of E, then d_{21} and d_{22} can't be placed into the matrix E, this is a prohibitive case. In the same way it can be proved that d_{11} can't be placed in the last column of E.

Summarizing the above inferences, we know that the d_{11} can be put in any of the places except to the first, second and the n-th rows and columns of E, while the enumeration for these possible places for d_{11} is just $(n-3)^2$. If the D_1 has been placed in the left lower, right upper or right lower corners then this enumeration result can also be proved in the same way.

Case 2. If D_1 has been placed on the bounds but corners of E, i.e.,

$$D_1 = \begin{bmatrix} E_{1i} & E_{1(i+1)} \\ E_{2i} & E_{2(i+1)} \end{bmatrix}, \quad \text{or } D_1 = \begin{bmatrix} E_{i(n-1)} & E_{in} \\ E_{(i+1)(n-1)} & E_{(i+1)n} \end{bmatrix},$$

$$\text{or } D_1 = \begin{bmatrix} E_{(n-1)i} & E_{(n-1)(i+1)} \\ E_{ni} & E_{n(i+1)} \end{bmatrix}, \quad \text{or } D_1 = \begin{bmatrix} E_{i1} & E_{i2} \\ E_{(i+1)1} & E_{(i+1)2} \end{bmatrix}$$

(where $2 \leqslant i \leqslant n-2$) then there are $(n-3)(n-4)$ possible places for D_2 in the matrix E.

Case 3. If D_1 has been put in the other places in E, then there are $(n-4)^2$ possible methods to place the D_2 into E.

The proof processes for Case 2 and Case 3 are the same as that for Case 1.

By now, with the consideration of order, the possible methods to place the D_1 and D_2 into E under the Case 1, Case 2 and Case 3 have been enumerated by $(n-1)^2(n-3)^2$, $(n-1)^2(n-3)(n-4)$ and $(n-1)^2(n-4)$ respectively. While all of these enumerations are bounded from ground and top by $(n-1)^2(n-4)^2$ and $(n-1)^2(n-3)^2$ respectively. In another aspect, because the D_1 and D_2 are the same sub-matrix, there will not be any essential change when the places of D_1 and D_2 are exchanged with each other. Therefore, the enumerations for the substantive different methods to place the D_1 and D_2 into E are bounded by $(1/2)(n-1)^2(n-4)^2$ and $(1/2)(n-1)^2(n-3)^2$ form ground and top respectively. After the D_1 and D_2 have been placed into E, we then put $(n-4)$ "1"s into E such that the resulting E is an n by n permutation matrix. It is clear that there are $(n-4)!$ such possible methods to place these $(n-4)$ "1"s into E. Now we have set up the following inequality:

$$(1/2)(n-1)^2(n-4)^2(n-4)! \leqslant N(B_{1n}) \leqslant (1/2)(n-1)^2(n-3)^2(n-4)!. \tag{8.61}$$

The above Eq. (8.61) can be relaxed by

$$(1/2)(n-k-3)^4(n-4)! \leqslant N(B_{1n}) \leqslant (1/2)n^4(n-4)!. \tag{8.62}$$

With the same analytical processes as those presented in Case 1–3, it can be proved that if the order is under consideration then the enumerations for the possible methods to place the two sub-matrices $H_1 = \text{diag}(1,0,\cdots,0,1) = B_i$ and $H_2 = \text{diag}(1,0,\cdots,0,1) = B_i$ into E are bounded by range from $(n-i)^2(n-i-3)^2$ to $(n-i)^2(n-i-2)^2$. In another aspect, because H_1 and H_2 are identical, therefore we have

$$(1/2)(n-i)^2(n-i-3)^2(n-4)! \leqslant N(B_{in}) \leqslant (1/2)(n-i)^2(n-i-2)^2(n-4)!. \tag{8.63}$$

Eq. (8.63) can also be relaxed by (for any $1 \leqslant i \leqslant k$)

$$(1/2)(n-k-3)^4(n-4)! \leqslant N(B_{in}) \leqslant (1/2)n^4(n-4)!. \tag{8.64}$$

Let $f_1(x) = x^4$ and $g_1(x) = (x-k-3)^4$, then Eq. (8.64) means that for any positive integer $i(1 \leqslant i \leqslant k)$ we can find two monic polynomials $f_1(x)$ and $g_1(x)$ of order 4 such that

$$(1/2)g_1(n)(n-4)! \leqslant N(B_{in}) \leqslant (1/2)f_1(n)(n-4)!, \tag{8.65}$$

where a polynomial is said to be a monic polynomial iff its highest coefficient equals to one. For example $f(x) = x+4$ is a monic polynomial of order one.

Now we turn to estimate the enumeration of $N(B_{1n} \cap B_{2n})$. Firstly we consider the lower bound. Denote

$$D_1 = \begin{bmatrix} a_{11} & a_{12} \\ a_{21} & a_{22} \end{bmatrix} = \begin{bmatrix} 1 & 0 \\ 0 & 1 \end{bmatrix}, \quad D_2 = \begin{bmatrix} b_{11} & b_{12} \\ b_{21} & b_{22} \end{bmatrix} = \begin{bmatrix} 1 & 0 \\ 0 & 1 \end{bmatrix},$$

$$D_3 = \begin{bmatrix} c_{11} & c_{12} & c_{13} \\ c_{21} & c_{22} & c_{23} \\ c_{31} & c_{32} & c_{33} \end{bmatrix} = \begin{bmatrix} 1 & 0 & 0 \\ 0 & 0 & 0 \\ 0 & 0 & 1 \end{bmatrix}, \quad D_4 = \begin{bmatrix} e_{11} & e_{12} & e_{13} \\ e_{21} & e_{22} & e_{23} \\ e_{31} & e_{32} & e_{33} \end{bmatrix} = \begin{bmatrix} 1 & 0 & 0 \\ 0 & 0 & 0 \\ 0 & 0 & 1 \end{bmatrix},$$

From the definition we know that $N(B_{1n} \cap B_{2n})$ is the number of possible methods to place the sub-matrices D_1, D_2, D_3, D_4 and some "1"s into an empty n by n matrix $E = (E_{ij}), (1 \leqslant i, j \leqslant n)$ such that the resulting matrix E is a permutation matrix of order n by n. Because the only restriction on the arrangement is that D_1 is disjointed with D_2 (i.e., a_{11}, a_{22}, a_{11} and b_{22} stay at different places in E) and D_3 is disjointed with D_4, therefore if we are asked to compute the exact value of $N(B_{1n} \cap B_{2n})$, then we should have to consider all of the cases of D_1 jointing with D_3, D_1 jointing with D_4, D_2 jointing with

D_3 or D_2 jointing with D_4 respectively. But at present time we need only the lower bound, therefore, if N denotes the number of possible permutation matrices containing D_1, D_2, D_3 and D_4 at the same time with all of the $D_i (1 \leqslant i \leqslant 4)$ disjointed with each other, then it is clear that

$$N(B_{1n} \cap B_{2n}) \geqslant N \tag{8.66}$$

In order to estimate the lower bound of N, let us put the D_1 into E (with $(n-1)^2$ possible methods) at first. After D_1 has been put into E, the possible methods to place D_2 into E may be different from each other depending upon the places where D_1 stays in E, but by the same inference process as described in the above three cases for the calculation of $N(B_{1n})$ it can be proved that there are at least $(n-4)^2$ possible methods to put the D_2 into E. When D_1 and D_2 have been placed into E, the possible methods to place D_3 into E are also different according to the places of D_1 and D_2 into E. But this enumeration is at least $(n-10)^2$. (This is because the c_{11} can be put in any place in E except in the n-th row, $(n-1)$-th row, n-th column, $(n-1)$-th column, the columns and rows where the b_{22}, b_{11}, a_{22} or a_{11} placed, the third row from the top and the third column from the left side in the places of a_{11}, a_{22}, b_{11} or b_{22}. While at most ten rows and columns have to be removed, hence there are at least $(n-10)^2$ possible methods to place D_3 into E.). After the D_1, D_2 and D_3 have been placed into E, the possible methods to place D_4 into E are also different according to the places of D_1, D_2 and D_3 in E. But using the same method as before, it can be proved that there are at least $(n-14)^2$ possible methods to place D_4 into E. Therefore under the consideration of the order, there are at least $(n-1)^2(n-4)^2(n-10)^2(n-14)^2$ possible methods to place D_1, D_2, D_3 and D_4 into E. Because D_1 and D_2 are identical and so are the D_3 and D_4, hence there will be no substantive change when D_1 is exchanged with D_2 or when D_3 is exchanged with D_4. By now we know that the number N is lower bounded by

$$N \geqslant (1/4)(n-1)^2(n-4)^2(n-10)^2(n-14)^2(n-8)!.$$

From this inequality and Eq. (8.66) it is clear that we can find a monic polynomial $g(x) = (x-1)^2(x-4)^2(x-10)^2(x-14)^2$ of order 8 such that

$$N(B_{1n} \cap B_{2n}) \geqslant (1/4)g(n)(n-8)!. \tag{8.67}$$

Then we begin to estimate the upper bound of $N(B_{1n} \cap B_{2n})$. By now we know that there are at most n^4 methods to place D_1 and D_2 into E. If the D_3 and D_4 have also been placed into E, then there will be the following seven possible cases.

Case 1. If D_1, D_2, D_3 and D_4 are disjointed with each other, then there are at most $n^8/4$ possible methods to place D_1, D_2, D_3 and D_4 into E. After the other $(n-8)$ "1"s are placed into E, it is clear that we can generate at most $(1/4)n^8(n-8)!$ such permutation matrices E.

Case 2. One of the D_3 and D_4 is jointed with one of the D_1 and D_2, but the other one in D_3 and D_4 is disjointed with D_1 and D_2. For example, D_3 is jointed with D_1, but D_4 is disjointed with D_1 and D_2. In this case there are at most 2 methods to place D_3 and n_2 methods to place D_4 into E. Then we have to put $(n-7)$ "1"s into E such that the resulting E is a permutation matrix. Therefore at this time the possible methods to form the permutation matrices are upper bounded by $2n^6(n-7)!$ which is the production of $(n-8)!$ and a polynomial of order 7.

Case 3. One of the D_3 and D_4 is jointed with one of the D_1 and D_2, and the other one in D_3 and D_4 is also jointed with the other one in D_1 and D_2. For example, D_3 is jointed with D_1 and D_4 is jointed with D_2. In this case there are at most 4 methods to place D_3 and D_4. Then we put the other $(n-6)$ "1"s into E such that the resulting E is a permutation matrix. It can be proved that the possible methods to form such permutation matrices E are upper bounded by $4n^4(n-6)!$ which is the production of $(n-8)!$ and a polynomial of order 6.

Case 4. D_3 and D_4 are jointed with one of the D_1 and D_2 at the same time. For example, D_3 and D_4 are jointed with D_1 at the same time. Using the same method as above, it can be proved that the possible methods to form such permutation matrices E are upper bounded by $2n^4(n-6)!$ which is also the production of $(n-8)!$ and a polynomial of order 6.

Case 5. One of the D_3 and D_4 is jointed with D_1 and D_2 at the same time, while the other one in D_3 and D_4 is disjointed with both D_1 and D_2. For example, D_3 is jointed with D_1 and D_2 at same time while D_4 is disjointed with both D_1 and D_2. In this case the place of D_2 is confined by that of D_1, in fact after D_1 has been placed there are only 2 possible places for D_2. The place of D_3 is completely determined by the places of D_1 and D_2. Because the possible places for D_1 and D_4 are upper bounded by n^2 and n^2 respectively, we know that after the other $(n-6)$ "1"s have been placed into E the possible methods to form such permutation matrices are upper bounded by $2n^4(n-6)!$ which is the production of $(n-8)!$ and a polynomial of order 6.

Case 6. One of the D_3 and D_4 is jointed with D_1 and D_2 at the same time, and the other one in D_3 and D_4 is also jointed with one of the D_1 and D_2. For example, D_3 is jointed with D_1 and D_2 at same time and D_4 is jointed with D_2. With the same method as before, it can be proved that there are at most $4n^2(n-5)!$ possible methods to place D_1, D_2, D_3 and D_4 into E. After the other $(n-5)$ "1"s have been placed into E to form a permutation matrix E, the possible methods are upper bounded by $4n^2(n-5)!$ which is the production of $(n-8)!$ and a polynomial of order 5.

Case 7. D_3 is jointed with D_1 and D_2, and D_4 is also jointed with D_1 and D_2 at the same time. In this case there are at most $8n^2$ possible methods to place D_1, D_2, D_3 and D_4 into E. After the other $(n-4)$ "1"s have been placed into E to form a permutation matrix E, the possible methods are upper bounded by $8n^2(n-4)!$ which is the production of $(n-8)!$ and a polynomial of order 6.

Summarizing all of the above seven cases, it is clear that we can find a monic polynomial $f(x)$ of order 8 such that the possible permutation matrices E formed by placing D_1, D_2, D_3 and D_4 and some "1"s into an empty n by n matrix are upper bounded by $(1/4)f(n)(n-8)!$. This means that the number $N(B_{1n}\cap B_{2n})$ (the number of permutation matrices containing at least two sub-matrices of the form $B_1 = \text{diag}(1,1)$ and $B_2 = \text{diag}(1,0,1)$) is upper bounded by

$$N(B_{1n}\cap B_{2n}) \leqslant (1/4)f(n)(n-8)!. \tag{8.68}$$

By combining Eq. (8.67) and Eq. (8.68), it has been proved that we can find two monic polynomials $f(x)$ and $g(x)$ of order 8 such that

$$(1/4)g(n)(n-8)! \leqslant N(B_{1n} \cap B_{2n}) \leqslant (1/4)f(n)(n-8)!.$$

In the same way, it can also be proved that we can find two monic polynomials $f_2(x)$ and $g_2(x)$ of order 8 such that for any $1 \leqslant i, j \leqslant k$,

$$(1/4)g_2(n)(n-8)! \leqslant N(B_{in} \cap B_{jn}) \leqslant (1/4)f_2(n)(n-8)!. \tag{8.69}$$

In general, it can be proved that for any $1 \leqslant i_1, i_2, \cdots, i_r \leqslant k$ we can find two monic polynomials $f_r(x)$ and $g_r(x)$ of order $4r$ such that

$$(1/2)^r g_r(n)(n-4r)! \leqslant N(B_{i_1 n} \cap \cdots \cap B_{i_r n}) \leqslant (1/2)^r f_r(n)(n-4r)!. \tag{8.70}$$

From Eq. (8.70) and Eq. (8.60), we know that

$$\begin{aligned} N(B_{1n} \cup \cdots \cup B_{kn}) \geqslant & \sum_{r=1,\ r \text{be odd}}^{k} (1/2)^r C_k^r g_r(n)(n-4r)! \\ & - \sum_{r=1,\ r \text{be even}}^{k} (1/2)^r C_k^r f_r(n)(n-4r)!. \end{aligned} \tag{8.71}$$

From Eq. (8.71) and Eq. (8.59), we have

$$\begin{aligned} C(n)/n! \leqslant 1 - 1/n! & \left[\sum_{r=1,\ r \text{be odd}}^{k} (1/2)^r C_k^r g_r(n)(n-4r)! \right. \\ & \left. - \sum_{r=1,\ r \text{be even}}^{k} (1/2)^r C_k^r f_r(n)(n-4r)! \right]. \end{aligned} \tag{8.72}$$

Because the integer k has been fixed in advance and $1 \leqslant r \leqslant k$ then

$$\lim_{n\to\infty} (1/n!)g_r(n)(n-4r)! = \lim_{n\to\infty} (1/n!)f_r(n)(n-4r)! = 1. \tag{8.73}$$

There two limiting equations are due to the fact that $f_r(x)$ and $g_r(x)$ are monic polynomials of order $4r$.

Let $n \to \infty$ in both sides of Eq. (8.72), then from Eq. (8.73) we have

$$\begin{aligned} \lim_{n\to\infty} C(n)/n! & \leqslant 1 - \sum_{r=1,\ r \text{be odd}}^{k} (1/2)^r C_k^r + \sum_{r=1,\ r \text{be even}}^{k} (1/2)^r C_k^r \\ & = \sum_{r=0}^{k} C_k^r (-1/2)^r = (1-1/2)^k = (1/2)^k. \end{aligned} \tag{8.74}$$

Because the integer k is arbitrarily fixed, we can let $k \to \infty$ in Eq. (8.74), then from the non-negative property of $C(n)$,

$$\lim_{n\to\infty} C(n)/n! = 0.$$

By now the Golomb-Taylor's fifth conjecture has been proved to be true.

A few construction schemes for the Costas arrays will be generalized in the following text.

Generalization of Welch's scheme Let p be a prime integer, α be one of the primitive elements in GF(p), u be a non-zero element in GF(p), v be an arbitrary element in GF(p). For any $1 \leqslant i \leqslant p-1, 1 \leqslant j \leqslant p-2$, if we let $A_{ij}=1$ iff $ui=\alpha^{j+v}$, otherwise $A_{ij}=0$, then the matrix $A=[A_{ij}]$ is a Costas array of order $(p-1)$ by $(p-1)$.

Obviously when $u=1$ and $v=0$, the above generalization is reduced to the original Welch's construction scheme. Now we show a short proof for this generalization. Because of $i=u-1\alpha^{j+v}$ and $j=\log(ui)-v$, from the equation $ui=\alpha^{j+v}$ it is obvious that the numbers i and j are uniquely determined by each other, i.e., A is indeed a permutation matrix. In addition let

$$c(r,s)=\sum_{i,j=1}^{n} A_{ij}A_{(i+r)(j+s)}.$$

If there is a non-zero pair (r,s) such that $c(r,s)\geqslant 2$, then there will be at least two sets (a,b) and (i,j) such that $A_{ij}=A_{(i+r)(j+s)}=1$ and $A_{ab}=A_{(a+r)(b+s)}=1$. This is equivalent to $ui=\alpha^{j+v}, u(i+r)=\alpha^{j+s+v}, ua=\alpha^{b+v}$ and $u(a+r)=\alpha^{b+s+v}$. If $r\neq 0$ then we have $(i+r)/i=\alpha^{v}$ (by the first two equations) and $(a+r)/a=\alpha^{v}$ (by the last two equations). Therefore we have $(a+r)/a=(i+r)/i$, i.e., $ar=ir$ and $i=a$. Hence $\alpha^{j+v}=ui=ua=\alpha^{b+v}$, and $j=b$. Then (i,j) and (a,b) become the same point, which is a contradiction.

If $s\neq 0$, then we have $ur=\alpha^{j+s+v}-\alpha^{j+v}=(\alpha^{s}-1)\alpha^{j+v}$ (by the first two equations) and $ur=(\alpha^{s}-1)\alpha^{b+v}$ (by the last two equations). Therefore $\alpha^{j+v}=\alpha^{b+v}$ and $j=b$. Again we have $a=i$ and $(i,j),(a,b)$ also become the same point too, which is also a contradiction.

Summarizing the above results we know that $c(r,s)\geqslant 2$ is impossible, hence $c(r,s)\leqslant 1$. That means that $A=[A_{ij}]$ is indeed a Costas array. The proof is finished.

Generalizations for the Lempel's and Golomb's schemes Let α be a primitive element of GF(q) (where $q=p^{k}$, p is a prime integer), a is any non-zero element in GF(q), m and n are two positive integers such that $\gcd(m,q-1)=\gcd(n,q-1)=1$, s is any integer, for any $1\leqslant i,j\leqslant q-2$ if we let $A_{ij}=1$ iff $\alpha^{mi+s}+\alpha^{nj+s}=a$, otherwise $A_{ij}=0$, then the matrix $A=[A_{ij}]$ is a Costas matrix of order $(q-2)$ by $(q-2)$.

If $m=n=1, \alpha=1, s=0$ then this generalization reduces to the Lempel's construction scheme $L_2^{[1]}$.

If $m=1, \alpha=1, s=0$ and n satisfies $\alpha^{n}=\beta$, then this generalization reduces to the Golomb's construction scheme $G_2^{[1]}$.

This generalization scheme can be proved in the same way as that of the above generalization.

8.2 Optical Orthogonal Codes

8.2.1 Parameters Bounds of Optical Orthogonal Codes[4]

Optical orthogonal code (OOC) is a special kind of code designed for a code-division multiple-access fiber optical channel[7]. The flexibility of the network systems can be strengthened by OOCs with the efficient transmission of information between asynchronous users. In addition, it has just been found that OOCs can also be widely used in mobile radio, frequency-hopping, spread-spectrum communications, radar and sonar systems and artificial neural networks.

Definition 8.2.1[7] *An $(n,\omega,\lambda_a,\lambda_c)$ optical orthogonal code C consists of a group of binary n-vectors $(x_0,\cdots,x_{n-1})$ such that the Hamming weight of any vector $(x_0,\cdots,x_{n-1})$ equals to ω and the auto- and cross-correlations satisfy the following two inequalities:*

$$\sum_{i=0}^{n-1} x_i x_{i+\tau} \leqslant \lambda_a, \quad 0<\tau<n$$

$$\sum_{i=0}^{n-1} x_i y_{i+\tau} \leqslant \lambda_a, \quad \textit{for any integer } \tau.$$

The above correlations are circular ones, i.e., $i+\tau$ means $(i+\tau) \bmod n$.

The number of codewords in an $(n,\omega,\lambda_a,\lambda_c)$ optical orthogonal code C (i.e., $|C|$) is called the volume of C. And the maximum volume of all such $(n,\omega,\lambda_a,\lambda_c)$ OOCs is denoted by $\varphi(n,\omega,\lambda_a,\lambda_c)$. When $\lambda_a=\lambda_c=\lambda$, the $(n,\omega,\lambda_a,\lambda_c)$ OOC is simply written as (n,ω,λ) and $\varphi(n,\omega,\lambda_a,\lambda_c)$ as $\varphi(n,\omega,\lambda)$.

A few known results about the maximum volume $\varphi(n,\omega,\lambda)$ of OOCs may be summarized as

(1) an upper bound (obtained from the constant weight codes);

(2) a lower bound (obtained with the greedy algorithm);

(3) if $n=(q^{d-1}-1)/(q-1), \omega=q+1$, where q is the power of a prime integer, d is even, then $\varphi(n,\omega,1)=\lfloor (n-1)/\omega(\omega-1)\rfloor$ (obtained with affine geometry);

(4) if $n \neq 2 \bmod 6$, then $\varphi(n,3,1)=\lfloor (n-1)/6\rfloor$(ditto).

Obviously, to determine the exact values of $\varphi(n,\omega,\lambda)$ for all possible n,ω,λ is very difficult. With the assumption of ω, some new exact values of $\varphi(n,\omega,\lambda)$ will be presented in this subsection.

Theorem 8.2.1 $\varphi(n,\omega,\omega)=\begin{pmatrix} n \\ \omega \end{pmatrix}=n!/[\omega!(n-\omega)!]$.

Proof For any 2 binary sequences $(x_0,\cdots,x_{n-1})$ and $(y_0,\cdots,y_{n-1})$ with Hamming weight ω, it is easy to see that

$$\sum_{i=0}^{n-1} x_i x_{i+\tau} \leqslant \omega \quad \text{and} \quad \sum_{i=0}^{n-1} x_i y_{i+\tau} \leqslant \omega.$$

This means that (n,ω,ω) OOC consists of all the binary n-vector with Hamming weight ω. And the number of such n-vectors is clearly $\begin{pmatrix} n \\ \omega \end{pmatrix}$. □

Definition 8.2.2 *Sequences $(x_0,\cdots,x_{n-1})$ and $(y_0,\cdots,y_{n-1})$ are said to be shift-equivalent with each other if $x_i=y_{i+t} (0\leqslant i\leqslant n-1)$ for some integer t, where $i+t$ means $(i+t) \bmod n$.*

Lemma 8.2.1 *Let $x=(x_0,\cdots,x_{n-1})$ and $y=(y_0,\cdots,y_{n-1})$ be two binary sequences with the same weight ω and not shift-equivalent to each other, then for any integer τ,*

$$\sum_{i=0}^{n-1} x_i y_{i+\tau} \leqslant \omega-1.$$

Proof Assume that there exists some s such that $\sum_{i=0}^{n-1} x_i y_{i+s} > \omega - 1$. But as the weights of x and y are all ω, we have

$$\sum_{i=0}^{n-1} x_i y_{i+s} \leqslant \omega.$$

This means, $\sum_{i=0}^{n-1} x_i y_{i+s} = \omega$. Again as the weights of x and y are all ω, we have $x_i = y_{i+s}(0 \leqslant i \leqslant n-1)$, i.e., x and y are shift-equivalent to each other. This is a contradiction. □

Lemma 8.2.2 *Let $x = (x_0, \cdots, x_{n-1})$ be a binary sequence with weight ω. Then, the autocorrelation of x satisfies $\sum_{i=0}^{n-1} x_i x_{i+\tau} \leqslant \omega - 1, (0 < \tau < n)$, if the minimum period of x is n (i.e., the minimum period of the infinite binary sequence $(x, x, \cdots)$ is n).*

Proof Necessity. Assume that the minimum period T of x is less than n, then T/n and there exists some $(a_0, \cdots, a_{T-1})$ such that $\underbrace{(a, a, \cdots, a)}_{n/T}$. So it is clear that $\sum_{i=0}^{n-1} x_i x_{i+T} = \omega > \omega - 1$, which is a contradiction.

Sufficiency. Assume that there exists some $0 < s < n$ such that $\sum_{i=0}^{n-1} x_i x_{i+s} \geqslant \omega$. In another aspect the weight ω of x leads to the inequality $\sum_{i=0}^{n-1} x_i x_{i+\tau} \leqslant \omega$. Hence we have $\sum_{i=0}^{n-1} x_i x_{i+s} = \omega$. Again from the weight ω of x, we have $x_i = x_{i+s}(0 \leqslant i \leqslant n-1)$, i.e., s is the period of x, which is in contradiction to the minimum period n. □

With the combination of Lemmas 8.2.1 and 8.2.2, we have

Lemma 8.2.3 *A set S of binary sequences with length n and Hamming weight ω forms an $(n, \omega, \omega - 1)$ optical orthogonal code, if S satisfies the following two conditions at the same time:* (1) *the minimum period of any $x \in S$ in n;* (2) *any two different $x, y \in S$ are not shift-equivalent to each other.*

Lemma 8.2.3 tells us that $\varphi(n, \omega, \omega-1)$ equals the maximum number of binary sequences that satisfy the conditions 1 and 2 at the same time. This fact is very useful for calculation of $\varphi(n, \omega, \omega - 1)$ in the following.

Theorem 8.2.2 *If $n = p$ is a prime integer, then $\varphi(n, \omega, \omega - 1) = \frac{1}{n}\binom{n}{\omega}$.*

Proof Because $n = p$ is prime, any binary sequence with length n and weight ω (altogether $\binom{n}{\omega}$ sequences) has minimum period n (in fact if the minimum period T is less than n, then T/n which is in contradiction to prime $n = p$). In the other aspect, there are n sequences which are shift-equivalent to one another. So the number of binary sequences with minimum period n and weight ω which are not shift-equivalent equals $\frac{1}{n}\binom{n}{\omega}$, i.e.,

$$\varphi(n, \omega, \omega - 1) = \frac{1}{n}\binom{n}{\omega}.$$

□

Theorem 8.2.3 *If $n = pq$ (p and q are different primes), then*

$$\varphi(n, \omega, \omega - 1) = \frac{1}{n}\left[\binom{n}{\omega} - \binom{p}{\omega p/n} - \binom{q}{\omega q/n}\right],$$

where $\binom{R}{m/n} = 0$ *if* $n \nmid m$.

Proof For $n = pq$, the minimum period T of any binary sequence with length n and weight ω (altogether $\binom{n}{\omega}$ sequences) may only be n,p or q, and the number of 1's in every minimum period equals $\omega T/n$. From the proof of Theorem 8.2.2, we know that if $T = p$ then there are $\binom{p}{\omega p/n}$ binary sequences with minimum period p and weight $\omega p/n$ in every minimum period. If $T = q$ then there are $\binom{q}{\omega q/n}$ binary sequences with minimum period q and weight $\omega q/n$ in every minimum period. So the number of binary sequences with minimum period n and weight ω equals $\binom{n}{\omega} - \binom{p}{\omega p/n} - \binom{q}{\omega q/n}$. In the other aspect, for any fixed x, there are n sequences that are shift-equivalent to one another. This means that, from Lemma 8.2.3,

$$\varphi(n, \omega, \omega - 1) = \frac{1}{n}\left[\binom{n}{\omega} - \binom{p}{\omega p/n} - \binom{q}{\omega q/n}\right]. \qquad \square$$

Theorem 8.2.4 *If $n = p^2$, p is a prime, then*

$$\varphi(n, \omega, \omega - 1) = \frac{1}{n}\left[\binom{n}{\omega} - \binom{p}{\omega p/n}\right].$$

Proof For $n = p$, the minimum period T of any binary sequence with length n and weight ω (altogether $\binom{n}{\omega}$ sequences) may be n or p and the number of 1's in every minimum period equals $\omega T/n$. From the proof of Theorem 8.2.2, we know that if $T = p$ then there are $\binom{p}{\omega p/n}$ binary sequences with minimum period p and Hamming weight $\omega p/n$ in every minimum period. So there are $\binom{n}{\omega} - \binom{p}{\omega p/n}$ binary sequences with minimum period n and weight ω. With the same reasoning as that in the proof of Theorem 8.2.3, it is clear that

$$\varphi(n, \omega, \omega - 1) = \frac{1}{n}\left[\binom{n}{\omega} - \binom{p}{\omega p/n}\right]. \qquad \square$$

By now, with Theorems 8.2.2, 8.2.3 and 8.2.4, we have new exact values of $\varphi(n, \omega, \omega - 1)$ for some special n. In order to show the generalized formula of $\varphi(n, \omega, \omega - 1)$ for any integer n, we restate some known definitions and results about the Mobius function $\mu(n)$.

The Mobius function $\mu(n)$ is defined as

$$\mu(n) = \begin{cases} 1, & \text{if} \quad n = 1, \\ 0, & \text{if} \quad n = p^2 m \quad (p \text{ prime}, m \text{ positive integer}), \\ -1, & \text{if} \quad n = p_1 p_2 \cdots p_r \quad (p_i's \text{ are different primes}). \end{cases}$$

Lemma 8.2.4 $\sum_{d|n} \mu(d) = \begin{cases} 1, & \textit{if} \quad n = 1, \\ 0, & \textit{if} \quad n > 1. \end{cases}$

Theorem 8.2.5 *For any positive integer n,*

$$\varphi(n,\omega,\omega-1)=\frac{1}{n}\sum_{d|n}\mu(d)\binom{n/d}{\omega/d},$$

where if $d\nmid\omega$, $\binom{m}{\omega/n}=0$, *as in Theorem* 8.2.3. *It is easy to verify that Theorems* 8.2.2, 8.2.3 *and* 8.2.4 *are special cases of Theorem* 8.2.5.

Proof Let $g(n,r)$ be the number of binary sequences with minimum period n and weight r in every minimum period. Let T be the possible minimum period of binary sequences with length n and weight ω (altogether $\binom{n}{\omega}$ sequences), then T/n and the weight of every minimum period equals $\omega T/n$. So, we have

$$\sum_{T|n}g\left(T,\frac{\omega T}{n}\right)=\binom{n}{\omega}$$

and

$$\begin{aligned}\binom{n/d'}{\omega/d'}&=\sum_{d|\frac{n}{d'}}g\left(d,\frac{\omega d}{d'}\frac{d'}{n}\right)\\&=\sum_{d|\frac{n}{d'}}g\left(d,\frac{\omega d}{n}\right)\sum_{d'|n}\mu(d')\binom{n/d'}{\omega/d'}\\&=\sum_{d'|n}\sum_{d|\frac{n}{d'}}g\left(d,\frac{\omega d}{n}\right)\mu(d')=\sum_{d|n}\sum_{d'|\frac{n}{d}}g\left(d,\frac{\omega d}{n}\right)\mu(d')\\&=\sum_{d|n}g\left(d,\frac{\omega d}{n}\right)\sum_{d'|\frac{n}{d}}\mu(d')=g(n,\omega).\end{aligned}$$

By now, we know that the number of binary sequences with minimum period n and weight ω is $g(n,\omega)=\sum_{d|n}\mu(d)\binom{n/d}{\omega/d}$.

Because for any fixed binary sequence x with length n, there are n shift-equivalent sequences. So there are $g(n,\omega)/n$ binary sequences that satisfy the conditions in Lemma 8.2.3, i.e.,

$$\varphi(n,\omega,\omega-1)=g(n,\omega)/n.$$

□

The $\varphi(n,\omega,\lambda)$ is one of the parameters about the OOC, which means the maximum volume of (n,ω,λ) OOC when n,ω and λ are fixed. It's not difficult to see that after n,ω,λ and the volume v are fixed, there may be more than one (n,ω,λ) OOCs $C_1,C_2,\cdots$. Let $\psi(v,n,\omega,\lambda)$ be the maximum number of possible (n,ω,λ) OOCs with the fixed volume v.

Theorem 8.2.6 *The following three relations hold true between* $\varphi(n,\omega,\lambda)$ *and* $\psi(v,n,\omega,\lambda)$:

(1) $\psi(v,n,\omega,\lambda)\geqslant\psi((v+1),n,\omega,\lambda)$.

(2) $\psi(\varphi(n,\omega,\lambda),n,\omega,\lambda)\leqslant\dfrac{\psi(1,n,\omega,\lambda)!}{\varphi(n,\omega,\lambda)![\psi(1,n,\omega,\lambda)-\varphi(n,\omega,\lambda)]!}$.

(3) $\psi(v,n,\omega,\lambda)\geqslant\dfrac{\varphi(n,\omega,\lambda)!}{V![\varphi(n,\omega,\lambda)-V]!}$.

Proof (1) is clear.

(2) Because every codeword in any (n, ω, λ) OOC forms an (n, ω, λ) OOC of volume 1, thus any (n, ω, λ) OOC of volume v is a set of special (n, ω, λ) OOCs of volume 1. In the other aspect, there are $\psi(1, n, \omega, \lambda)$ (n, ω, λ) OOCs of volume 1, thus, we have at most $\binom{\psi(1,n,\omega,\lambda)}{v}$ subsets of volume v. Therefore $\psi(v, n, \omega, \lambda) \leqslant \binom{\psi(1,n,\omega,\lambda)}{v}$. The result (2) follows from this inequality by letting $v = \varphi(n, \omega, \lambda)$.

(3) If C is an (n, ω, λ) OOC of maximum volume $\varphi(n, \omega, \lambda)$, then any v codewords in C form an (n, ω, λ) OOC of volume v. Hence we have at least $\binom{\varphi(n,\omega,\lambda)}{v}$ (n, ω, λ) OOCs of volume v, i.e. $\psi(v, n, \omega, \lambda) \leqslant \binom{\varphi(n,\omega,\lambda)}{v}$. □

Now let us consider $\psi(1, n, \omega, 1)$ and $\psi(1, n, \omega, 2)$.

Let $x = (x_0, \cdots, x_{n-1})$ be a binary sequence of length n and weight ω. The map $g(x)$ of x is defined by $g(x) = (i_1, \cdots, i_\omega)$, if $x_i = 1$ for $i = i_1, i_2, \cdots, i_\omega$ and $x_j = 0$ for other integer j in $[0, n-1]$. It is clear that x and $g(x)$ are uniquely determined by each other. $g(x) = (i_1, \cdots, i_\omega)$ is called the characteristic sequence of x.

Lemma 8.2.5 (1) *Let* $x = (x_0, \cdots, x_{n-1})$ *and its* $g(x) = (i_1, \cdots, i_\omega)$*, Then, the* $x = (x_0, \cdots, x_{n-1})$ *forms an* $(n, \omega, 1)$ OOC *of volume 1, if for any* $(p, q) \neq (r, s)$ *hold* $(i_p - i_q) \bmod n \neq (i_r - i_s) \bmod n$ *(where* $1 \leqslant p, q, r, s \leqslant \omega$*).*

(2) *Let* $x = (x_0, \cdots, x_{n-1})$ *and its* $g(x) = (i_1, \cdots, i_\omega)$ *and let the set* D *be* $D = \{(i_p - i_q) \bmod n : 1 \leqslant p, q \leqslant \omega\}$*. If there are only two equal elements in* D *and the other elements are different from each other, then,* x *forms an* $(n, \omega, 2)$ OOC *of volume* 1.

Proof For (1). Sufficiency. If x does not form an (n, ω, λ) OOC of volume 1, there should be some $0 < s < n$ such that $\sum_{j=0}^{n-1} x_j x_{j+s} \geqslant 2$. Hence we can find at least two integers u and v such that $x_u = x_{u+s} = 1$ and $x_v = x_{v+s} = 1$ which is contradictory to the condition $(i_p - i_q) \bmod n \neq (i_r - i_s) \bmod n$.

Necessity. If there exists some $(a, b) \neq (c, d)$ such that $(i_a - i_b) \bmod n = (i_c - i_d) \bmod n$ and for simplicity we assume $i_b < i_a < i_d < i_c$ (the other cases may be studied similarly). Then, $x_{i_a} = x_{i_b} = x_{i_c} = x_{i_d} = 1$ and $i_a - i_b = i_c - i_d$. If $\tau_0 = i_d - i_b$, then $\sum_{j=0}^{n-1} x_j x_{j+\tau_0} \geqslant x_{i_{q^0}} x_{i_{q^0}} + \tau_0 + x_{i_{p^0}} x_{i_{p^0}} + \tau_0 = x_{i_{q^0}} x_{i_{\tau^0}} + x_{i_{p^0}} x_{i_{\tau^0}} = 2$. This is contradictory to $\sum_{j=0}^{n-1} x_i x_{i+\tau} \leqslant 1$.

For (2), the result may be proved similarly as the sufficiency of (1). □

From Lemma 8.2.5, we have

Theorem 8.2.7 (1) *Let* $E_1 = \{(i_1, \cdots, i_\omega) : 1 \leqslant i_1 < \cdots < i_\omega \leqslant n-1$*, and* $(i_p - i_q) \bmod n \neq (i_r - i_s) \bmod n$ *for any* $(p, q) \neq (r, s)\}$*, then,* $\psi(1, n, \omega, 1) = |E_1|$ *(where* $|x|$ *means the cardinality of the set* x*).*

(2) *Let* $E_2 = \{(i_1, \cdots, i_\omega)$ *and there are only two equal elements in* D *and all the other elements are different from each other*$\}$*. Then,* $\psi(1, n, \omega, 2) \geqslant |E_2|$.

The calculation of $|E_1|$ and $|E_2|$ is also very difficult. We will only derive a few upper bounds.

Let $E_1^* = \{(i_1, \cdots, i_\omega) : 1 \leqslant i_1 < \cdots < i_\omega \leqslant n-1$,and $k_1 = i_2 - i_1, k_2 = i_3 - i_2, \cdots, k_{\omega-1} = i_\omega - i_{\omega-1}$ and $k_\omega = n + i_1 - i_\omega$ are different from each other$\}$, then $E_1 \subseteq E_1^*$ and $|E_1| \leqslant |E_1^*|$. In the other aspect, $|E_1^*|$ is the number of possible ways to choose ω

integers $k_1, k_2, \cdots, k_\omega$ from the set of $1, \cdots, n-1$ such that $k_1 + k_2 + k_\omega = n$. Hence $|E_1^*|$ equals the coefficient of x^n in the polynomial

$$f_\omega(x) = \sum_{1 \leqslant j_1, \cdots, j_\omega \leqslant n-1 \text{ all} j's \text{ differ}} x^{j_1 + \cdots + j_\omega},$$

$$\text{i.e., } |E_1^*| = \left.\frac{\partial^{(n)} f_\omega(x)}{\partial x^n}\right|_{x=0},$$

let $E_2^* = \{(i_1, \cdots, i_\omega) : 1 \leqslant i_1 < \cdots < i_\omega \leqslant n-1$, and at most two of $k_1 = i_2 - i_1, k_2 = i_3 - i_2, \cdots, k_{\omega-1} = i_\omega - i_{\omega-1}$ and $k_\omega = n + i_1 - i_\omega$ are identical $\}$. In the same way we have $E_2 \subseteq E_2^* \cup E_1^*$ and $|E_2^*|$ equals the coefficient of x^n in the polynomial

$$g_\omega(x) = \sum_{1 \leqslant j_1, \cdots, j_\omega \leqslant n-1 \text{ only} 2j's \text{ iden}} x^{j_1 + \cdots + j_\omega}$$

$$= \binom{\omega}{2} \sum_{1 \leqslant j_1, \cdots, j_\omega \leqslant n-1 \text{ all} j's \text{ differ}} x^{2j_2 + \cdots + j_\omega}.$$

The last equation is due to mathematical symmetry of the $j_1, j_2 \cdots, j_\omega$. Therefore, $|E_2| \leqslant |E_2^*| + |E_1^*| = \frac{\partial^{(n)} g_\omega(x)}{\partial x^n}|_{x=0} + \frac{\partial^{(n)} f_\omega(x)}{\partial x^n}|_{x=0}$ up to here, we have

Theorem 8.2.8

$$\psi(1, n, \omega, 1) \leqslant \left.\frac{\partial^{(n)} f_\omega(x)}{\partial x^n}\right|_{x=0},$$

$$\psi(1, n, \omega, 2) \leqslant \left.\frac{\partial^{(n)} g_\omega(x)}{\partial x^n}\right|_{x=0} + \left.\frac{\partial^{(n)} f_\omega(x)}{\partial x^n}\right|_{x=0}.$$

In the common channel the cross-correlation between binary sequence $x = (x_0, \cdots, x_{n-1})$ and $y = (y_0, \cdots, y_{n-1})$ is defined by $R(\tau) = \sum_{i=0}^{n-1}(-1)^{x_i y_{i+\tau}}$. But in the fiber optical channel the cross-correlation should be defined as $r(\tau) = \sum_{i=0}^{n-1} x_i y_{i+\tau}$. Perfect codes designed for different channels can't be replaced by each other. Thus most of results obtained for the known signals can't be directly used in the construction of OOCs.

Lemma 8.2.6 *Let $a = (a_0, \cdots, a_{n-1})$ and $b = (b_0, \cdots, b_{n-1})$ be two binary sequences of Hamming weight ω, then*

$$\sum_{i=0}^{n-1} a_i b_{i+\tau} = \omega - \frac{1}{4}\left[n - \sum_{i=0}^{n-1}(-1)^{a_i + b_{i+\tau}}\right].$$

Therefore $r(\tau) = \omega - \frac{1}{4}[n - R(\tau)]$.

Proof The identity $a \oplus b = a + b - 2ab$ is true for any binary variable a and b (where $\oplus$ means the summation over mod 2)

$$\sum_{i=0}^{n-1} a_i b_{i+\tau} = \frac{1}{2}\left[\sum_{i=0}^{n-1} a_i + \sum_{i=0}^{n-1} b_{i+\tau} - \sum_{i=0}^{n-1}(a_i \oplus b_{i+\tau})\right] = \frac{1}{2}\left[2\omega - \sum_{i=0}^{n-1}(a_i \oplus b_{i+\tau})\right].$$

In the other aspect,

$$\sum_{i=0}^{n-1}(-1)^{a_i b_{i+\tau}} = \left[n - \sum_{i=0}^{n-1}(a_i \oplus b_{i+\tau})\right] - \sum_{i=0}^{n-1}(a_i \oplus b_{i+\tau}) = n - 2\sum_{i=0}^{n-1}(a_i \oplus b_{i+\tau}).$$

The proof is completed by combining the above two equations. □

In the relation between $r(\tau)$ and $R(\tau)$ shown in Lemma 8.2.6 and some known results for perfect discrete signals, we can easily obtain the following conclusion.

Theorem 8.2.9 (1) *Any m-sequence of length* $2^n - 1$ *form a* $(2^n - 1, 2^{n-1}, 2^{n-2})$ OOC.

(2) *Let a and b be two m-sequence of length* $2^n - 1$, *and their primitive elements are* α *and* α^q *respectively (where* $q = 2^{k+1}$ *or* $2^{2k} - 2^k + 1$, *k is such an integer that* $e = \gcd(n, k)$ *and* n/e *is an odd integer). Then, a and b form a* $(2^n - 1, 2^{n-1}, 2^{n-2} + 2^{(n+e-1)/2})$ OOC *of volume* 2.

(3) *Let a and b be two m-sequence with period* $2^n - 1$ *and their generator elements are* α *and* $\alpha^{-1+(2n+2)/2}$ *respectively, then a and b form a* $(2^n - 1, 2^{n-1}, 2^{n-2} + 2^{n/2})$ OOC *of volume* 2.

(4) *Let a and b are generated by the primitive elements* α *and* α^{-1} *respectively, then they form a* $(2^n - 1, 2^{n-1}, 2^{n-2} + 2^{n/2})$ OOC *of volume* 2.

(5) *Any subset with weight* ω *in the families of Gold sequences,* GL *sequences,* DBCH *sequences or* LK *sequences forms a* $(2^n - 1, \omega, \omega - 2^{n-2} + 2^{(n-2)/2})$ OOC.

(6) *Any subset with weight* ω *in the family* SK *sequences, forms a* $(2^n - 1, \omega, \omega - 2^{n-2} + 2^{n/2-2})$ OOC.

(7) *Any subset with weight* ω *in the families of McElience sequences forms a* $(2^n - 1, \omega, \omega - 2^{n-2} + 2^{n/4+1})$ OOC.

In addition, some other perfect codes can also be used for the construction of OOCs. But it should be noted that the OOCs designed in this way are not the best ones.

8.2.2 Truncated Costas Optical Orthogonal Codes

Let $A = [A(i,j)]$ and $B = [B(i,j)]$, $0 \leqslant i \leqslant N-1$, $0 \leqslant j \leqslant M-1$, be two arrays of the same size $N \times M$. The following three arrays $R_{AB} = [R_{AB}(s,t)]$, $0 \leqslant s \leqslant N-1$, $0 \leqslant t \leqslant M-1$; $T_{AB} = [T_{AB}(u,v)]$, $-N+1 \leqslant u \leqslant N-1$, $-M+1 \leqslant v \leqslant M-1$; and $C_{AB} = [C_{AB}(u,t)]$, $0 \leqslant t \leqslant M-1$, $-N+1 \leqslant u \leqslant N-1$, are called periodic, non-periodic and mixed cross-correlation arrays between A and B, respectively, if

$$R_{AB}(s,t) = \sum_{i=0}^{N-1}\sum_{j=0}^{M-1} A(i,j)B(i \oplus_M s, j \oplus_M t),$$
$$0 \leqslant s \leqslant N-1,\ 0 \leqslant t \leqslant M-1, \tag{8.75}$$

$$T_{AB}(u,v) = \sum_{i=0}^{N-1-u}\sum_{j=0}^{M-1-v} A(i,j)B(i+u, j+v),$$
$$-N+1 \leqslant u \leqslant N-1, -M+1 \leqslant v \leqslant M-1, \tag{8.76}$$

$$C_{AB}(u,t) = \sum_{i=0}^{N-1-u}\sum_{j=0}^{M-1} A(i,j)B(i+u, j \oplus_M t),$$
$$0 \leqslant t \leqslant M-1, -N+1 \leqslant u \leqslant N-1, \tag{8.77}$$

where $N-1 \geqslant i \oplus_N s \equiv (i+s) \bmod N \geqslant 0$, $M-1 \geqslant j \oplus_M t \equiv (j+t) \bmod M \geqslant 0$, and $B(i+u, j+V)=0$, if $i+u$ exceeds the range $[-N+1, N-1]$ or $j+V$ exceeds $[-M+1, M-1]$, and $B(i+u, j \oplus_M t)=0$, if $i+u$ exceeds $[-N+1, N-1]$. If A=B, then $R_A =: R_{AA} = [R_{AA}(s,t)]$, $T_A =: T_{AA} = [T_{AA}(u,t)]$, and $C_A =: C_{AA} = [C_{AA}(u,t)]$ are called the periodic, non-periodic, and mixed autocorrelation arrays of A, respectively. These correlations are called 1-dimensional(1D), if $N=1$; and 2-dimensional(2D) if $N>1$.

The relationships among the periodic, non-periodic, and mixed correlations are revealed by the following theorem.

Theorem 8.2.10 *If all the elements in $A=[A(i,j)]$ and $B=[B(i,j)]$ are non-negative, especially 0 or 1, then*

$$\max_{u,v} T_{AB}(u,v) \leqslant \max_{u,t} C_{AB}(u,t) \leqslant \max_{s,t} R_{AB}(s,t), \tag{8.78}$$

$$\max_{u,t} C_{AB}(u,t) \leqslant 2 \max_{u,v} T_{AB}(u,v), \tag{8.79}$$

$$\max_{s,t} R_{AB}(s,t) \leqslant 2 \max_{u,t} C_{AB}(u,t). \tag{8.80}$$

A permutation matrix $A=[A(i,j)]$, $0 \leqslant i,j \leqslant N-1$, of order $N \times N$ is called a Costas array if and only if its out-of-phase non-periodic autocorrelation is upper bounded by 1, i.e., $T_A(u,v) \leqslant 1$ for all $-N+1 \leqslant u,v \leqslant N-1$ and $(u,v) \neq (0,0)$.

Welch-Costas Arrays $A=[A(i,j)]$, $0 \leqslant i \leqslant p-2$, $0 \leqslant j \leqslant p-2$, with

$$A(i,j) = \begin{cases} 1, & \text{if } i+1=\alpha^j \text{ in GF}(p), \\ 0, & \text{otherwise}, \end{cases} \tag{8.81}$$

where p is an odd prime integer, α a primitive element in the finite field GF(p). The size of a Welch-Costas array is $N \times N$, $N=p-1$.

Golomb-Costas Arrays $A=[A(i,j)]$,$0 \leqslant i,j \leqslant q-3$, with

$$A(i,j) = \begin{cases} 1, & \text{if } \alpha^{j+1}+\beta^{j+1}=1, \\ 0, & \text{otherwise}, \end{cases} \tag{8.82}$$

where $q=p^r$ is a prime power, α and β two primitive elements in the field GF(q). The size of a Golomb-Costas array is $N \times N$, $N=q-2$.

Taylor's Variant Golomb-Costas Arrays $A=[A(i,j)]$,$0 \leqslant i,j \leqslant q-2$, with $A(0,0)=1$, and for other $0 \leqslant m,n \leqslant q-2$, $A(n,0)=A(0,m)=0$,

$$A(n,m) = \begin{cases} 1, & \text{if } \alpha^n+\beta^m=1, \\ 0, & \text{otherwise}, \end{cases} \tag{8.83}$$

where $q=p^r$ is a prime power and $p \neq 2$, α and β two primitive elements in the field GF(q). The size of a Taylor's Variant Golomb-Costas array is $N \times N$, $N=q-1$.

In Theorem 8.2.10, the out-of-phase mixed and periodic autocorrelations of a general Costas array are upper bounded by 2 and 4, respectively. While the following two results show that, in some special cases, these bounds can be improved to 1 and 2, respectively.

Theorem 8.2.11 *The out-of-phase mixed autocorrelation of a Welch-Costas array A is upper bounded by* 1*, i.e.,* $\max_{(u,t)\neq(0,0)} C_A(u,t) \leqslant 1$.

Corollary 8.2.1 *The out-of-phase periodic autocorrelation of a Taylor's Variant Golomb-Costas array A is upper bounded by 2, i.e.,* $\max_{(s,t)\neq(0,0)} R_A(s,t) \leqslant 2$.

Remark 8.2.1 *The Taylor's Variant Golomb-Costas array considered here is generated from a Golomb-Costas array by adding a corner dot ("1") on the position* $(0,0)$. *Other kinds of Taylor's Variant Golomb-Costas arrays of size* $(q-1)\times(q-1)$ *can be produced by adding a corner dot ("1") at the position* $(0,q-1)$, $(q-1,0)$, *or* $(q-1,q-1)$. *Corollary* 8.2.1 *and the following Corollary* 8.2.3 *are valid for the other Taylor's Variant Golomb-Costas arrays.*

Similar to the autocorrelation cases, the cross-correlations of Welch-Costas arrays and Taylor's Variant Golomb-Costas arrays are also better than the general Costas arrays. Precisely, we have the following corollaries.

Corollary 8.2.2 *Let* $A=[A(i,j)]$ *and* $B=[B(i,j)]$, $0\leqslant i,j\leqslant p-2$, *be two Welch-Costas arrays defined by the primitive elements* α *and* β, *respectively. Then the mixed cross-correlation between* A *and* B *is upper bounded by* $l=\log_a(\beta)$, *the discrete logarithm of* β *under the base* α.

Remark 8.2.2 *Discrete logarithm* $\log_a(b)$ *is a function in finite field* $\mathrm{GF}(q)$, q *a prime power. Let* α *be a primitive element,* b *non-zero in* $\mathrm{GF}(q)$, *then* $\log_a(b)=k$ *if and only if* $0\leqslant k\leqslant p-2$ *and* $\alpha^k=b$ *in* $\mathrm{GF}(q)$.

Corollary 8.2.3 *Let* $A=[A(i,j)]$ *and* $B=[B(i,j)]$, $0\leqslant i,j\leqslant p-2$, *be two Taylor's Variant Golomb-Costas arrays generated by* (α,β) *and* (η,σ), *respectively, then the periodic cross-correlation between* A *and* B *is upper bounded by* $1+\max(m,l)$, *where* $m=\log_\alpha(\eta)$ *and* $l=\log_\beta(\sigma)$.

The following two theorems are useful for the construction of TC OOC.

Theorem 8.2.12 *Let* $A=[A(i,j)]$, $B=[B(i,j)]$, $0\leqslant i\leqslant N-1$, $0\leqslant i\leqslant M-1$, *be two arrays with non-negative elements.*

(1) *If* C *and* D *are the subarrays consisting of the first* $n(n\leqslant N)$ *rows and* $m\,(m\leqslant M)$ *columns of* A *and* B, *respectively, then the maximum non-periodic cross-correlation between* C *and* D *is upper bounded by that between* A *and* B, *i.e.,* $\max_{u,v} T_{CD}(u,v)\leqslant \max_{u,v} T_{AB}(u,v)$.

(2) *If* E *and* F *are the subarrays consisting of the first* $n(n\leqslant N)$ *rows (columns have not been cut!) of* A *and* B, *respectively, then the maximum mixed cross-correlation between* E *and* F *is upper bounded by that between* A *and* B, *i.e.,* $\max_{u,t} C_{EF}(u,t)\leqslant\max_{u,t} C_{AB}(u,t)$ *(this claim will be false if the column was cut).*

Theorem 8.2.13 *Let* $A=[A(i,j)]$, $B=[B(i,j)]$, $0\leqslant i\leqslant N-1$, $0\leqslant j\leqslant M-1$,*be two arrays with non-negative elements, and let 0 be an all-zero array of size* $(N-1)\times M$. *If*

$$C=\begin{pmatrix} A \\ \vdots \\ 0 \end{pmatrix},\quad D=\begin{pmatrix} B \\ \vdots \\ 0 \end{pmatrix}$$

are the arrays produced by putting 0 under A *and* B, *respectively, then the maximum periodic cross-correlation between* C *and* D *is upper bounded by the maximum mixed cross-correlation* *between the original arrays A and B, i.e.,* $\max_{s,t} R_{CD}(s,t) \leqslant \max_{u,t} C_{AB}(u,t)$.

Remark 8.2.3 *Besides Theorem 8.2.12 and Theorem 8.2.13, there are some other approaches for the construction of arrays with better correlation. For example, let A and B be two arrays with non-negative elements. If G and H are two arrays produced by replacing some rows (or columns) of A and B, respectively, by all-zero rows (columns), then the non-periodic, mixed, and periodic cross-correlations between G and H are smaller than or equal to that between A and B, respectively, i.e.,* $\max_{u,v} T_{GH}(u,v) \leqslant \max_{u,v} T_{AB}(u,v)$, $\max_{u,t} C_{GH}(u,t) \leqslant \max_{u,t} C_{AB}(u,t)$, *and* $\max_{s,t} R_{GH}(s,t) \leqslant \max_{s,t} R_{AB}(s,t)$.

Definition 8.2.3 *Let $A = [A(i,j)]$, $0 \leqslant i \leqslant N-1$, $0 \leqslant j \leqslant M-1$, be an array of size $N \times M$, the concatenated sequence of A is a 1-dimensional sequence $x = (x(k))$, $0 \leqslant k \leqslant MN-1$, of length MN which is formed by concatenating the columns of A one after another. Mathematically, for each $0 \leqslant k \leqslant MN-1$, if $k = mN+n$, $0 \leqslant n \leqslant N-1$, then the k-th element $x(k)$ of the concatenated sequence is defined by $x(k) = A(n,m)$ (note that every integer k, $0 \leqslant k \leqslant MN-1$, can be uniquely defined by $k = mN+n$, $0 \leqslant n \leqslant N-1$, $0 \leqslant m \leqslant M-1$, where m and n are the quotient and residue, respectively, of k divided by N).*

For example, if A is the following 3×5 array

$$A = \begin{bmatrix} A(0,0) & A(0,1) & A(0,2) & A(0,3) & A(0,4) \\ A(1,0) & A(1,1) & A(1,2) & A(1,3) & A(1,4) \\ A(2,0) & A(2,1) & A(2,2) & A(2,3) & A(2,4) \end{bmatrix},$$

then its concatenated sequence is

$$\begin{aligned} x = \; & \overbrace{A(0,0), A(1,0), A(2,0),}^{\text{1st column}} \overbrace{A(0,1), A(1,1), A(2,1),}^{\text{2st column}} \\ & \overbrace{A(0,2), A(1,2), A(2,2),}^{\text{3st column}} \overbrace{A(0,3), A(1,3), A(2,3),}^{\text{4st column}} \\ & \overbrace{A(0,4), A(1,4), A(2,4),}^{\text{5st column}} \end{aligned}$$

which is a 1-dimensional sequence of length $15 (= 3 \times 5)$.

Theorem 8.2.14 *Let $A = [A(i,j)]$, $B = [B(i,j)]$, $0 \leqslant i \leqslant N-1$, $0 \leqslant j \leqslant M-1$, be two arrays with non-negative elements, and let $x = (x(k))$, $y = (y(k))$, $0 \leqslant k \leqslant MN-1$ be the concatenated sequences of A and B respectively. Then the maximum (1D) periodic cross-correlation between x and y is upper bounded by the doubled maximum mixed cross-correlation between the original arrays A and B, i.e.,* $\max_r R_{xy}(r) \leqslant 2\max_{u,t} C_{AB}(u,t)$.

Definition 8.2.4 *An $(n, \omega, \lambda_a, \lambda_c)$ OOC is an array of binary $(0,1)$ sequences of length n and Hamming weight ω such that (1) the out-of-phase periodic autocorrelation of each member sequence is uniformly upper bounded by λ_a, and (2) the periodic cross-correlation between each member sequences pair is uniformly upper bounded by λ_c. The number of member sequences in this family is called the volume of the OOC.*

Based on Theorem 8.2.14 and Definition 8.2.4, we have the following general construction for OOCs.

Construction 8.2.1 *If $X_1, X_2, \cdots, X_r$ are r binary $(0,1)$ arrays of size $N \times M$ such that*

(1) *Each X_i has the same Hamming weight ω;*

(2) *The out-of-phase mixed autocorrelation of each* $0 \leqslant i \leqslant r$, *is upper bounded by* λ_a;

(3) *The mixed cross-correlation between each pair* X_i *and* X_j, $1 \leqslant i \neq j \leqslant r$, *is upper bounded by* λ_c.

Then the sequences family consisting of the r concatenated sequences x_i of X_i, $1 \leqslant i \leqslant r$, is an $(N, \omega, 2\lambda_a, 2\lambda_c)$ OOC (note that both the auto- and cross-correlation upper bounds have been doubled).

Theorem 8.2.15 *Let* $A = [A(i,j)]$, $B = [B(i,j)]$, $0 \leqslant i \leqslant N-1$, $0 \leqslant j \leqslant M-1$, *be two arrays with non-negative elements, and let 0 be two arrays with non-negative elements, and let 0 be an all-zero array of size* $(N-1) \times M$, *and let*

$$C = \begin{pmatrix} A \\ \vdots \\ 0 \end{pmatrix}, \quad D = \begin{pmatrix} B \\ \vdots \\ 0 \end{pmatrix}.$$

If $x = (x(k))$ *and* $y = (y(k))$, $0 \leqslant k \leqslant (2N-1)M-1$, *are the concatenated sequences of* C *and* D, *respectively, then the maximum periodic cross-correlation between* x *and* y *is upper bounded by the maximum mixed cross-correlation between the original* A *and* B, *i.e.,* $\max_r R_{xy}(r) \leqslant \max_{u,t} C_{AB}(u,t)$.

Remark 8.2.4 *Notice that, different from Theorem* 8.2.14, *the* $\max_{u,t} C_{AB}\ (u,t)$ *is not doubled in Theorem* 8.2.15. *There is clearly a trade-off between the upper bound of periodic cross-correlation and the number of zero elements contained in the sequences* x *and* y.

Based on Theorem 8.2.15 and Definition 8.2.4, we get another construction for OOCs:

Construction 8.2.2 *If* $X_1, X_2, \cdots, X_r$ *are* r *binary* $(0,1)$ *arrays of size* $N \times M$ *such that*

(1) *Each* X_i *has the same Hamming weight* ω;

(2) *The out-of-phase mixed autocorrelation of each* X_i, $0 \leqslant i \leqslant r$, *is upper bounded by* λ_a;

(3) *The mixed cross-correlation between each pair* X_i *and* X_j, $1 \leqslant i \neq j \leqslant r$, *is upper bounded by* λ_c.

Then the sequences family consisting of the r concatenated sequences y_i of $C = \begin{pmatrix} X_i \\ \vdots \\ 0 \end{pmatrix}$, $1 \leqslant i \leqslant r$, is an $(M(2N-1), \omega, \lambda_a, \lambda_c)$ OOC, where 0 is an all-zero array of size $(N-1) \times M$.

In the following, we concentrate on the problem of designing arrays with good auto- and cross-correlation properties and then use them to construct new OOCs by Construction 8.2.1 and Construction 8.2.2.

We know that it is an important problem to find two or more Costas arrays such that their non-periodic cross-correlation is ideal ($\leqslant 1$) when the allowable Doppler shift is constrained. Mathematically, this problem is to find a family of Costas arrays of size $N \times N$ such that for each pair members A and B:

$$T_{AB}(u,v) \leqslant 1, \quad |u| \leqslant N-1, \quad |v| \leqslant m \leqslant N-1, \tag{8.84}$$

or equivalently,

$$T_{AB}(u,v) \leqslant 1, \quad |u| \leqslant n \leqslant N-1, \quad |v| \leqslant N-1. \tag{8.85}$$

The first family of Costas arrays satisfying Eq. (8.84) (or equivalently Eq. (8.85)) was proposed by Chang and Scarbrough. Now, we will present the second family of such good Costas arrays.

Note that every $M \times N$ permutation matrix can be represented by an integer sequence $a_0, a_1, \cdots, a_{M-1}$ chosen from the set $\{0, 1, 2, \cdots, M-1\}$ in such a way that the j-th column of the matrix has a "1" in the a_j-th row. If two $M \times M$ arrays A and B are represented by $a_0, a_1, \cdots, a_{M-1}$ and $b_0, b_1, \cdots, b_{M-1}$, respectively, then the condition

$$T_{AB}(u, v) \geqslant 2 \tag{8.86}$$

implies that for some integers k and h, $0 \leqslant k \leqslant M-1$, $0 \leqslant h \leqslant M-1$,

$$a_k - a_{k+h} = b_{k-u} - b_{k+h-u}. \tag{8.87}$$

In fact, Eq. (8.87) means that the "1"s of array A on the k-th and $(k+h)$-th columns overlap with the "1"s of array B on the $(k-u)$-th and $(k+h-u)$-th columns, respectively, when B is horizontally shifted by u and vertically shifted by some v.

Let $M = q-2$, α and β be two primitive elements in the finite field GF(q), $q = p^r$ a prime power. Then by the definition the integers $\log_\alpha(1-\beta)$, $\log_\alpha(1-\beta^2), \cdots, \log_\alpha(1-\beta^M)$ are all distinct and exhaust all elements in the set $\{1, 2, \cdots, M\}$.

Define integers N and Q satisfying

$$NQ \leqslant M. \tag{8.88}$$

Let $d_j = N_j + 1$, $0 \leqslant j \leqslant Q-1$, for each j, $0 \leqslant j \leqslant Q-1$, the following sequence

$$A_j \sim [\log_\alpha(1-\beta^{d_j}) - 1, \log_\alpha(1-\beta^{d_j+1}) - 1, \cdots, \log_\alpha(1-\beta^{d_j+M-1}) - 1] \tag{8.89}$$

represents a permutation array, denoted by A_j. Moreover, we have the following important theorem.

Theorem 8.2.16 *The arrays set $A_0, A_1, \cdots, A_{Q-1}$, defined by Eq. (8.89), is a family of Costas arrays satisfying*

$$T_{A_iA_j}(u, v) \leqslant 1, \quad |u| \leqslant N-1, |v| \leqslant M-1 \tag{8.90}$$

for all $0 \leqslant i, j \leqslant Q-1$, where N and Q are determined by Eq. (8.88).

Proof For each fixed j, $0 \leqslant j \leqslant Q-1$, the array A_j is clearly a permutation matrix. Thus it is needed to prove that (1) A_j is Costas array, and (2) Eq. (8.90) is satisfied by each pair A_i and A_j, $0 \leqslant i \neq j \leqslant Q-1$.

(1) Let $A_i = [A_j(n, m)]$, $0 \leqslant n, m \leqslant M-1$, by the definition, $A_j(n, m) = 1$ if $n = \log_\alpha(1-\beta^{d_j+m}) - 1$. Thus

$$\begin{aligned} T_{A_j}(u, v) &= \sum_{m,n=0}^{M-1} A_j(n, m) A_j(n+u, m+v) \\ &= \sum_{m=0}^{M-1} A_j(\log_\alpha(1-\beta^{d_j+m}) - 1, m) A_j \\ &\quad \cdot \log_\alpha(1-\beta^{d_j+m} - 1 + u, m+v) \\ &= \sum_{m=0}^{M-1} A_j(\log_\alpha(1-\beta^{d_j+m}) - 1 + u, m+v). \end{aligned} \tag{8.91}$$

Because of that

$$\begin{aligned}
&A_j(\log_\alpha(1-\beta^{d_j+m})-1+u, m+v)=1\\
\Longleftrightarrow &\log_\alpha(1-\beta^{d_j+m})-1+u=\log_\alpha(1-\beta^{d_j+m+v})-1\\
\Longrightarrow &\alpha^n(1-\beta^{d_j+m})=1-\beta^{d_j+m+v}\\
\Longleftrightarrow &\beta^m(\alpha^n\beta^{d_j}-\beta^{d_j+v})=\alpha^n-1,
\end{aligned}$$

which has exactly one solution m if $\alpha^n \neq \beta^v$, or has no solution m, if $\alpha^n = \beta^v$. Hence $T_{A_j}(u,v) \leqslant 1$, i.e., A_j is indeed a Costas array. In fact, A_j is produced by periodically shifting a Golomb-Costas array by $d_j - 1$ units in horizontal direction.

(2) In order to prove Eq. (8.90), for $0 \leqslant i \neq j \leqslant Q-1$, we assume on the contrary that $T_{A_iA_j}(u,v) \geqslant 2$ for some $|u| \leqslant N-1$, $|v| \leqslant M-1$. Then from Eq. (8.86) and Eq. (8.87), there exists integers k and h, $0 \leqslant k \leqslant M-1$, $0 \leqslant h \leqslant M-1$, such that the following equality holds

$$\begin{aligned}
&\log_\alpha(1-\beta^{d_j+k})-\log_\alpha(1-\beta^{d_j+k+h})\\
&=\log_\alpha(1-\beta^{d_j+k-u})-\log_\alpha(1-\beta^{d_j+k+h-u})\\
\longrightarrow &(1-\beta^{d_j+k})(1-\beta^{d_j+h+k-u})=(1-\beta^{d_j+k-u})(1-\beta^{d_j+h+k})\\
\Longleftrightarrow &\beta^{d_j+k-u}(\beta^h-1)=\beta^{d_j+k}(\beta^h-1)\\
\Longleftrightarrow &\beta^{d_j+k-u}=\beta^{d_j+k}, \text{ for } h\geqslant 1 \Longleftrightarrow d_i-u\equiv d_j(\text{mod}(M+1)), \text{ for } \beta^{M+1}=\beta^{q-1}=1,
\end{aligned}$$

which is impossible for $i \neq j$ and $|u| \leqslant N-1$. In fact, $d_i - d_j = N(i-j)$. Using the condition $i \neq j$ and $0 \leqslant i,j \leqslant Q-1$, one obtains $N \leqslant |d_i - d_j| \leqslant N(Q-1) \leqslant M-N$. If $|u| \leqslant N-1$, it follows that $1 \leqslant |d_i - d_j - u| \leqslant M$. This implies that $d_i - d_j - u \equiv 0 \bmod(M+1)$, i.e., the equation $d_i - u \equiv d_j \bmod(M+1)$ is impossible. □

Summarizing Theorem 8.2.16 and its proof, we have the following construction, which produces a family of Q Costas arrays of size $M \times M$ such that $T_{AB}(u,v) \leqslant 1$, for all $|u| \leqslant N-1$, $|v| \leqslant M-1$, and for each member arrays pair A and B, where $M = q-2$, q a prime power, and $NQ \leqslant M$.

Construction 8.2.3 *Costas arrays with small non-periodic cross-correlation.*

Step 1. Construct an $M \times M$ Golomb-Costas array using a pair of primitive elements α and β in GF(q). Call it A_0 (i.e., $A_0(i,j) = 1$ iff $\alpha^{i+1} + \beta^{j+1} = 1$, $A_0(i,j) = 0$, otherwise, $0 \leqslant i,j \leqslant q-3$).

Step 2. Extend the above Golomb-Costas array horizontally to the right by periodic extension with period M.

Step 3. From this periodically extended array, take an $M \times M$ subarray starting at the N-th column and call it A_1. Similarly, A_j is defined to be the $M \times M$ subarray starting at the Nj-th column, for $0 \leqslant j \leqslant Q-1$.

Step 4. The arrays $A_0, A_1, \cdots, A_{Q-1}$ provide a family of Costas arrays satisfying Eq. (8.85), or equivalently, $A'_0, A'_1, \cdots, A'_{Q-1}$ provide a family of Costas arrays satisfying Eq. (8.84), where A' denotes the transpose of A.

By using Construction 8.2.3, we can generate a family of binary arrays $A_0, A_1, \cdots, A_{Q-1}$ of size $M \times M$ such that $T_{A_iA_j}(u,v) \leqslant 1$, for $|u| \leqslant N-1$, $|v| \leqslant M-1$. Let X_i, $0 \leqslant i \leqslant Q-1$, be the subarray of A_i consisting of its first N rows, then $X_0, X_1, \cdots, X_{Q-1}$

provide a family of Q arrays of size $N \times M$ such that both their non-periodic cross- and out-of-phase autocorrelations are upper bounded by 1. Consequently, by using Theorem 8.2.10, their mixed cross- and out-of-phase autocorrelations are upper bounded by 2. Applying Construction 8.2.2 to these arrays $X_0, X_1, \cdots, X_{Q-1}$, we get an $(M(2N-1), N, 2, 2)$ OOC.

Precisely, the first TC OOC is constructed by the following steps.

Construction 8.2.4 (The first TC OOC).

Step A.1. For a given prime power q, primitive elements α and β in GF(q), construct $a(q-2) \times (q-2)$ Golomb-Costas array $A_0 = [A_0(i,j)]$, i.e., $A_0(i,j) = 1$ iff $\alpha^{i+1} + \beta^{j+1}$, $0 \leqslant i, j \leqslant q-3$.

Step A.2 (New Costas arrays in Theorem 8.2.16). Select integers N and Q satisfying $1 \leqslant NQ \leqslant q-2$. In order to obtain Q different $(q-2) \times (q-2)$ Costas arrays, first periodically (in horizontal direction) extend the array A_0 obtained in Step A.1 and take as separate $(q-2) \times (q-2)$ arrays, the arrays starting at (Nj)-th column of the periodically extended array for $j = 0, 1, 2, \cdots, Q-1$. These arrays are denoted by $A_0, A_1, \cdots, A_{Q-1}$.

Step A.3. Truncate the arrays $A_0, A_1, \cdots, A_{Q-1}$ keeping in each array the first N rows to get Q arrays $X_0, X_1, \cdots, X_{Q-1}$, respectively, of size $N \times (q-2)$.

Step A.4. (Applying Construction 2 to $X_0, X_1, \cdots, X_{Q-1}$). Extend each X_i by putting under it an all-zero array 0 of size $(N-1) \times (q-2)$ to get $Y_i = \begin{pmatrix} X_i \\ \vdots \\ 0 \end{pmatrix}$, $0 \leqslant i \leqslant Q-1$,

The concatenated sequence y_i of Y_i, $0 \leqslant i \leqslant Q-1$, provides the first new TC OOC, which is a $((q-2)(2N-1), N, 2, 2)$ optical orthogonal codes.

Example 8.2.1 Let $q = 8$, $Q = 2$, $N = 3$, α be a root of the primitive polynomial $f(x) = x^3 + x + 1$, in GF(2), and $\beta = \alpha^2$. Thus α and β are two primitive elements in the finite field GF(2^3).

These Costas arrays A_0, A_1 in Step A.2 are

$$A_0 = \begin{bmatrix} 0&0&0&0&1&0 \\ 0&0&1&0&0&0 \\ 0&0&0&1&0&0 \\ 0&0&0&0&0&1 \\ 0&1&0&0&0&0 \\ 1&0&0&0&0&0 \end{bmatrix}, \quad A_1 = \begin{bmatrix} 0&1&0&0&0&0 \\ 0&0&0&0&0&1 \\ 1&0&0&0&0&0 \\ 0&0&1&0&0&0 \\ 0&0&0&0&1&0 \\ 0&0&0&1&0&0 \end{bmatrix}.$$

The truncated arrays X_0, X_1 in Step A.3 are

$$X_0 = \begin{bmatrix} 0&0&0&0&1&0 \\ 0&0&1&0&0&0 \\ 0&0&0&1&0&0 \end{bmatrix}, \quad X_1 = \begin{bmatrix} 0&1&0&0&0&0 \\ 0&0&0&0&0&1 \\ 1&0&0&0&0&0 \end{bmatrix}.$$

The extended arrays Y_0, Y_1 and their corresponding concatenated sequences y_0, y_1 are

$$Y_0 = \begin{bmatrix} 0&0&0&0&1&0 \\ 0&0&1&0&0&0 \\ 0&0&0&1&0&0 \\ 0&0&0&0&0&0 \\ 0&0&0&0&0&0 \end{bmatrix}, \quad Y_1 = \begin{bmatrix} 0&1&0&0&0&0 \\ 0&0&0&0&0&1 \\ 1&0&0&0&0&0 \\ 0&0&0&0&0&0 \\ 0&0&0&0&0&0 \end{bmatrix},$$

$$y_0 = (000000000000100000100100000000),$$

$$y_1 = (001001000000000000000000001000).$$

The sequences y_0, y_1 provide a (30,3,2,2) OOC.

The second new TC OOC is constructed by the following steps.

Construction 8.2.5 (The second TC OOC).

Step B.1. For a given prime integer p, and a primitive element α in GF(p). Construct a $(p-1)\times(p-1)$ Welch-Costas array $B_0 = [B_0(i,j)]$, i.e., $B_0(i,j) = 1$ iff $i+1 = \alpha^j$, $0 \leqslant i,j \leqslant p-2$.

Step B.2. Select integers N and Q satisfying $1 \leqslant NQ \leqslant p-1$. In order to obtain Q different $(p-1)\times(p-1)$ Costas arrays, first periodically (in horizontal direction) extend the array B_0 obtained in Step B.1 and take as separate $(p-1)\times(p-1)$ arrays the arrays starting at (Nj)-th column of the periodically extended array for $j = 0,1,2,\cdots,Q-1$. These arrays are denoted by $B_0, B_1, \cdots, B_{Q-1}$.

Step B.3. Truncate the arrays $B_0, B_1, \cdots, B_{Q-1}$ keeping in each array the first N rows to get Q arrays $X_0^*, X_1^*, \cdots, X_{Q-1}^*$, respectively, of size $N\times(q-2)$.

Step B.4 (Applying Construction 2 to $X_0^*, X_1^*, \cdots, X_{Q-1}^*$). Extend each X_i^* by putting under it an all-zero array 0 of size $(N-1)\times(p-1)$ to get $Y_i^* = \begin{pmatrix} X_i^* \\ \vdots \\ 0 \end{pmatrix}$, $0 \leqslant i \leqslant Q-1$. The concatenated sequence y_i^* (concatenating the columns one after another, see Definition 8.2.3) of Y_i^*, $0 \leqslant i \leqslant Q-1$, provides the second new TC OOC, which is a $((p-1)(2N-1), N, 1, 2)$ optical orthogonal code.

In fact, it is easy to see that the non-periodic cross-correlations between each array pair in $X_0^*, X_1^*, \cdots, X_{Q-1}^*$ are upper bounded by 1, thus from Theorem 8.2.10, their mixed cross-correlations are upper bounded by 2. Therefore, from Theorem 8.2.15, the periodic cross-correlation upper bound λ_c of the second new TC OOC is 2.

On the other hand, from Theorem 8.2.11, the out-of-phase mixed autocorrelation of each Welch-Costas array is upper bounded by 1. Therefore, from Theorem 8.2.15, the out-of-phase periodic autocorrelation upper bound λ_a of the second new TC OOC is 1.

Example 8.2.2 Let $p = 7$, $Q = 2$, $N = 3$, and $\alpha = 3$. The Costas arrays B_0, B_1 in Step B.2 are

$$B_0 = \begin{bmatrix} 1&0&0&0&0&0 \\ 0&0&1&0&0&0 \\ 0&1&0&0&0&0 \\ 0&0&0&0&1&0 \\ 0&0&0&0&0&1 \\ 1&0&0&1&0&0 \end{bmatrix}, \quad B_1 = \begin{bmatrix} 0&0&0&1&0&0 \\ 0&0&0&0&0&1 \\ 0&0&0&0&1&0 \\ 0&1&0&0&0&0 \\ 0&0&1&0&0&0 \\ 1&0&0&0&0&0 \end{bmatrix}.$$

The truncated arrays X_0^* and X_1^* in Step B.3 are

$$X_0^* = \begin{bmatrix} 1&0&0&0&0&0 \\ 0&0&1&0&0&0 \\ 0&1&0&0&0&0 \end{bmatrix}, \quad X_1^* = \begin{bmatrix} 0&0&0&1&0&0 \\ 0&0&0&0&0&1 \\ 0&0&0&0&1&0 \end{bmatrix}.$$

The extended arrays Y_0^* and Y_1^* and their corresponding concatenated sequences y_0^* and y_1^* are

$$Y_0^* = \begin{bmatrix} 1 & 0 & 0 & 0 & 0 & 0 \\ 0 & 0 & 1 & 0 & 0 & 0 \\ 0 & 1 & 0 & 0 & 0 & 0 \\ 0 & 0 & 0 & 0 & 0 & 0 \\ 0 & 0 & 0 & 0 & 0 & 0 \end{bmatrix}, \quad Y_1^* = \begin{bmatrix} 0 & 0 & 0 & 1 & 0 & 0 \\ 0 & 0 & 0 & 0 & 0 & 1 \\ 0 & 0 & 0 & 0 & 1 & 0 \\ 0 & 0 & 0 & 0 & 0 & 0 \\ 0 & 0 & 0 & 0 & 0 & 0 \end{bmatrix},$$

$$y_0^* = (1000000100010000000000000000000),$$

$$y_1^* = (10000000000000001000010001000),$$

y_0^* and y_1^* provide a $(30, 3, 1, 2)$ OCC.

Remark 8.2.5 *If Construction 1 is applied to the arrays $X_0, X_1, \cdots, X_{Q-1}$ in Step* A.3 *and to the arrays $X_0^*, X_1^*, \cdots, X_{Q-1}^*$ in Step* B.3, *then we can get another two families of* TC OOCs *with parameters $(N(q-2), N, 4, 4)$ and $(N(p-1), N, 2, 4)$, respectively. The details are omitted here because their λ_a and λ_c are not good.*

Construction 8.2.6 (The third TC OOCs).

Step C.1. For a given prime p, construct a $(p-1) \times (p-1)$ Welch-Costas array $A = [A(i,j)], 1 \leqslant i, j \leqslant p-1, A(i,j) = 1$ if $i = \alpha^j; A(i,j) = 0$, otherwise, where α is a primitive element in the field GF(p).

Step C.2. Select integers ω and Q satisfying $1 \leqslant \omega Q \leqslant p-1$. Periodically (in a horizontal direction) extend the Costas array A obtained in Step C.1 and take as separate $(p-1) \times (p-1)$ arrays starting at the $(\omega(q-1)+1)$-th column of the periodically extended array for $q = 1, 2, \cdots, Q$. Rotate the resulting Q arrays counterclockwise by 90°.

Step C.3. Truncate the arrays, keeping in each array the first (in the direction down-to-up) ω rows.

Step C.4. Extend the obtained arrays by an all-zero auxiliary $\omega \times (p-2)$ array. Each member sequence of the TC OOC's is formed from the elements of each truncated Costas array by shearing the rows of the array and placing them in a sequence.

Example 8.2.3 Let $p = 11$, $\omega = 4$, and $Q = 2$, thus $\omega Q = 4 \times 2 = 8 \leqslant p - 1 = 10$. The Welch-Costas array A defined by the primitive element $\alpha = 2$ is

$$A = \begin{bmatrix} 0 & 0 & 0 & 0 & 1 & 0 & 0 & 0 & 0 & 0 \\ 0 & 0 & 0 & 0 & 0 & 1 & 0 & 0 & 0 & 0 \\ 0 & 0 & 1 & 0 & 0 & 0 & 0 & 0 & 0 & 0 \\ 0 & 0 & 0 & 0 & 0 & 0 & 1 & 0 & 0 & 0 \\ 0 & 0 & 0 & 0 & 0 & 0 & 0 & 0 & 1 & 0 \\ 0 & 0 & 0 & 1 & 0 & 0 & 0 & 0 & 0 & 0 \\ 0 & 1 & 0 & 0 & 0 & 0 & 0 & 0 & 0 & 0 \\ 0 & 0 & 0 & 0 & 0 & 0 & 0 & 1 & 0 & 0 \\ 1 & 0 & 0 & 0 & 0 & 0 & 0 & 0 & 0 & 0 \\ 0 & 0 & 0 & 0 & 0 & 0 & 0 & 0 & 0 & 1 \end{bmatrix}. \tag{8.92}$$

By periodically extending the Costas array A in a horizontal direction, we get the following B, shown in Eq. (8.93).

$$B = \begin{bmatrix} 0&0&0&0&1&0&0&0&0&0&0&0&0&0&1&0&0&0&0&0 \\ 0&0&0&0&0&1&0&0&0&0&0&0&0&0&0&1&0&0&0&0 \\ 0&0&1&0&0&0&0&0&0&0&0&0&1&0&0&0&0&0&0&0 \\ 0&0&0&0&0&0&1&0&0&0&0&0&0&0&0&0&1&0&0&0 \\ 0&0&0&0&0&0&0&0&1&0&0&0&0&0&0&0&0&0&1&0 \\ 0&0&0&1&0&0&0&0&0&0&0&0&0&1&0&0&0&0&0&0 \\ 0&1&0&0&0&0&0&0&0&0&0&1&0&0&0&0&0&0&0&0 \\ 0&0&0&0&0&0&0&1&0&0&0&0&0&0&0&0&0&1&0&0 \\ 1&0&0&0&0&0&0&0&0&0&1&0&0&0&0&0&0&0&0&0 \\ 0&0&0&0&0&0&0&0&0&1&0&0&0&0&0&0&0&0&0&1 \end{bmatrix}. \quad (8.93)$$

We start at the first $(\omega(1-1)+1) = 1$ and fifth columns of $(\omega(2-1)+1) = 5$ of the extended array to get the following two separate $(p-1) \times (p-1)$ arrays C and D:

$$C = \begin{bmatrix} 0&0&0&0&1&0&0&0&0&0 \\ 0&0&0&0&0&1&0&0&0&0 \\ 0&0&1&0&0&0&0&0&0&0 \\ 0&0&0&0&0&0&1&0&0&0 \\ 0&0&0&0&0&0&0&0&1&0 \\ 0&0&0&1&0&0&0&0&0&0 \\ 0&1&0&0&0&0&0&0&0&0 \\ 0&0&0&0&0&0&0&1&0&0 \\ 1&0&0&0&0&0&0&0&0&0 \\ 0&0&0&0&0&0&0&0&0&1 \end{bmatrix}. \quad (8.94)$$

$$D = \begin{bmatrix} 1&0&0&0&0&0&0&0&0&0 \\ 0&1&0&0&0&0&0&0&0&0 \\ 0&0&0&0&0&0&0&0&1&0 \\ 0&0&1&0&0&0&0&0&0&0 \\ 0&0&0&0&1&0&0&0&0&0 \\ 0&0&0&0&0&0&0&0&0&1 \\ 0&0&0&0&0&0&0&1&0&0 \\ 0&0&0&1&0&0&0&0&0&0 \\ 0&0&0&0&0&0&1&0&0&0 \\ 0&0&0&0&0&1&0&0&0&0 \end{bmatrix}. \quad (8.95)$$

Rotating the arrays C and D counterclockwise by 90°, we get the following arrays E and F:

$$E = \begin{bmatrix} 0&0&0&0&0&0&0&0&0&1 \\ 0&0&0&0&1&0&0&0&0&0 \\ 0&0&0&0&0&0&0&1&0&0 \\ 0&0&0&1&0&0&0&0&0&0 \\ 0&1&0&0&0&0&0&0&0&0 \\ 1&0&0&0&0&0&0&0&0&0 \\ 0&0&0&0&0&1&0&0&0&0 \\ 0&0&1&0&0&0&0&0&0&0 \\ 0&0&0&0&0&0&1&0&0&0 \\ 0&0&0&0&0&0&0&0&1&0 \end{bmatrix}, \quad (8.96)$$

$$F = \begin{bmatrix} 0&0&0&0&0&1&0&0&0&0 \\ 0&0&1&0&0&0&0&0&0&0 \\ 0&0&0&0&0&0&1&0&0&0 \\ 0&0&0&0&0&0&0&0&1&0 \\ 0&0&0&0&0&0&0&0&0&1 \\ 0&0&0&0&1&0&0&0&0&0 \\ 0&0&0&0&0&0&0&1&0&0 \\ 0&0&0&1&0&0&0&0&0&0 \\ 0&1&0&0&0&0&0&0&0&0 \\ 1&0&0&0&0&0&0&0&0&0 \end{bmatrix}. \quad (8.97)$$

Truncating the arrays E and F and keeping in each array the first $\omega = 4$ rows (in the direction of down-to-up) we get the following arrays G and H:

$$G = \begin{bmatrix} 0 & 0 & 0 & 0 & 0 & 1 & 0 & 0 & 0 & 0 \\ 0 & 0 & 1 & 0 & 0 & 0 & 0 & 0 & 0 & 0 \\ 0 & 0 & 0 & 0 & 0 & 0 & 1 & 0 & 0 & 0 \\ 0 & 0 & 0 & 0 & 0 & 0 & 0 & 0 & 1 & 0 \end{bmatrix}, \tag{8.98}$$

$$H = \begin{bmatrix} 0 & 0 & 0 & 0 & 0 & 0 & 1 & 0 & 0 & 0 \\ 0 & 0 & 0 & 1 & 0 & 0 & 0 & 0 & 0 & 0 \\ 0 & 1 & 0 & 0 & 0 & 0 & 0 & 0 & 0 & 0 \\ 1 & 0 & 0 & 0 & 0 & 0 & 0 & 0 & 0 & 0 \end{bmatrix}. \tag{8.99}$$

Extending the arrays G and H by an all-zero auxiliary $\omega \times (p-2) = 4 \times 9$ array, we get the arrays I and J shown in (8.100) and (8.101).

$$I = \begin{bmatrix} 0 & 0 & 0 & 0 & 0 & 1 & 0 & 0 & 0 & 0 & 0 & 0 & 0 & 0 & 0 & 0 & 0 & 0 & 0 \\ 0 & 0 & 1 & 0 & 0 & 0 & 0 & 0 & 0 & 0 & 0 & 0 & 0 & 0 & 0 & 0 & 0 & 0 & 0 \\ 0 & 0 & 0 & 0 & 0 & 0 & 1 & 0 & 0 & 0 & 0 & 0 & 0 & 0 & 0 & 0 & 0 & 0 & 0 \\ 0 & 0 & 0 & 0 & 0 & 0 & 0 & 0 & 1 & 0 & 0 & 0 & 0 & 0 & 0 & 0 & 0 & 0 & 0 \end{bmatrix}, \tag{8.100}$$

$$J = \begin{bmatrix} 0 & 0 & 0 & 0 & 0 & 0 & 1 & 0 & 0 & 0 & 0 & 0 & 0 & 0 & 0 & 0 & 0 & 0 & 0 \\ 0 & 0 & 0 & 1 & 0 & 0 & 0 & 0 & 0 & 0 & 0 & 0 & 0 & 0 & 0 & 0 & 0 & 0 & 0 \\ 0 & 1 & 0 & 0 & 0 & 0 & 0 & 0 & 0 & 0 & 0 & 0 & 0 & 0 & 0 & 0 & 0 & 0 & 0 \\ 1 & 0 & 0 & 0 & 0 & 0 & 0 & 0 & 0 & 0 & 0 & 0 & 0 & 0 & 0 & 0 & 0 & 0 & 0 \end{bmatrix}. \tag{8.101}$$

The two sequences of TC OOC are formed by shearing the rows of the arrays I and J and placing them in sequences a and b, respectively. Thus

$$a = \left\{ \begin{matrix} 0 & 0 & 0 & 0 & 0 & 0 & 0 & 0 \\ 1 & 0 & 0 & 0 & 0 & 0 & 0 & 0 \\ 0 & 0 & 0 & 0 & 0 & 0 & 0 & 0 \\ 0 & 1 & 0 & 0 & 0 & 0 & 0 & 0 \\ 0 & 0 & 0 & 0 & 0 & 0 & 0 & 0 \\ 1 & 0 & 0 & 0 & 0 & 0 & 0 & 0 \\ 0 & 0 & 0 & 0 & 0 & 0 & 0 & 0 \\ 0 & 0 & 0 & 0 & 0 & 0 & 1 & 0 \\ 0 & 0 & 0 & 0 & 0 & 0 & 0 & 0 \\ 0 & 0 & 0 & 0 & & & & \end{matrix} \right\}$$

and

$$b = \left\{ \begin{matrix} 1 & 0 & 0 & 0 & 0 & 0 & 0 & 0 \\ 0 & 0 & 0 & 0 & 0 & 0 & 0 & 0 \\ 0 & 0 & 0 & 0 & 1 & 0 & 0 & 0 \\ 0 & 0 & 0 & 0 & 0 & 0 & 0 & 0 \\ 0 & 0 & 0 & 0 & 0 & 0 & 0 & 0 \\ 0 & 1 & 0 & 0 & 0 & 0 & 0 & 0 \\ 0 & 0 & 0 & 0 & 0 & 0 & 0 & 0 \\ 0 & 0 & 0 & 0 & 0 & 0 & 0 & 0 \\ 1 & 0 & 0 & 0 & 0 & 0 & 0 & 0 \\ 0 & 0 & 0 & 0 & & & & \end{matrix} \right\}.$$

The periodic autocorrelation of the first sequence a equals 2 at two out-of-phase shifts. In fact the periodic autocorrelation of the a sequence is

$$R_a = \left\{ \begin{array}{cccccccc} 4 & 0 & 0 & 0 & 0 & 0 & 0 & 0 \\ 0 & 0 & 0 & 0 & 0 & 0 & 0 & 1 \\ 0 & 1 & 0 & 0 & 0 & 0 & 2 & 0 \\ 0 & 0 & 0 & 0 & 0 & 0 & 0 & 0 \\ 1 & 0 & 0 & 0 & 0 & 1 & 0 & 1 \\ 0 & 0 & 0 & 0 & 1 & 0 & 0 & 0 \\ 0 & 0 & 0 & 0 & 0 & 0 & 2 & 0 \\ 0 & 0 & 0 & 1 & 0 & 1 & 0 & 0 \\ 0 & 0 & 0 & 0 & 0 & 0 & 0 & 0 \\ 0 & 0 & 0 & 0 & & & & \end{array} \right\}.$$

The periodic cross-correlation between the first sequence a and the second sequence b, which equals 2 at one shift. In fact, the periodic cross-correlation between a and b is

$$R_{ab} = \left\{ \begin{array}{cccccccc} 0 & 1 & 1 & 0 & 0 & 0 & 0 & 0 \\ 0 & 0 & 0 & 0 & 1 & 0 & 1 & 0 \\ 1 & 0 & 0 & 0 & 0 & 0 & 0 & 0 \\ 1 & 0 & 0 & 0 & 0 & 0 & 0 & 0 \\ 0 & 1 & 1 & 0 & 1 & 0 & 0 & 1 \\ 0 & 0 & 0 & 0 & 0 & 0 & 0 & 0 \\ 0 & 0 & 0 & 1 & 0 & 0 & 0 & 1 \\ 2 & 0 & 0 & 0 & 0 & 0 & 0 & 0 \\ 0 & 0 & 0 & 0 & 1 & 0 & 0 & 1 \\ 0 & 0 & 0 & 0 & & & & \end{array} \right\}.$$

Now we will prove that the third TC OOC's are $(\omega(2p-3), \omega, 2, 2)$ optical orthogonal codes, i.e., both the periodic auto- and cross-correlations are upper bounded by 2.

In fact, let

$$X' = [X'(i,j)] \text{ and } Y' = [Y'(i,j)], \quad 0 \leqslant i \leqslant \omega - 1, 0 \leqslant j \leqslant p-2$$

represent two arrays produced prior to Step C.3. Extending X' and Y' by an all-zero auxiliary $\omega \times (p-2)$ array, we obtain the two following $\omega \times (2p-3)$ arrays:

$$X = [X(i,j)] \text{ and } Y = [Y(i,j)], \quad 0 \leqslant i \leqslant \omega - 1, 0 \leqslant j \leqslant 2p-4.$$

By this construction, we have the following lemma.

Lemma 8.2.7 *The two-dimensional (2-D) nonperiodic cross-correlation between X' and Y' (of course, X and Y) are upper bounded by 1.*

The two TC OOC sequences

$$x' = (x(k)) \text{ and } y' = (y(k)), \quad 0 \leqslant k \leqslant \omega(2p-3) - 1 = N - 1$$

produced by Step C.4 from X and Y are defined by

$$x(m(2p-3)+n) = X(m,n) \text{ and}$$
$$y(m(2p-3)+n) = Y(m,n), \quad 0 \leqslant m \leqslant \omega - 1,\ 0 \leqslant n \leqslant 2p-4,$$

respectively. Thus, at the shift

$$0 \leqslant \tau = e(2p-3) + f \leqslant N-1, \quad 0 \leqslant e \leqslant \omega - 1, 0 \leqslant f \leqslant 2p-4,$$

the periodic cross-correlation between the sequences x and y is equivalent to

$$\begin{aligned} R_{xy}(\tau) &= \sum_{i=0}^{N-1} x(i)y((i+\tau) \bmod N) \\ &= \sum_{m=0}^{\omega-1}\sum_{n=0}^{2p-4} x(m(2p-3)+n)y(((m+e)(2p-3)+n+f) \bmod N) \\ &= \sum_{m=0}^{\omega-1}\sum_{n=0}^{p-1} X(m,n)y(((m+e)(2p-3)+n+f) \bmod N), \end{aligned} \tag{8.102}$$

where the last equation is due to the fact that $X(m,n) = 0$ for $n \geqslant p$. Equation (8.102) is considered in the following two cases.

Case 1. If $0 \leqslant f \leqslant p-3$, Eq. (8.102) becomes

$$\begin{aligned} &\sum_{m=0}^{\omega-1}\sum_{n=0}^{p-1} X(m,n)Y((m+e) \bmod \omega, n+f) \\ &= \sum_{m=0}^{\omega-e-1}\sum_{n=0}^{p-1} X(m,n)Y(m+e, n+f) \\ &\quad + \sum_{m=\omega-e}^{\omega-1}\sum_{n=0}^{p-1} X(m,n)Y(m+e-\omega, n+f). \end{aligned} \tag{8.103}$$

Note that the first and second terms of Eq. (8.103) are parts of nonperiodic cross-correlations between X and Y at 2-D shifts (e, f) and $(e-\omega, f)$, respectively. Thus, they are upper bounded by 1 (from Lemma 8.2.7). Consequently, the periodic cross-correlation between x and y are upper bounded by 2 (in Case 1).

Case 2. If $p-2 \leqslant f \leqslant 2p-4$, Eq. (8.102) becomes

$$\begin{aligned} &\sum_{m=0}^{\omega-1}\sum_{n=0}^{p-1} X(m,n)Y((m+e-1) \bmod \omega, n+f-(2p-3)) \\ &= \sum_{m=0}^{\omega-e}\sum_{n=0}^{p-1} X(m,n)Y(m+e-1, n+f-(2p-3)) \\ &\quad + \sum_{m=\omega-e+1}^{\omega-1}\sum_{n=0}^{p-1} X(m,n)Y(m+e-1-\omega, n+f-(2p-3)). \end{aligned} \tag{8.104}$$

Note that the first and second terms of Eq. (8.104) are parts of nonperiodic cross-correlations between X and Y at 2-D shifts $(e-1, f-(2p-3))$ and $(e-\omega-1, f-(2p-3))$, respectively. Thus, they are upper bounded by 1 (from Lemma 8.2.7). Consequently, the periodic cross-correlation between x and y are upper bounded by 2 (in Case 2).

Thus far, we have proven the following:

Theorem 8.2.17 *The* TC OOC's *constructed by Steps* C.1–C.4 *are* $(\omega(2p-3), \omega, 2, 2)$ *optical orthogonal codes.*

Bibliography

[1] Golomb S W and Taylor H. Constructions and properties of Costas arrays. *Proc. Of IEEE*, 1984, 72(9): 1143–1163.

[2] Yixian Yang. On Costas arrays. *J. of Electronics*, 1992, 9(1): 17–25.

[3] Yixian Yang. Algebraically constructed Costas arrays with small number of cross-coincidences. *System Sciences and Mathematical Sciences*, 2000, 13(3): 242–249.

[4] Yixian Yang. Optical Orthogonal Codes. *Chinese J. of Electronics*, 1991, 1(1): 72–78.

[5] Yixian Yang, Niu X X and Xu C Q. Counterexample of truncated Costas optical orthogonal codes. *IEEE Trans. on Communications.*, 1997, 45(6): 640–643.

[6] Yixian Yang. New Costas arrays with small cross-coincidences and new truncated Costas optical orthogonal codes. *Chinese Journal of Electronics*, 1997, 6(3): 54–62.

[7] Salehi J and Brackett C. Codes division multiple access techniques in optical fiber networks. Part I and II. *IEEE Trans. On Com.*, 1989, 1(37): 824–842.

Concluding Questions

The topic of (higher-) dimensional Hadamard matrices is a developing one. As the first book concerning the topic, this book has investigated many problems of constructions, existences, enumeration, transforms and fast algorithms, whilst there are still many open problems in both the areas of theory and practice. In order to motivate more research we list here some interesting research problems.

1. (Hadamard Conjecture) There exists an Hadamard matrix of order $4t$ for each positive integer t.

2. (Wallis Conjecture) There exists a weighting matrix $W(4t, k)$, $0 \leqslant k \leqslant 4t$, for each positive integer t (this conjecture is clearly a generalization of the Hadamard conjecture).

3. It is conjectured that there exist amicable Hadamard matrices of each order $n \equiv 0 \pmod{4}$.

4. (Conjecture) Let x, y, q, m, and t be non-negative integers satisfying $m = 2^t q$ and $0 \leqslant x + y \leqslant m$. When t is sufficiently large, there always exists an orthogonal design of type $(x, y, m - x - y)$ and order m.

5. Prove or disprove the existence of three-dimensional Hadamard matrices of orders $4k + 2 \neq 2.3^b$, $k > 1$, $b \geqslant 0$.

6. Construct more three-dimensional Hadamard matrices of orders $4k + 2$, $k > 1$.

7. The three-dimensional Hadamard matrices can be constructed from two-dimensional ones. Could the two-dimensional Hadamard matrices be constructed by three- or higher-dimensional Hadamard matrices? If the answer is positive, it is possible that more new two-dimensional Hadamard matrices would be discovered. In general, could the lower-dimensional Hadamard matrices be constructed by decomposing the higher-dimensional Hadamard matrices?

8. What is the enumeration of n-dimensional Hadamard matrices of order two (or equivalently the H-Boolean functions in n-variables)?

9. What is the enumeration of the Boolean functions satisfying the SAC of order $k \leqslant n - 4$? Construct more Boolean functions satisfying the SAC of order k.

10. Construct and enumerate Bent functions.

11. Construct and enumerate Boolean functions satisfying propagation criterion of degree k and order m.

12. Let $n \geqslant 4$. Is there an n-dimensional Hadamard matrix of order $2t$, for each $t \geqslant 1$?

13. What are the necessary and sufficient conditions for the existence of PBA$(a_1, \cdots, a_n)$? How can one construct as many PBAs as possible?

14. What are the necessary and sufficient conditions for the existence of higher-dimensional

orthogonal designs (HDOD)? How can one construct as many HDODs as possible?

15. How can one use the higher-dimensional arrays introduced in this book to construct families of secure crypto-systems?

16. Try to develop more applications of higher-dimensional Walsh and Hadamard matrices engineering areas, e.g., signal processing, telecommunications (especially in mobile, optical, and data communications), image processing, EMS and so forth.

Remark *After the manuscript of this book was finished, the author learned of many new wonderful works on higher-dimensional Hadamard and perfect binary arrays by Yuqing Chen, De Launey, Horadam and Cantian Lin et al.*

Index

A

absolutely improper, 58, 102, 184
absolutely proper three-dimensional Hadamard matrix, 58
absolutely proper, 58, 183
ACS, 333
ACSP, 333
additive Abelian group, 46
amicable Hadamard matrices, 40, 45, 419
amicable orthogonal designs, 40, 42, 45
antipodal, 125
anti-symmetric, 39
aperiodic ambiguity function, 353
autocorrelation, 69

B

balance, 251
Bent functions, 158
Bent sequences, 160
binary supplementary quadruple, 217
block design, 23
BNN, 340
Boolean functions, 340
BSPT, 315

C

circulant matrix, 228
column-wise quasi-perfect, 71
complete orthogonal, 13
complex Hadamard matrix, 51
correlational immune, 247
Costas arrays, 369
cross-coincidences, 384
cross-correlation, 389
cyclic difference sets, 369
cyclic group, 228
cyclic Williamson matrices, 32
cyclotomic classes, 46

D

DFP, 332
difference set, 320
direct multiplication, 43, 60
doppler shift, 383
double proximity shells, 125
doubly quasi-perfect binary arrays, 76
DSP, 320
dyadic shift, 161
d-form sequences, 306

E

entropy dropping, 272
enumerate, 419
enumeration, 247
error correcting code, 3

F

fast Hadamard transform (FHT), 14
fast Walsh transform (FWT), 14
fast Walsh-Hadamard transforms, 101
folding, 195
Fourier transforms, 14

G

generalized 2-dimensional Walsh matrix, 87
generalized Hadamard matrix, 187
generalized perfect array, 201
geometric sequences, 276
GMW sequence, 304
Gold-sequence, 277
Golomb-Costas, 373

H

Hadamard conjecture, 19, 419
Hadamard design, 24
Hadamard matrices, 3
Hadamard matrix, 7, 19
Hadamard-ordered Walsh function, 7

Hadamard-ordered Walsh matrices, 20
Hadamard-ordered Walsh-Hadamard transform, 16
Hamming weights, 130
H-Boolean function, 126
higher dimensional Walsh-Hadamard transforms, 112
higher-dimensional Hadamard matrices, 166
higher-dimensional orthogonal transforms, 114
hyperplane, 277

I

image processing, 99
improper, 58
incomplete, 13
inner product, 222
inverse matrix, 107

K

Kronecker production, 34

L

Lagrange's theorem, 372
Lemple-Costas, 380
linear complexity, 267
linear kernel, 247
LUT, 239

M

Menon difference set, 195
minimal, 125
mismatched sequence, 331
Mobius function, 399
m-sequences, 179
multiplication, 102

N

n-cube, 169
n-dimensional Hadamard matrices, 167
n-dimensional Hadamard matrix, 103
n-dimensional Hadamard matrix of order 2, 125
n-dimensional matrices, 103
nonaffine, 239
nondegeneracy, 239

O

OOC, 396
optical orthogonal code, 397
orthogonal Arrays, 256
orthogonal complete functions, 3
orthogonal designs, 42, 211, 420
orthogonal matrix, 44
orthogonal Split, 255
orthogonal, 12
orthogonality, 221

P

Paley-ordered Walsh matrix, 5
Paley-ordered Walsh functions, 4
Parseval theorem, 100
partial enumeration, 148
Partially Bent, 252
PCS, 332
PCSP, 332
perfect binary arrays, 67
periodic cross-correlation, 282
petrie polygon, 125
propagation characteristics, 158
PSP, 316

Q

QBR attack, 267
quadratic characteristic function, 37
quasi-perfect binary arrays, 72

R

Rademacher functions, 3
random matrices, 169
recursive constructions, 68
Reed–Muller codes, 162
resilient functions, 254
row-wise quasi-perfect, 71

S

sampling, 194
second type linear, 246
sequency-ordered Walsh function, 6
shift registers, 261

skew Hadamard matrices, 40
skew Hadamard matrix, 22
spectral characterization, 156
square circulant matrices, 41
stream-key generator, 261
strict avalanche criterion (SAC), 153
supplementary difference sets, 47
supplementary quadruple, 217
symmetric conference matrix, 48
symmetric cryptosystem, 153
symmetric Hadamard matrix, 21
symmetry, 239

T

TC OOC, 405
tensor products, 222
TFHC-EQC, 353
three-dimensional Hadamard matrices, 58, 419
trigonometric functions, 13
triple correlation, 359
truncated Costas, 412

U

uncorrelated binary arrays, 70
uncorrelatedness, 239
unit matrix, 107

W

Wallis conjecture, 419
Walsh functions, 3
Walsh matrices, 3
Walsh-Hadamard transform, 15, 100
Walsh-ordered Walsh-Hadamard transforms, 15
Walsh-spectrum, 157
weighting matrix, 33
Welch-Costas, 374
Williamson matrices, 28, 221

***

ε-correlation immune, 268
2-dimensional Walsh-Hadamard transforms, 99
3-dimensional Pan-Walsh matrix, 91
3-dimensional Walsh matrices, 86
4-dimensional Hadamard matrix, 102
4-variable's H-Boolean function, 140
5-variable's H-Boolean functions, 147